Lecture Notes in Computer Science   1993
Edited by G. Goos, J. Hartmanis and J. van Leeuwen

Springer
*Berlin
Heidelberg
New York
Barcelona
Hong Kong
London
Milan
Paris
Singapore
Tokyo*

Eckart Zitzler   Kalyanmoy Deb
Lothar Thiele   Carlos A. Coello Coello
David Corne (Eds.)

Evolutionary Multi-Criterion Optimization

First International Conference, EMO 2001
Zurich, Switzerland, March 7-9, 2001
Proceedings

Volume Editors

Eckart Zitzler
Lothar Thiele
Swiss Federal Institute of Technology, Department of Electrical Engineering
Computer Engineering and Networks Laboratory
ETH Zentrum, Gloriastrasse 35, 8092 Zurich, Switzerland
E-mail: {zitzler/thiele}@tik.ee.ethz.ch

Kalyanmoy Deb
Indian Institute of Technology, Department of Mechanical Engineering
Kanpur Genetic Algorithms Laboratory, Kanpur, UP 208 016, India
E-mail: deb@iitk.ac.in

Carlos Artemio Coello Coello
CINVESTAV-IPN, Electrical Engineering Department, Computer Science Section
Av. Instituto Politecnico Nacional No. 2508
Col. San Pedro Zacatenco, Mexico City 07300, Mexico
E-mail: ccoello@cs.cinvestav.mx

David Corne
University of Reading, Department of Computer Science
P.O. Box 225, Whiteknights, Reading RG6 6AY, UK
E-mail: d.w.corne@reading.ac.uk

Cataloging-in-Publication Data applied for

Die Deutsche Bibliothek - CIP-Einheitsaufnahme

Evolutionary multi-criterion optimization : first international conference ;
proceedings / EMO 2001, Zurich, Switzerland, March 7 - 9, 2001.
Eckart Zitzler ... (ed.). - Berlin ; Heidelberg ; New York ; Barcelona ;
Hong Kong ; London ; Milan ; Paris ; Singapore ; Tokyo : Springer, 2001
  (Lecture notes in computer science ; Vol. 1993)
  ISBN 3-540-41745-1

CR Subject Classification (1998): F.2, G.1.6, G.1.2, I.2.8

ISSN 0302-9743
ISBN 3-540-41745-1 Springer-Verlag Berlin Heidelberg New York

This work is subject to copyright. All rights are reserved, whether the whole or part of the material is
concerned, specifically the rights of translation, reprinting, re-use of illustrations, recitation, broadcasting,
reproduction on microfilms or in any other way, and storage in data banks. Duplication of this publication
or parts thereof is permitted only under the provisions of the German Copyright Law of September 9, 1965,
in its current version, and permission for use must always be obtained from Springer-Verlag. Violations are
liable for prosecution under the German Copyright Law.

Springer-Verlag Berlin Heidelberg New York
a member of BertelsmannSpringer Science+Business Media GmbH
http://www.springer.de
© Springer-Verlag Berlin Heidelberg 2001
Printed in Germany

Typesetting: Camera-ready by author, data conversion by PTP-Berlin, Stefan Sossna
Printed on acid-free paper   SPIN: 10782086   06/3142   5 4 3 2 1 0

Preface

Multi-criterion optimization deals with multiple, often conflicting objectives which naturally arise in a real-world scenario. The field of multiple criteria decision making (MCDM) is well established, investigated by many researchers and scientists, and widely applied in practice. Unlike in single-objective optimization, a multi-criterion optimization problem gives rise to a number of optimal solutions, known as Pareto-optimal solutions, of which none can be said to be better than the others with respect to all objectives. Thus, one of the primary goals in multi-criterion optimization is to find or to approximate the set of Pareto-optimal solutions. Since evolutionary algorithms work with a population of solutions, they have been used in multi-criterion optimization for more than a decade. To date, there exist a number of evolutionary approaches and application case studies, demonstrating the usefulness and efficiency of evolutionary multi-criterion optimization (EMO). Due to the growing interest in EMO, the general chairs envisaged organizing this first-ever international conference covering all aspects of the intersection of evolutionary computation and classical MCDM. The aim was to promote and share research activities in this promising field.

The first international conference on evolutionary multi-criterion optimization (EMO 2001) was held in Zürich at the Swiss Federal Institute of Technology (ETH) on March 7–9, 2001. This event included two keynote speeches, one delivered by Ralph E. Steuer on current state-of-the-art methodology and the other delivered by Ian C. Parmee on real-world applications of evolutionary techniques. Furthermore, two extended tutorials were presented, one on classical multiple criteria decision making methodologies by Kaisa Miettinen and another one on evolutionary algorithms by Carlos A. Coello Coello.

In response to the call for papers, 87 papers from 27 countries were submitted, each of which was independently reviewed by at least three members of the program committee. This volume presents a selection of 45 of the refereed papers, together with contributions based on the invited talks and tutorials.

We would like to express our appreciation to the keynote speakers who accepted our invitation, to the tutorial organizers, to all authors who submitted papers to EMO 2001, and to Marco Laumanns and Monica Fricker for their invaluable help in organizing the conference.

March 2001   Eckart Zitzler, Kalyanmoy Deb, Lothar Thiele,
Carlos A. Coello Coello, and David Corne

Organization

EMO 2001 took place from March 7th to 9th, 2001 at the Swiss Federal Institute of Technology (ETH) Zürich, Switzerland, and was organized in cooperation with ACM/SIGART, IEEE Neural Network Council, and the International Society for Genetic and Evolutionary Computation (ISGEC).

General Chairs

Kalyanmoy Deb   IIT Kanpur, India
Lothar Thiele   ETH Zürich, Switzerland
Eckart Zitzler   ETH Zürich, Switzerland

Executive Program Committee

Hojjat Adeli   Ohio State University, USA
Carlos A. Coello Coello   CINVESTAV-IPN, Mexico
David Corne   University of Reading, UK
Carlos Fonseca   Universidade do Algarve, Portugal
David E. Goldberg   University of Illinois at Urbana-Champaign, USA
Jeffrey Horn   Northern Michigan University, USA
Sourav Kundu   Mastek Limited, Japan
Gary B. Lamont   Air Force Institute of Technology, USA
Shigeru Obayashi   Tohoku University, Japan
Ian C. Parmee   University of Plymouth, UK
Carlo Poloni   University of Trieste, Italy
Günter Rudolph   University of Dortmund, Germany
J. David Schaffer   Phillips Research, USA
Hans-Paul Schwefel   University of Dortmund, Germany
El-ghazali Talbi   Université des Sciences et Technologies de Lille, France

Program Committee

Enrique Baeyens   University de Valladolid, Spain
Tapan P. Bagchi   IIT Kanpur, India
Peter J. Bentley   University College London, UK
Jürgen Branke   University of Karlsruhe, Germany
Nirupam Chakraborti   IIT Kharagpur, India
William A. Crossley   Purdue University, USA

Dragan Cvetkovic	Soliton Associates Ltd., Canada
Nicole Drechsler	University of Freiburg, Germany
Rolf Drechsler	Siemens AG, Germany
Peter Fleming	University of Sheffield, UK
Kary Främling	Helsinki University of Technology, Finland
Antonio Gaspar-Cunha	University of Minho, Portugal
Prabhat Hajela	Rensselaer Polytechnic Institute, USA
Thomas Hanne	Institute for Techno- and Economathematics, Germany
Alberto Herreros	University of Valladolid, Spain
Evan J. Hughes	Cranfield University, UK
Hisao Ishibuchi	Osaka Prefecture University, Japan
Andrzej Jaszkiewicz	Poznan University of Technology, Poland
Joshua D. Knowles	University of Reading, UK
Petros Koumoutsakos	ETH Zürich, Switzerland
Rajeev Kumar	IIT Kharagpur, India
Bill Langdon	University College London, UK
Marco Laumanns	ETH Zürich, Switzerland
Mark Sh. Levin	Ben-Gurion University, Israel
Daniel H. Loughlin	North Carolina State University, USA
Filippo Menczer	University of Iowa, USA
Martin Middendorf	University of Karlsruhe, Germany
Tadahiko Murata	Ashikaga Institute of Technology, Japan
Pedro Oliveira	Universidade do Minho, Portugal
Andrzej Osyczka	Cracow University of Technology, Poland
S. Ranji Ranjithan	North Carolina State University, USA
Katya Rodriguez-Vazquez	IIMAS-UNAM, Mexico
Carlos Mariano Romero	IMTA, Mexico
Ralf Salomon	University of Zürich, Switzerland
Marc Schoenauer	Ecole Polytechnique, France
Pratyush Sen	University of Newcastle, UK
Hisashi Tamaki	Kobe University, Japan
Kay Chen Tan	National University, Singapore
Dirk Thierens	Utrecht University, The Netherlands
Mark Thompson	Sheffield Hallam University, UK
Thanh Binh To	Institute of Automation and Communication Magdeburg, Germany
Marco Tomassini	University of Lausanne, Switzerland
David A. Van Veldhuizen	US Air Force, USA

Table of Contents

Tutorials

Some Methods for Nonlinear Multi-objective Optimization 1
    Kaisa Miettinen

A Short Tutorial on Evolutionary Multiobjective Optimization 21
    Carlos A. Coello Coello

Invited Talks

An Overview in Graphs of Multiple Objective Programming 41
    Ralph E. Steuer

Poor-Definition, Uncertainty, and Human Factors – Satisfying Multiple
Objectives in Real-World Decision-Making Environments 52
    I. C. Parmee

Algorithm Improvements

Controlled Elitist Non-dominated Sorting Genetic Algorithms for Better
Convergence 67
    Kalyanmoy Deb, Tushar Goel

Specification of Genetic Search Directions in Cellular Multi-objective
Genetic Algorithms 82
    Tadahiko Murata, Hisao Ishibuchi, Mitsuo Gen

Adapting Weighted Aggregation for Multiobjective Evolution Strategies .. 96
    Yaochu Jin, Tatsuya Okabe, Bernhard Sendhoff

Incrementing Multi-objective Evolutionary Algorithms: Performance
Studies and Comparisons 111
    K. C. Tan, T. H. Lee, E. F. Khor

A Micro-Genetic Algorithm for Multiobjective Optimization 126
    Carlos A. Coello Coello, Gregorio Toscano Pulido

Evolutionary Algorithms for Multicriteria Optimization with Selecting a
Representative Subset of Pareto Optimal Solutions 141
    Andrzej Osyczka, Stanislaw Krenich

Multi-objective Optimisation Based on Relation *Favour* 154
    Nicole Drechsler, Rolf Drechsler, Bernd Becker

Performance Assessment and Comparison

Comparison of Evolutionary and Deterministic Multiobjective Algorithms
for Dose Optimization in Brachytherapy 167
    Natasa Milickovic, Michael Lahanas, Dimos Baltas,
    Nikolaos Zamboglou

On The Effects of Archiving, Elitism, and Density Based Selection in
Evolutionary Multi-objective Optimization 181
    Marco Laumanns, Eckart Zitzler, Lothar Thiele

Global Multiobjective Optimization with Evolutionary Algorithms:
Selection Mechanisms and Mutation Control 197
    Thomas Hanne

Inferential Performance Assessment of Stochastic Optimisers and the
Attainment Function 213
    Viviane Grunert da Fonseca, Carlos M. Fonseca, Andreia O. Hall

A Statistical Comparison of Multiobjective Evolutionary Algorithms
Including the MOMGA-II 226
    Jesse B. Zydallis, David A. Van Veldhuizen, Gary B. Lamont

Performance of Multiple Objective Evolutionary Algorithms on a
Distribution System Design Problem – Computational Experiment 241
    Andrzej Jaszkiewicz, Maciej Hapke, Paweł Kominek

Constraint Handling and Problem Decomposition

An Infeasibility Objective for Use in Constrained Pareto Optimization 256
    Jonathan Wright, Heather Loosemore

Reducing Local Optima in Single-Objective Problems by
Multi-objectivization 269
    Joshua D. Knowles, Richard A. Watson, David W. Corne

Constrained Test Problems for Multi-objective Evolutionary Optimization 284
    Kalyanmoy Deb, Amrit Pratap, T. Meyarivan

Constraint Method-Based Evolutionary Algorithm (CMEA) for
Multiobjective Optimization... 299
    S. Ranji Ranjithan, S. Kishan Chetan, Harish K. Dakshina

Uncertainty and Noise

Pareto-Front Exploration with Uncertain Objectives..................... 314
    Jürgen Teich

Evolutionary Multi-objective Ranking with Uncertainty and Noise 329
    Evan J. Hughes

Hybrid and Alternative Methods

Tabu-Based Exploratory Evolutionary Algorithm for Effective
Multi-objective Optimization 344
    E. F. Khor, K. C. Tan, T. H. Lee

Bi-Criterion Optimization with Multi Colony Ant Algorithms 359
    Steffen Iredi, Daniel Merkle, Martin Middendorf

Multicriteria Evolutionary Algorithm with Tabu Search for
Task Assignment 373
    Jerzy Balicki, Zygmunt Kitowski

A Hybrid Multi-objective Evolutionary Approach to Engineering Shape
Design 385
    Kalyanmoy Deb, Tushar Goel

Fuzzy Evolutionary Hybrid Metaheuristic for Network Topology Design ... 400
    Habib Youssef, Sadiq M. Sait, Salman A. Khan

A Hybrid Evolutionary Approach for Multicriteria Optimization Problems:
Application to the Flow Shop 416
    El-Ghazali Talbi, Malek Rahoual, Mohamed Hakim Mabed,
    Clarisse Dhaenens

The Supported Solutions Used as a Genetic Information in a Population
Heuristics 429
    Xavier Gandibleux, Hiroyuki Morita, Naoki Katoh

Scheduling

Multi-objective Flow-Shop: Preliminary Results 443
    C. Brizuela, N. Sannomiya, Y. Zhao

Pareto-Optimal Solutions for Multi-objective Production
Scheduling Problems 458
    Tapan P. Bagchi

Comparison of Multiple Objective Genetic Algorithms for Parallel Machine
Scheduling Problems 472
    W. Matthew Carlyle, Bosun Kim, John W. Fowler, Esma S. Gel

Applications

A Bi-Criterion Approach for the Airlines Crew Rostering Problem 486
    Walid El Moudani, Carlos Alberto Nunes Cosenza, Marc de Coligny,
    Félix Mora-Camino

Halftone Image Generation with Improved Multiobjective
Genetic Algorithm 501
    Hernán E. Aguirre, Kiyoshi Tanaka, Tatsuo Sugimura, Shinjiro Oshita

Microchannel Optimization Using Multiobjective Evolution Strategies 516
    Ivo F. Sbalzarini, Sibylle Müller, Petros Koumoutsakos

Multi-objective Optimisation of Cancer Chemotherapy Using Evolutionary
Algorithms 531
    Andrei Petrovski, John McCall

Application of Multi Objective Evolutionary Algorithms to
Analogue Filter Tuning 546
    Mark Thompson

Multiobjective Design Optimization of Real-Life Devices in
Electrical Engineering: A Cost-Effective Evolutionary Approach 560
    P. Di Barba, M. Farina, A. Savini

Application of Multiobjective Evolutionary Algorithms for Dose
Optimization Problems in Brachytherapy 574
    Michael Lahanas, Natasa Milickovic, Dimos Baltas,
    Nikolaos Zamboglou

Multiobjective Optimization in Linguistic Rule Extraction from Numerical
Data 588
    Hisao Ishibuchi, Tomoharu Nakashima, Tadahiko Murata

Determining the Color-Efficiency Pareto Optimal Surface for Filtered Light
Sources 603
    Neil H. Eklund, Mark J. Embrechts

Multi-objective Design Space Exploration of Road Trains with
Evolutionary Algorithms 612
    Nando Laumanns, Marco Laumanns, Dirk Neunzig

Multiobjective Optimization of Mixed Variable Design Problems 624
    Johan Andersson, Petter Krus

Aerodynamic Shape Optimization of Supersonic Wings by Adaptive Range
Multiobjective Genetic Algorithms 639
    Daisuke Sasaki, Masashi Morikawa, Shigeru Obayashi,
    Kazuhiro Nakahashi

Accurate, Transparent, and Compact Fuzzy Models for Function
Approximation and Dynamic Modeling through Multi-objective
Evolutionary Optimization 653
    Fernando Jiménez, Antonio F. Gómez-Skarmeta, Hans Roubos,
    Robert Babuška

Multi-objective Evolutionary Design of Fuzzy Autopilot Controller 668
   Anna L. Blumel, Evan J. Hughes, Brian A. White

The Niched Pareto Genetic Algorithm 2 Applied to the Design of
Groundwater Remediation Systems.................................... 681
   Mark Erickson, Alex Mayer, Jeffrey Horn

MOLeCS: Using Multiobjective Evolutionary Algorithms for Learning 696
   Ester Bernadó i Mansilla, Josep M. Garrell i Guiu

Author Index 711

Some Methods for Nonlinear Multi-objective Optimization

Kaisa Miettinen*

University of Jyväskylä, Department of Mathematical Information Technology,
P.O. Box 35 (Agora), FIN-40351 Jyväskylä, Finland
miettine@mit.jyu.fi
WWW home page: http://www.mit.jyu.fi/miettine/engl.html

Abstract. A general overview of nonlinear multiobjective optimization methods is given. The basic features of several methods are introduced so that an appropriate method could be found for different purposes. The methods are classified according to the role of a decision maker in the solution process. The main emphasis is devoted to interactive methods where the decision maker progressively provides preference information so that the most satisfactory solution can be found.

1 Introduction

Multiple criteria decision making (MCDM) problems form an extensive field where the best possible compromise should be found by evaluating several conflicting objectives. There is a good reason to classify such problems on the basis of their different characteristics. Here we concentrate on problems involving continuous nonlinear functions with deterministic values. We present versatile methods for solving such problems.

The solution process usually requires the participation of a human decision maker (DM) who can give preference information related to conflicting goals. Here we assume that a single DM is involved.

Methods are divided into four classes according to the role of the DM. Either no DM takes part in the solution process or (s)he expresses preference relations before, after or during the process. The last-mentioned, interactive, methods form the most extensive class of methods.

Multiobjective optimization problems are usually solved by scalarization. *Scalarization* means that the problem is converted into one single or a family of single objective optimization problems. This new problem has a real-valued objective function that possibly depends on some parameters and it can be solved using single objective optimizers.

Further information about the methodology of deterministic multiobjective optimization can be found, e.g., in the monographs [6,12,14,20,27,30]. For a more detailed presentation of the methods treated here as well as other related topics we refer to [20] and references therein.

---

* This research was supported by the Academy of Finland, grant #65760.

2 Concepts and Background

A *multiobjective optimization problem* is of the form

$$\begin{aligned}\text{minimize} \quad & \{f_1(\boldsymbol{x}), f_2(\boldsymbol{x}), \ldots, f_k(\boldsymbol{x})\} \\ \text{subject to} \quad & \boldsymbol{x} \in S\end{aligned} \quad (1)$$

involving k (≥ 2) conflicting *objective functions* $f_i : \mathbb{R}^n \to \mathbb{R}$ that we want to minimize simultaneously. The *decision (variable) vectors* $\boldsymbol{x} = (x_1, x_2, \ldots, x_n)^T$ belong to the (nonempty) *feasible region* $S \subset \mathbb{R}^n$. The feasible region is formed by *constraint functions* but we do not fix them here.

We denote the image of the feasible region by $Z \subset \mathbb{R}^k$ and call it a *feasible objective region*. The elements of Z are called *objective vectors* and they consist of *objective (function) values* $\boldsymbol{f}(\boldsymbol{x}) = (f_1(\boldsymbol{x}), f_2(\boldsymbol{x}), \ldots, f_k(\boldsymbol{x}))^T$. Note that if f_i is to be maximized, it is equivalent to minimize $-f_i$.

In what follows, a function is called *nondifferentiable* if it is locally Lipschitzian (and not necessarily continuously differentiable).

Definition 1. *When all the objective and the constraint functions are linear, the problem is called* linear *or an* MOLP *problem. If at least one of the functions is nonlinear, the problem is a* nonlinear multiobjective optimization problem. *Correspondingly, the problem is* nondifferentiable *is some of the functions is nondifferentiable and* convex *if all the objective functions and the feasible region are convex.*

Because of the contradiction and possible incommensurability of the objective functions, it is not possible to find a single solution that would optimize all the objectives simultaneously. In multiobjective optimization, vectors are regarded as optimal if their components cannot be improved without deterioration to at least one of the other components. This is usually called Pareto optimality.

Definition 2. *A decision vector $\boldsymbol{x}^* \in S$ is* Pareto optimal *if there does not exist another $\boldsymbol{x} \in S$ such that $f_i(\boldsymbol{x}) \leq f_i(\boldsymbol{x}^*)$ for all $i = 1, \ldots, k$ and $f_j(\boldsymbol{x}) < f_j(\boldsymbol{x}^*)$ for at least one index j. An objective vector is Pareto optimal if the corresponding decision vector is Pareto optimal.*

There are usually a lot (infinite number) of Pareto optimal solutions and they form a set of Pareto optimal solutions or a *Pareto optimal set*. This set can be nonconvex and nonconnected.

Definition 2 introduces *global Pareto optimality*. Another important concept is local Pareto optimality defined in a small environment of the point considered. Naturally, any globally Pareto optimal solution is locally Pareto optimal. The converse is valid for convex problems. To be more specific, if the feasible region is convex and the objective functions are quasiconvex with at least one strictly quasiconvex function, then locally Pareto optimal solutions are also globally Pareto optimal.

Other related optimality concepts are weak and proper Pareto optimality. The properly Pareto optimal set is a subset of the Pareto optimal set which is a subset of the weakly Pareto optimal set.

A vector is *weakly Pareto optimal* if there does not exist any other feasible vector for which all the components are better. In other words, when compared to Definition 2, all the inequalities are strict. Weakly Pareto optimal solutions are often relevant from a technical point of view because they are sometimes easier to generate than Pareto optimal points.

Pareto optimal solutions can be divided into improperly and properly Pareto optimal ones according to whether unbounded trade-offs between objectives are allowed or not. Proper Pareto optimality can be defined in several ways (see, e.g., [20]). According to [9] a solution is properly Pareto optimal if there is at least one pair of objectives for which a finite decrement in one objective is possible only at the expense of some reasonable increment in the other objective.

Mathematically, all the Pareto optimal points are equally acceptable solutions of the multiobjective optimization problem. However, it is generally desirable to obtain one point as a solution. Selecting one out of the set of Pareto optimal solutions calls for a *decision maker (DM)*. (S)he is a person who has better insight into the problem and who can express preference relations between different solutions.

Finding a solution to (1) is called a *solution process*. It usually means the co-operation of the DM and an analyst. An *analyst* is a person or a computer program responsible for the mathematical side of the solution process. The analyst generates information for the DM to consider and the solution is selected according to the preferences of the DM.

By solving a multiobjective optimization problem we here mean finding a feasible decision vector such that it is Pareto optimal and satisfies the DM. Assuming such a solution exists, it is called a *final solution*.

The ranges of the Pareto optimal set provide valuable information for the solution process if the objective functions are bounded over the feasible region. The components $z_i^\star$ of the *ideal objective vector* $\boldsymbol{z}^\star \in \mathbb{R}^k$ are obtained by minimizing each of the objective functions individually subject to the constraints. The ideal objective vector is not feasible because of the conflict among the objectives. From the ideal objective vector we obtain the lower bounds of the Pareto optimal set. Note that in nonconvex problems we need a global optimizer for calculating the ideal objective vector.

The upper bounds of the Pareto optimal set, that is, the components of a *nadir objective vector* $\boldsymbol{z}^{\mathrm{nad}}$, are usually rather difficult to obtain. They can be estimated from a payoff table (see, e.g., [20]) but this is not a reliable way as can be seen, e.g., in [15,31].

For nonlinear problems, there is no constructive method for calculating the nadir objective vector. Nonetheless, the payoff table may be used as a rough estimate as long as its robustness is kept in mind. Because of the above-described difficulty of calculating the actual nadir objective vector, $\boldsymbol{z}^{\mathrm{nad}}$ is usually an approximation.

Sometimes we need a vector that is strictly better than every Pareto optimal solution. Such a vector is called a *utopian objective vector* $z^{\star\star} \in \mathbb{R}^k$ and its components are formed by decreasing the components of $z^\star$ by a positive scalar.

It is often assumed that the DM makes decisions on the basis of an underlying function. This function representing the preferences of the DM is called a *value function* $U : \mathbb{R}^k \to \mathbb{R}$ (see [14]). In many methods, the value (or utility) function is assumed to be known implicitly.

Value functions are important in the development of solution methods and as a theoretical background. Generally, the value function is assumed to be strongly decreasing. This means that the preference of the DM will increase if the value of an objective function decreases while all the other objective values remain unchanged (i.e., less is preferred to more). In this case, the maximum of U is Pareto optimal. Regardless of the existence of a value function, it is usually assumed that less is preferred to more by the DM.

Instead of as a maximum of the value function, a final solution can be understood as a satisficing one. *Satisficing decision making* means that the DM does not intend to maximize any value function but tries to achieve certain aspirations. A solution which satisfies all the aspirations of the DM is called a *satisficing solution*.

During solution processes, various kinds of information are solicited from the DM. *Aspiration levels* $\bar{z}_i$, $i = 1, \ldots, k$, are such desirable or acceptable levels in the objective function values that are of special interest and importance to the DM. The vector $\bar{z} \in \mathbb{R}^k$ is called a *reference point*.

According to the definition of Pareto optimality, moving from one Pareto optimal solution to another necessitates trading off. This is one of the basic concepts in multiobjective optimization. A *trade-off* reflects the ratio of change in the values of the objective functions concerning the increment of one objective function that occurs when the value of some other objective function decreases. For details, see, e.g., [6,20].

It is said that two feasible solutions are situated on the same *indifference curve* if the DM finds them equally desirable. For any two Pareto optimal solutions on the same indifference curve there is a trade-off involving a certain increment in one objective function value that the DM can tolerate in exchange for a certain amount of decrement in some other objective function while the preferences of the two solutions remain the same. This is called the *marginal rate of substitution* $m_{ij}(x^\star)$ $(i, j = 1, \ldots, k,\ i \neq j)$.

To conclude this section, let us have a look at how the Pareto optimality of a feasible decision vector can be tested. This topic is investigated, e.g., in [1,20]. A decision vector $x^\star \in S$ is Pareto optimal if and only if the problem

$$\begin{array}{ll} \text{minimize} & \sum_{i=1}^{k} \varepsilon_i \\ \text{subject to} & f_i(x) + \varepsilon_i = f_i(x^\star) \quad \text{for all} \ \ i = 1, \ldots, k, \\ & \varepsilon_i \geq 0 \quad \text{for all} \ \ i = 1, \ldots, k, \\ & x \in S \end{array} \qquad (2)$$

has an optimal objective function value of zero, where both $x \in \mathbb{R}^n$ and $\varepsilon \in \mathbb{R}^k_+$ are variables. On the other hand, if (2) has a finite nonzero optimal objective function value obtained at $\hat{x}$, then $\hat{x}$ is Pareto optimal. Note that the equalities in (2) can be replaced with inequalities.

3 Methods

Mathematically, the multiobjective optimization problem is considered to be solved when the Pareto optimal set is found. This is also known as *vector optimization*. However, this is not always enough. Instead, we want to obtain one final solution. This means that we must find a way to order the Pareto optimal solutions and here we need a DM and her/his preferences.

In what follows, we present several methods for finding a final solution. We cannot cover every existing method but we introduce several philosophies and ways of approaching the problem.

The methods can be classified in many ways. Here we apply the classification presented in [12] based on the participation of the DM in the solution process. The classes are *no-preference methods*, *a posteriori methods*, *a priori methods* and *interactive methods*. Note that no classification can be complete and overlapping and combinations of classes are possible.

In addition, we consider an alternative way of classification into ad hoc and non ad hoc methods. This division, suggested in [29], is based on the existence of an underlying value function. Even if one knew the DM's value function, one would not exactly know how to respond to the questions posed by an *ad hoc* algorithm. On the other hand, in *non ad hoc* methods the responses can be determined or at least confidently simulated with the help of a value function.

In no-preference methods, the opinions of the DM are not taken into consideration. Thus, the problem is solved using some relatively simple method and the solution is presented to the DM who may either accept or reject it. For details if this class see, e.g., [20]. Next we introduce examples of a posteriori, a priori and interactive methods.

4 A Posteriori Methods

A posteriori methods could also be called *methods for generating Pareto optimal solutions*. After the Pareto optimal set (or a part of it) has been generated, it is presented to the DM, who selects the most preferred solution. The inconveniences here are that the generation process is usually computationally expensive and sometimes in part, at least, difficult. On the other hand, it is hard for the DM to select from a large set of alternatives. An important question related to this is how to display the alternatives to the DM in an illustrative way.

If there are only two objective functions, the Pareto optimal set can be generated parametrically (see, e.g., [2,8]). The problem becomes more complicated with more objectives.

4.1 Weighting Method

In the weighting method (see, e.g. [8,35]), we solve the problem

$$\begin{array}{ll} \text{minimize} & \sum_{i=1}^{k} w_i f_i(\boldsymbol{x}) \\ \text{subject to} & \boldsymbol{x} \in S \;, \end{array} \qquad (3)$$

where $w_i \geq 0$ for all $i = 1, \ldots, k$ and $\sum_{i=1}^{k} w_i = 1$. The solution of (3) is weakly Pareto optimal and it is Pareto optimal if $w_i > 0$ for all $i = 1, \ldots, k$ or the solution is unique.

The weakness of the weighting method is that not all of the Pareto optimal solutions can be found unless the problem is convex. The conditions under which the whole Pareto optimal set can be generated by the weighting method with positive weights are presented in [5].

Systematic ways of perturbing the weights to obtain different Pareto optimal solutions are suggested, e.g., in [6] (pp. 234–236). In addition, an algorithm for generating different weights automatically for convex problems to produce an approximation of the Pareto optimal set is proposed in [4].

The method has several weaknesses. On the one hand, a small change in the weights may cause big changes in the objective vectors. On the other hand, dramatically different weights may produce nearly similar objective vectors. In addition, an evenly distributed set of weights does not necessarily produce an evenly distributed representation of the Pareto optimal set.

4.2 ε-Constraint Method

In the ε-constraint method, introduced in [11], one of the objective functions is optimized in the form

$$\begin{array}{ll} \text{minimize} & f_\ell(\boldsymbol{x}) \\ \text{subject to} & f_j(\boldsymbol{x}) \leq \varepsilon_j \ \text{ for all } \ j = 1, \ldots, k, \ j \neq \ell, \\ & \boldsymbol{x} \in S \;, \end{array} \qquad (4)$$

where $\ell \in \{1, \ldots, k\}$ and ε_j are upper bounds for the objectives $j \neq \ell$.

The solution of (4) is weakly Pareto optimal. On the other hand, $\boldsymbol{x}^* \in S$ is Pareto optimal if and only if it solves (4) for every $\ell = 1, \ldots, k$, where $\varepsilon_j = f_j(\boldsymbol{x}^*)$ for $j = 1, \ldots, k, j \neq \ell$. In addition, the unique solution of (4) is Pareto optimal for any upper bounds. Thus, finding any Pareto optimal solution does not necessitate convexity.

In order to ensure Pareto optimality in this method, we have to either solve k different problems or obtain a unique solution. In general, uniqueness is not necessarily easy to verify. Systematic ways of perturbing the upper bounds to obtain different Pareto optimal solutions are suggested in [6] (pp. 283–295).

4.3 Method of Weighted Metrics

In the method of weighted metrics, the distance between some reference point and the feasible objective region is minimized. A common way is to use the ideal objective vector and L_p-metrics. We can produce different solutions by weighting the metrics. This method is also sometimes called *compromise programming*.

The solution obtained depends greatly on the value chosen for p. For $1 \leq p < \infty$ we have a problem

$$\begin{array}{ll} \text{minimize} & \left(\sum_{i=1}^{k} w_i |f_i(\boldsymbol{x}) - z_i^\star|^p\right)^{1/p} \\ \text{subject to} & \boldsymbol{x} \in S \ . \end{array} \quad (5)$$

The exponent $1/p$ can be dropped. For $p = \infty$ we have a *weighted Tchebycheff problem*

$$\begin{array}{ll} \text{minimize} & \max_{i=1,\ldots,k} \left[w_i |f_i(\boldsymbol{x}) - z_i^\star| \right] \\ \text{subject to} & \boldsymbol{x} \in S \ . \end{array} \quad (6)$$

Notice that no absolute values are needed if we know the global ideal objective vector. The solution of (5) is Pareto optimal if either the solution is unique or all the weights are positive. Furthermore, the solution of (6) is weakly Pareto optimal for positive weights. Finally, (6) has at least one Pareto optimal solution.

Convexity of the problem is needed in order to guarantee that every Pareto optimal solution can be found by (5). On the other hand, any Pareto optimal solution can be found by (6) when $\boldsymbol{z}^{\star\star}$ is used as a reference point.

Weakly Pareto optimal solutions can be avoided in (6) by giving a slight slope to the contour of the metric (see [27]). The augmented problem to be solved is

$$\begin{array}{ll} \text{minimize} & \max_{i=1,\ldots,k} \left[w_i |f_i(\boldsymbol{x}) - z_i^{\star\star}| \right] + \rho \sum_{i=1}^{k} |f_i(\boldsymbol{x}) - z_i^{\star\star}| \\ \text{subject to} & \boldsymbol{x} \in S \ , \end{array} \quad (7)$$

where ρ is a sufficiently small positive scalar. In this case, it may be impossible to find every Pareto optimal solution. Instead, (7) generates properly Pareto optimal solutions and any properly Pareto optimal solution can be found.

Let us mention that different connections between the weighting method, the ε-constraint method and the method of weighted metrics are presented in [17].

4.4 Achievement Scalarizing Function Approach

Scalarizing functions of a special type are called *achievement (scalarizing) functions*. They have been introduced by Wierzbicki, e.g., in [32,33]. These functions are of the form $s_{\bar{z}} : Z \to \mathbb{R}$, where $\bar{z} \in \mathbb{R}^k$ is an arbitrary reference point. Because we do not know Z explicitly, in practice we minimize the function $s_{\bar{z}}(\boldsymbol{f}(\boldsymbol{x}))$ subject to $\boldsymbol{x} \in S$.

We can define so-called order-representing and order-approximating achievement functions. Then we have the following properties for any reference point: If the achievement function is order-representing, then its solution is weakly Pareto optimal and if the function is order-approximating, then its solution is Pareto optimal. On the other hand, any (weakly) Pareto optimal solution can be found if the achievement function is order-representing. Thus, weakly Pareto optimal or Pareto optimal solutions can be obtained by moving the reference point only.

There are many achievement functions satisfying the above-presented conditions. An example of order-representing functions is $s_{\bar{z}}(z) = \max_{i=1,\ldots,k}[w_i(z_i - \bar{z}_i)]$, where w is some fixed positive weighting vector. An example of order-approximating achievement functions is

$$s_{\bar{z}}(z) = \max_{i=1,\ldots,k}[w_i(z_i - \bar{z}_i)] + \rho \sum_{i=1}^{k} w_i(z_i - \bar{z}_i) , \qquad (8)$$

where w is as above and $\rho > 0$.

5 A Priori Methods

In a priori methods, the DM must specify her/his preferences, hopes and opinions before the solution process. Unfortunately, the DM does not necessarily know beforehand what it is possible to attain in the problem and how realistic her/his expectations are.

5.1 Value Function Method

The value function approach was already mentioned earlier. It is an excellent method if the DM happens to know an explicit mathematical formulation for the value function and if that function represents wholly her/his preferences. Unfortunately, it may be difficult, if not impossible, to get that mathematical expression. On the other hand, it can be difficult to optimize because of its possible complicated nature.

Note that the DM's preferences must satisfy certain conditions so that a value function can be defined on them. The DM must, e.g., be able to specify consistent preferences.

5.2 Lexicographic Ordering

In lexicographic ordering, the DM must arrange the objective functions according to their absolute importance. This ordering means that a more important objective is infinitely more important than a less important objective. After ordering, the most important objective function is minimized subject to the original constraints. If this problem has a unique solution, it is the final one. Otherwise, the second most important objective function is minimized. Now, a new constraint is added to guarantee that the most important objective function preserves its optimal value. If this problem has a unique solution, it is the final one. Otherwise, the process goes on.

The solution of lexicographic ordering is Pareto optimal. The method is quite simple and people usually make decisions successively. However, the DM may have difficulties in specifying an absolute order of importance. Besides, it is very likely that the process stops before less important objective functions are taken into consideration.

Note that lexicographic ordering does not allow a small increment of an important objective function to be traded off with a great decrement of a less important objective. Yet, trading off might often be appealing to the DM.

5.3 Goal Programming

Goal programming is one of the first methods expressly created for multiobjective optimization. The DM is asked to specify aspiration levels $\bar{z}_i$ ($i = 1, \ldots, k$) for the objective functions and deviations from these aspiration levels are minimized. An objective function jointly with an aspiration level forms a *goal*. For minimization problems, goals are of the form $f_i(\boldsymbol{x}) \leq \bar{z}_i$. Here the aspiration levels are assumed to be selected so that they are not achievable simultaneously. Next, the *overachievements* δ_i of the objective function values are minimized.

The method has several variants. In the *weighted* approach, see [7], the weighted sum of the deviational variables is minimized. This means that in addition to the aspiration levels, the DM must specify positive weights. Then we have a problem

$$
\begin{array}{ll}
\text{minimize} & \sum_{i=1}^{k} w_i \delta_i \\
\text{subject to} & f_i(\boldsymbol{x}) - \delta_i \leq \bar{z}_i \text{ for all } i = 1, \ldots, k, \\
& \delta_i \geq 0 \text{ for all } i = 1, \ldots, k, \\
& \boldsymbol{x} \in S ,
\end{array}
\qquad (9)
$$

where $\boldsymbol{x} \in \mathbb{R}^n$ and δ_i ($i = 1, \ldots, k$) are the variables and $\delta_i = \max[0, f_i(\boldsymbol{x}) - \bar{z}_i]$.

In the *lexicographic* approach, the DM must specify a lexicographic order for the goals in addition to the aspiration levels. After the lexicographic ordering, the problem with the deviational variables as objective functions is solved subject to the constraints of (9) as explained in Sect. 5.2.

A combination of the weighted and the lexicographic approaches is quite popular. In this case, several objective functions may belong to the same class of importance in the lexicographic order. In each priority class, a weighted sum of the deviational variables is minimized.

The solution of a weighted or a lexicographic goal programming problem is Pareto optimal if either the aspiration levels form a Pareto optimal reference point or all the deviational variables δ_i have positive values at the optimum. In other words, if the aspiration levels are all feasible, the solution is equal to the reference point that is not necessarily Pareto optimal.

Goal programming is a very widely used and popular solution method. Goal-setting is an understandable and easy way of making decisions. The specification of the weights or the lexicographic ordering may be more difficult. It may also be hard to specify weights because they have no direct physical meaning.

6 Interactive Methods

The extensive interest devoted to interactive methods can be explained by the fact that assuming the DM has enough time and capabilities for co-operation, interactive methods can be presumed to produce the most satisfactory results. Many of the weak points of the methods in the other method classes are overcome. Namely, only part of the Pareto optimal points has to be generated and evaluated, the DM is not overloaded with information, and the DM can specify and correct her/his preferences and selections as the solution process continues and (s)he gets to know the problem and its potentialities better. This also means that the DM does not have to know any global preference structure. In addition, the DM can be assumed to have more confidence in the final solution since (s)he is involved throughout the solution process.

In interactive methods, the DM works together with an analyst or an interactive computer program. One can say that the analyst tries to determine the preference structure of the DM in an interactive way. After every iteration, some information is given to the DM and (s)he is asked to answer some questions or provide some other type of information.

Interactive methods differ from each other by the form in which information is given to the DM, by the form in which information is provided by the DM, and how the problem is transformed into a single objective optimization problem. It is always important that the DM finds the method worthwhile and acceptable and is able to use it properly.

There are three main stopping criteria in interactive methods. Either the DM finds a desirable solution and is convinced that it is preferred to all the other Pareto optimal solutions (see [16]), some algorithmic stopping or convergence rule is fulfilled or the DM gets tired of the solution process.

The number of interactive methods is large. Here we briefly describe some of them. In all the methods, less is assumed to be preferred to more by the DM. For more details, see [20] and references therein.

6.1 Geoffrion-Dyer-Feinberg Method

The Geoffrion-Dyer-Feinberg (GDF) method, proposed in [10], is one of the most well-known interactive methods and it is based on the maximization of the underlying (implicitly known) value function. The objective functions are assumed to be continuously differentiable and the feasible region S must be compact and convex.

Marginal rates of substitution specified by the DM at the current point $\boldsymbol{x}^h$ are here used to approximate the direction of steepest ascent of the value function. We have $m_{ij}(\boldsymbol{x}^h) = m_i = \frac{\partial U(\boldsymbol{f}(\boldsymbol{x}^h))}{\partial f_j} / \frac{\partial U(\boldsymbol{f}(\boldsymbol{x}^h))}{\partial f_i}$. Then the approximation is optimized by the method of Frank and Wolfe by solving the problem

$$\begin{aligned} \text{minimize} \quad & \left(\sum_{i=1}^{k} -m_i \nabla_x f_i(\boldsymbol{x}^h) \right)^T \boldsymbol{y} \\ \text{subject to} \quad & \boldsymbol{y} \in S \end{aligned} \qquad (10)$$

with $\boldsymbol{y} \in \mathbb{R}^n$ being the variable.

The basic phases of the GDF algorithm are the following.
1. Ask the DM to specify a reference function.
2. Ask the DM to specify marginal rates of substitution between the reference function and the other objectives at the current solution point.
3. Solve (10). Set the search direction as the difference between the old (i.e. current) and the new solution.
4. Determine with the help of the DM the appropriate step-size to be taken in the direction.
5. If the DM wants to continue, go to step (2). Otherwise, stop.

When determining the step-size, the DM is asked to select the most preferred objective vector obtained with different step-sizes taken in the search direction. Note that the alternatives are not necessarily Pareto optimal. It is obvious that the task of selection becomes more difficult for the DM as the number of objective functions increases.

The GDF method can be characterized to be a non ad hoc method. If one knows the value function, it is easy to specify the marginal rates of substitution and select the best alternative. In spite of the plausible theoretical foundation of the GDF method, it is not so convincing and powerful in practice. The most important difficulty for the DM is the determining of the $k - 1$ marginal rates of substitution at each iteration. Even more difficult is to give consistent and correct marginal rates of substitution at every iteration.

6.2 Tchebycheff Method

The Tchebycheff method, presented in [27] (pp. 419–450) and refined in [28], is an interactive weighting vector space reduction method where value functions are not used. The method has been designed to be user-friendly for the DM and, thus, complicated information is not required. It is assumed that the objective functions are bounded (from below).

To start with, a utopian objective vector is established. Then the distance from the utopian objective vector to the feasible objective region, measured by the weighted Tchebycheff metric, is minimized. Different solutions are obtained with well dispersed positive weighting vectors in the metric, as introduced in Sect. 4.3.

At the first iteration, a sample of the whole Pareto optimal set is generated. The solution space is reduced by tightening the upper and the lower bounds for the weights. Then a concentrated group of weighting vectors centred about the selected one is formed. Thus, the idea is to develop a sequence of progressively smaller subsets of the Pareto optimal set until a final solution is located.

Every Pareto optimal solution of can be found by solving the weighted Tchebycheff problem with $z^{\star\star}$ but some of the solutions may be weakly Pareto optimal. Here this weakness is overcome by formulating the distance minimization problem in a lexicographic form:

$$\text{lex minimize} \quad \max_{i=1,\ldots,k}\left[w_i(f_i(\boldsymbol{x}) - z_i^{\star\star})\right], \sum_{i=1}^{k}(f_i(\boldsymbol{x}) - z_i^{\star\star}) \quad (11)$$
$$\text{subject to} \quad \boldsymbol{x} \in S \ .$$

The solution of (11) is Pareto optimal and any Pareto optimal solution can be found.

The number of the alternative objective vectors to be presented to the DM is denoted by P. It may be fixed or different at each iteration. We can now present the main steps of the Tchebycheff algorithm.

1. Specify values for the set size P and the number of iterations H. Construct the utopian objective vector. Set $h = 1$.
2. Form the current weighting vector space and generate $2P$ dispersed weighting vectors.
3. Solve (11) for each of the $2P$ weighting vectors.
4. Present the P most different of the resulting objective vectors to the DM and let her/him choose the most preferred among them.
5. If $h = H$, stop. Otherwise, gather information for reducing the weighting vector space, set $h = h + 1$ and go to step (2).

The predetermined number of iterations is not necessarily conclusive. The DM can stop iterating when (s)he obtains a satisfactory solution or continue the solution process longer if necessary.

All the DM has to do in the Tchebycheff method is to compare several Pareto optimal objective vectors and select the most preferred one. The ease of the comparison depends on the magnitude of P and on the number of objective functions. This can be characterized as a non ad hoc method. If the value function is known, it is easy to select the alternative maximizing the value function.

The flexibility of the method is reduced by the fact that the discarded parts of the weighting vector space cannot be restored if the DM changes her/his mind. Thus, some consistency is required. The weakness of the Tchebycheff method is that a great deal of calculation is needed at each iteration and many of the results are discarded. For large and complex problems, the Tchebycheff method is not a realistic choice. On the other hand, parallel computing can be utilized.

6.3 Reference Point Method

As its name suggests, the reference point method (see, e.g., [32]) is based on a reference point of aspiration levels. The reference point is used to derive achievement scalarizing functions as introduced in Sect. 4.4. No specific assumptions are set on the problem to be solved. The reference point idea has been utilized in several methods in different ways. Wierzbicki's reference point method (to be discussed here) was among the first of them.

Before the solution process starts, some information is given to the DM about the problem. If possible, the ideal objective vector and the (approximated) nadir objective vector are presented to illustrate the ranges of the Pareto optimal set. Another possibility is to minimize and maximize the objective functions individually in the feasible region (if it is bounded). An appropriate form for the achievement function must also be selected.

The basic steps of the reference point method are the following:

1. Present information about the problem to the DM.
2. Ask the DM to specify a reference point.
3. Minimize the achievement function. Present the solution to the DM.
4. Calculate a number of k other (weakly) Pareto optimal solutions by minimizing the achievement function with perturbed reference points.
5. Present the alternatives to the DM. If (s)he finds one of the $k+1$ solutions satisfactory, stop. Otherwise, go to step (2).

The reference point method can be characterized as an ad hoc method or a method having both non ad hoc and ad hoc features. Alternatives are easy to compare if the value function is known. On the other hand, a reference point cannot be directly defined with the help of the value function. However, it is possible to test whether a new reference point has a higher value function value than the earlier solutions.

The reference point method is quite easy for the DM to understand. The DM only has to specify appropriate aspiration levels and compare objective vectors. What has earlier been said about the comparison of alternatives is also valid here. The solutions are weakly or Pareto optimal depending on the achievement function employed.

The freedom of the DM has both positive and negative aspects. On the one hand, the DM can direct the solution process and is free to change her/his mind during the process. On the other hand, there is no clear strategy to produce the final solution since the method does not help the DM to find improved solutions. A software family called DIDAS (Dynamic Interactive Decision Analysis and Support) has been developed on the basis of the reference point ideas (see [34] for details).

6.4 GUESS Method

The GUESS method is a simple interactive method related to the reference point method. The method is also sometimes called a *naïve method* and it is presented in [3]. The ideal objective vector $z^\star$ and the nadir objective vector z^{nad} are required to be available.

The method proceeds as follows. The DM specifies a reference point (or a guess) below the nadir objective vector and the minimum weighted deviation from the nadir objective vector is maximized. Then the DM specifies a new reference point and the iteration continues until the DM is satisfied with the solution produced.

The problem to be solved is

$$\begin{aligned}&\text{minimize}\quad \min_{i=1,\ldots,k}\left[\frac{z_i^{\mathrm{nad}}-f_i(\boldsymbol{x})}{z_i^{\mathrm{nad}}-\bar{z}_i}\right]\\&\text{subject to}\quad \boldsymbol{x}\in S\ .\end{aligned} \qquad (12)$$

The solution of (12) is weakly Pareto optimal and any Pareto optimal solution can be found. The algorithm can be formulated as follows.

1. Calculate the ideal objective vector and the nadir objective vector and present them to the DM.
2. Let the DM specify upper or lower bounds to the objective functions if (s)he so desires. Update (12), if necessary.
3. Ask the DM to specify a reference point between the ideal and the nadir objective vectors.
4. Solve (12) and present the solution to the DM.
5. If the DM is satisfied, stop. Otherwise, go to step (2).

The only stopping rule is the satisfaction of the DM. No guidance is given to the DM in setting new aspiration levels. This is typical of many reference point-based methods. The GUESS method is an ad hoc method. The existence of a value function would not help in determining new reference points or upper or lower bounds for the objective functions.

The weakness of the GUESS method is its heavy reliance on the availability of the nadir objective vector. As mentioned earlier, the nadir objective vector is not easy to determine and it is usually only an approximation.

An interesting practical observation is mentioned in [3]. Namely, DMs are easily satisfied if there is a small difference between the reference point and the solution obtained. Somehow they feel a need to be satisfied when they have almost achieved what they wanted. In this case they may stop iterating 'too early.' The DM is naturally allowed to stop the solution process if the solution really is satisfactory. But, the coincidence of setting the reference point near an attainable solution may unnecessarily increase the DM's satisfaction.

6.5 Satisficing Trade-Off Method

The satisficing trade-off method (STOM) (see, e.g., [25]) is based on ideas similar to the two earlier methods with emphasis on finding a satisficing solution. The differentiating factor is the trade-off information utilized.

The functioning of STOM is the following. After a (weakly) Pareto optimal solution has been obtained by optimizing a scalarizing function, it is presented to the DM. On the basis of this information (s)he is asked to classify the objective functions into three classes. The classes are the unacceptable objective functions whose values (s)he wants to improve, the acceptable objective functions whose values (s)he agrees to relax (impair) and the acceptable objective functions whose values (s)he accepts as they are.

The objective and the constraint functions are assumed to be twice continuously differentiable. Under some additional special assumptions, trade-off information can be obtained from the KKT multipliers related to the scalarizing function. With this information, appropriate upper bounds can be determined for the functions to be relaxed. Thus, the DM only has to specify aspiration levels for functions to be improved. This is called *automatic trade-off*. Next, a modified scalarizing function is minimized and the DM is asked to classify the objective functions at the new solution.

Different scalarizing functions have been suggested for use in STOM. In the original formulation, the weighted Tchebycheff metric is used and the weights are set as $w_i = \frac{1}{\bar{z}_i - z_i^{\star\star}}$ for $i = 1, \ldots, k$, where $\bar{z}$ is a reference point and $z^{\star\star}$ is a utopian objective vector so that $\bar{z} > z^{\star\star}$. If weakly Pareto optimal solutions are to be avoided, the scalarizing function can be augmented as described in Sect. 4.3.

Even if a value function existed, it could not be directly used to determine the functions to be decreased and increased or the amounts of change. Thus the method is characterized as an ad hoc method. If automatic trade-off is not used, the method resembles the GUESS method.

6.6 Light Beam Search

Light beam search, described in [13], combines the reference point idea and tools of multiattribute decision analysis. The achievement function (8) is used with weights only in the maximum part. The reference point is here assumed to be an infeasible objective vector.

It is assumed that the objective and the constraint functions are continuously differentiable and the ideal and the nadir objective vectors are available. In addition, none of the objective functions is allowed to be more important than all the others together.

In the light beam search, the learning process of the DM is supported by providing additional information about the Pareto optimal set at each iteration. This means that other solutions in the neighbourhood of the current solution (based on the reference point) are displayed. However, an attempt is made to avoid frustration on the part of the DM caused, e.g., by indifference between the alternatives.

Concepts used in ELECTRE methods (see, e.g., [26]) are here employed. The idea is to establish *outranking relations* between alternatives. It is said that one alternative outranks the other if it is at least as good as the latter. In the light beam search, additional alternatives near the current solution are generated so that they outrank the current one. Incomparable or indifferent alternatives are not shown to the DM.

To be able to compare alternatives and to define outranking relations, we need several thresholds from the DM. The DM is asked to provide *indifference thresholds* for each objective function describing intervals where indifference prevails. Furthermore, the line between indifference and preference does not have to be sharp. The hesitation between indifference and preference can be expressed by *preference thresholds*. One more type of threshold, namely a *veto threshold* can be defined. It prevents a good performance in some objectives from compensating for poor values on some other objectives.

Let us now outline the light beam algorithm.

1. If the DM wants to or can specify the best and the worst values for each objective function, save them. Alternatively calculate $z^\star$ and z^{nad}. Set $z^\star$ as

a reference point. Ask the DM to specify indifference thresholds. If desired, (s)he can also specify preference and veto thresholds.
2. Minimize the achievement function.
3. Present the solution to the DM. Calculate k Pareto optimal characteristic neighbours and present them as well to the DM. If the DM wants to see alternatives between any two of the $k+1$ alternatives displayed, set their difference as a search direction, take different steps in this direction and project them onto the Pareto optimal set before showing them to the DM. If desired, save the current solution.
4. If desired, let the DM revise the thresholds and go to step (3). Otherwise, if the DM wants to give another reference point, go to step (2). If, on the other hand, the DM wants to select one of the alternatives displayed or saved as a current solution, go to step (3). Finally, if one of the alternatives is satisfactory, stop.

Characteristic neighbours are new alternative objective vectors that outrank the current solution. See [13] for details. The option of saving desirable solutions increases the flexibility of the method. The DM can explore different directions and select the best among different trials.

The light beam search can be characterized as an ad hoc method. If a value function were available, it could not directly determine new reference points. It could, however, be used in comparing the set of alternatives. Yet, the thresholds are important in the method and they must come from the DM. Specifying different thresholds is a new aspect when compared to the methods presented earlier. This may be demanding for the DM. Anyway, it is positive that the thresholds are not assumed to be global but can be altered at any time.

The light beam search is a rather versatile solution method where the DM can specify reference points, compare a set of alternatives and affect the set of alternatives in different ways. Thresholds are used to try to make sure that the alternatives generated are not worse than the current solution. In addition, the alternatives must be different enough to be compared and comparable on the whole. This should decrease the burden on the DM.

6.7 NIMBUS Method

NIMBUS (Nondifferentiable Interactive Multiobjective BUndle-based optimization System), presented in [20,21,22], is an interactive multiobjective optimization method designed especially to be able to handle nondifferentiable functions efficiently. For this reason, it is capable of solving complicated real-world problems. It is assumed that the objective and the constraint functions are locally Lipschitzian (if a nondifferentiable solver is used) and the ideal objective vector is available.

NIMBUS is based on the classification of the objective functions where the DM can easily indicate what kind of improvements are desirable and what kind of impairments are tolerable. The DM examines at iteration h the values of the objective functions calculated at the current solution x^h and divides the objective functions into up to five classes; functions f_i whose values

- should be decreased ($i \in I^<$),
- should be decreased to a certain aspiration level $\bar{z}_i < f_i(\boldsymbol{x}^h)$ ($i \in I^{\leq}$),
- are satisfactory at the moment ($i \in I^=$),
- are allowed to increase to a certain upper bound $\varepsilon_i > f_i(\boldsymbol{x}^h)$ ($i \in I^>$) and
- are allowed to change freely ($i \in I^\circ$),

where $I^< \cup I^{\leq} \neq \emptyset$ and $I^= \cup I^> \cup I^\circ \neq \emptyset$.

The difference between the classes $I^<$ and $I^{\leq}$ is that the functions in $I^<$ are to be minimized as far as possible but the functions in $I^{\leq}$ only as far as the aspiration level. The classification is the core of NIMBUS. However, the DM can specify optional positive weights w_i summing up to one.

After the DM has classified the objective functions, a subproblem

$$\begin{aligned}\text{minimize} \quad & \max_{\substack{i \in I^< \\ j \in I^{\leq}}} \left[w_i(f_i(\boldsymbol{x}) - z_i^\star), w_j \max\left[f_j(\boldsymbol{x}) - \bar{z}_j, 0 \right] \right] \\ \text{subject to} \quad & f_i(\boldsymbol{x}) \leq f_i(\boldsymbol{x}^h) \quad \text{for all} \quad i \in I^< \cup I^{\leq} \cup I^=, \\ & f_i(\boldsymbol{x}) \leq \varepsilon_i \quad \text{for all} \quad i \in I^>, \\ & \boldsymbol{x} \in S \end{aligned} \qquad (13)$$

is formed, where $z_i^\star$ ($i \in I^<$) are components of the ideal objective vector. The solution of (13) is weakly Pareto optimal if the set $I^<$ is nonempty. On the other hand, any Pareto optimal solution can be found with an appropriate classification.

If the DM does not like the solution of (13) for some reason, (s)he can explore other solutions between the old and this new one. Then we calculate a search direction as a difference of these two solutions and provide more solutions by taking steps of different sizes in this direction.

The NIMBUS algorithm is given below. Note that the DM must be ready to give up something in order to attain improvement for some other objective functions.

1. Ask the DM to classify the objective functions at the current point.
2. Solve the subproblem and present the solution to the DM. If (s)he wants to see different alternatives between the old and the new solution, go to step (3). If the DM prefers either of the two solutions and want to continue from it, go to step (1). Otherwise, go to step (4).
3. Ask the DM to specify the desired number of alternatives P and calculate alternative vectors. Present their Pareto optimal counterparts to the DM and let her/him choose the most preferred one among them. If the DM wants to continue, go to step (1).
4. Check Pareto optimality and stop.

In NIMBUS, the DM expresses iteratively her/his desires. Unlike some other methods based on classification, the success of the solution process does not depend entirely on how well the DM manages in specifying the classification and the appropriate parameter values. It is important that the classification is not irreversible. Thus, no irrevocable damage is caused in NIMBUS if the solution obtained is not what was expected. The DM is free to go back or explore intermediate points. (S)he can easily get to know the problem and its possibilities by specifying, e.g., loose upper bounds and examining intermediate solutions.

In NIMBUS, the DM can explore the (weakly) Pareto optimal set and change her/his mind if necessary. The DM can also extract undesirable solutions from further consideration.

The method is ad hoc in nature, since the existence of a value function would not directly advise the DM how to act to attain her/his desires. A value function could only be used to compare different alternatives.

The method has been implemented as a WWW-NIMBUS system on the Internet (see [23]). Via the Internet we can centralize the computing to one server computer (at the University of Jyväskylä) and the WWW is a way of distributing the graphical user interface to the computers of each individual user. Besides, the user always has the latest version of NIMBUS available.

The most important aspect of WWW-NIMBUS is that it is easily accessible and available to any Internet user (http://nimbus.mit.jyu.fi/). No special tools, compilers or software besides a WWW browser are needed. The user saves the trouble of installing any software and the system is independent of the computer and the operating system used.

The system contains both a nondifferentiable local solver (proximal bundle method) (see [19], pp. 112–143) and a global solver (genetic algorithms) for the subproblem. When the first version of WWW-NIMBUS was implemented in 1995 it was a pioneering interactive optimization system on the Internet.

7 Conclusions

As has been stressed, a large variety of methods exists for multiobjective optimization problems and none of them can be claimed to be superior to the others in every aspect. When selecting a solution method, the specific features of the problem to be solved must be taken into consideration. In addition, the opinions of the DM are important. One can say that selecting a multiobjective optimization method is a problem with multiple objectives itself.

The theoretical properties of the methods can rather easily be compared. A comparative table summarizing some of the features of interactive methods is given in [20]. However, in addition to theoretical properties, practical applicability also plays an important role in the selection of an appropriate method for the problem to be solved. The difficulty is that practical applicability is hard to determine without experience and experimentation.

The features of the problem and the capabilities of the DM have to be charted before a solution method can be chosen. Some methods may suit some problems and some DMs better than others. A decision tree is provided in [20] for easing the selection.

As far as the future is concerned, the obvious conclusion in the development of methods is the importance of continuing in the direction of user-friendliness. Methods must be even better able to correspond to the characteristics of the DM. If the aspirations of the DM change during the solution process, the algorithm must be able to cope with this situation.

Computational tests have confirmed the idea that DMs want to feel in control of the solution process, and consequently they must understand what is happening. However, sometimes the DM simply needs support, and this should be available as well. Thus, the aim is to have methods that support learning so that guidance is given whenever necessary. The DM can be supported by using visual illustrations and further development of such tools is essential. In addition to bar charts, value paths and petal diagrams of alternatives, we may use 3D slices of the feasible objective region (see [18]) and other tools. Specific methods for different areas of application that take into account the characteristics of the problems are also important.

References

1. Benson, H.P.: Existence of Efficient Solutions for Vector Maximization Problems. Journal of Optimization Theory and Applications **26** (1978) 569–580
2. Benson, H.P.: Vector Maximization with Two Objective Functions. Journal of Optimization Theory and Applications (1979) 253–257
3. Buchanan, J.T.: A Naïve Approach for Solving MCDM Problems: The GUESS Method. Journal of the Operational Research Society **48** (1997) 202–206
4. Caballero, R., Rey, L., Ruiz, F., González, M.: An Algorithmic Package for the Resolution and Analysis of Convex Multiple Objective Problems. In: Fandel, G., Gal, T. (eds.): Multiple Criteria Decision Making: Proceedings of the Twelfth International Conference, Hagen (Germany). Springer-Verlag (1997) 275–284
5. Censor, Y.: Pareto Optimality in Multiobjective Problems. Applied Mathematics and Optimization **4** (1977) 41–59
6. Chankong, V., Haimes, Y.Y.: Multiobjective Decision Making Theory and Methodology. Elsevier Science Publishing Co., Inc., (1983)
7. Charnes, A., Cooper, W.W.: Goal Programming and Multiple Objective Optimization; Part 1. European Journal of Operational Research **1** (1977) 39–54
8. Gass, S., Saaty, T.: The Computational Algorithm for the Parametric Objective Function. Naval Research Logistics Quarterly **2** (1955) 39–45
9. Geoffrion, A.M.: Proper Efficiency and the Theory of Vector Maximization. Journal of Mathematical Analysis and Applications **22** (1968) 618–630
10. Geoffrion, A.M., Dyer, J.S., Feinberg, A.: An Interactive Approach for Multi-Criterion Optimization, with an Application to the Operation of an Academic Department. Management Science **19** (1972) 357–368
11. Haimes, Y.Y., Lasdon, L.S., Wismer, D.A.: On a Bicriterion Formulation of the Problems of Integrated System Identification and System Optimization. IEEE Transactions on Systems, Man, and Cybernetics **1** (1971) 296–297
12. Hwang, C.-L., Masud, A.S.M.: Multiple Objective Decision Making – Methods and Applications: A State-of-the-Art Survey. Springer-Verlag (1979)
13. Jaszkiewicz, A., Slowiński, R.: The 'Light Beam Search' Approach – An Overview of Methodology and Applications. European Journal of Operational Research **113** (1999) 300–314
14. Keeney, R.L., Raiffa, H.: Decisions with Multiple Objectives: Preferences and Value Tradeoffs. John Wiley & Sons, Inc. (1976)
15. Korhonen, P., Salo, S., Steuer, R.E.: A Heuristic for Estimating Nadir Criterion Values in Multiple Objective Linear Programming. Operations Research **45** (1997) 751–757
16. Korhonen, P., Wallenius, J.: Behavioural Issues in MCDM: Neglected Research Questions. Journal of Multi-Criteria Decision Analysis **5** (1996) 178–182

17. Li, D., Yang, J.-B., Biswal, M.P.: Quantitative Parametric Connections between Methods for Generating Noninferior Solutions in Multiobjective Optimization. European Journal of Operational Research **117** (1999) 84–99
18. Lotov, A.V., Bushenkov, V., Chernykh, O: Multi-Criteria DSS for River Water Quality Planning. Microcomputers in Civil Engineering **12** (1997) 57–67
19. Mäkelä, M.M., Neittaanmäki, P.: Nonsmooth Optimization: Analysis and Algorithms with Applications to Optimal Control. World Scientific Publishing Co. (1992)
20. Miettinen, K.: Nonlinear Multiobjective Optimization. Kluwer Academic Publishers (1999)
21. Miettinen, K., Mäkelä, M.M.: Interactive Bundle-Based Method for Nondifferentiable Multiobjective Optimization: NIMBUS. Optimization **34** (1995) 231–246
22. Miettinen, K., Mäkelä, M.M.: Comparative Evaluation of Some Interactive Reference Point-Based Methods for Multi-Objective Optimisation. Journal of the Operational Research Society **50** (1999) 949–959
23. Miettinen, K., Mäkelä, M.M.: Interactive Multiobjective Optimization System WWW-NIMBUS on the Internet. Computers & Operations Research **27** (2000) 709–723
24. Miettinen, K., Mäkelä, M.M, Männikkö, T.: Optimal Control of Continuous Casting by Nondifferentiable Multiobjective Optimization. Computational Optimization and Applications **11** (1998) 177–194
25. Nakayama, H.: Aspiration Level Approach to Interactive Multi-Objective Programming and Its Applications. In: Pardalos, P.M., Siskos, Y., Zopounidis, C. (eds.): Advances in Multicriteria Analysis. Kluwer Academic Publishers (1995) 147–174
26. Roy, B.: The Outranking Approach and the Foundations of ELECTRE Methods. In: Bana e Costa, C.A. (ed.): Readings in Multiple Criteria Decision Aid. Springer-Verlag (1990) 155–183
27. Steuer, R.E.: Multiple Criteria Optimization: Theory, Computation, and Applications. John Wiley & Sons, Inc. (1986)
28. Steuer, R.E.: The Tchebycheff Procedure of Interactive Multiple Objective Programming. In: Karpak, B., Zionts, S. (eds.): Multiple Criteria Decision Making and Risk Analysis Using Microcomputers. Springer-Verlag (1989) 235–249
29. Steuer, R.E., Gardiner, L.R.: On the Computational Testing of Procedures for Interactive Multiple Objective Linear Programming. In: Fandel, G., Gehring, H. (eds.): Operations Research. Springer-Verlag (1991) 121–131
30. Vincke, P.: Multicriteria Decision-Aid. John Wiley & Sons, Inc. (1992)
31. Weistroffer, H.R.: Careful Usage of Pessimistic Values is Needed in Multiple Objective Optimization. Operations Research Letters **4** (1985) 23–25
32. Wierzbicki, A.P.: A Mathematical Basis for Satisficing Decision Making. Mathematical Modelling **3** (1982) 391–405
33. Wierzbicki, A.P.: On the Completeness and Constructiveness of Parametric Characterizations to Vector Optimization Problems. OR Spektrum **8** (1986) 73–87
34. Wierzbicki, A.P., Granat, J.: Multi-Objective Modeling for Engineering Applications: DIDASN++ System. European Journal of Operational Research **113** (1999) 372–389
35. Zadeh, L.: Optimality and Non-Scalar-Valued Performance Criteria. IEEE Transactions on Automatic Control **8** (1963) 59–60

A Short Tutorial on Evolutionary Multiobjective Optimization

Carlos A. Coello Coello*

CINVESTAV-IPN
Depto. de Ingeniería Eléctrica
Sección de Computación
Av. Instituto Politécnico Nacional No. 2508
Col. San Pedro Zacatenco
México, D. F. 07300
ccoello@cs.cinvestav.mx

Abstract. This tutorial will review some of the basic concepts related to evolutionary multiobjective optimization (i.e., the use of evolutionary algorithms to handle more than one objective function at a time). The most commonly used evolutionary multiobjective optimization techniques will be described and criticized, including some of their applications. Theory, test functions and metrics will be also discussed. Finally, we will provide some possible paths of future research in this area.

1 Introduction

Most real-world engineering optimization problems are multiobjective in nature, since they normally have several (possibly conflicting) objectives that must be satisfied at the same time. The notion of "optimum" has to be re-defined in this context and instead of aiming to find a single solution, we will try to produce a set of good compromises or "trade-offs" from which the decision maker will select one.

Over the years, the work of a considerable amount of operational researchers has produced an important number of techniques to deal with multiobjective optimization problems [46]. However, it was until relatively recently that researchers realized of the potential of evolutionary algorithms in this area.

The potential of evolutionary algorithms in multiobjective optimization was hinted by Rosenberg in the 1960s [52], but this research area, later called Evolutionary Multi-Objective Optimization (EMOO for short) remained unexplored for almost twenty five years. However, researchers from many different disciplines have shown an increasing interest in EMOO in recent years. The considerable amount of research related to EMOO currently reported in the literature (over 630 publications[1]) is a clear reflection of such interest.

---

* This work was done while the author was at the Laboratorio Nacional de Informática Avanzada, Rébsamen 80, Xalapa, Veracruz 91090, México.
[1] The author maintains a repository on Evolutionary Multiobjective Optimization at: http://www.lania.mx/~ccoello/EMOO/ with a mirror at http://www.jeo.org/emo/

This paper will provide a short tutorial on EMOO, including a review of the main existing approaches (a description of the technique, together with its advantages and disadvantages and some of its applications) and of the most significant research done in theory, test functions and metrics. We will finish with a short review of two promising areas of future research.

2 Basic Definitions

Multiobjective optimization (also called multicriteria optimization, multiperformance or vector optimization) can be defined as the problem of finding [49]:

> a vector of decision variables which satisfies constraints and optimizes a vector function whose elements represent the objective functions. These functions form a mathematical description of performance criteria which are usually in conflict with each other. Hence, the term "optimize" means finding such a solution which would give the values of all the objective functions acceptable to the designer.

Formally, we can state it as follows:

Find the vector $\boldsymbol{x}^* = [x_1^*, x_2^*, \ldots, x_n^*]^T$ which will satisfy the m inequality constraints:

$$g_i(\boldsymbol{x}) \geq 0 \quad i = 1, 2, \ldots, m \tag{1}$$

the p equality constraints

$$h_i(\boldsymbol{x}) = 0 \quad i = 1, 2, \ldots, p \tag{2}$$

and optimizes the vector function

$$\boldsymbol{f}(\boldsymbol{x}) = [f_1(\boldsymbol{x}), f_2(\boldsymbol{x}), \ldots, f_k(\boldsymbol{x})]^T \tag{3}$$

where $\boldsymbol{x} = [x_1, x_2, \ldots, x_n]^T$ is the vector of decision variables.

In other words, we wish to determine from among the set $\mathcal{F}$ of all numbers which satisfy (1) and (2) the particular set $x_1^*, x_2^*, \ldots, x_k^*$ which yields the optimum values of all the objective functions.

It is rarely the case that there is a single point that simultaneously optimizes all the objective functions. Therefore, we normally look for "trade-offs", rather than single solutions when dealing with multiobjective optimization problems. The notion of "optimum" is therefore, different. The most commonly adopted notion of optimality is that originally proposed by Francis Ysidro Edgeworth [22], and later generalized by Vilfredo Pareto [50]. Although some authors call *Edgeworth-Pareto optimum* to this notion (see for example Stadler [61]), we will use the most commonly accepted term: *Pareto optimum*.

We say that a vector of decision variables $\boldsymbol{x}^* \in \mathcal{F}$ is *Pareto optimal* if there does not exist another $\boldsymbol{x} \in \mathcal{F}$ such that $f_i(\boldsymbol{x}) \leq f_i(\boldsymbol{x}^*)$ for all $i = 1, \ldots, k$ and $f_j(\boldsymbol{x}) < f_j(\boldsymbol{x}^*)$ for at least one j.

In words, this definition says that x^* is Pareto optimal if there exists no feasible vector of decision variables $x \in \mathcal{F}$ which would decrease some criterion without causing a simultaneous increase in at least one other criterion. Unfortunately, this concept almost always gives not a single solution, but rather a set of solutions called the *Pareto optimal set*. The vectors x^* correspoding to the solutions included in the Pareto optimal set are called *nondominated*. The plot of the objective functions whose nondominated vectors are in the Pareto optimal set is called the *Pareto front*.

2.1 An Example

Let us analyze a simple example of a multiobjective optimization problem, that has been studied by Stadler & Dauer [62]. We want to design the four-bar plane truss shown in Figure 1. We will consider two objective functions: minimize the volume of the truss (f_1) and minimize its joint displacement Δ (f_2). The mathematical definition of the problem is:

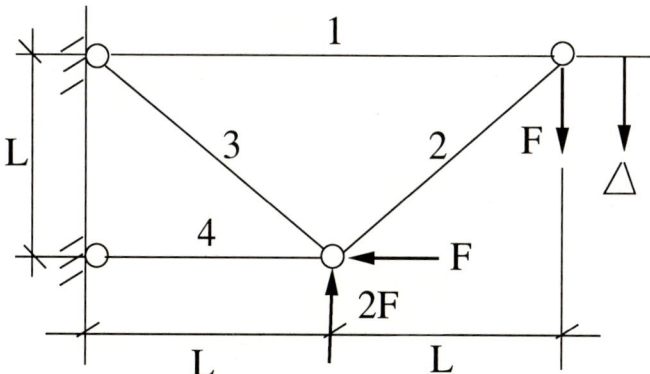

Fig. 1. A four-bar plane truss.

$$\text{Minimize} \begin{cases} f_1(x) = L\left(2x_1 + \sqrt{2}x_2 + \sqrt{x_3} + x_4\right) \\ f_2(x) = \frac{FL}{E}\left(\frac{2}{x_1} + \frac{2\sqrt{2}}{x_2} - \frac{2\sqrt{2}}{x_3} + \frac{2}{x_4}\right) \end{cases} \quad (4)$$

such that:

$$\begin{aligned} (F/\sigma) &\leq x_1 \leq 3(F/\sigma) \\ \sqrt{2}(F/\sigma) &\leq x_2 \leq 3(F/\sigma) \\ \sqrt{2}(F/\sigma) &\leq x_3 \leq 3(F/\sigma) \\ (F/\sigma) &\leq x_4 \leq 3(F/\sigma) \end{aligned} \quad (5)$$

where $F = 10$ kN, $E = 2 \times 10^5$ kN/cm^2, $L = 200$ cm, $\sigma = 10$ kN/cm^2.

The global Pareto front of this problem can be obtained by enumeration. The process consists of iterating on the four decision variables (with a reasonable granularity) to get a set of points representing the search space. Then, we apply the concept of Pareto optimality previously defined to the points generated. The result of this procedure, plotted on objective function space is shown in Figure 2. This is the true (or global) Pareto front of the problem.

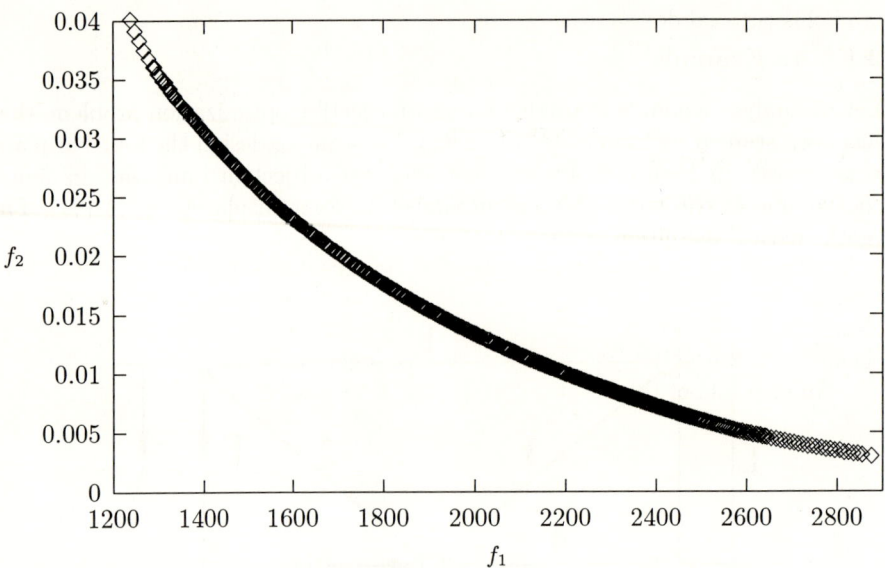

Fig. 2. True Pareto front of the four-bar plane truss problem.

3 Why Evolutionary Algorithms?

The first implementation of an EMOO approach was Schaffer's *Vector Evaluation Genetic Algorithm* (VEGA), which was introduced in the mid-1980s, mainly intended for solving problems in machine learning [57,58,59].

Schaffer's work was presented at the *First International Conference on Genetic Algorithms* in 1985 [58]. Interestingly, his simple unconstrained two-objective functions became the usual test suite to validate most of the evolutionary multiobjective optimization techniques developed during the following years [60,38].

Evolutionary algorithms seem particularly suitable to solve multiobjective optimization problems, because they deal simultaneously with a set of possible solutions (the so-called population). This allows us to find several members of the Pareto optimal set in a single run of the algorithm, instead of having to perform a series of separate runs as in the case of the traditional mathematical programming techniques [5]. Additionally, evolutionary algorithms are less susceptible to the shape or continuity of the Pareto front (e.g., they can easily deal with discontinuous or concave Pareto fronts), whereas these two issues are a real concern for mathematical programming techniques.

4 Reviewing EMOO Approaches

There are several detailed surveys of EMOO reported in the literature [5,27,64] and this tutorial does not intend to produce a new one. Therefore, we will limit ourselves to a short discussion on the most popular EMOO techniques currently in use, including two recent approaches that look very promising.

4.1 Aggregating Functions

A genetic algorithm relies on a scalar fitness function to guide the search. Therefore, the most intuitive approach to deal with multiple objectives would be to combine them into a single function. The approach of combining objectives into a single (scalar) function is normally denominated aggregating functions, and it has been attempted several times in the literature with relative success in problems in which the behavior of the objective functions is more or less well-known.

An example of this approach is a sum of weights of the form:

$$\min \sum_{i=1}^{k} w_i f_i(\boldsymbol{x}) \tag{6}$$

where $w_i \geq 0$ are the weighting coefficients representing the relative importance of the k objective functions of our problem. It is usually assumed that

$$\sum_{i=1}^{k} w_i = 1 \tag{7}$$

Since the results of solving an optimization model using (6) can vary significantly as the weighting coefficients change, and since very little is usually known about how to choose these coefficients, a necessary approach is to solve the same problem for many different values of w_i.

Advantages and Disadvantages. This approach does not require any changes to the basic mechanism of a genetic algorithm and it is therefore very simple, easy to implement and efficient. The approach can work properly in simple multiobjective optimization problems with few objective functions and convex search spaces. One obvious problem of this approach is that it may be difficult to generate a set of weights that properly scales the objectives when little is known about the problem. However, its most serious drawback is that it cannot generate proper members of the Pareto optimal set when the Pareto front is concave regardless of the weights used [13].

Sample Applications

- Truck packing problems [30].
- Real-time scheduling [47].
- Structural synthesis of cell-based VLSI circuits [1].

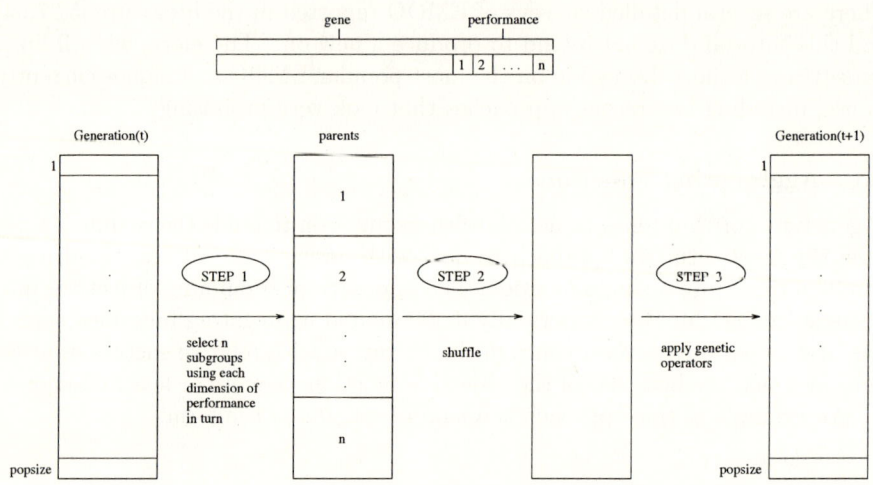

Fig. 3. Schematic of VEGA selection.

4.2 VEGA

Schaffer [58] proposed an approach that he called the *Vector Evaluated Genetic Algorithm* (VEGA), and that differed of the simple genetic algorithm (GA) only in the way in which selection was performed. This operator was modified so that at each generation a number of sub-populations was generated by performing proportional selection according to each objective function in turn. Thus, for a problem with k objectives and a population size of M, k sub-populations of size M/k each would be generated. These sub-populations would be shuffled together to obtain a new population of size M, on which the GA would apply the crossover and mutation operators in the usual way. This process is illustrated in Figure 3.

The solutions generated by VEGA are locally nondominated, but not necessarily globally nondominated. VEGA presents the so-called "speciation" problem (i.e., we could have the evolution of "species" within the population which excel on different objectives). This problem arises because this technique selects individuals who excel in one objective, without looking at the others. The potential danger doing that is that we could have individuals with what Schaffer [58] called "middling" performance[2] in all dimensions, which could be very useful for compromise solutions, but that will not survive under this selection scheme, since they are not in the extreme for any dimension of performance (i.e., they do not produce the best value for any objective function, but only moderately good values for all of them). Speciation is undesirable because it is opposed to our goal of finding compromise solutions.

Advantages and Disadvantages. Since only the selection mechanism of the GA needs to be modified, the approach is easy to implement and it is quite efficient. However, the "middling" problem prevents the technique from finding the compromise solutions that we normally aim to produce. In fact, if proportional selection is used with VEGA (as Schaffer did), the shuffling and merging of all the sub-populations corresponds to averaging the fitness components associated with each of the objectives [51]. In other words, under these conditions, VEGA behaves as an aggregating approach and therefore, it is subject to the same problems of such techniques.

Sample Applications

- Optimal location of a network of groundwater monitoring wells [4].
- Combinational circuit design [8].
- Design multiplierless IIR filters [71].

4.3 MOGA

Fonseca and Fleming [25] proposed the *Multi-Objective Genetic Algorithm* (MOGA). The approach consists of a scheme in which the rank of a certain individual corresponds to the number of individuals in the current population by which it is dominated. All nondominated individuals are assigned rank 1, while dominated ones are penalized according to the population density of the corresponding region of the trade-off surface.

Fitness assignment is performed in the following way [25]:

1. Sort population according to rank.
2. Assign fitness to individuals by interpolating from the best (rank 1) to the worst (rank $n \leq M$) in the way proposed by Goldberg [29] (the so-called Pareto ranking assignment process), according to some function, usually linear, but not necessarily.
3. Average the fitnesses of individuals with the same rank, so that all of them will be sampled at the same rate. This procedure keeps the global population fitness constant while maintaining appropriate selective pressure, as defined by the function used.

---

[2] By "middling", Schaffer meant an individual with acceptable performance, perhaps above average, but not outstanding for any of the objective functions.

Since the use of a blocked fitness assignment scheme as the one indicated before is likely to produce a large selection pressure that might produce premature convergence [29], the authors proposed the use of a niche-formation method to distribute the population over the Pareto-optimal region [20]. Sharing is performed on the objective function values, and the authors provided some guidelines to compute the corresponding niche sizes. MOGA also uses mating restrictions.

Advantages and Disadvantages. The main strengths of MOGA is that is efficient and relatively easy to implement [11]. Its main weakness is that, as with all the other Pareto ranking techniques[3], its performance is highly dependent on an appropriate selection of the sharing factor.

MOGA has been a very popular EMOO technique (particularly within the control community), and it normally exhibits a very good overall performance [11].

Some Applications

- Fault diagnosis [45].
- Control system design [3,69,21].
- Wing planform design [48].
- Design of multilayer microwave absorbers [68].

4.4 NSGA

The *Nondominated Sorting Genetic Algorithm* (NSGA) was proposed by Srinivas and Deb [60], and is based on several layers of classifications of the individuals. Before selection is performed (stochastic remainder proportionate selection was used), the population is ranked on the basis of domination (using Pareto ranking): all nondominated individuals are classified into one category (with a dummy fitness value, which is proportional to the population size). To maintain the diversity of the population, these classified individuals are shared (in decision variable space) with their dummy fitness values. Then this group of classified individuals is removed from the population and another layer of nondominated individuals is considered (i.e., the remainder of the population is re-classified). The process continues until all individuals in the population are classified. Since individuals in the first front have the maximum fitness value, they always get more copies than the rest of the population. This allows us to search for nondominated regions, and results in convergence of the population toward such regions. Sharing, on its part, helps to distribute the population over this region. Figure 4 (taken from Srinivas and Deb [60]) shows the general flow chart of this approach.

---

[3] The use of a ranking scheme based on the concept of Pareto optimality was originally proposed by Goldberg [29].

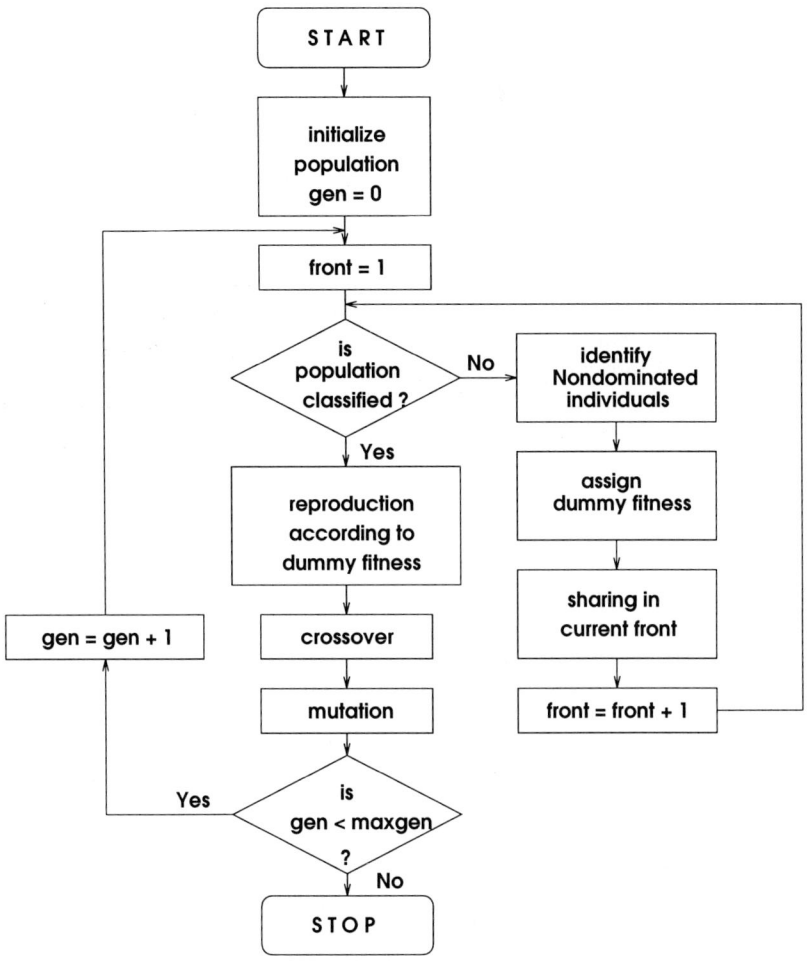

Fig. 4. Flowchart of the Nondominated Sorting Genetic Algorithm (NSGA).

Advantages and Disadvantages. Some researchers have reported that NSGA has a lower overall performance than MOGA (both computationally and in terms of quality of the Pareto fronts produced), and it seems to be also more sensitive to the value of the sharing factor than MOGA [11]. However, Deb et al. [18,19] have recently proposed a new version of this algorithm, called NSGA-II, which is more efficient (computationally speaking), uses elitism and a crowded comparison operator that keeps diversity without specifying any additional parameters. The new approach has not been extensively tested yet, but it certainly looks promising.

Sample Applications

- Airfoil shape optimization [43].
- Scheduling [2].
- Minimum spanning tree [73].

4.5 NPGA

Horn et al. [38] proposed the *Niched Pareto Genetic Algorithm*, which uses a tournament selection scheme based on Pareto dominance. Instead of limiting the comparison to two individuals (as normally done with traditional GAs), a higher number of individuals is involved in the competition (typically around 10% of the population size). When both competitors are either dominated or nondominated (i.e., when there is a tie), the result of the tournament is decided through fitness sharing in the objective domain (a technique called *equivalent class sharing* was used in this case) [38].

The pseudocode for Pareto domination tournaments assuming that all of the objectives are to be maximized is presented below [37]. S is an array of the N individuals in the current population, *random_pop_index* is an array holding the N indices of S, in a random order, and t_{dom} is the size of the comparison set.

function selection /* Returns an individual from the current population S */
begin
    shuffle(random_pop_index); /* Re-randomize random index array */
    candidate_1 = random_pop_index[1];
    candidate_2 = random_pop_index[2];
    candidate_1_dominated = **false**;
    candidate_2_dominated = **false**;
    for comparison_set_index = 3 to t_{dom} + 3 **do**
    /* Select t_{dom} individuals randomly from S */
        begin
            comparison_individual = random_pop_index[comparison_set_index];
            if S[comparison_individual] dominates S[candidate_1]
                then candidate_1_dominated = **true**;
            if S[comparison_individual] dominates S[candidate_2]
                then candidate_2_dominated = **true**;
    end /* end for loop */
    if (candidate_1_dominated **AND** ¬ candidate_2_dominated)
        then return candidate_2;
    else if (¬ candidate_1_dominated **AND** candidate_2_dominated)
        then return candidate_1;
    else
        do sharing;

end

This technique normally requires population sizes considerably larger than usual with other approaches, so that the noise of the selection method can be tolerated by the emerging niches in the population [26].

Advantages and Disadvantages. Since this approach does not apply Pareto ranking to the entire population, but only to a segment of it at each run, its main strength are that it is faster than MOGA and NSGA[4]. Furthermore, it also produces good nondominated fronts that can be kept for a large number of generations [11]. However, its main weakness is that besides requiring a sharing factor, this approach also requires an additional parameter: the size of the tournament.

Sample Applications

- Automatic derivation of qualitative descriptions of complex objects [55].
- Feature selection [24].
- Optimal well placement for groundwater containment monitoring [37,38].
- Investigation of feasibility of full stern submarines [63].

4.6 Target Vector Approaches

Under this name we will consider approaches in which the decision maker has to assign targets or goals that wishes to achieve for each objective. The GA in this case, tries to minimize the difference between the current solution found and the vector of goals (different metrics can be used for that purpose). The most popular techniques included here are hybrids with: Goal Programming [16,70], Goal Attainment [71,72] and the min-max approach [32,9].

Advantages and Disadvantages. The main strength of these methods is their efficiency (computationally speaking) because they do not require a Pareto ranking procedure. However, their main weakness is the definition of the desired goals which requires some extra computational effort (normally, these goals are the optimum of each objective function, considered separately). Furthermore, these techniques will yield a nondominated solution only if the goals are chosen in the feasible domain, and such condition may certainly limit their applicability.

Some Applications

- Truss design [56,7].
- Design of a robot arm [10].
- Synthesis of low-power operational amplifiers [72].

---

[4] Pareto ranking is $O(kM^2)$, where k is the number of objectives and M is the population size.

4.7 Recent Approaches

Recently, several new EMOO approaches have been developed. We consider important to discuss briefly at least two of them: PAES and SPEA.

The *Pareto Archived Evolution Strategy* (PAES) was introduced by Knowles & Corne [42]. This approach is very simple: it uses a (1+1) evolution strategy (i.e., a single parent that generates a single offspring) together with a historical archive that records all the nondominated solutions previously found (such archive is used as a comparison set in a way analogous to the tournament competitors in the NPGA). PAES also uses a novel approach to keep diversity, which consists of a crowding procedure that divides objective space in a recursive manner. Each solution is placed in a certain grid location based on the values of its objectives. A map of such grid is maintained, indicating the amount of solutions that reside in each grid location. Since the procedure is adaptive, no extra parameters are required (except for the number of divisions of the objective space). Furthermore, the procedure has a lower computational complexity than traditional niching methods. PAES has been used to solve the off-line routing problem [41] and the adaptive distributed database management problem [42].

The *Strength Pareto Evolutionary Algorithm* (SPEA) was introduced by Zitzler & Thiele [78]. This approach was conceived as a way of integrating different EMOO techniques. SPEA uses an archive containing nondominated solutions previously found (the so-called external nondominated set). At each generation, nondominated individuals are copied to the external nondominated set. For each individual in this external set, a strength value is computed. This strength is similar to the ranking value of MOGA, since it is proportional to the number of solutions to which a certain individual dominates. The fitness of each member of the current population is computed according to the strengths of all external nondominated solutions that dominate it. Additionally, a clustering technique is used to keep diversity. SPEA has been used to explore trade-offs of software implementations for DSP algorithms [76] and to solve 0/1 knapsack problems [78].

5 Theory

The most important theoretical work related to EMOO has concentrated on two main issues:

– Studies of convergence towards the Pareto optimum set [53,54,33,34,65].
– Ways to compute appropriate sharing factors (or niche sizes) [36,35,25].

Obviously, a lot of work remains to be done. It would be very interesting to study, for example, the structure of fitness landscapes in multiobjective optimization problems [40,44]. Such study could provide some insights regarding the sort of problems that are particularly difficult for an evolutionary algorithm and could also provide clues regarding the design of more powerful EMOO techniques.

Also, there is a need for detailed studies of the different aspects involved in the parallelization of EMOO techniques (e.g., load balancing, impact on Pareto convergence, performance issues, etc.), including new algorithms that are more suitable for parallelization than those currently in use.

6 Test Functions

The design of test functions that are appropriate to evaluate EMOO approaches was disregarded in most of the early research in this area. However, in recent years, there have been several interesting proposals. Deb [14,15] proposed ways to create controllable test problems for evolutionary multiobjective optimization techniques using single-objective optimization problems as a basis. He proposed to transform deceptive and massively multimodal problems into very difficult multiobjective optimization problems. More recently, his proposal was extended to constrained multiobjective optimization problems [17] (in most of the early papers on EMOO techniques, only unconstrained test functions were used).

Van Veldhuizen and Lamont [66,67] have also proposed some guidelines to design a test function suite for evolutionary multiobjective optimization techniques, and have included in a technical report some sample test problems (mainly combinatorial optimization problems) [66]. In this regard, the literature on multiobjective combinatorial optimization can be quite useful [23]. The benchmarks available for problems like the multiobjective 0/1 knapsack can be used to validate EMOO approaches. Such idea has been explored by a few EMOO researchers (for example [78,39]), but more work in this direction is still necessary.

7 Metrics

Assuming that we have a set of test functions available, the next issue is how to compare different EMOO techniques. The design of metrics has been studied recently in the literature. The main proposals so far are the following:

- Van Veldhuizen and Lamont [65] proposed the so-called *generational distance*, which is a measure of how close our current Pareto front is from the true Pareto front (assuming we know where it lies).
- Srinivas and Deb [60] proposed the use of an statistical measure (the chi-square distribution) to estimate the spread of the population on the Pareto front with respect to the sharing factor used.
- Zitzler and Thiele [77] proposed two measures: the first concerns the size of the objective value space which is covered by a set of nondominated solutions and the second compares directly two sets of nondominated solutions, using as a metric the fraction of the Pareto front covered by each of them. Several other similar metrics have been also suggested recently by Zitzler et al. [75].

- Fonseca and Fleming [28] proposed the definition of certain (arbitrary) goals that we wish the GA to attain; then we can perform multiple runs and apply standard non-parametric statistical procedures to evaluate the quality of the solutions (i.e. the Pareto fronts) produced by the EMOO technique under study, and/or compare it against other similar techniques.

There are few comparative studies of EMOO techniques where these metrics have been used and more comprehensive comparisons are still lacking in the literature [75,64,74]. Also, it is important to consider that most of the previously mentioned metrics assume that the user can generate the global Pareto front of the problem under study (using, for example, an enumerative approach), and that will not be possible in most real-world applications.

8 Promising Areas of Future Research

There are at least two areas of future research that deserve more attention in the next few years:

- **Incorporation of preferences**: We should not ignore the fact that the solution of a multiobjective optimization problem really involves three stages: measurement, search, and decision making. Most EMOO research tends to concentrate on issues related to the search of nondominated vectors. However, these nondominated vectors do not provide any insight into the process of decision making itself (the decision maker still has to choose manually one of the several alternatives produced), since they are really a useful generalization of a utility function under the conditions of minimum information (i.e., all attributes are considered as having equal importance; in other words, the decision maker does not express any preferences of the attributes). Thus, the issue is how to incorporate the decision maker's preferences into an EMOO approach as to guide the search only to the regions of main interest. There are a few recent proposals in this area [12,6], but more research is still needed. Issues such as scalability of the preferences' handling mechanism and capability of the approach to incorporate preferences from several decision makers deserve special attention.
- **Emphasis on efficiency**: Efficiency has been emphasized in EMOO research until recently, mainly regarding the number of comparisons performed for ranking the population [18], ways to maintain diversity [42], and procedures to reduce the computational cost involved in evaluating several (expensive) objective functions [21]. However, more work is still needed. For example, EMOO researchers have paid little attention to the use of efficient data structures. In contrast, operational researchers have used, for example, domination-free quad trees where a nondominated vector can be retrieved from the tree very efficiently. Checking if a new vector is dominated by the vectors in one of these trees can also be done very efficiently [31]. It is therefore necessary to pay more attention to efficiency issues in the design of new EMOO approaches, to make them more suitable for real-world applications.

9 Conclusions

This paper has attempted to provide a short tutorial of evolutionary multiobjective optimization. Our discussion has covered the main EMOO approaches currently in use, their advantages and disadvantages, and some of their applications reported in the literature.

We have also discussed briefly the theoretical work done in this area, as well as some of the research that has attempted to produce benchmarks that are appropriate to validate EMOO approaches. We also discussed another problem related to this last issue: the definition of appropriate metrics that allow us to compare several EMOO techniques. Such metrics should evaluate the capability of an EMOO approach to produce a sufficient amount of elements of the Pareto optimal set of the problem as well as to spread them appropriately.

Our discussion finishes with a short description of two possible areas of future research in EMOO: mechanisms that facilitate the incorporation of user's preferences and the search for efficient procedures and algorithms for evolutionary multiobjective optimization and to keep diversity.

Acknowledgements. The author gratefully acknowledges support from CONACyT through project 34201-A.

References

1. T. Arslan, D. H. Horrocks, and E. Ozdemir. Structural Synthesis of Cell-based VLSI Circuits using a Multi-Objective Genetic Algorithm. *IEE Electronic Letters*, 32(7):651–652, March 1996.
2. Tapan P. Bagchi. *Multiobjective Scheduling by Genetic Algorithms*. Kluwer Academic Publishers, Boston, 1999.
3. A. J. Chipperfield and P. J. Fleming. Gas Turbine Engine Controller Design using Multiobjective Genetic Algorithms. In A. M. S. Zalzala, editor, *Proceedings of the First IEE/IEEE International Conference on Genetic Algorithms in Engineering Systems : Innovations and Applications, GALESIA'95*, pages 214–219, Halifax Hall, University of Sheffield, UK, September 1995. IEEE.
4. Scott E. Cieniawski, J. W. Eheart, and S. Ranjithan. Using Genetic Algorithms to Solve a Multiobjective Groundwater Monitoring Problem. *Water Resources Research*, 31(2):399–409, February 1995.
5. Carlos A. Coello Coello. A Comprehensive Survey of Evolutionary-Based Multiobjective Optimization Techniques. *Knowledge and Information Systems. An International Journal*, 1(3):269–308, August 1999.
6. Carlos A. Coello Coello. Handling Preferences in Evolutionary Multiobjective Optimization: A Survey. In *2000 Congress on Evolutionary Computation*, volume 1, pages 30–37, Piscataway, New Jersey, July 2000. IEEE Service Center.
7. Carlos A. Coello Coello. Treating Constraints as Objectives for Single-Objective Evolutionary Optimization. *Engineering Optimization*, 32(3):275–308, 2000.
8. Carlos A. Coello Coello, Arturo Hernández Aguirre, and Bill P. Buckles. Evolutionary Multiobjective Design of Combinational Logic Circuits. In Jason Lohn, Adrian Stoica, Didier Keymeulen, and Silvano Colombano, editors, *Proceedings of the Second NASA/DoD Workshop on Evolvable Hardware*, pages 161–170, Los Alamitos, California, July 2000. IEEE Computer Society.

9. Carlos A. Coello Coello and Alan D. Christiansen. Two New GA-based methods for multiobjective optimization. *Civil Engineering Systems*, 15(3):207–243, 1998.
10. Carlos A. Coello Coello, Alan D. Christiansen, and Arturo Hernández Aguirre. Using a New GA-Based Multiobjective Optimization Technique for the Design of Robot Arms. *Robotica*, 16(4):401–414, July–August 1998.
11. Carlos Artemio Coello Coello. *An Empirical Study of Evolutionary Techniques for Multiobjective Optimization in Engineering Design*. PhD thesis, Department of Computer Science, Tulane University, New Orleans, LA, April 1996.
12. Dragan Cvetković. *Evolutionary Multi–Objective Decision Support Systems for Conceptual Design*. PhD thesis, School of Computing, University of Plymouth, Plymouth, UK, November 2000.
13. Indraneel Das and John Dennis. A Closer Look at Drawbacks of Minimizing Weighted Sums of Objectives for Pareto Set Generation in Multicriteria Optimization Problems. *Structural Optimization*, 14(1):63–69, 1997.
14. Kalyanmoy Deb. Multi-Objective Genetic Algorithms: Problem Difficulties and Construction of Test Problems. Technical Report CI-49/98, Dortmund: Department of Computer Science/LS11, University of Dortmund, Germany, 1998.
15. Kalyanmoy Deb. Evolutionary Algorithms for Multi-Criterion Optimization in Engineering Design. In Kaisa Miettinen, Marko M. Mäkelä, Pekka Neittaanmäki, and Jacques Periaux, editors, *Evolutionary Algorithms in Engineering and Computer Science*, chapter 8, pages 135–161. John Wiley & Sons, Ltd, Chichester, UK, 1999.
16. Kalyanmoy Deb. Solving Goal Programming Problems Using Multi-Objective Genetic Algorithms. In *1999 Congress on Evolutionary Computation*, pages 77–84, Washington, D.C., July 1999. IEEE Service Center.
17. Kalyanmoy Deb. An Efficient Constraint Handling Method for Genetic Algorithms. *Computer Methods in Applied Mechanics and Engineering*, 2000. (in Press).
18. Kalyanmoy Deb, Samir Agrawal, Amrit Pratab, and T. Meyarivan. A Fast Elitist Non-Dominated Sorting Genetic Algorithm for Multi-Objective Optimization: NSGA-II. KanGAL report 200001, Indian Institute of Technology, Kanpur, India, 2000.
19. Kalyanmoy Deb, Samir Agrawal, Amrit Pratab, and T. Meyarivan. A Fast Elitist Non-Dominated Sorting Genetic Algorithm for Multi-Objective Optimization: NSGA-II. In *Proceedings of the Parallel Problem Solving from Nature VI Conference*, pages 849–858. Springer, 2000.
20. Kalyanmoy Deb and David E. Goldberg. An Investigation of Niche and Species Formation in Genetic Function Optimization. In J. David Schaffer, editor, *Proceedings of the Third International Conference on Genetic Algorithms*, pages 42–50, San Mateo, California, June 1989. George Mason University, Morgan Kaufmann Publishers.
21. N.M. Duarte, A. E. Ruano, C.M. Fonseca, and P.J. Fleming. Accelerating Multi-Objective Control System Design Using a Neuro-Genetic Approach. In *2000 Congress on Evolutionary Computation*, volume 1, pages 392–397, Piscataway, New Jersey, July 2000. IEEE Service Center.
22. F. Y. Edgeworth. *Mathematical Physics*. P. Keagan, London, England, 1881.
23. Matthias Ehrgott and Xavier Gandibleux. An Annotated Bibliography of Multiobjective Combinatorial Optimization. Technical Report 62/2000, Fachbereich Mathematik, Universitat Kaiserslautern, Kaiserslautern, Germany, 2000.
24. C. Emmanouilidis, A. Hunter, and J. MacIntyre. A Multiobjective Evolutionary Setting for Feature Selection and a Commonality-Based Crossover Operator. In *2000 Congress on Evolutionary Computation*, volume 1, pages 309–316, Piscataway, New Jersey, July 2000. IEEE Service Center.

25. Carlos M. Fonseca and Peter J. Fleming. Genetic Algorithms for Multiobjective Optimization: Formulation, Discussion and Generalization. In Stephanie Forrest, editor, *Proceedings of the Fifth International Conference on Genetic Algorithms*, pages 416–423, San Mateo, California, 1993. University of Illinois at Urbana-Champaign, Morgan Kauffman Publishers.
26. Carlos M. Fonseca and Peter J. Fleming. An Overview of Evolutionary Algorithms in Multiobjective Optimization. Technical report, Department of Automatic Control and Systems Engineering, University of Sheffield, Sheffield, U. K., 1994.
27. Carlos M. Fonseca and Peter J. Fleming. An Overview of Evolutionary Algorithms in Multiobjective Optimization. *Evolutionary Computation*, 3(1):1–16, Spring 1995.
28. Carlos M. Fonseca and Peter J. Fleming. On the Performance Assessment and Comparison of Stochastic Multiobjective Optimizers. In Hans-Michael Voigt, Werner Ebeling, Ingo Rechenberg, and Hans-Paul Schwefel, editors, *Parallel Problem Solving from Nature—PPSN IV*, Lecture Notes in Computer Science, pages 584–593, Berlin, Germany, September 1996. Springer-Verlag.
29. David E. Goldberg. *Genetic Algorithms in Search, Optimization and Machine Learning*. Addison-Wesley Publishing Company, Reading, Massachusetts, 1989.
30. Pierre Grignon, J. Wodziack, and G. M. Fadel. Bi-Objective optimization of components packing using a genetic algorithm. In *NASA/AIAA/ISSMO Multidisciplinary Design and Optimization Conference*, pages 352–362, Seattle, Washington, September 1996. AIAA-96-4022-CP.
31. W. Habenicht. Quad trees, A data structure for discrete vector optimization problems. In *Lecture notes in economics and mathematical systems*, volume 209, pages 136–145, 1982.
32. P. Hajela and C. Y. Lin. Genetic search strategies in multicriterion optimal design. *Structural Optimization*, 4:99–107, 1992.
33. T. Hanne. On the convergence of multiobjective evolutionary algorithms. *European Journal of Operational Research*, 117(3):553–564, September 2000.
34. Thomas Hanne. Global Multiobjective Optimization Using Evolutionary Algorithms. *Journal of Heuristics*, 6(3):347–360, August 2000.
35. Jeffrey Horn. Multicriterion Decision Making. In Thomas Bäck, David Fogel, and Zbigniew Michalewicz, editors, *Handbook of Evolutionary Computation*, volume 1, pages F1.9:1 – F1.9:15. IOP Publishing Ltd. and Oxford University Press, 1997.
36. Jeffrey Horn. *The Nature of Niching: Genetic Algorithms and the Evolution of Optimal, Cooperative Populations*. PhD thesis, University of Illinois at Urbana Champaign, Urbana, Illinois, 1997.
37. Jeffrey Horn and Nicholas Nafpliotis. Multiobjective Optimization using the Niched Pareto Genetic Algorithm. Technical Report IlliGAl Report 93005, University of Illinois at Urbana-Champaign, Urbana, Illinois, USA, 1993.
38. Jeffrey Horn, Nicholas Nafpliotis, and David E. Goldberg. A Niched Pareto Genetic Algorithm for Multiobjective Optimization. In *Proceedings of the First IEEE Conference on Evolutionary Computation, IEEE World Congress on Computational Intelligence*, volume 1, pages 82–87, Piscataway, New Jersey, June 1994. IEEE Service Center.
39. Andrzej Jaszkiewicz. On the performance of multiple objective genetic local search on the 0/1 knapsack problem. a comparative experiment. Technical Report RA-002/2000, Institute of Computing Science, Poznan University of Technology, Poznań, Poland, July 2000.

40. S. Kaufmann. Adaptation on rugged fitness landscapes. In D. Stein, editor, *Lectures in the Sciences of Complexity*, pages 527–618. Addison-Wesley, Reading, Massachusetts, 1989.
41. Joshua D. Knowles and David W. Corne. The Pareto Archived Evolution Strategy: A New Baseline Algorithm for Multiobjective Optimisation. In *1999 Congress on Evolutionary Computation*, pages 98–105, Washington, D.C., July 1999. IEEE Service Center.
42. Joshua D. Knowles and David W. Corne. Approximating the Nondominated Front Using the Pareto Archived Evolution Strategy. *Evolutionary Computation*, 8(2):149–172, 2000.
43. R. Mäkinen, P. Neittaanmäki, J. Périaux, and J. Toivanen. A genetic Algorithm for Multiobjective Design Optimization in Aerodynamics and Electromagnetics. In K. D. Papailiou et al., editor, *Computational Fluid Dynamics '98, Proceedings of the ECCOMAS 98 Conference*, volume 2, pages 418–422, Athens, Greece, September 1998. Wiley.
44. Bernard Manderick, Mark de Weger, and Piet Spiessens. The Genetic Algorithm and the Structure of the Fitness Landscape. In Richard K. Belew and Lashon B. Booker, editors, *Proceedings of the Fourth International Conference on Genetic Algorithms*, pages 143–150, San Mateo, California, 1991. Morgan Kaufmann.
45. Teodor Marcu. A multiobjective evolutionary approach to pattern recognition for robust diagnosis of process faults. In R. J. Patton and J. Chen, editors, *IFAC Symposium on Fault Detection, Supervision and Safety for Technical Processes: SAFEPROCESS'97*, pages 1183–1188, Kington Upon Hull, United Kingdom, August 1997.
46. Kaisa M. Miettinen. *Nonlinear Multiobjective Optimization*. Kluwer Academic Publishers, Boston, Massachusetts, 1998.
47. David Montana, Marshall Brinn, Sean Moore, and Garrett Bidwell. Genetic Algorithms for Complex, Real-Time Scheduling. In *Proceedings of the 1998 IEEE International Conference on Systems, Man, and Cybernetics*, pages 2213–2218, La Jolla, California, October 1998. IEEE.
48. S. Obayashi, S. Takahashi, and Y. Takeguchi. Niching and Elitist Models for MOGAs. In A. E. Eiben, M. Schoenauer, and H.-P. Schwefel, editors, *Parallel Problem Solving From Nature — PPSN V*, pages 260–269, Amsterdam, Holland, 1998. Springer-Verlag.
49. Andrzej Osyczka. Multicriteria optimization for engineering design. In John S. Gero, editor, *Design Optimization*, pages 193–227. Academic Press, 1985.
50. Vilfredo Pareto. *Cours D'Economie Politique*, volume I and II. F. Rouge, Lausanne, 1896.
51. Jon T. Richardson, Mark R. Palmer, Gunar Liepins, and Mike Hilliard. Some Guidelines for Genetic Algorithms with Penalty Functions. In J. David Schaffer, editor, *Proceedings of the Third International Conference on Genetic Algorithms*, pages 191–197, George Mason University, 1989. Morgan Kaufmann Publishers.
52. R. S. Rosenberg. *Simulation of genetic populations with biochemical properties*. PhD thesis, University of Michigan, Ann Harbor, Michigan, 1967.
53. Günter Rudolph. On a Multi-Objective Evolutionary Algorithm and Its Convergence to the Pareto Set. In *Proceedings of the 5th IEEE Conference on Evolutionary Computation*, pages 511–516, Piscataway, New Jersey, 1998. IEEE Press.
54. Günter Rudolph and Alexandru Agapie. Convergence Properties of Some Multi-Objective Evolutionary Algorithms. In *Proceedings of the 2000 Conference on Evolutionary Computation*, volume 2, pages 1010–1016, Piscataway, New Jersey, July 2000. IEEE Press.

55. Enrique H. Ruspini and Igor S. Zwir. Automated Qualitative Description of Measurements. In *Proceedings of the 16th IEEE Instrumentation and Measurement Technology Conference*, Venice, Italy, 1999.
56. Eric Sandgren. Multicriteria design optimization by goal programming. In Hojjat Adeli, editor, *Advances in Design Optimization*, chapter 23, pages 225–265. Chapman & Hall, London, 1994.
57. J. David Schaffer. *Multiple Objective Optimization with Vector Evaluated Genetic Algorithms*. PhD thesis, Vanderbilt University, 1984.
58. J. David Schaffer. Multiple Objective Optimization with Vector Evaluated Genetic Algorithms. In *Genetic Algorithms and their Applications: Proceedings of the First International Conference on Genetic Algorithms*, pages 93–100. Lawrence Erlbaum, 1985.
59. J. David Schaffer and John J. Grefenstette. Multiobjective Learning via Genetic Algorithms. In *Proceedings of the 9th International Joint Conference on Artificial Intelligence (IJCAI-85)*, pages 593–595, Los Angeles, California, 1985. AAAI.
60. N. Srinivas and Kalyanmoy Deb. Multiobjective Optimization Using Nondominated Sorting in Genetic Algorithms. *Evolutionary Computation*, 2(3):221–248, Fall 1994.
61. W. Stadler. Fundamentals of multicriteria optimization. In W. Stadler, editor, *Multicriteria Optimization in Engineering and the Sciences*, pages 1–25. Plenum Press, New York, 1988.
62. W. Stadler and J. Dauer. Multicriteria optimization in engineering: A tutorial and survey. In *Structural Optimization: Status and Future*, pages 209–249. American Institute of Aeronautics and Astronautics, 1992.
63. Mark W. Thomas. *A Pareto Frontier for Full Stern Submarines via Genetic Algorithm*. PhD thesis, Ocean Engineering Department, Massachusetts Institute of Technology, Cambridge, MA, june 1998.
64. David A. Van Veldhuizen. *Multiobjective Evolutionary Algorithms: Classifications, Analyses, and New Innovations*. PhD thesis, Department of Electrical and Computer Engineering. Graduate School of Engineering. Air Force Institute of Technology, Wright-Patterson AFB, Ohio, May 1999.
65. David A. Van Veldhuizen and Gary B. Lamont. Evolutionary Computation and Convergence to a Pareto Front. In John R. Koza, editor, *Late Breaking Papers at the Genetic Programming 1998 Conference*, pages 221–228, Stanford University, California, July 1998. Stanford University Bookstore.
66. David A. Van Veldhuizen and Gary B. Lamont. Multiobjective Evolutionary Algorithm Research: A History and Analysis. Technical Report TR-98-03, Department of Electrical and Computer Engineering, Graduate School of Engineering, Air Force Institute of Technology, Wright-Patterson AFB, Ohio, 1998.
67. David A. Van Veldhuizen and Gary B. Lamont. Multiobjective Evolutionary Algorithm Test Suites. In Janice Carroll, Hisham Haddad, Dave Oppenheim, Barrett Bryant, and Gary B. Lamont, editors, *Proceedings of the 1999 ACM Symposium on Applied Computing*, pages 351–357, San Antonio, Texas, 1999. ACM.
68. D. S. Weile, E. Michielssen, and D. E. Goldberg. Genetic algorithm design of pareto optimal broad-band microwave absorbers. Technical Report CCEM-4-96, Electrical and Computer Engineering Department, Center for Computational Electromagnetics, University of Illinois at Urbana-Champaign, May 1996.
69. J. F. Whidborne, D.-W. Gu, and I. Postlethwaite. Algorithms for the method of inequalities – a comparative study. In *Procedings of the 1995 American Control Conference*, pages 3393–3397, Seattle, Washington, 1995.

70. P. B. Wienke, C. Lucasius, and G. Kateman. Multicriteria target optimization of analytical procedures using a genetic algorithm. *Analytical Chimica Acta*, 265(2):211–225, 1992.
71. P. B. Wilson and M. D. Macleod. Low implementation cost IIR digital filter design using genetic algorithms. In *IEE/IEEE Workshop on Natural Algorithms in Signal Processing*, pages 4/1–4/8, Chelmsford, U.K., 1993.
72. R. S. Zebulum, M. A. Pacheco, and M. Vellasco. A multi-objective optimisation methodology applied to the synthesis of low-power operational amplifiers. In Ivan Jorge Cheuri and Carlos Alberto dos Reis Filho, editors, *Proceedings of the XIII International Conference in Microelectronics and Packaging*, volume 1, pages 264–271, Curitiba, Brazil, August 1998.
73. Gengui Zhou and Mitsuo Gen. Genetic Algorithm Approach on Multi-Criteria Minimum Spanning Tree Problem. *European Journal of Operational Research*, 114(1), April 1999.
74. Eckart Zitzler. *Evolutionary Algorithms for Multiobjective Optimization: Methods and Applications*. PhD thesis, Swiss Federal Institute of Technology (ETH), Zurich, Switzerland, November 1999.
75. Eckart Zitzler, Kalyanmoy Deb, and Lothar Thiele. Comparison of Multiobjective Evolutionary Algorithms: Empirical Results. *Evolutionary Computation*, 8(2):173–195, Summer 2000.
76. Eckart Zitzler, Jürgen Teich, and Shuvra S. Bhattacharyya. Multidimensional Exploration of Software Implementations for DSP Algorithms. *VLSI Signal Processing Systems*, 1999. (To appear).
77. Eckart Zitzler and Lothar Thiele. An Evolutionary Algorithm for Multiobjective Optimization: The Strength Pareto Approach. Technical Report 43, Computer Engineering and Communication Networks Lab (TIK), Swiss Federal Institute of Technology (ETH), Zurich, Switzerland, May 1998.
78. Eckart Zitzler and Lothar Thiele. Multiobjective Evolutionary Algorithms: A Comparative Case Study and the Strength Pareto Approach. *IEEE Transactions on Evolutionary Computation*, 3(4):257–271, November 1999.

An Overview in Graphs of Multiple Objective Programming

Ralph E. Steuer

Terry College of Business
University of Georgia
Athens, Georgia 30602-6253 USA

Abstract. One of the keys to getting one's arms around multiple objective programming is to understand its geometry. With this in mind, the purpose of this paper is to function as a short tutorial on multiple objective programming that is accomplished maximally with graphs, and minimally with text.

1 Introduction

Consider the multiple objective program

$$max \{f_1(x) = z_1\} \tag{1}$$

$$\vdots$$

$$max \{f_k(x) = z_k\}$$
$$s.t. \quad x \in S$$

where k is the number of objectives, the z_i are *criterion values*, and S is the feasible region in *decision space*. Let $Z \subset R^k$ denote the feasible region in *criterion space* where $z \in Z$ iff there exists an $x \in S$ such that $z = (f_1(x), \ldots, f_k(x))$. Let $K = \{1, \ldots, k\}$. Criterion vector $\bar{z} \in Z$ is *nondominated* iff there does not exist another $z \in Z$ such that $z_i \geq \bar{z}_i$ for all $i \in K$ and $z_i > \bar{z}_i$ for at least one $i \in K$. The set of all nondominated criterion vectors is designated N and is called the *nondominated set*. A point $\bar{x} \in S$ is *efficient* iff its criterion vector $\bar{z} = (f_1(\bar{x}), \ldots, f_k(\bar{x}))$ is nondominated. The set of all efficient points is designated E and is called the *efficient set*.

Letting $U : R^k \to R$ be the utility function of the decision maker (DM), any $z^0 \in Z$ that maximizes U over Z is an *optimal criterion vector* and any $x^0 \in S$ such that $(f_1(x^0), \ldots, f_k(x^0)) = z^0$ is an *optimal solution*. We are interested in the efficient and nondominated sets because if U is *coordinatewise increasing* (i.e., more is always better that less of each $f_i(x)$), $z^0 \in N$ and any *inverse image* $x^0 \in E$. Thus, instead of searching all of Z, we need only find the best criterion vector in N to locate an optimal solution to a multiple objective problem. To overview the nature of multiple objective programming problems and methods for solving them, we have the following tutorial topics:

1. Decision Space vs. Criterion Space
2. Ideal Way to Solve a Multiple Objective Program?
3. Graphically Detecting Nondominated Criterion Vectors
4. Reference Criterion Vectors
5. Size of the Nondominated Set
6. Weighted Tchebycheff Metrics
7. Points on Smallest Intersecting Contours
8. Lexicographic Weighted Tchebycheff Sampling Program
9. T-Vertex λ-Vectors
10. Wierzbicki's Aspiration Criterion Vector Method
11. Tchebycheff Method
12. Why Not Weighted-Sums Approach?
13. Other Interactive Procedures

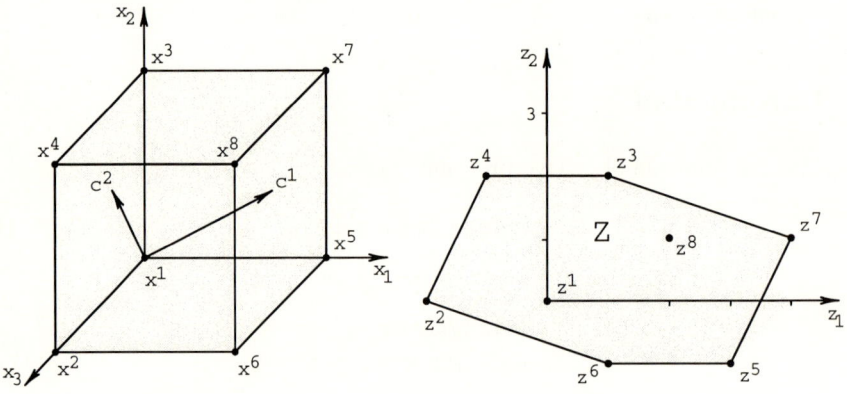

Fig. 1. S in Decision Space   **Fig. 2.** Z in Criterion Space

2 Tutorial Topics

1. Decision Space vs. Criterion Space (Figs. 1–2). Whereas single objective programming is typically studied in decision space, multiple objective programming is mostly studied in criterion space. To illustrate, consider the two-objective multiple objective linear program (MOLP) of Fig. 1 in which $c^1 = (3, 1, -2)$ and $c^2 = (-1, 2, 0)$ are the gradients of the objective functions, and S, the feasible region in decision space, is the unit cube in R^3. Fig. 2 shows the feasible region Z in criterion space in which the eight z^i are the images of the eight extreme points of S. Note that (a) not all extreme points of S map into extreme points of Z, (b) Z is at most of dimensionality k, and (c) Z is not necessarily confined to the nonnegative orthant.

2. Ideal Way to Solve a Multiple Objective Program? (Fig. 3). While one might consider maximizing U over Z as in Fig. 3 to be an ideal way to solve a multiple objective program, this approach does not work in practice because (a) of the impossibility of obtaining U, (b) the approach does not give the user a feel for the nondominated set N, and (c) the approach does not allow for "learning" during the course of the solution process. Therefore, without *explicit* knowledge of U, the emphasis is on methods that only rely on *implicit* information (e.g., extracting from the DM answers to questions such as which in a group of solutions is the most preferred, or which objectives can be relaxed to enable more achievement in others).

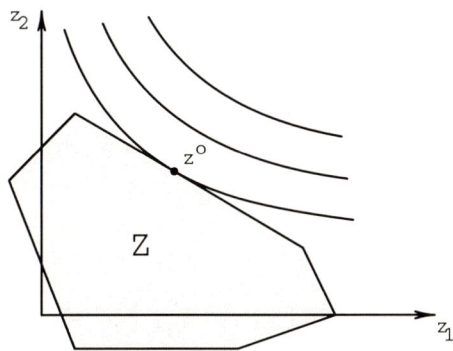

Fig. 3. Maximizing U over Z

3. Graphically Detecting Nondominated Criterion Vectors (Figs. 4–7). Let R^+ be the nonnegative orthant in R^k. To determine graphically whether a $\bar{z} \in Z$ is dominated or nondominated, translate R^+ to $\bar{z}$. This forms the set called the *translated nonnegative orthant at* $\bar{z}$, designated $\{\bar{z}\} \oplus R^+$ where $\oplus$ denotes *set addition*.

Theorem 1. *Criterion vector* $\bar{z} \in Z$ *is nondominated iff* $(\{\bar{z}\} \oplus R^+) \cap Z = \{\bar{z}\}$.

In other words, a $\bar{z} \in Z$ is nondominated if and only if, aside from $\bar{z}$, no other points reside in the *intersection* between the nonnegative orthant translated to $\bar{z}$ and Z except for $\bar{z}$ itself. In the integer multiple objective program of Fig. 4 in which $Z = \{z^i \mid 1 \leq i \leq 6\}$, $N = \{z^2, z^3, z^6\}$. For instance, z^3 is nondominated because if there were a $z^i \in Z$ dominating z^3, it would have to be in the translated nonnegative orthant. On the other hand, z^5 is dominated because z^6 is in the intersection. In the MOLP of Fig. 5, $N = \text{bls}[z^1 \to z^2]$, where the notation designates the *boundary line segment* in the *clockwise* direction from z^1 to z^2. For instance, z^4 is nondominated because the intersection is empty other than for z^4. On the other hand, z^3 is dominated. With regard to the extreme points of Z, of the five, two are nondominated. In the multiple objective program of Fig. 6, $N = \{z^1\} \cup \text{bls}(z^2 \to z^3]$, where the left parenthesis in the "bls" signifies that z^2 is an *open* endpoint. This occurs because z^2 is dominated by z^1 (i.e., $z^1 \in (\{z^2\} \oplus R^+))$. In Fig. 7, $N = \text{bls}[z^1 \to z^2) \cup \text{bls}[z^3 \to z^4] \cup \text{bls}(z^5 \to z^6]$.

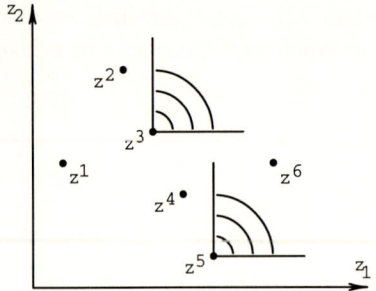

Fig. 4. Nondominance in an Integer Case

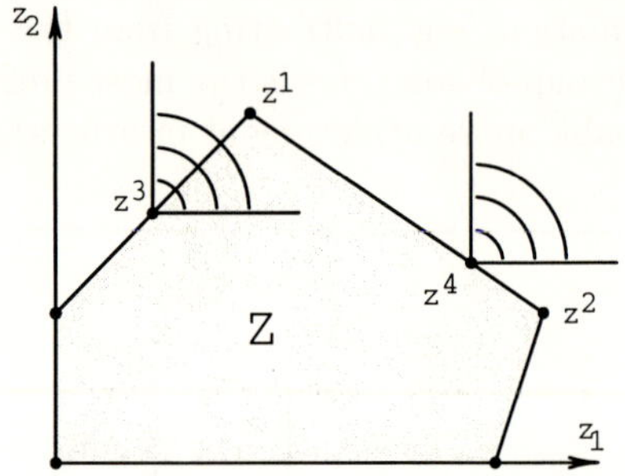

Fig. 5. Nondominance in an MOLP

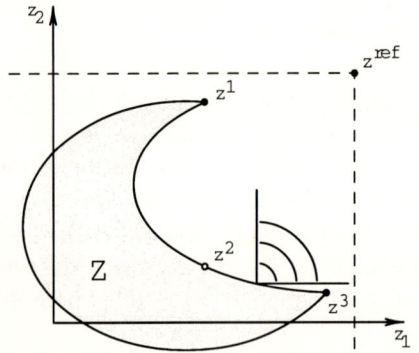

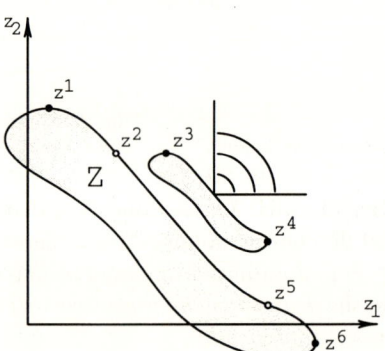

Fig. 6. Nondominance in a Nonlinear Case

Fig. 7. Nondominance in Another Nonlinear Case

4. Reference Criterion Vectors (Fig. 8). Let $K = 1, \ldots, k$ and let $z^{ref} \in R^k$ be a *reference criterion vector* whose components are given by

$$z_i^{ref} = max\,\{f_i(x) \mid x \in S\} + \epsilon_i$$

where the ϵ_i need only be small positive values. An often convenient scheme is to use values for ϵ_i that raise each z_i^{ref} to the smallest integer greater than

$max\{f_i(x) \mid x \in S\}$ as in Fig. 8. A z^{ref} serves two purposes. One is to define the *domain* of the problem $D = \{z \in R^k \mid z_i \leq z_i^{ref}\}$, the region to the "lower left" of the dashed lines in which everything relevant happens. The other, since z^{ref} dominates all points in N, is to function as point from which *downward probes* can be made to sample N.

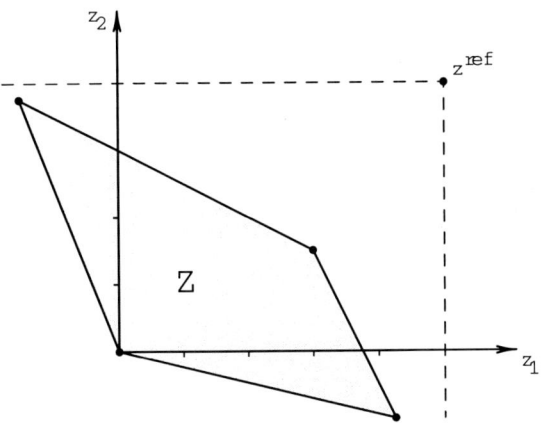

Fig. 8. Construction of a z^{ref} Reference Point

5. Size of the Nondominated Set (Table 1). Now a few facts about the nondominated set. While N is always connected in the linear case, N can be disconnected in integer and nonlinear cases. Also, N is always a portion of the surface of Z. However, since the portion is often quite large, finding a most preferred point in N is often not a trivial task. From computational results gleaned from Steuer [10], Table 1 indicates the numbers of nondominated extreme points typically possessed by MOLPs by size. While N grows with the number of variables and constraints, it grows most substantially with the number of objectives.

Table 1. Nondominated Extreme Point Indications of Size of N

MOLP Problem Size	Typical Number of Nondominated Extreme Points
$3 \times 30 \times 25$	100
$3 \times 50 \times 50$	500
$4 \times 30 \times 50$	1000
$5 \times 40 \times 25$	3000

6. Weighted Tchebycheff Metric (Fig. 9).

To compute a distance between a $z \in Z$ and z^{ref}, it is often useful to employ a λ-*weighted Tchebycheff metric*

$$\| z - z^{ref} \|_\infty^\lambda = \max_{i \in K} \{ \lambda_i \mid z_i - z_i^{ref} \mid \}$$

where (a) $\lambda \in \Lambda = \{\lambda \in R^k \mid \lambda_i \in (0,1), \sum_{i \in K} \lambda_i = 1\}$ and (b) associated with each λ-weighted Tchebycheff metric is a *probing ray* emanating from z^{ref} in the downward direction $-(\frac{1}{\lambda_1}, \ldots, \frac{1}{\lambda_k})$. The *contours* (of points in R^k equidistant from z^{ref}) of a given λ-weighted Tchebycheff metric form a family of rectangles centered at z^{ref}. Moreover, in the domain D of a problem, the vertices of all of the rectangles lie along the probing ray. In Fig. 9 with the probing ray in the direction $-(2,1)$, the rectangles are contours of the $\lambda = (\frac{1}{3}, \frac{2}{3})$ weighted Tchebycheff metric. With this metric, z^1 is closest to z^{ref} because it lies on the smallest rectangle, z^2 is the next closest, and so forth. Note that by changing the λ-vector, we change the direction of the probing ray and thus the shape of the rectangular contours.

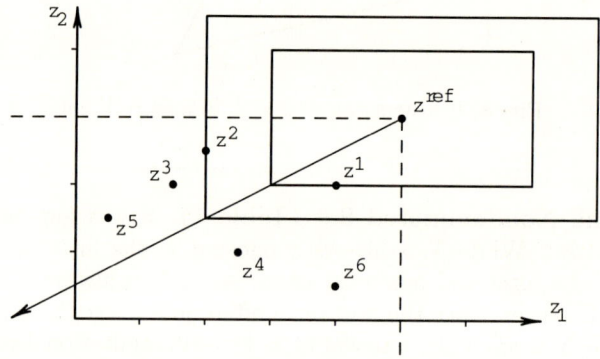

Fig. 9. Determining Points Closest to z^{ref}

7. Points on Smallest Intersecting Contours (Figs. 10–12).

In Fig. 10, $N = \text{bls}[z^1 \to z^2]$. With the probing ray as drawn, it is only necessary to show the portion of the smallest intersecting rectangular contour in D to see that z^3 is the point in Z closest to z^{ref}. Because the portion of any rectangular contour in D also portrays a translated nonnegative orthant, we further observe that z^3 is nondominated. Several questions now arise. What happens if

  (a) the point on the smallest intersecting contour does not occur at the vertex of the contour as in Fig. 11?
  (b) there is more than one point on the smallest intersecting contour as in Fig. 12? Are they all nondominated?

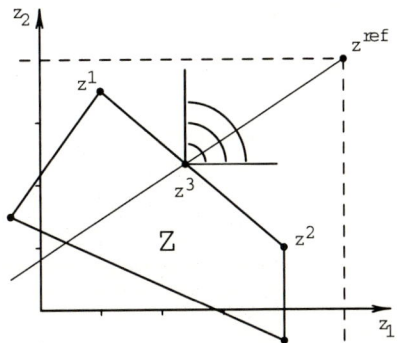
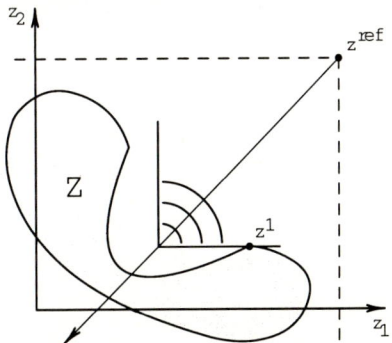

Fig. 10. Point Ecountered with Vertex   **Fig. 11.** Point Encountered on Side

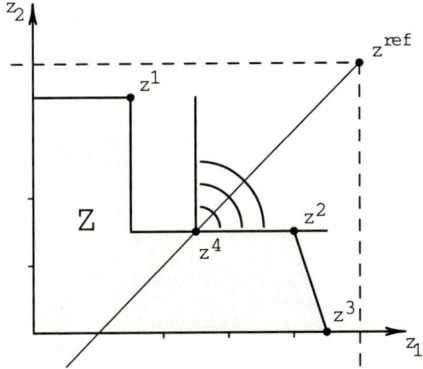

Fig. 12. Ties on Smallest Intersecting Contour

As long as the point on the smallest intersecting contour is unique, by Theorem 2 (a generalization of Theorem 1), the point is nondominated.

Theorem 2. *Let $\bar{z} \in Z$ and let $z^v \in R^k$ function as the vertex of a translated nonnegative orthant such that $(\{z^v\} \oplus R^+) \cap Z = \{\bar{z}\}$. Then $\bar{z}$ in nondominated.*

In answer to (b), let $Z_\lambda \subset Z$ be the set of points on the smallest intersecting contour. By Theorem 3, at least one point in Z_λ is nondominated.

Theorem 3. *Let $\lambda \in \Lambda$. Then $\bar{z}$ is nondominated if it is a point in Z_λ that is closest to a z^{ref} according to a L_1-metric.*

In Fig. 12 with $Z_\lambda = \text{bls}[z^4 \to z^2]$, z^2 is seen to be nondominated as it is the point in Z_λ that minimizes the sum of the coordinate deviations between it and z^{ref}, or in other words, z^2 is the point that solves

$$\min_{z \in Z_\lambda} \left[\sum_{i=1}^{k} (z_i^{ref} - z_i) \right]$$

8. Lexicographic Weighted Tchebycheff Sampling Program. With regard to the ability of a weighted Tchebycheff metric to *generate* a nondominated criterion vector, we have a two-stage process. In the first stage, we compute Z_λ, the set of points in Z closest to z^{ref} according to the λ-weighted Tchebycheff metric. If Z_λ is a singleton set, the point in Z_λ is the nondominated point generated. If Z_λ contains more than one point, the second stage is called into action to compute, as the nondominnated point generated, a point in Z_λ closest to z^{ref} according to the L_1-metric. Incorporating this geometry into an optimization problem, we have the two-stage *lexicographic weighted Tchebycheff sampling program* for

$$\text{lex min}\{\alpha, \sum_{i=1}^{k}(z_i^{ref} - z_i)\} \qquad (2)$$

$$\text{s.t.} \quad \alpha \geq \lambda_i(z_i^{ref} - z_i)$$
$$f_i(x) = z_i$$
$$x \in S$$

$$0 \leq \alpha \in R, \; z \in R^k \text{unrestricted}$$

generating, not only a nondominated criterion vector, but an inverse image (i.e., an efficient point) from S that goes with it. Note that the first optimization stage minimizes the scalar α to implement the λ-weighted Tchebycheff metric, and that the second stage, when invoked, minimizes $\sum_{i \in K}(z_i^{ref} - z_i)$ to implement the L_1-metric. It is called a "sampling" program because in this way, with a group of dispersed representatives from Λ, and solving (2) for each of them, we have a strategy for sampling points from N.

9. T-Vertex λ-Vectors. Note that in some cases, many different λ-vectors could cause the lexicographic weighted Tchebycheff sampling program to generate the same nondominated point. However, out of each such set, one is special and it is called the *T-vertex* λ-vector. The T-vertex λ-vector is the one that causes the smallest intersecting contour to hit the nondominated point in question head-on with its vertex. For a given $z \in N$ and z^{ref}, the coordinates of a T-vertex λ-vector are given by

$$\lambda_i = \frac{1}{z_i^{ref} - z_i} \left[\sum_{j \in K}^{k} \frac{1}{z_j^{ref} - z_j} \right]^{-1} \qquad (3)$$

10. Wierzbicki's Aspiration Criterion Vector Method (Figs. 13–14). As an introduction to the world of interactive procedures of multiple objective programming, we first discuss Wierzbicki's Aspiration Criterion Vector method [12]. The purpose of the procedure is to find at each iteration h the point $z^{(h)} \in N$ that is closest to that iteration's *aspiration criterion vector* $q^{(h)} < z^{ref}$, where $q^{(h)}$'s purpose is to capture the DM's criterion value hopes, expectations, and aspirations of the moment.

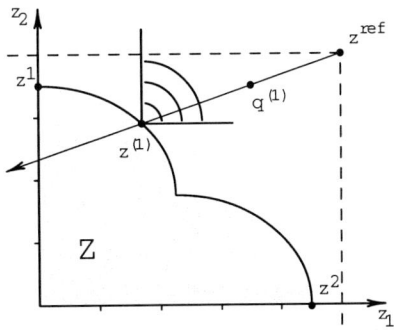
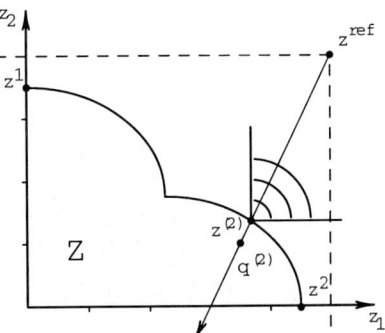

Fig. 13. Wierzbicki 1-st Iteration   **Fig. 14.** Wierzbicki 2-nd Iteration

1. $h = 0$. Establish a $z^{ref} \in R^k$.
2. $h = h + 1$. Specify a $q^{(h)}$ aspiration criterion vector.
3. Compute the $\lambda^{(h)} \in \Lambda$ that causes the probing ray to pass through $q^{(h)}$. Equation (3) can be used by substituting $q^{(h)}$ for z^{ref}.
4. Using $\lambda^{(h)}$ in (2), obtain the nondominated point $z^{(h)}$ closest to $q^{(h)}$ as computed by the lexicographic weighted Tchebycheff sampling program.
5. If after examining $z^{(h)}$ the DM decides to continue with another iteration, go to Step 2. Otherwise, terminate with $z^{(h)}$ as the *final solution*.

Consider the feasible region Z in Figs. 13 and 14 with $z^{ref} = (5,4)$. On the first iteration, let the DM's aspiration criterion vector be $q^{(1)} = (3\frac{1}{2}, 3\frac{1}{2})$. Then $\lambda^{(1)} = (\frac{1}{4}, \frac{3}{4})$ and the nondominated point generated by (2) is $z^{(1)}$. Continuing with another iteration, the DM specifies $q^{(2)} = (3,1)$. Then $\lambda^{(2)} = (\frac{3}{5}, \frac{2}{5})$ and the nondominated point generated by (2) is $z^{(2)}$. This iteration is interesting because, recognizing that $q^{(2)}$ is dominated, the method produces a superior nondominating point. Lacking a formal stopping criterion, the method continues as long as the DM is willing to specify new aspiration criterion vectors.

11. Tchebycheff Method (Figs. 15–16). Instead of generating only one solution at each iteration, the Tchebycheff Method conducts *multiple probes* by sampling each in a sequence of progressively small subsets of N. Letting P be the number of solutions presented to the DM at each iteration, we begin by selecting P dispersed λ-vectors from $\Lambda^{(1)} = \Lambda$ to form a group of dispersed probing rays as in Fig. 15. Then the lexicographic weighted Tchebycheff program (2) is solved for each of the λ-vectors. From the P resulting nondominated criterion vectors, the DM selects the most preferred, designating it $z^{(1)}$. Now, about the T-vertex λ-vector defined by $z^{(1)}$ and z^{ref}, a reduced subset $\Lambda^{(2)} \subset \Lambda$ is centered. Then P dispersed λ-vectors are selected from $\Lambda^{(2)}$ to form a group of more concentrated probing rays as in Fig. 16, and (2) is solved for each of them. From the P resulting nondominated criterion vectors, the DM selects the most preferred, designating it $z^{(2)}$. Now, about the T-vertex λ-vector defined by $z^{(2)}$ and z^{ref}, a further reduced subset $\Lambda^{(3)} \subset \Lambda$ is centered. Then P dispersed λ-vectors are selected from $\Lambda^{(3)}$, and so forth.

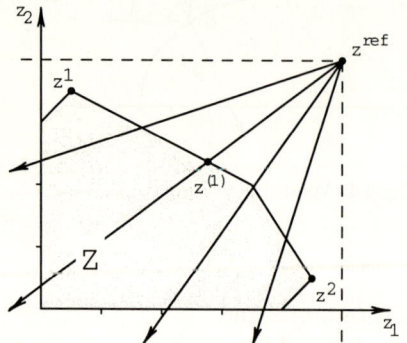

Fig. 15. Tchebycheff 1-st Iteration   **Fig. 16.** Tchebycheff 2-nd Iteration

12. Why Not Weighted-Sums Approach? One might ask why not to considered assigning a λ_i-weight to each objective function and solve the *weighted-sums* program

$$\max\{\sum_{i \in K}^{k} \lambda_i f_i(x) \mid x \in S\}$$

The reason is that the weighted-sums program only computes points that support hyperplanes and is unable to compute *unsupported* nondominated criterion vectors. A $\bar{z} \in N$ is unsupported if and only if there exists a convex combination of other nondominated criterion vectors that dominates $\bar{z}$. Otherwise, $\bar{z}$ is *supported*. Let N^u and N^s designate the sets of unsupported and supported nondominated criterion vectors, respectively. For example in Fig. 4, $N^u = \{z^3\}$ and $N^s = \{z^2, z^6\}$. In Fig. 6, $N^u = \text{bls}(z^2 \to z^3)$ and $N^s = \{z^1, z^3\}$, and so forth. If an unsupported point were optimal, the weighted-sums program would be unable to compute it. In contrast, the lexicographic weighted Tchebycheff program can compute *any* nondominated criterion vector without reservation.

13. Other Interactive Procedures. In addition to the Aspiration Criterion Vector and Tchebycheff methods, there are other procedures of multiple objective programming including STEM [1], Global Shooting [2], TRIMAP [3], Light Beam Search [5], Pareto Race [6], Bi-Reference Procedure [7], Fuzzy Satisficing [8], PSI [9], and FFANN [11], among others. While embodying various philosophies, most of these procedures nonetheless use variants of the lexicographic weighted Tchebycheff program [4] to probe N in different ways.

3 Conclusion

Because the lexicographic weighted Tchebycheff sampling program and its variants in other procedures can be configured for solution using conventional single criterion mathematical programming software, the procedures of multiple objective programming can generally address problems with as many constraints and variables as in the single criterion case. However, there is a limit to the number of objectives. While problems with up to 5–6 objectives can generally be accommodated, above this number gets us into uncharted territory where future research is needed.

References

[1] Benayoun, R., J. de Montgolfier, J. Tergny, and O. Larichev (1972). "Linear Programming with Multiple Objective Functions: Step Method (STEM)," *Mathematical Programming*, **1**, 366-375.

[2] Benson, H. P. and S. Sayin (1997). "Towards Finding Global Representations of the Efficient Set in Multiple Objective Mathematical Programming," *Naval Research Logistics*, **44**, 47-67.

[3] Climaco, J. and C. Antunes (1989). "Implementation of a User-Friendly Software Package—A Guided Tour of TRIMAP," *Mathematical and Computer Modelling*, **12**, 1299-1309.

[4] Gardiner, L. R. and R. E. Steuer (1994). "Unified Interactive Multiple Objective Programming," *European Journal of Operational Research*, **74**, 391-406.

[5] Jaszkiewicz, A. and R. Slowinski (1999). "The Light Beam Search Approach: An Overview of Methodology and Applications," *European Journal of Operational Research*, **113**, 300-314.

[6] Korhonen, P. and J. Wallenius (1988). "A Pareto Race," *Naval Research Logistics*, **35**, 277-287.

[7] Michalowski, W. and T. Szapiro (1992). "A Bi-Reference Procedure for Interactive Multiple Criteria Programming," *Operations Research*, **40**, 247-258.

[8] Sakawa, M. and H. Yano (1990). "An Interactive Fuzzy Satisficing Method for Generalized Multiobjective Programming Problems with Fuzzy Parameters," *Fuzzy Sets and Systems*, **35**, 125-142.

[9] Statnikov, R. B. and J. B. Matusov (1995). *Multicriteria Optimization and Engineering*, Chapman & Hall, New York.

[10] Steuer, R. E. (1994). "Random Problem Generation and the Computation of Efficient Extreme Points in Multiple Objective Linear Programming," *Computational Optimization and Applications*, **3**, 333-347.

[11] Sun, M., A. Stam and R. E. Steuer (1996). "Solving Multiple Objective Programming Problems Using Artificial Feed-Forward Neural Networks: The Interactive FFANN Procedure," *Management Science*, **42**, 835-849.

[12] Wierzbicki, A. P. (1986). "On the Completeness and Constructiveness of Parametric Characterizations to Vector Optimization Problems," *OR Spektrum* **8**, 73-87.

Poor-Definition, Uncertainty, and Human Factors – Satisfying Multiple Objectives in Real-World Decision-Making Environments

I. C. Parmee

Advanced Computational Technologies, Exeter, UK.
iparmee@ad-comtech.co.uk

Abstract. Ill-definition, uncertainty and multiple objectives are primary characteristics of real-world decision-making processes. During the initial stages of such processes little knowledge appertaining to the problem at hand may be available. A primary task relates to improving problem definition in terms of variables, constraint and both quantitative and qualitative objectives. The problem space develops with information gained in a dynamical process where optimisation plays a secondary role following the establishment of a well-defined problem domain. The paper speculates upon the role of evolutionary computing, complementary computational intelligence techniques and interactive systems that support such problem definition where multi-objective satisfaction plays a major role.

1 Introduction

The author's area of interest has primarily been within the field of evolutionary engineering design particularly relating to the higher levels of the design process where problem conceptualisation can represent a highly complex human-centred activity supported by a range of relatively basic machine-based models of the problem domain.

The process generally consists of search across an ill-defined space of possible solutions using fuzzy objective functions and vague concepts of the structure of the final solution. Solutions and partial solutions are explored and assessed in terms of their feasibility with regard to those constraints and objectives considered relevant at that time. Heuristics, approximation and experimentation play a major role with a high degree of flexibility evident in the establishment of domain bounds, objectives and constraints. The design environment itself will evolve with the solutions as the designer/design team gain understanding of the functional requirements and the resulting structures. Simple human and computer-based models which may be largely qualitative in nature are utilised in order to establish initial direction. The decision-making environment is characterised by uncertainty in terms of lack of available data and a poorly defined initial specification. Discovery and the accumulation of knowledge appertaining to problem definition and objective preferences are prevalent in this highly dynamical human / machine-based process The following quote [1] relating to creative design captures these aspects:

...problem formulation and reformulation are integral parts of creative design. Designers' understanding of a problem typically evolves during creative design processing. This evolution of problem understanding may lead to (possibly radical) changes in the problem and solution representations. [....] in creative design, knowledge needed to address a problem typically is not available in a form directly applicable to the problem. Instead, at least some of the needed knowledge has to be acquired from other knowledge sources, by analogical transfer from a different problem for example. [...] creativity in design may occur in degrees, where the degree of creativity may depend upon the extent of problem and solution reformulation and the transfer of knowledge from different knowledge sources to the design problem.

2 Changing Objectives during Decision-Making

Discovery and knowledge accumulation aspects of problem formulation are common across decision-making as a whole. Whether we are designing an engineering system, developing a financial strategy or establishing a business process, exploration involving initial variables, constraints and objectives will likely result in re-formulation of the problem domain through iterative search and user analysis of identified solutions.

This can be illustrated relatively simply through a decision-making process familiar to most which, although unrelated to industrial or commercial activities, could be seen as analogous in terms of discovery, knowledge accumulation, problem-reformulation and the eventual identification of a best compromise solution.

For illustrative purposes let us therefore consider a job-related relocation to a new city and the daunting problem of finding a family home. Initial investigation will likely relate to identifying appropriate districts based upon criteria relating to quality of local schools; safety / security issues; proximity to places of work, transport, highway networks, shopping and leisure facilities etc. plus average price and type of housing and overall environment. Other criteria relate directly to the ideal property such as maximum cost, number of bedrooms, garden, garage, parking etc. Several of the above criteria would be considered hard constraints (i.e. maximum cost) in the first instance.

The decision-making team is the family who would all probably rate the relative importance of the above criteria in a slightly different manner and whose opinions will carry a varying degree of influence. It is likely that initially there is a pretty clear vision of what the ideal property will look like and the preferred location.

Initial information gathering will provide quantitative and qualitative data relating to location from a wide variety of sources some reliable and some based upon hearsay. Gradually an overall picture will be established which will result in possible elimination of some options and the inclusion of new possibilities. New possible locations will be discovered during explorative trips to those already identified.

As details of properties are gathered it will likely become apparent that the ideal solution is hard to find and the concept of compromise becomes a reality. Hard constraints may soften whereas objective preferences will constantly be discussed and re-defined in the light of accumulated knowledge regarding districts and property availability within them. Particular characteristics of areas initially thought to be unsuitable may suddenly appear attractive. Search concentration may shift as it is discovered that such areas have suitable properties within the pre-set price range. Alternatively, the initial hard constraint relating to maximum price may soften as close to ideal properties in favoured locations become available. Other possible compromises are investigated in an attempt to accommodate increased costs.

The whole decision-making process becomes an uncertain mix of subjective / objective decisions as goal-posts move, objectives rapidly change in nature and external pressures relating to time constraints begin to take precedence. At the end of the day it is quite probable that the chosen home differs significantly from the one first envisaged. That guest bedroom may have been sacrificed and the garden may be far smaller but the location is ideal. Alternatively, the route to work may be longer and more tortuous but a property close to ideal at the right price in an up-and-coming neighbourhood has been found.

Although a seemingly simple problem the overall search process is highly complex and uncertainty, compromise and problem re-definition are inherent features. Although a little far removed from commercial and industrial decision-making scenarios analogies are apparent. It is suggested that we can learn much from such everyday decision-making scenarios and utilise this knowledge when designing interactive evolutionary search environments that can support more complex decision-making processes.

3 Knowledge Generation and Extraction

A machine-based search and exploration environment has been proposed that provides relevant problem information to the designer / decision-making team [2,3]. The intention is that such information can be processed and subsequent discussion can result in the recognition of similarities with other problem areas and the discovery of possible alternative approaches. One of the major characteristics of population-based evolutionary search is the generation of a large amount of possibly relevant information most of which may be discarded through the actions of various operators. The development of interactive systems may support the capture of such information and its utilisation in the subsequent re-formulation of the problem through the application and integration of experiential knowledge. It can be argued that such problem re-formulation captures this knowledge which then plays a significant role in further evolutionary search relating to the re-defined problem. The re-definition of objectives and objective preferences is an important aspect of this evolution of the problem space.

It is apparent that a core activity within the decision-making process relates to the gathering of information relating to diverse aspects of the problem space. It has been proposed [4] that a primary role of evolutionary machine-based search and exploration processes can provide a central role in the generation of such information. This moves the utilisation of EC away from application over a set number of generations or until some convergence criteria is met to a more continuous exploratory process where changes to objective weightings, variable ranges and constraint based upon information generated results in a moving, evolving problem space. The primary task of such an approach is the design of an optimal problem space as opposed to the identification of an optimal solution.

This theme has been central to much of the author's previous work where the development of EC strategies relating to the higher levels of the design process has related to the identification of high performance regions of complex conceptual design space (vmCOGAs, [5,6]) or the identification of optimal alternative system configurations through the utilisation of dual-agent strategies for search across mixed discrete / continuous design hierarchies [7,8]. Other work relates to the concurrent satisfaction of both quantitative and qualitative criteria through the integration of fuzzy rule bases with evolutionary search [9].

More recent work has led to the establishment of an experimental interactive evolutionary design system (IEDS) that supports a relatively continuous iterative user / evolutionary search process that involves EC, agent-based approaches and a number of other complementary techniques.

An overview of this research that illustrates the manner in which EC-based information gathering can support decision-making in complex multi-objective design environments follows. A much more detailed treatment can be found in the referenced text whereas the author's research as a whole is presented in the book 'Evolutionary and Adaptive Computing in Engineering Design' [10].

4 Experimental Approaches

4.1 The Qualitative Evaluation System (QES)

Early work related to the development of a Qualitative Evaluation System. The QES strategy provides support to the designer when attempting to determine trade-offs between both quantitative and qualitative criteria. This support utilises a linguistic rule base (generated in close collaboration with Rolls Royce turbine engineers) which resides within a fuzzy expert system. The rules relate to the comparative effectiveness of a GA-generated design solution in terms of manufacturability, choice of materials and a number of special preferences relating to in-house capabilities. Quantitative aspects of a design are combined with qualitative ratings to generate a measure of the overall fitness of the solutions. Domain knowledge concerning variable preferences and heuristics is utilised and combined using a concept of compromise [11,12].

The work concerns the preliminary design of gas turbine engine cooling hole geometries. The primary objective is to minimise the mass flow through the radial cooling hole passage. Adaptive Restricted Tournament Selection (ARTS [13]) identifies a number of single high performance solutions from the design space. The qualitative evaluation system receives all the design variable values of each of these solutions as inputs and develops an overall qualitative rating concerning the effectiveness of the design.

The ARTS multi-modal GA is utilised to first identify a number of 'good' quantitative (i.e. minimal mass coolant flow) design solutions. These "good" solutions are next evaluated by the QES which takes the variable values of each solution as inputs and outputs a qualitative rating for the design. The QES has three components, the fuzzifier, the fuzzy inference engine and the defuzzifier. Designer knowledge provided by Rolls Royce engineers is stored in a static fuzzy rule base. During the fuzzification stage each variable range is divided into five sub-ranges, and expressed using linguistic terms. A crisp value for the effectiveness is obtained through centre-of-gravity type defuzzification.

The knowledge base for the system is developed using fuzzy rules and facts embodying qualitative aspects of the design problem in terms of manufacturability, choice of materials and some designer's special preferences. The knowledge base is presented in three categories: *Inter Variable Knowledge* which relates to the relative importance of each variable in terms of the objective function; *Intra Variable Knowledge* which relates to preferred subsets of each variables' range e.g. blade wall thickness needs to be *low* in terms of material cost but *high* in terms of stress considerations; *Heuristics* which mostly concern specific cases where there is no uncertainty concerning the conclusion.

The inter and intra variable knowledge is then integrated using a concept of compromise which is implemented to reduce the severity of qualitative ratings. Inter variable knowledge determines the degree of compromise possible on every variable (*slight compromise, less compromise, compromise* and *more compromise*). A more complete definition of the fuzzy rule base can be found in [9 & 11].

Having evaluated a solution both *quantitatively* via the ARTS GA process and *qualitatively* via the QES the results can be combined and presented in a graphical manner that facilitates overall understanding of the major aspects of the problem. Four solutions for each of three internal cooling-hole geometries (plane, ribbed and pedestal) are presented in figure 1with their relative quantitative fitness plainly shown by the major bars of the chart. The qualitative ratings for each solution are then shown as a series of embedded, shaded bars. Such a representation contains much information in a relatively transparent manner. It is apparent, for instance, that although the plane cast internal geometry provides a low quantitative fitness the solutions are relatively robust in terms of the qualitative criteria. The qualitative bars relating to the other two geometries show a much greater degree of fluctuation. This can perhaps provide insight into the problem and criteria characteristics which aid the designer in terms of both modelling the system and determining preference rankings for each criteria. For instance, if the priority is for a solution that can be considered low risk in terms of possible problems relating to manufacturing, material and special preferences aspects and losses relating to quantitative performance can be made up within other areas of the turbine design then a compromise can be made and a plane cast internal geometry can be chosen. If, however, quantitative performance is paramount then a pedestal geometry that best satisfies preferences relating to the three qualitative criteria may be considered appropriate.

The QES provides a good indication of the relative merits of high-performance solutions in terms of a number of qualitative criteria. A possible problem area here however is the flexibility of the rule-base. There is a requirement for on-line rule changes that should be easy to implement by users with no knowledge of fuzzy inference technologies. A high degree of flexibility in terms of the representation of objectives and their weightings is a major requirement during these higher levels of the design process. It is possible that the major utility offered by the QES relates to more routine design tasks where problem definition is already high.

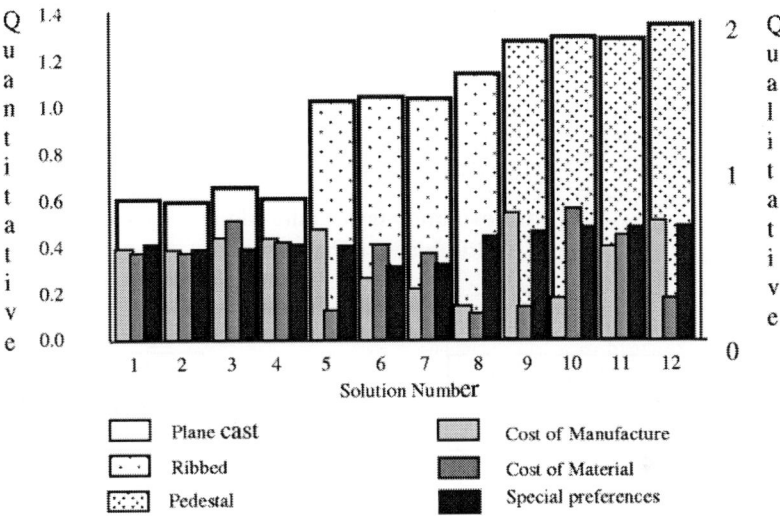

Fig. 1. Graphical representation of ARTS and the QES

4.2 The Interactive Design System

4.2.1 Introduction

The QES illustrates the manner in which qualitative criteria can be combined with quantitative evaluation but it does not readily support on-line interaction. The requirement for a system that supports the on-line extraction of information that can be presented to the user in a succinct manner thereby supporting easily implemented change has led to an investigation of various techniques that can be combined within an overall architecture.

The satisfaction of multiple objectives (i.e.> 10) is a major requirement and such objectives must be entirely flexible in terms of preferences / weightings to allow adequate exploration of the problem domain and to support a better understanding of the complex interactions between variable space and objective space.

The developed system involves a number of machine-based processes that can communicate as shown simply in figure 2. The user is an integral part of the system accepting, analysing and processing information from the system, introducing objective change via the Preferences component and variable parameter changes directly into the Evolutionary component. The Evolutionary component can operate in several modes:
1. Single evolutionary process
2. Multiple individual evolutionary processes – no co-operation
3. Multiple co-evolving processes

Mode 1 relates to problem decomposition aspects where the Cluster-oriented Genetic Algorithms of the Information Gathering component extract information relating to variable interaction, variable redundancy and the setting of appropriate variable parameter ranges. This is describe fully in references [3 & 4] and is not included in the following text as it does not directly relate to multi-objective aspects.

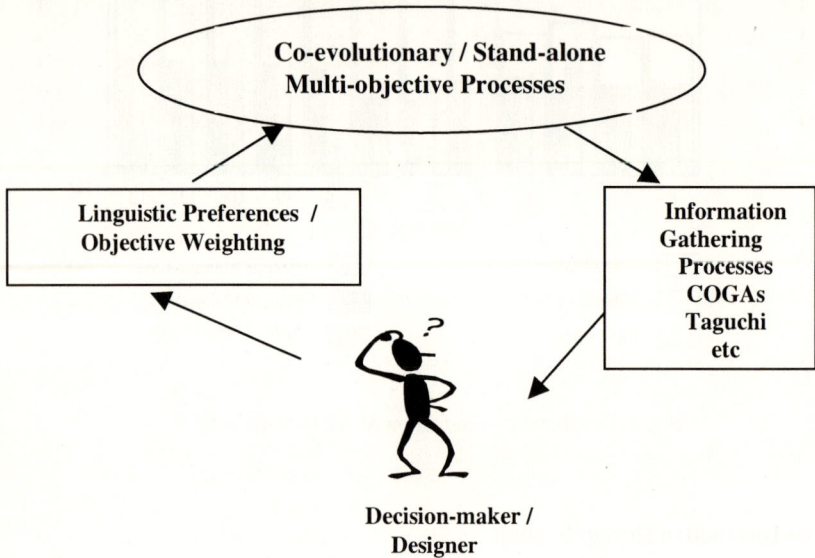

Fig. 2. Components of the Interactive Evolutionary Design System

4.2.2 COGAs as Multi-objective Information Gatherers

Mode 2 concerns the concurrent identification of high-performance (HP) regions of the problem space relating to individual objectives and the subsequent definition of common HP regions where best compromise solutions may be found. Again, the COGA techniques of the Information Gathering component are involved. The basic structures of COGAs and the associated adaptive filter (AF) have been described in a number of papers [4,5,6]. Their function relates to the rapid decomposition of complex, multi-variate design space into regions of high-performance. At the COGA core is an explorative genetic algorithm (GA). Exploration has initially been promoted through variable mutation regimes (vmCOGA) and more recently through the integration of various sampling techniques [13]. The AF mechanism extracts and scales populations (in terms of fitness) from a continuous GA evolutionary process only allowing solutions that lie above a filter threshold to pass into a final clustering set. The design exploration capabilities are well described in Parmee and Bonham [4] along with extensive discussion relating to possible interactive utilisation.

Of particular interest here is the identification of high-performance regions relating to differing objectives from consecutive runs of COGA. Identified HP regions can be overlaid upon selected two-dimensional hyperplanes described by pairs of variables selected from the set of variable parameters that describe the problem space.

The utilisation of the Evolutionary and Information Gathering components is demonstrated within a design domain relating to the preliminary design of military aircraft. This is a complex design domain characterised by uncertain requirements and fuzzy objectives relating to the long gestation periods between initial design brief and realisation of the product. Changes in operational requirements in addition to technological advances cause a demand for a responsive, highly flexible strategy where design change and compromise are inherent features for much of the design period. Design exploration leading to innovative and creative activity must be supported. The ability to introduce rapid change to satisfy the many operational, engineering and marketing considerations as they themselves change is essential. In this case the COGA software is manipulating the BAE Systems mini-CAPS model. This model is a much condensed version of the CAPS (Computer-aided Project Studies) suite of software for conceptual and preliminary airframe design. Mini-CAPS maintains many of the characteristics of the overall suite especially in terms of multiple objectives. The nine input variables that define the problem space can generate up to thirteen outputs,

Figure 3a shows high-performance regions relating to three mini-CAPS objectives: attained turn rate (ATR), specific excess power (SEP) and Ferry Range (FR) plotted on the gross wing plan area / wing aspect ratio variable parameter hyperplane. As can be seen regions relating to ATR and FR overlap forming a region containing HP compromise solutions. There is no mutually inclusive HP region relating to SEP however. By returning to COGA and reducing the severity of the adaptive filter in relation to SEP solutions the SEP region can be expanded as shown in figure 3band 3c until a mutually inclusive region involving all objectives is identified. This relaxing of the adaptive filter threshold allows lower performance SEP solutions through to the final clustering set. This could be considered equivalent to a lessening of the relative importance of this objective.

The technique allows the projection of objective space onto variable space. This gives a visual appreciation of the interaction between the various objectives and supports the user in the determination of initial preferences concerning their relative performance. Current work is investigating the relationship of the solutions within the mutually inclusive regions to the non-dominated solutions of the Pareto frontier. This work is indicating that a good approximation to the Pareto front is contained within these regions which can be realised by identifying the non-dominated solutions contained within the high performance regions describing all objectives.

It is stressed that this visual representation is an indication only. Even if mutually inclusive compromise regions are apparent in all variable parameter hyperplanes there still exists a possibility that these regions do not exist to the extent suggested in the graphical representations. This could be due to highly convoluted high-performance regions. Current work relates to the development of agent-based systems that check the validity of compromise regions across all dimensions. These agents will then inform the designer of any possible problem. This is not a complex procedure as all solution vectors describing the regions are available.

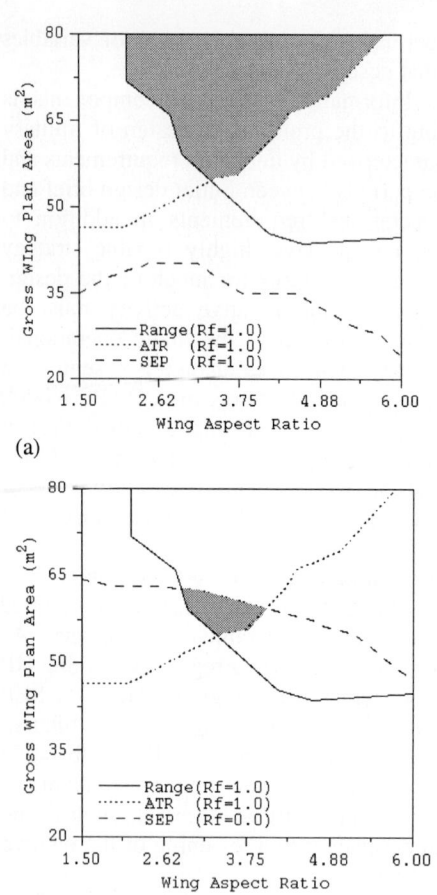

(a)

(c)

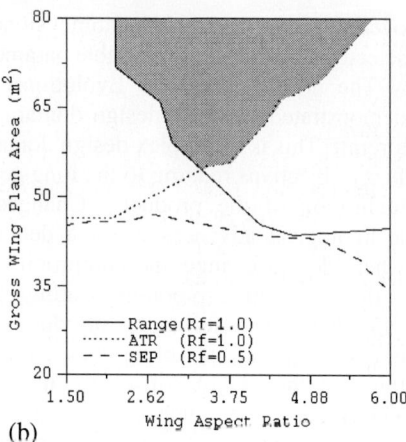

(b)

Fig. 3. Identification of compromise high-performance regions relating through filter threshold relaxation

(a) A common region for Ferry Range and Turn Rate has been identified but Specific Excess Power objectives cannot be satisfied.

(b) Relaxing the SEP filter threshold allows lower fitness solutions through and boundary moves

(c) Further relaxation results in the identification of a common region for all objectives.

The aim of this work, however, is to support a better understanding of objective interaction and conflict through graphical representation rather than providing a succinct and accurate representation of compromise regions or of the Pareto frontier. In this sense, the technique again supports the generation of information appertaining to the problem at hand where variables and objectives and can vary as problem knowledge expands. The approach therefore takes into consideration the uncertainties and ill-definition inherent in the utilisation of preliminary design models and of the degree of initial understanding of the problem domain. A visual representaion of the degree of possible conflict and the manner in which a changing of objective preferences can result in a relaxation of such conflict is considered more viable than the utilisation of more sophisticated techniques that identify optimal non-dominated solutions that lie upon the true Pareto frontier at this stage. The basic notion of 'rubbish in, rubbish out' must be taken into consideration. Much time could be spent upon more definitive analysis to identify Pareto optimal points that prove erroneous upon the introduction of more definitive problem models. The strategy therefore indicates a probable best way forward rather than the global solution to the problem which, at this stage, is poorly defined.

4.2.2 The Preference Component

Having gained a better understanding of objective interactions and of problems relating to the degree of difficulty likely to be encountered in the satisfaction of initial objective preferences a highly flexible interface for the introduction of such preferences to a more definitive multi-objective search process is required.

A methodology that supports on-line variation of design preferences has been developed. Efficient exploration across the many different design variants that the designer wishes to assess is of more interest than the identification of single optimal solutions. The system should therefore be able to further support such exploration whilst also suggesting best design direction.

It is generally accepted that it is easier for the decision-maker to give qualitative ratings to objectives i.e. 'Objective A is much more important than objective B' than to set the weight w_A of objective A to, say, 0.1 or to 0.09. The method of fuzzy preferences [14] and induced preference order can be utilised to introduce such rule-based preference representation for transformation into appropriate objective weightings. The following predicates can be introduced [15]:

Table 1. Rule-based preference representation

relation	intended meaning	relation	intended meaning
≈	is much less important	<	is less important
<<	is equally important	#	don't know or don't care
¬	is not important	!	is important

These together with the complementary relations > and >>, can help build the relationship matrix R necessary for a 'words to numbers' transformation. For this transformation, concepts of 'leaving score' [14] amongst other techniques, can be employed.

It is first necessary for the engineer to rank the objectives in terms of relative importance but numeric weightings are not required. If transitivity is assumed, then the number of required questions to establish overall preference ratings is reduced. The preference algorithm has been described fully in a number of publications [2,15] and it is not intended to reproduce it here. Examples from the graphical user interface follow which illustrate the preference procedure. The problem domain again relates to BAE preliminary airframe design.

The user first selects those objectives that require investigation. In this case Take-off Distance, Landing Speed, Ferry Range and Mass at Take-off have been selected from a possible thirteen outputs from the mini-CAPS model. Having selected the objectives it is necessary to establish equivalence classes which classify the objectives in terms of relative performance. In the example (figure 4), two of the objectives are considered to be equally important, and two further objectives are considered to have different levels of importance. Three different levels of importance are therefore processed.

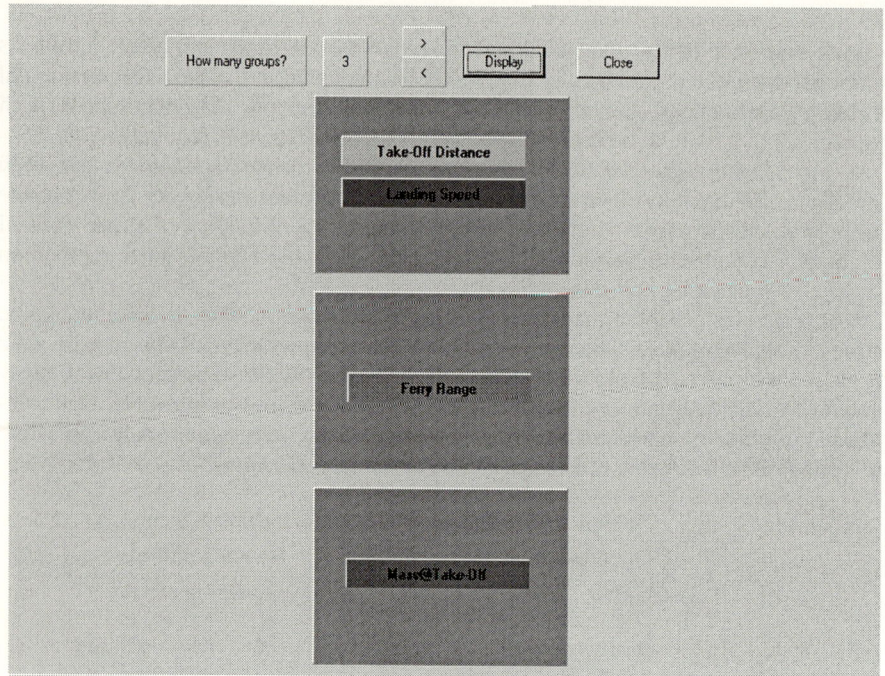

Fig. 4. Establishing equivalence classes

The interface facilitates the on-line change of objective preferences which allows further exploratory runs providing information relating to possible problem redefinition. Having selected objectives, established objective preferences and performed the machine-based 'words-to-numbers' transformation the numeric weightings can be passed to the Evolutionary process module.

4.2.3 The Co-evolutionary Multi-objective Approach (Mode 3)

The Preference component has been linked with the core co-evolutionary processes. The goal is to explore high-performance solutions relating to several objectives whilst providing maximum information concerning: appropriate regions of complex, multi-dimensional, Pareto surfaces; single objective optimal solutions and a number of solutions that best satisfy a range of ideal scenarios. This approach is an alternative to the generation of n-dimensional trade-off surfaces comprising very large numbers of non-dominated solutions that can be achieved using standard EC-based Pareto approaches. Again, the intention is to generate information that supports a better understanding of the multiple criteria aspects of the problem rather than identity Pareto-optimal solutions.

The distributed method involves individual GAs for the optimisation of each objective. Search relating to an individual objective takes place upon each evolutionary process. Subsequently, through the application of penalties, the co-evolving processes are drawn into that region of the overall space that offers best compromise relating to all objectives and their pre-set preferences (Parmee I. et al 2000).

During the co-evolutionary run the fitness for each objective is normalised relative to the maximum and minimum values found for each GA with constant adjustment as new upper and lower limits are identified. In each generation, the variables of solutions relating to each objective are compared with those of the best individual from the other co-evolving GA populations. If a variable is outside a range defined by a range constraint map it is adjusted by a penalty function. The range constraint map reduces maximum allowable distances between variables at each generation. Initially the map must allow each GA to produce a good solution based on its own specified objective. As the run progresses inflicted penalties increasingly reduce variable diversity to draw all concurrent GA searches from their separate objectives towards a single compromise design region where all objectives are best satisfied. This process is illustrated in figure 5(b) where the individual evolution of each objective is projected upon the Ferry Range objective hyperplane. In this case all objectives are of equal importance.

The machine-generated numeric weightings resulting from the preference ranking introduced by the user can now modify the penalties inflicted. A heavy penalty inflicted upon a 'much more important' objective is therefore moderated in order to allow some influence upon the co-evolutionary search whereas such penalties upon objectives of lesser importance may not change or may be modified to take into account objective ranking. The effect of varying the relative importance of the Ferry Range objective via the Preference component is shown in figure 5(a), 5(b) and 5(c).

In most real decision-making situations variables will have differing degrees of influence upon any given objective. An on-line sensitivity analysis which ranks variables according to their influence upon each objective has been introduced. This design sensitivity ranking is then used to adjust the fitness of each solution to ensure that the values of the most influential variables are within the range defined by the constraint map. Solutions are assigned the highest fitness penalty where their most influential variables lie outside of the current constraint map range. This ensures that populations contain high levels of compromise solutions in terms of the most influential variables and relatively redundant variables have little or no effect on overall solution fitness. The Taguchi method has been selected to determine the sensitivity of each input [16].

Again, the concentration is upon information-gathering and visual representation rather than the identification of Pareto optimal points. The co-evolutionary multi-objective approach provides the following information within one run of the process: high-performance solutions relating to the individual objectives; evolutionary 'tracks' that trace the Pareto surface to some extent; the bounds of a compromise region where all objectives will likely be best satisfied and the identification of influential and redundant variables relating to each objective. A more in-depth description of the process with results can be found in reference [17].

4.2.4 Agent-Based Support

Stochastic population-based search generates a mass of information much of which is discarded. The intention is to introduce agents that monitor solutions generated from single or co-evolutionary processes and identify those that may be considered 'interesting' by the decision-maker. The notion of 'interesting' may relate to: a good solution with a large Hamming or Euclidean distance from the majority of the population; a good solution that may satisfy the majority of constraints or objectives but is not satisfactory in a few; a not particularly high-performance solution where the constituent variable values lie within preferred ranges.

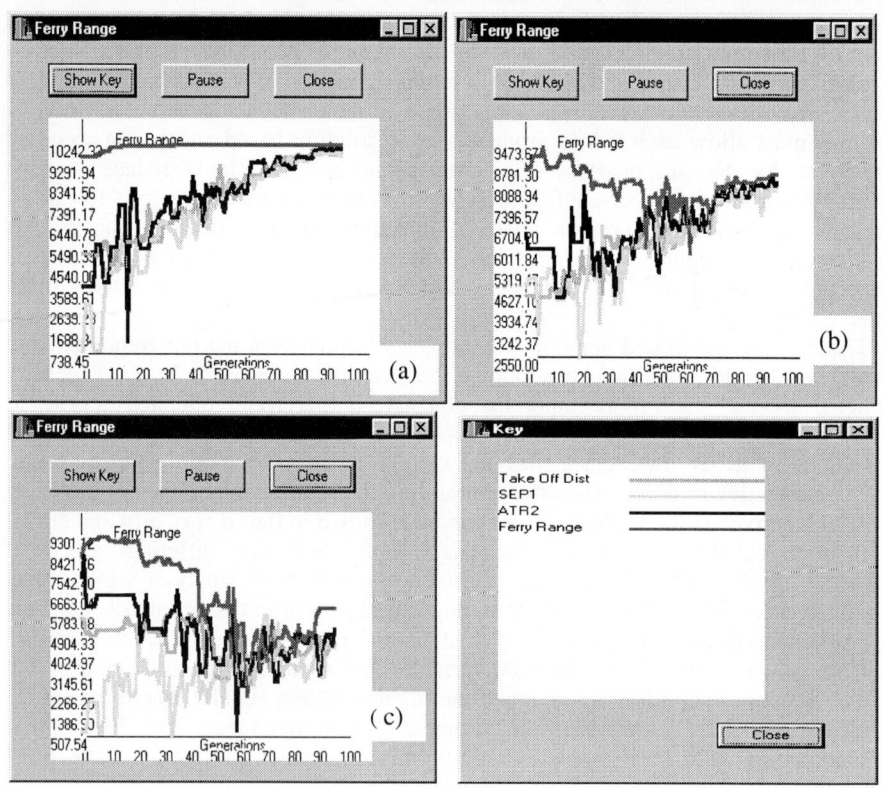

Fig. 5. (a) Ferry Range is much more important
(b) All objectives are of equal importance
(c) Ferry Range is much less important

The concept of 'interesting' may be largelay subjective which suggest that a degree of machine-learning may be appropriate where the responsible agents 'learn' from user reaction to possible interesting solutions presented to them. This is obviously an area requiring extensive further research.

Experimental negotiating agent systems utilising the rule-based preferences have been established for the identification of solutions that satisfy a range of design scenarios relating to multiple objectives and ideal variable values. For instance, the designer is likely to have several ideal scenarios such as: 'I would like objective A to be greater than 0.6 and objective C to be less than 83.5; objectives B, D, E should be maximised; variable 2 should have a value of between 128.0 and 164.5; a value greater than 0.32 is prefered for variable 7...etc'. The Incremental Agent operates as follows: use designer's original preferences for both objectives and scenarios and run the optimisation process; if some scenarios are not fulfilled, the agent suggests an increase in the importance of these scenarios; if some scenarios are still not fulfilled even when classed as 'most important' agent suggests change to variable ranges in scenario; if some scenarios are still not fulfilled the agent reports to designer and asks for assistance.

The Incremental Agent strategies have been integrated with the Preferences and Co-evolutionary Multi-objective component on an experimental basis. Both the Scenario and Incremental Agent components sit between the designer and the Preference module drawing information from both. For a more detailed description of the processes and initial results based upon a miniCAPS example the reader is directed to Cvetkovic D. [18].

5 Discussion

The paper discusses complexities relating to real-world multi-objective decision-making processes where the problem domain develops with information gained from evolutionary search and exploration. The intention has been to illustrate the manner in which EC can support such processes through the establishment of highly interactive systems where generated information provides problem insights which result in problem reformulation. An initial framework has been briefly outlined that supports the concept of an Interactive Evolutionary Design System (IEDS). Much further research and development is required to achieve relatively seamless development of the problem space where the decision-maker's knowledge becomes embedded within an iterative human / evolutionary computational process.

The entire concept moves away from the identification of non-dominated solutions and the generation of an n-dimensional Pareto frontier. The reasoning is that the inherent uncertainties and human-centred aspects of complex decision-making environments renders such approaches less viable although their utility is well-founded in more well-defined problem areas.

The intention is also to move away from the identification of solutions through the short-term application of evolutionary search techniques. The goal is a continuous, dynamic explorative process primarily made possible by the search and exploration capabilities of iterative designer / evolutionary systems. Although ambitious, it is suggested that such a concept could best utilise the processing capabilities of present and future computing technology during complex human / machine-based decision-making activities.

The proposed architecture and IEDS concept provides an indication of what is possible in terms of interaction, information gathering and problem reformulation relating to variables and objectives. Further research may result in a much modified structure where agent technologies play a major and, to some extent, autonomous role to ensure appropriate communication and information processing capabilities. It is hoped that the concepts presented here will stimulate interest and such further research.

References

1. Goel A. K.: Design, Analogy and Creativity. IEEE Expert, Intelligent Systems and their Applications, 12 (3). (1997) 62 – 70
2. Parmee I. C., Cvetkovic C., Watson A. H., Bonham C. R.: Multi-objective Satisfaction within an Interactive Evolutionary Design Environment. Evolutionary Computation. **8** (2), (2000) 197 – 222.
3. Parmee I. C., Cvetkovic C., A. H., Bonham C. R., Packham I.: Introducing Prototype Interactive Evolutionary Systems for Ill-defined Design Environments. To be published in Journal of Advances in Engineering Software, Elsevier, (2001).
4. Parmee I. C., Bonham C. R.: Towards the Support of Innovative Conceptual Design Through Interactive Designer / Evolutionary Computing Strategies. Artificial Intelligence for Engineering Design, Analysis and Manufacturing Journal; Cambridge University Press, 14, (1999) 3 – 16.
5. Parmee I. C.: The Maintenance of Search Diversity for Effective Design Space Decomposition using Cluster-Orientated Genetic Algorithms (COGAs) and Multi-Agent Strategies (GAANT). Proceedings of 2nd International Conference on Adaptive Computing in Engineering Design and Control, PEDC, University of Plymouth; (1996) 128 – 138.
6. Parmee I. C.: Cluster Oriented Genetic Algorithms (COGAs) for the identification of High Performance Regions of Design Spaces. *First International Conference on Evolutionary Computation and its Applications, EvCA 96*, Presidium of the Russian Academy of Sciences, Moscow; (1996) 66-75.
7. Parmee, I.C.: The Development Of A Dual-Agent Strategy For Efficient Search Across Whole System Engineering Design Hierarchies. Proceedings of Parallel Problem Solving from Nature. (PPSN IV), Lecture notes in Computer Science No. 1141; Springer-Verlag, Berlin (1996) 523-532.
8. Parmee I. C.: Evolutionary and Adaptive Strategies for Efficient Search Across Whole System Engineering Design Hierarchies. Journal of Artificial Intelligence for Engineering Design, Analysis and Manufacturing; 12 (1998) 431 – 435.
9. Roy R, Parmee I. C, Purchase G.: Integrating the Genetic Algorithm with the Preliminary Design of Gas Turbine Cooling Systems. Proceedings of 2nd International Conference on Adaptive Computing in Engineering Design and Control, PEDC, University of Plymouth (1996).
10. Parmee I. C.: Evolutionary and Adaptive Computing in Engineering Design. Springer Verlag, London (2001).
11. Roy R., Parmee I. C., Purchase G.: Qualitative Evaluation of Engineering Designs using Fuzzy Logic. Proceedings of ASME Design Engineering Technical Conferences and Computers in Engineering Conference, Irvine, California; (1996) 96-DETC/DAC-1449.
12. Roy R., Parmee I. C.: Adaptive Restricted Tournament Selection for the Identification of Multiple Sub-optima in a Multi-modal Function. Lecture Notes in Computer Science, Evolutionary Computing; Springer-Verlag, (1996) 236-256.
13. Bonham C. R., Parmee I. C.: An Investigation of Exploration and Exploitation in Cluster-oriented Genetic Algorithms. Proceedings of the Genetic and Evolutionary Computation Conference, Orlando, Florida, USA; (1999) 1491 – 1497.
14. Fodor J., Roubens M.: Fuzzy Preference Modelling and Multi-criteria Decision Support. System Theory, Knowledge Engineering and Problem Solving, **14**; Kluwer Academic Publishers (1994).
15. Cvetkovic D., Parmee I. C.: Designer's Preferences and Multi-objective Preliminary Design Processes. Evolutionary Design and Manufacture: Proceedings of the Fourth International Conference on Adaptive Computing in Design and Manufacture. Springer-Verlag (2000) 249 – 260.
16. Peace G. S.: *Taguchi Methods*. Addison Wesley, Reading, M. A. (1992).
17. Parmee I. C., Watson A. W.: Preliminary Airframe Design using Co-evolutionary Multi-objective Genetic Algorithms. Proceedings of the Genetic and Evolutionary Computation Conference, Orlando, Florida, USA; (1999) 1657 - 1665.
18. Cvetkovic D.: Evolutionary Multi-objective Decision Support Systems for Conceptual Design. PhD Thesis, University of Plymouth (2000).

Controlled Elitist Non-dominated Sorting Genetic Algorithms for Better Convergence

Kalyanmoy Deb and Tushar Goel

Kanpur Genetic Algorithms Laboratory (KanGAL)
Indian Institute of Technology Kanpur
Kanpur, PIN 208 016, India
{deb,tusharg}@iitk.ac.in
http://www.iitk.ac.in/kangal

Abstract. Preserving elitism is found to be an important issue in the study of evolutionary multi-objective optimization (EMO). Although there exists a number of new elitist algorithms, where elitism is introduced in different ways, the extent of elitism is likely to be an important matter. The desired extent of elitism is directly related to the so-called exploitation-exploration issue of an evolutionary algorithm (EA). For a particular recombination and mutation operators, there may exist a selection operator with a particular extent of elitism that will cause a smooth working of an EA. In this paper, we suggest an approach where the extent of elitism can be controlled by fixing a user-defined parameter. By applying an elitist multi-objective EA (NSGA-II) to a number of difficult test problems, we show that the NSGA-II with controlled elitism has much better convergence property than the original NSGA-II. The need for a controlled elitism in evolutionary multi-objective optimization, demonstrated in this paper should encourage similar or other ways of implementing controlled elitism in other multi-objective evolutionary algorithms.

1 Introduction

It is now well established through a number of studies [1,11] that elitist multi-objective evolutionary algorithms (MOEAs) have better convergence characteristics than non-elitist MOEAs. Motivated by these studies, researchers and practitioners now concentrate on developing and using elitist MOEAs. This have resulted in a number of elitist MOEAs, such as strength Pareto EA or SPEA [13], Pareto-archived evolution strategy or PAES[8], and others. After the early suggestion of a non-elitist multi-objective genetic algorithm (NSGA) [9], the first author and his students have also suggested an elitist, fast, and parameter-free multi-objective GA (NSGA-II) [2,4]. All these elitist algorithms are interesting and have shown tremendous potential in solving multi-objective optimization problems.

In the context of single objective optimization, the elitism is introduced in a number of different ways. Preserving elitism means emphasizing the currently-best solutions in the subsequent generations. In some implementations, this is achieved by simply carrying over the best $\epsilon\%$ of the population in the next generation. In other implementations, the parent and child populations are combined together and the best 50% of the population is retained. This way previously-found good solutions are given a chance to carry over to subsequent generations.

In the context of multi-objective optimization, the meaning of elite solutions is different from that in single objective optimization. Here, all solutions that belong to the currently-best non-dominated front are best solutions in the population and are all equally important. Thus, all these solutions are elite solutions. In many occasions, a population may be mostly comprised of currently-best non-dominated solutions. When this happens, the preservation of elitism means acceptance of all such solutions. In such a scenario, not many new solutions can be accepted in the population. As a result, the search process may stagnate or prematurely converge to a suboptimal solution set. Thus, there is a need of introducing elitism in a controlled manner, in the context of multi-objective optimization.

In this paper, we address the issue of controlling elitism from the point of maintaining a balance between underlying exploitation and exploration issues. Thereafter, we argue that our earlier elitist implementation NSGA-II can have uncontrolled elitism in tackling certain problems. We suggest a controlled elitist approach, where only a certain portion of the population is allowed to contain the currently-best non-dominated solutions. The dominated solutions are purposefully kept in the population to reduce the elitism effect. By introducing a parameter to control the extent of elitism, we study its effect on five complex test problems. In all simulation results, it is clear that the controlled NSGA-II has a better convergence property than the original NSGA-II with uncontrolled elitism.

2 Elitist Non-dominated Sorting Genetic Algorithm (NSGA-II)

The details of NSGA-II algorithm appear elsewhere [4]. Essentially, NSGA-II differs from our original NSGA implementation [9] in a number of ways. Firstly, NSGA-II uses an elite-preserving mechanism, thereby assuring preservation of previously found good solutions. Secondly, NSGA-II uses a fast non-dominated sorting procedure. Thirdly, NSGA-II does not require any tunable parameter, thereby making the algorithm independent of the user.

Initially, a random parent population P_0 is created. The population is sorted based on the non-domination. A special book-keeping procedure is used in order to reduce the computational complexity to $O(MN^2)$. Each solution is assigned a fitness equal to its non-domination level. Binary tournament selection, recombination, and mutation operators are used to create a child population Q_0 of size N. Thereafter, we use the algorithm shown in Figure 1 in every generation.

First, a combined population $R_t = P_t \cup Q_t$ is formed. This allows parent solutions to be compared with the child population, thereby ensuring elitism. The population R_t is of size $2N$. Then, the population R_t is sorted according to

```
R_t = P_t ∪ Q_t
F = fast-non-dominated-sort(R_t)
P_{t+1} = ∅ and i = 1
until |P_{t+1}| + |F_i| ≤ N
    P_{t+1} = P_{t+1} ∪ F_i
    crowding-distance-assignment(F_i)
    i = i + 1
Sort(F_i, ≺_n)
P_{t+1} = P_{t+1} ∪ F_i[1 : (N − |P_{t+1}|)]
Q_{t+1} = make-new-pop(P_{t+1})
t = t + 1
```

Fig. 1. NSGA-II algorithm is shown.

non-domination [10] and different non-dominated fronts F_1, F_2, and so on are found. The algorithm is illustrated in the following:

The new parent population P_{t+1} is formed by adding solutions from the first front F_1 and continuing to other fronts successively till the size exceeds N. Individuals of each front are used to calculate the crowding distance—the distance between the neighboring solutions [4]. Thereafter, the solutions of the last accepted front are sorted according to a *crowded comparison criterion* and a total of N points are picked. Since the diversity among the solutions is important, the crowded comparison criterion uses a relation $\prec_n$ as follows:

Definition 1 *Solution i is better than solution j in relation $\prec_n$ if $(i_{rank} < j_{rank})$ or $((i_{rank} = j_{rank})$ and $(i_{distance} > j_{distance}))$.*

That is, between two solutions with differing non-domination ranks we prefer the point with the lower rank. Otherwise, if both the points belong to the same front then we prefer the point which is located in a region with smaller number of points (or with larger crowded distance). This way solutions from less dense regions in the search space are given importance in deciding which solutions to choose from R_t. This constructs the population P_{t+1}. This population of size N is now used for selection, crossover and mutation to create a new population Q_{t+1} of size N. We use a binary tournament selection operator but the selection criterion is now based on the crowded comparison operator $\prec_n$. The above procedure is continued for a specified number of generations.

It is clear from the above description that NSGA-II uses (i) a faster non-dominated sorting approach, (ii) an elitist strategy, and (ii) no niching parameter. It has been shown elsewhere [2] that the above procedure has $O(MN^2)$ computational complexity.

2.1 Exploitation Versus Exploration Issue

The above description of NSGA-II raises an important issue relating to EA research: the issue of exploitation and exploration [7]. Let us imagine that at a generation, we have a population R_t where most of the members lie on the non-dominated front of rank one and this front is distant from the true Pareto-optimal front. This will happen in the case of multi-modal problems, where a population can get attracted to a local Pareto-optimal front. Since most members belong to the current best non-dominated front, the elitism operation will result in deleting most members belonging to other fronts. In the above NSGA-II algorithm, elitist solutions are emphasized in two occasions. Once in the make-new-pop(P_{t+1}) operation and again during the elitism operation in the until loop. The former operation involves the crowded tournament selection operator, which emphasizes the elitist solutions (the currently-best non-dominated solutions). In the latter case, solutions are selected starting from the currently-best non-dominated solutions till all population slots are filled. This way, the elitist solutions also get emphasized. This dual emphasis of elitist solutions will cause a rapid deletion of solutions belonging to non-elitist fronts. Although the crowding tournament operator will ensure diversity along the current non-dominated front, lateral diversity will be lost. In many problems, when this happens the search slows down, simply because there may be a lack of diversity in certain decision variables left to push the search towards better regions of optimality. Thus, in order to ensure better convergence, a search algorithm may need diversity in both directions—along the Pareto-optimal front and lateral to the Pareto-optimal front, as shown in Figure 2. In the test problems suggested elsewhere [5], the lateral diversity is

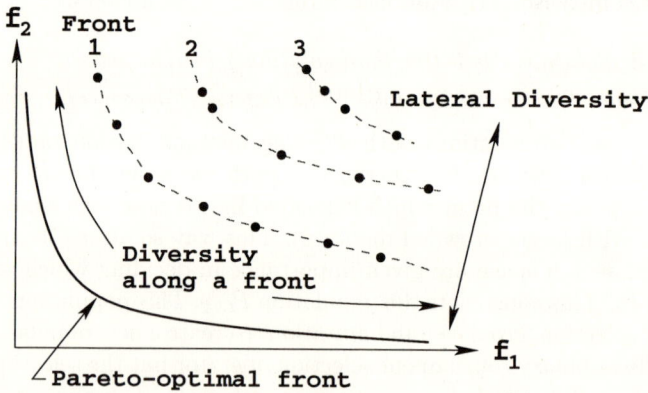

Fig. 2. Controlled elitism procedure is illustrated.

ensured by the functional $g()$ with $(n-1)$ decision variables. Solutions converging to any local Pareto-optimal front will make all these decision variables to take an identical value. The dual elitism will reduce the variability in these solutions and eventually prevent the algorithm to move towards the true Pareto-optimal front.

The difficulty arises because of the way elitism is implemented in NSGA-II. There is no control on the exploitation of the currently best non-dominated solutions. In the above discussion, it is clear that in certain complex functions, NSGA-II, in the absence of a lateral diversity-preserving operator such as mutation, causes too much exploitation of currently-best non-dominated solutions. In order to counteract this excessive selection pressure, an adequate exploration by means of the search operators must be used. A proper balance of these two issues is not possible to achieve with the uncontrolled elitism mechanism in NSGA-II. In an earlier study [2], we have shown that in the test problem ZDT4 with Rastrigin's multi-modal function as the g functional, NSGA-II could not converge to the global Pareto-optimal front. However, when a mutation operator with a larger mutation strength is used, NSGA-II succeeds in converging to the global Pareto-optimal front. Increasing variability through mutation enhances the exploration power of NSGA-II and a balance between enhanced exploitation of NSGA-II and the modified exploration can be maintained.

Although many researchers have adopted an increased exploration requirement by using a large mutation strength in the context of single objective EAs, the extent of needed exploration is always problem dependent. In the following section, instead of concentrating on changing the search operators, we suggest a controlled elitism mechanism which will control the extent of exploitation rather than controlling the extent of exploration.

3 Controlled Approach

In the proposed controlled NSGA-II, we restrict the number of individuals in the currently best non-dominated front adaptively. We attempt to maintain a pre-defined distribution of number of individuals in each front. Specifically, we use a geometric distribution for this purpose:

$$n_i = rn_{i-1}, \tag{1}$$

where n_i is the maximum number of allowed individuals in the i-th front and r (< 1) is the reduction rate. Although the parameter r is user-defined, the procedure is adaptive as follows.

First, the population $R_t = P_t \cup Q_t$ is sorted for non-domination. Let us say that the number of non-dominated fronts in the population (of size $2N$) is K. Thus, according to the geometric distribution, the maximum number of individual allowed in the i-th front ($i = 1, 2, \ldots, K$) in the new population of size N is

$$n_i = N \frac{1-r}{1-r^K} r^{i-1}. \tag{2}$$

Since $r < 1$, the maximum allowable number of individuals in the first front is highest. Thereafter, each front is allowed to have an exponentially reducing number of solutions. This exponential distribution considered above is an assumption and must be tried with other distributions such as an arithmetic distribution or a harmonic distribution. But the principle in all these approaches is the same—forcibly allow solutions from all non-dominated fronts to co-exist in the population.

Although equation 2 denotes the maximum allowable number of individuals n_i in each front i, in a population, there may not exist exactly n_i individuals in front i. We resolve this problem by starting a procedure from the first front. The number of individuals in the first front is counted. Let us say that there are n_1^t individuals. If $n_1^t > n_1$ (that is, there are more solutions than allowed), we only choose n_1 solutions using the crowded tournament selection. This way, exactly n_1 solutions that are residing in less crowded region are selected. On the other hand, if $n_1^t < n_1$ (that is, there are less solutions in the population than allowed), we choose all n_1^t solutions and count the number of remaining slots $\rho_1 = n_1 - n_1^t$. The maximum allowed number of individuals in the second front is now increased to $n_2 \rightarrow n_2 + \rho_1$. Thereafter, the actual number of solutions n_2^t present in the second front is counted and is compared with n_2 as above. This procedure is continued till all N individuals are selected. Figure 3 shows that a population of size $2N$ (having four non-dominated fronts with topmost subpopulation representing front one and so on) is reduced to a new population P_{t+1} of size N using the above procedure. In the transition shown in the right, all four fronts have representative solutions. Besides this elitism procedure the rest of the procedure is kept the same as that in NSGA-II. The figure also shows

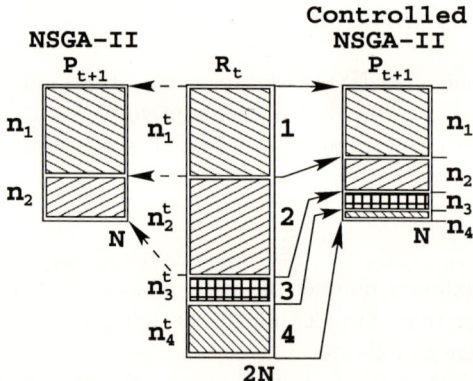

Fig. 3. Controlled elitism procedure is illustrated.

the new population P_{t+1} which would have obtained using the usual NSGA-II procedure. It is clear that the new population under the controlled NSGA-II is more diverse than that under NSGA-II.

Since the population is halved, it is likely that in each front there would be more solutions than allowed. However, there could be some situations where after all $2N$ solutions are processed as above, there are still some slots left in the new population to be filled. This may happen particularly when r is large. In such cases, we make another pass with the left-out individuals from the first front, continuing to other fronts, and start including them till we fill up the remaining slots.

3.1 Discussions

As mentioned earlier, keeping individuals from many non-dominated fronts in the population help to the recombination operator to create diverse solutions. NSGA-II and many other successful MOEAs thrive at maintaining diversity among solutions of independent non-dominated fronts. The controlled elitism procedure suggested above will help to maintain diversity in solutions across the non-dominated fronts. In solving difficult multi-objective optimization problems, this additional feature may be helpful in progressing towards the true Pareto-optimal front.

It is intuitive that the parameter r is important in maintaining the correct balance between the exploitation and exploration issue discussed earlier. This parameter sets up the extent of exploration allowed in an MOEA. If r is small, the extent of exploration is large and vice versa. In general, the optimal value of r will depend on the problem and will be difficult to find theoretically. In the following section, we present simulation results on a number of difficult problems to investigate if there exists any range of r, where NSGA-II performs well.

4 Simulation Results

In order to demonstrate the working of the proposed methodology, we use a set of test problems. The first four problems have multiple Pareto-optimal fronts, among those many fronts, one is the globally Pareto-optimal set. Next, we consider one constrained test problem, which has infeasible regions restricting the search towards the true Pareto-optimal front.

In order to investigate the effect of the controlled elitism alone, we do not use the mutation operator. For controlled NSGA-II runs, we do not use the selection operator in make_new_pop() to create the child population. For all problems, we use a population size of 100, a crossover probability of 0.95, and spread factor for the SBX operator [6] of 20. All simulation runs are continued until 200 generations are over.

4.1 Multi-objective Rastrigin's Problem

This problem is identical to the ZDT4 function, introduced elsewhere [12]:

$$\text{ZDT4}: \begin{cases} \text{Minimize } f_1(\mathbf{x}) = x_1 \\ \text{Minimize } f_2(\mathbf{x}) = g(\mathbf{x})\left[1 - \sqrt{x_1/g(\mathbf{x})}\right] \\ g(\mathbf{x}) = 91 + \sum_{i=2}^{10}\left[x_i^2 - 10\cos(4\pi x_i)\right] \\ x_1 \in [0,1], \quad x_i \in [-5,5], \ i = 2,\ldots,10. \end{cases} \quad (3)$$

It was also observed in another study [2] that NSGA-II has difficulties in converging to the global Pareto-optimal front. However, when the mutation strength is increased, NSGA-II could converge to the correct front.

Figure 4 shows that a simulation run with NSGA-II could not converge to the true Pareto-optimal front (the plot shown with diamonds). Since mutation is not

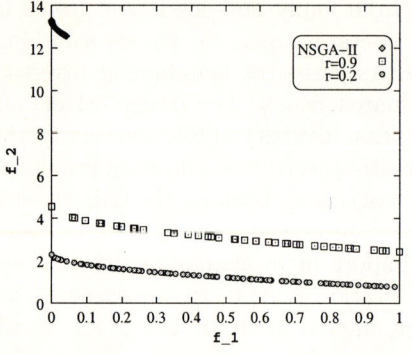 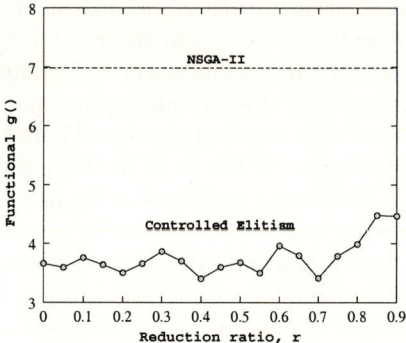

Fig. 4. Distribution of obtained set of solutions for ZDT4.

Fig. 5. Average $g()$ for controlled and uncontrolled NSGA-IIs for ZDT4.

used, NSGA-II also faced difficulty in maintaining a diverse set of solutions in the obtained front. The figure also shows two other simulations with controlled elitism ($r = 0.2$ and $r = 0.9$), starting from the same initial population. The figure clearly shows that when controlled elitism is used, existence of multiple fronts ensures better convergence and better diversity among obtained solutions. To make a thorough investigation, we have used 19 controlled elitist NSGA-IIs in the range $r \in [0.05, 0.9]$ with an interval of 0.05 and compared the results with the original NSGA-II. For each NSGA-II, we have performed 25 simulation studies, each starting with a different initial population. But for all NSGA-IIs, 25 identical populations are used. The average value of the best functional $g()$ in 200 generations is calculated and plotted in Figure 5. Equation 3 reveals that all solutions having $g = 1$ lie in the global Pareto-optimal front. Thus, an MOEA with a smaller value of functional $g()$ can be considered to have done well in solving the above problem. The figure clearly shows that all controlled elitism runs have done much better than the original NSGA-II. The average $g()$ value is much closer to one than NSGA-II. Although there is some fluctuations in the obtained $g()$ values for different setting of the reduction rate, all controlled NSGA-IIs are better.

In order to investigate the composition of population with controlled elitism, we count the number of solutions in the first and the second fronts in 25 different NSGA-II and controlled NSGA-II with $r = 0.5$ runs. The average of these numbers is plotted in Figures 6 and 7. Figure 6 shows that in the case of original

NSGA-II, the number of solutions is in the first front grows very rapidly after a certain number of generations. Within about 90 generations, all population members belong to the best non-dominated front. Thereafter, NSGA-II works by keeping all population members in the first front. The only way a new solution will be accepted, if it lies in the current best non-dominated front and if it resides in a less crowded area. On the other hand, with controlled NSGA-II (with $r = 0.5$) the population is never filled with currently-best non-dominated solutions alone. With $r = 0.5$, only about 50% population members belong to the best non-dominated front. Moreover, the rate of increase of solutions in the best non-dominated front is also small and importantly, the growth of solutions in this front begins much later than that with the original NSGA-II. Figure 7

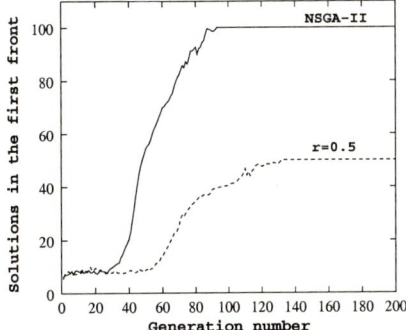
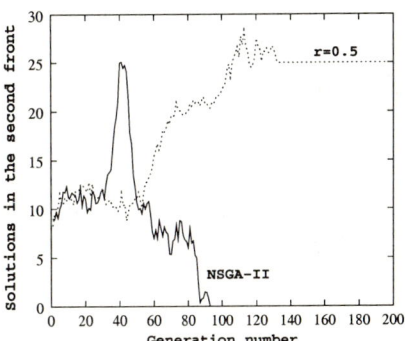

Fig. 6. Average number of solutions in the first (best) non-dominated front.

Fig. 7. Average number of solutions in the second non-dominated front.

shows the number of solutions in the second front. The figure clearly shows that in the original NSGA-II, there exists no member in the second front after about 90 generations, meaning that all solutions belong to the first non-dominated front. But, in the case of controlled NSGA-II, about 25% population belongs to the second front. The presence of so many solutions in the second front and subsequent fronts provides a lateral diversity needed for a better convergence to the global Pareto-optimal front.

We would like to distinguish the original NSGA-II algorithm with controlled NSGA-II having $r = 0$. With $r = 0$, the allowed number of solutions in the best non-dominated front is N (the population size). Thus, there is no pressure to maintain other fronts in the population. This algorithm may apparently look the same as the original NSGA-II, but there is a difference. In the $r = 0$ controlled NSGA-II, the tournament selection in the make_new_pop() operator is eliminated. This way, the selection pressure for currently-best elitist solutions is smaller than that in the original NSGA-II. Although this simple change in NSGA-II exhibits a better performance, the controlled NSGA-IIs with $r > 0$ are generic. Moreover, NSGA-IIs with some values of $r > 0$ show better performance than NSGA-II with $r = 0$. It is a matter of future research to investigate if a dynamically changing r with generation makes the convergence better than a fixed r. However, the effect of r is evident in the following problem.

4.2 Biased Test Function

Next, we construct a test problem using equation 3, with the funcitonal $g()$ as given by the following equation [5].

$$g(\mathbf{x}) = 1 + 9 \left(\frac{\sum_{i=2}^{30} x_i - \sum_{i=2}^{30} x_i^{\min}}{\sum_{i=2}^{30} x_i^{\max} - \sum_{i=2}^{30} x_i^{\min}} \right)^{\gamma}. \quad (4)$$

where $\gamma = 0.25$. Here, the variables x_i are allowed to vary in the limits $[0, 1]$. The function has the property of having more solutions away from the Pareto-optimal front, thus making the search more and more difficult to come to the Pareto-optimal front. This function has the globally Pareto-optimal front with the functional $g(\mathbf{x}) = 1$.

Figure 8 shows a typical distribution on the Pareto-optimal front for NSGA-II runs and different controlled elitist NSGA-II on this function without mutation. Here, the weak definition of the dominance is used. It is clear from the Figure 9

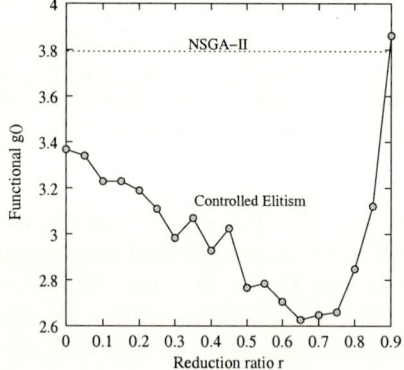
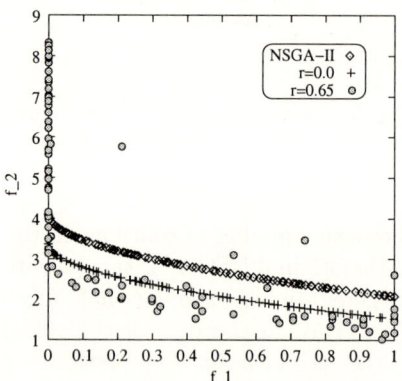

Fig. 8. Distribution of obtained set of solutions for ZDT4 with a biased function.

Fig. 9. Average $g()$ for controlled and uncontrolled NSGA-IIs for ZDT4 with a biased function.

that controlling elitism is useful. The best performance of the MOEA is achieved at $r = 0.65$. The performance of NSGA-II improves with r, but after a critical r the performance deteriorates. For small r, not enough fronts exist in the population, thereby slowing the performance. On the other hand, when r is large, not much selection pressure is assigned to the current best non-dominated solutions. Figure 8 shows the lateral diversity for a typical run for $r = 0$, $r = 0.65$ and the original NSGA-II runs. It is clear that diversity on the front is maintained in all the fronts. There exist a number of points in the run for $r = 0.65$ which are not on the best non-dominated front. These are the points which exist on the different fronts and provide the needed diversity to help proceed NSGA-II towards the Pareto-optimal region. The figure confirms the concept of surviving different fronts simultaneously to achieve the better convergence.

4.3 Multi-objective Griewangk Problem

Next, we construct a problem using equation 3, except that the functional $g()$ is now different ($g()$ is a Greigwank's function [5]):

$$g(\mathbf{x}) = 2 + \sum_{i=2}^{10} x_i^2/4000 - \Pi_{i=2}^{10} \cos(x_i/\sqrt{i}). \qquad (5)$$

Here, the variables x_2 until x_{10} are all allowed to vary in $[-512, 511]$. This function has 163^9 local Pareto-optimal fronts, of which only one is globally Pareto-optimal front having $g(\mathbf{x}) = 1$.

Figure 10 shows a typical performance plot of NSGA-II and controlled elitist NSGA-IIs on this function without mutation. It is clear that controlled elitism

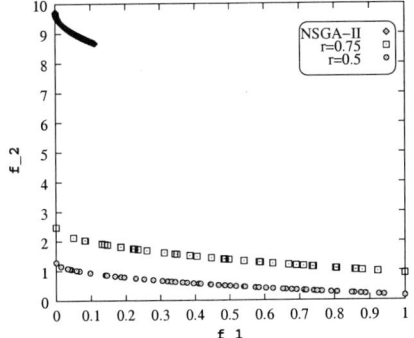
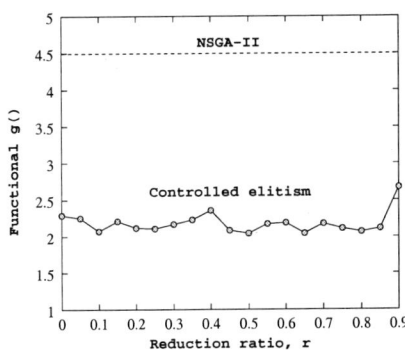

Fig. 10. Distribution of obtained set of solutions for ZDT4 with Griewangk's function.

Fig. 11. Average $g()$ for controlled and uncontrolled NSGA-IIs for ZDT4 with Griewangk's function.

runs are better converged near the true Pareto-optimal front. Because of lateral diversity present in the controlled NSGA-II runs, diversity in the obtained set of solutions also comes as a bonus. Figure 11 shows a detailed comparative study. Once again, the figure confirms that the controlled NSGA-II results are better.

4.4 Nonuniform Distribution

We now use the problem ZDT6, which was used in another study [12].

$$\text{ZDT6}: \begin{cases} \text{Minimize } f_1(\mathbf{x}) = 1 - \exp(-4x_1)\sin^6(6\pi x_1) \\ \text{Minimize } f_2(\mathbf{x}) = g(\mathbf{x})\left[1 - (x_1/g(\mathbf{x}))^2\right] \\ g(\mathbf{x}) = 1 + 9\left[(\sum_{i=2}^{10} x_i)/9\right]^{0.25} \\ x_1 \in [0,1], \quad x_i \in [-5,5], \quad i = 2, \ldots, 10. \end{cases} \quad (6)$$

The above problem should not cause much difficulty in converging to the correct front ($g(\mathbf{x}) = 1$), but causes difficulty in maintaining a uniform spread of solutions in the front.

Even in this problem, we observe from Figures 12 and 13 that the original NSGA-II without the help of mutation cannot come closer to the Pareto-optimal front. Moreover, the distribution of solutions in the obtained set is also non-

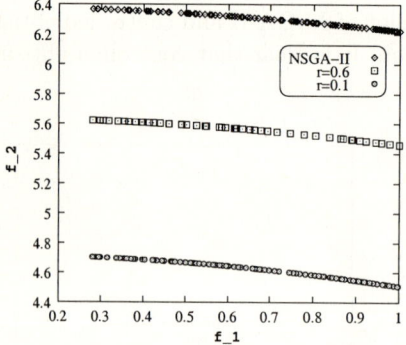 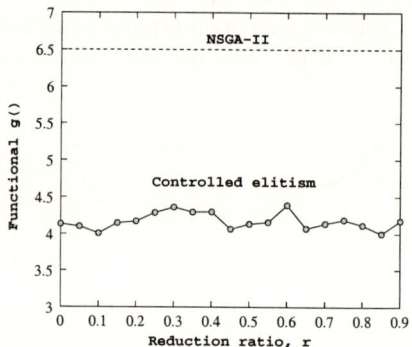

Fig. 12. Distribution of obtained set of solutions for ZDT6.

Fig. 13. Average $g()$ for controlled and uncontrolled NSGA-IIs for ZDT6.

uniform. Controlled NSGA-IIs performs better than the original NSGA-II in both converging near the true Pareto-optimal front and maintaining a uniform-like distribution. Figure 12 shows that NSGA-II's crowding mechanism is good enough to maintain a good distribution of solutions in this problem. But, NSGA-II's large selection pressure through uncontrolled elitism is detrimental. Once again, the performance of controlled NSGA-IIs is more or less insensitive to the chosen r value.

4.5 Constrained Test Problem (CTP7)

Finally, we attempt to solve a constrained test problem shown below [3]:

$$\text{CTP7}: \begin{cases} \text{Minimize } f_1(\mathbf{x}) = x_1 \\ \text{Minimize } f_2(\mathbf{x}) = g(\mathbf{x})\left(1 - \frac{f_1(\mathbf{X})}{g(\mathbf{X})}\right) \\ \text{Subject to } c(\mathbf{x}) \equiv \cos(\theta)(f_2(\mathbf{x}) - e) - \sin(\theta)f_1(\mathbf{x}) \geq \\ \qquad a\left|\sin\left(b\pi\left(\sin(\theta)(f_2(\mathbf{x}) - e) + \cos(\theta)f_1(\mathbf{x})\right)^c\right)\right|^d. \end{cases} \quad (7)$$

The decision variable x_1 is restricted to $[0, 1]$. The functional $g()$ is the Rastrigin's function used in equation 3. But, here we use five variables, instead of 10 used in ZDT4. Other parameter values are as follows:

$$\theta = -0.05\pi, \quad a = 40, \quad b = 5, \quad c = 1, \quad d = 6, \quad e = 0.$$

The feasible search space and the corresponding disconnected Pareto-optimal regions are shown in Figure 14.

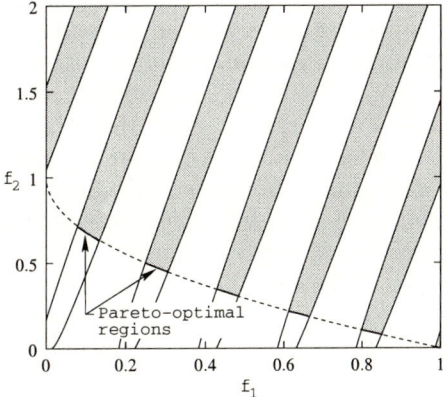

Fig. 14. Constrained test problem CTP7.

In the absence of a mutation operator, we observe that NSGA-II's strong elitism prevents the algorithm to come closer to the true Pareto-optimal front (Figure 15). The outcome of 25 runs suggests that the controlled elitism can overcome this problem somewhat and help NSGA-II to converge closer to the true Pareto-optimal front (Figure 16). The absence of mutation in all NSGA-II runs prohibits them to maintain a diverse set of population members in the obtained set.

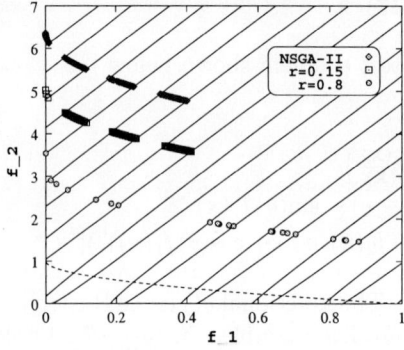 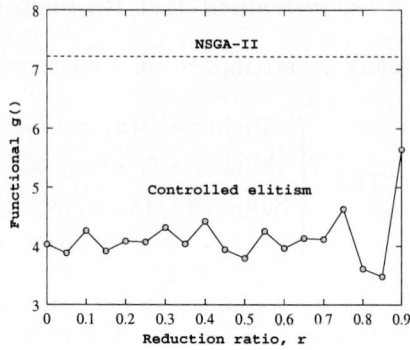

Fig. 15. Distribution of obtained set of solutions for CTP7.

Fig. 16. Average $g()$ for controlled and uncontrolled NSGA-IIs for CTP7.

5 Conclusion

In this paper, we have discussed the need to make a balance between exploitation offered by a selection operator along with an elite-preserving mechanism with exploration offered by a recombination operator. Although NSGA-II has been largely successful in many problems, in some difficult problems, it showed difficulty in converging to the Pareto optimal front. Here, we have clearly shown that the uncontrolled elitism present in NSGA-II produces a large selection pressure for currently-best non-dominated solutions. In difficult multi-modal problems and other problems, this forces NSGA-II without any mutation operator to have a premature convergence to a suboptimal front.

By forcing a number of non-dominated fronts to co-exist in an NSGA-II, we have shown that the selection pressure offered by NSGA-II's elitism operator can be controlled. On five difficult test problems, including a difficult constraint-handling problem, we have shown that the controlled NSGA-II has a better convergence property than the original NSGA-II. Since diverse non-dominated fronts are also maintained in the population, controlled NSGA-II also produces a better distribution of solutions than the original NSGA-II.

The controlled elitism mechanism implemented here with NSGA-II is generic and can also be implemented with other elitist MOEAs. It would be a matter of future research to investigate whether the convergence property of other MOEAs improves with such a controlled elitism operator.

Acknowledgements. Authors acknowledge the support provided by Department of Science and Technology (DST) for this study.

References

1. Corne, D. W., Knowles, J. D., and Oates, M. J. (2000). The Pareto envelope-based selection algorithm for multiobjective optimization. *Proceedings of the Parallel Problem Solving from Nature VI Conference*, pp. 839–848.
2. Deb, K., Pratap, A., Agrawal, S. and Meyarivan, T. (2000). A fast and elitist multi-objective genetic algorithm: NSGA-II. Technical Report No. 2000001. Kanpur: Indian Institute of Technology Kanpur, India.
3. Deb, K., Mathur, A. P., Meyarivan, T. (2000). Constrained test problems for multi-objective evolutionary optimization. Technical Report No. 200002. Kanpur: Kanpur Genetic Algorithms Laboratory, IIT Kanpur, India.
4. Deb, K., Agrawal, S., Pratap, A., Meyarivan, T. (2000). A Fast Elitist Non-dominated sorting genetic algorithm for multi-objective optimization: NSGA-II. *Proceedings of the Parallel Problem Solving from Nature VI Conference*, pp. 849–858.
5. Deb, K. (1999) Multi-objective genetic algorithms: Problem difficulties and construction of test functions. *Evolutionary Computation, 7*(3), 205–230.
6. Deb, K. and Agrawal, R. B. (1995) Simulated binary crossover for continuous search space. *Complex Systems, 9* 115–148.
7. Goldberg, D. E. (1989). *Genetic algorithms in search, optimization, and machine learning*. Reading: Addison-Wesley.
8. Knowles, J. and Corne, D. (1999) The Pareto archived evolution strategy: A new baseline algorithm for multiobjective optimisation. *Proceedings of the 1999 Congress on Evolutionary Computation*, Piscataway: New Jersey: IEEE Service Center, 98–105.
9. Srinivas, N. and Deb, K. (1995). Multi-objective function optimization using non-dominated sorting genetic algorithms. *Evolutionary Computation*(2), 221–248.
10. Steuer, R. E. (1986). *Multiple criteria optimization: Theory, computation, and application*. New York: Wiley.
11. Zitzler, E. (1999). Evolutionary algorithms for multiobjective optimization: Methods and applications. Doctoral thesis ETH NO. 13398, Zürich: Swiss Federal Institute of Technology (ETH), Aachen, Germany: Shaker Verlag.
12. Zitzler, E., Deb, K., and Thiele, L. (2000). Comparison of multiobjective evolutionary algorithms: Empirical results. *Evolutionary Computation Journal*, 8(2), 173–196.
13. Zitzler, E. and Thiele, L. (1998). An evolutionary algorithm for multiobjective optimization: The strength Pareto approach. *Technical Report No. 43 (May 1998)*. Zürich: Computer Engineering and Networks Laboratory, Switzerland.

Specification of Genetic Search Directions in Cellular Multi-objective Genetic Algorithms

Tadahiko Murata [1], Hisao Ishibuchi [2], and Mitsuo Gen [1]

[1] Department of Industrial and Information Systems Engineering,
Ashikaga Institute of Technology,
268 Omae-cho, Ashikaga 326-8558, Japan
{murata, gen}@ashitech.ac.jp

[2] Department of Industrial Engineering, Osaka Prefecture University,
1-1 Gakuen-cho, Sakai, Osaka 599-8531, Japan
hisaoi@ie.osakafu-u.ac.jp

Abstract. When we try to implement a multi-objective genetic algorithm (MOGA) with variable weights for finding a set of Pareto optimal solutions, one difficulty lies in determining appropriate search directions for genetic search. In our MOGA, a weight value for each objective in a scalar fitness function was randomly specified. Based on the fitness function with the randomly specified weight values, a pair of parent solutions are selected for generating a new solution by genetic operations. In order to find a variety of Pareto optimal solutions of a multi-objective optimization problem, weight vectors should be distributed uniformly on the Pareto optimal surface. In this paper, we propose a proportional weight specification method for our MOGA and its variants. We apply the proposed weight specification method to our MOGA and a cellular MOGA for examining its effect on their search ability.

1 Introduction

Genetic algorithms have been successfully applied to various optimization problems [1]. The extension of GAs to multi-objective optimization was proposed in several manners (for example, Schaffer [2], Kursawe [3], Horn *et al.* [4], Fonseca & Fleming [5], Murata & Ishibuchi [6], Zitzler & Thiele [7]). The aim of these algorithms is to find a set of Pareto-optimal solutions of a multi-objective optimization problem. Another issue in multi-objective optimization is to select a single final solution from Pareto-optimal solutions. Many studies on multi-objective GAs did not address this issue because the selection totally depends on the decision maker's preference. In this paper, we also concentrate our attention on the search for finding a set of Pareto-optimal solutions.

In this paper, we try to improve the search ability of our multi-objective genetic algorithm (MOGA) in [6] and its variants (i.e., extensions of our MOGA). Fig. 1 shows some extended algorithms in our previous studies [8-10]. By hybridizing our MOGA with a local search procedure, we have already extended it to a multi-objective genetic local search algorithm (MOGLS [8]). We have also extended our MOGA to a cellular multi-objective genetic algorithm (C-MOGA [9]) by introducing a cellular structure. We have employed a local search procedure and a cellular structure in a cellular MOGLS [10]. Furthermore we have extended the cellular algorithms by introducing a relocation procedure (i.e., a kind of immigration) in [10]. Each individual is relocated to a cell at every generation based on the values of multiple objectives (i.e., the location in the multi-dimensional objective space). Those extended algorithms, which are based on the cellular structure and the immigration procedure, are referred to as Cellular Immigrative ("CI-") algorithms in Fig. 1.

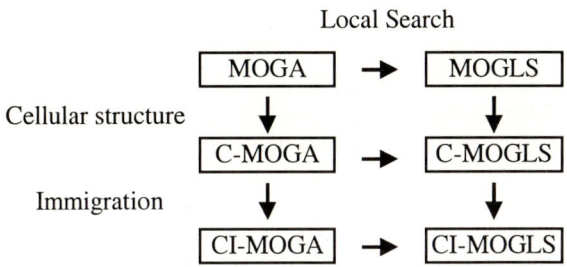

Fig. 1. Extensions of our MOGA by introducing local search, cellular structures and immigration procedures

When we try to implement our MOGA and its variants with variable weights, one difficulty lies in determining appropriate search directions for genetic search. In those algorithms, the weight value for each objective in a scalar fitness function is randomly specified. Based on the scalar fitness function with the randomly specified weight values, a pair of parent solutions are selected for generating a new solution by genetic operations. In order to find a variety of Pareto optimal solutions of a multi-objective optimization problem, a proportional weight specification method is more desirable than the random specification method. In this paper, we propose a proportional weight specification method for our MOGA and its variants. We apply the proposed weight specification method to our MOGA and a cellular MOGA for examining its effect on their search ability.

The concept of cellular genetic algorithms was proposed by Whitley [11]. In cellular genetic algorithms, each individual (i.e. a chromosome) resides in a cell of a spatially structured space. Genetic operations for generating new individuals are locally performed in the neighborhood of each cell. While the term "cellular genetic algorithm" was introduced by Whitley, such algorithms had already been proposed by Manderik & Spiessens [12]. A similar concept was also studied in evolutionary ecology in the framework of "structured demes" (Wilson [13], Dugatkin & Mesterton-Gibbons [14]). The effect of spatial structures on the evolution of cooperative behavior has also been examined in many studies (e.g., Nowak & May [15], Wilson *et al.* [16], Oliphant [17], Grim [18], and Ishibuchi *et al.* [19]) where each individual was located in a cell of single-dimensional or two-dimensional grid-worlds. The concept for generating a grid world on an n-objective space is also employed in the Pareto Archived Evolution Strategy (PAES) [20], where each individual is located in a grid on the objective space. The PAES employs a grid world in order to avoid introducing a niche size in the algorithm.

2 Multi-objective Optimization

Let us consider the following multi-objective optimization problem with n objectives:

$$\text{Maximize } f_1(\mathbf{x}), f_2(\mathbf{x}), ..., f_n(\mathbf{x}), \tag{1}$$

where $f_1(\cdot), f_2(\cdot), ..., f_n(\cdot)$ are n objectives. When the following inequalities hold between two solutions $\mathbf{x}$ and $\mathbf{y}$, the solution $\mathbf{y}$ is said to dominate the solution $\mathbf{x}$:

$$\forall i : f_i(\mathbf{x}) \leq f_i(\mathbf{y}) \text{ and } \exists j : f_j(\mathbf{x}) < f_j(\mathbf{y}). \tag{2}$$

If a solution is not dominated by any other solutions of the multi-objective optimization problem, that solution is said to be a Pareto-optimal solution. The task of multi-objective algorithms in this paper is not to select a single final solution but to find all Pareto-optimal solutions of the multi-objective optimization problem in (1). When we use heuristic search algorithms such as taboo search, simulated annealing, and genetic algorithms for finding Pareto-optimal solutions, we usually can not confirm the optimality of obtained solutions. We only know that each of the obtained solutions is not dominated by any other solutions examined during the execution of those algorithms. Therefore obtained solutions by heuristic algorithms are referred to as "nondominated" solutions. For a large-scale multi-objective optimization problem, it is impossible to find all Pareto-optimal solutions. Thus our task is to find many near-optimal nondominated solutions in a practically acceptable computational time. The performance of different multi-objective algorithms is compared based on several quality measures of obtained nondominated solutions.

3 Multi-objective Genetic Algorithms (MOGA)

In this section, we explain our MOGA [6], which is the basic algorithm of the C-MOGA (See Fig. 1). In our MOGA, the weighted sum of the n objectives is used as a fitness function:

$$f(\mathbf{x}) = w_1 f_1(\mathbf{x}) + w_2 f_2(\mathbf{x}) + ... + w_n f_n(\mathbf{x}), \tag{3}$$

where $w_1, ..., w_n$ are nonnegative weights for the n objectives, which satisfy the following relations:

$$w_i \geq 0 \text{ for } i = 1, 2, ..., n, \tag{4}$$

$$w_1 + w_2 + \cdots + w_n = 1. \tag{5}$$

This fitness function is utilized when a pair of parent solutions are selected for generating a new solution by crossover and mutation. One characteristic feature of our MOGA is to randomly specify weight values whenever a pair of parent solutions are selected. That is, each selection (i.e., the selection of two parents) is performed based on a different weight vector. This means that each of newly generated solutions by the genetic operations has its own weight vector. The other characteristic feature of our MOGA is preserving all nondominated solutions which are obtained during the execution of the algorithm. We describe these characteristic features in the following subsections.

3.1 Selection Operation

When a pair of parent solutions are to be selected from a current population in a selection operation for generating an offspring by genetic operations, first the n weight values ($w_1, w_2, ..., w_n$) are randomly specified as follows:

$$w_i = random_i / (random_1 + \cdots + random_n), \; i = 1, 2, ..., n, \quad (6)$$

where $random_i$ is a nonnegative random real number. For example, when N pairs of parent solutions are selected for generating a new population, N different weight vectors are specified by (6). This means that N search directions are utilized in a single generation. In other words, each selection (i.e., the selection of two parents) is governed by a different fitness function.

3.2 Elitist Strategy

Our MOGA separately stores two different sets of solutions: a current population and a tentative set of nondominated solutions. After genetic operations are applied to the current population, it is replaced with newly generated solutions. At the same time, the tentative set of nondominated solutions is updated. That is, if a newly generated solution is not dominated by any other solutions in the current population and the tentative set of nondominated solutions, this solution is added to the tentative set. Then all solutions dominated by the added one are removed from the tentative set. In this manner, the tentative set of nondominated solutions is updated at every generation in our MOGA.

From the tentative set of nondominated solutions, a few solutions are randomly selected and added to the current population (see Fig. 2). The randomly selected nondominated solutions may be viewed as elite solutions because they are added to the current population with no modification.

When a multi-objective optimization problem has a non-convex Pareto front, weighted sum approaches with constant weights fail to find its entire Pareto solutions. This is because those algorithms try to find a single optimal solution with respect to the fixed weights by their single trial. Our approach remedies such a difficulty by using variable weights and storing the tentative set of nondominated solutions. This set is updated by examining the Pareto optimality of every solution generated by genetic operations during the execution of the algorithm. It was shown in [6, 8] that our approach found nondominated solutions on a non-convex Pareto front of a two-objective continuous optimization problem. In [6, 8], our approach was also applied to two-objective flowshop scheduling problems with non-convex Pareto fronts.

3.3 Algorithm

Let us denote the population size by N_{pop}. We also denote the number of nondominated solutions added to the current population by N_{elite} (i.e., N_{elite} is the number of elite solutions, see Fig. 2). Using these notations, our MOGA can be written as follows.

Step 0) Initialization: Randomly generate an initial population of N_{pop} solutions.
Step 1) Evaluation: Calculate the values of the n objectives for each solution in the current population. Then update the tentative set of nondominated solutions.
Step 2) Selection: Repeat the following procedures to select ($N_{pop} - N_{elite}$) pairs of parent solutions.
  a) Randomly specify the weight values $w_1, w_2, ..., w_n$ in the fitness function (3) by (6).
  b) According to the following selection probability $P(\mathbf{x})$, select a pair of parent solutions from the current population Ψ.

$$P(\mathbf{x}) = \frac{f(\mathbf{x}) - f_{min}(\Psi)}{\sum_{\mathbf{y} \in \Psi}\{f(\mathbf{y}) - f_{min}(\Psi)\}}, \quad (7)$$

where $f_{min}(\Psi)$ is the minimum fitness value in the current population Ψ.
Step 3) Crossover and Mutation: Apply a crossover operator to each of the selected ($N_{pop} - N_{elite}$) pairs of parent solutions. A new solution is generated from each pair of parent solutions. Then apply a mutation operator to the generated new solutions.
Step 4) Elitist Strategy: Randomly select N_{elite} solutions from the tentative set of nondominated solutions, and add the selected N_{elite} solutions to the ($N_{pop} - N_{elite}$) solutions generated in Step 3 to construct a population of N_{pop} solutions.
Step 5) Termination Test: If a prespecified stopping condition is satisfied, end the algorithm. Otherwise, return to Step 1.

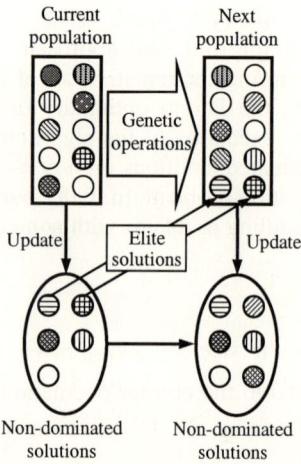

Fig. 2. Illustration of the elitist strategy in the MOGA

4 Cellular Algorithms

4.1 Relation between Cell Location and Weight Vector

In cellular algorithms, each individual (i.e. a solution) resides in a cell in a spatially structured space (e.g., two-dimensional grid-world). For utilizing a cellular structure in our MOGA, we assign a different weight vector to each cell. For our n-objective optimization problem, cells are structured in an n-objective weight space. Fig. 3 shows an example of structured cells for a two-dimensional optimization problem where the two weights w_1 and w_2 are used for the calculation of the fitness function $f(\mathbf{x})$ as $f(\mathbf{x}) = w_1 f_1(\mathbf{x}) + w_2 f_2(\mathbf{x})$. In this figure, the population size is eleven because an individual exists in each cell. As shown in Fig. 3, the location of each cell corresponds to its weight vector. In order to allocate cells on uniformly distributed weight vectors, we generate weight vectors systematically (not randomly). For example, weight vectors in Fig. 3 are (1.0, 0.0), (0.9, 0.1), ..., (0.0, 1.0).

As shown in Fig. 3, we can easily generate uniform weight vectors on the two-dimensional weight space. In order to generate uniformly distributed weight vectors for multi-objective optimization problems with three or more objectives, we propose a weight specification method on an n-dimensional grid world. Let us consider weight vectors satisfying the following conditions.

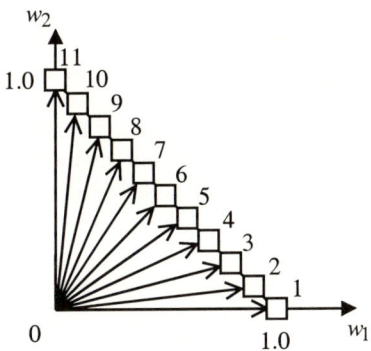

Fig. 3. Location of each cell in the two-dimensional weight space

These conditions show that weight vectors are generated by combining n non-negative integers with the sum of d. In our cellular algorithm, a cell is located on every weight vector satisfying the above conditions. Thus the number of cells (i.e., the population size) is equal to the total number of weight vectors satisfying the above conditions. This means that the population size is determined by d. For example, when we specify d as $d = 10$ in (8) for the case of two-objective problems, we will have eleven weight vectors (10, 0), (9, 1), ..., (0, 10). Each of these weight vectors has the same direction as the corresponding weight vector in Fig. 3.

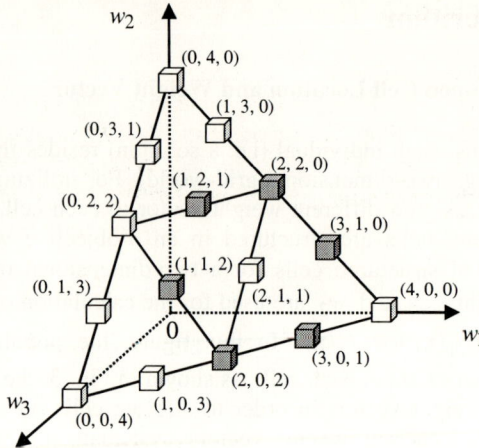

Fig. 4. Location of each cell in the three-dimensional weight space by the proposed method

$$w_1 + w_2 + \cdots + w_n = d, \tag{8}$$

$$w_i \in \{0,1,2,...,d\}. \tag{9}$$

This weight specification method is easily extended to the case with three or more objectives. For example, Fig. 4 shows an example of the three-objective case where d is specified as $d = 4$. From Fig. 4, we can observe that the value of d can be considered as the number of partitions of the edge between two extreme points (e.g., (0,4,0) and (4,0,0)). By this weight specification method, we can uniformly distribute cells on the n-dimensional space. We can calculate the number of cells generated for n-objective problems as follows:

$$N_2(d) = d+1 \approx O(d), \tag{10}$$

$$N_3(d) = \sum_{i=0}^{d} N_2(i) = \sum_{i=0}^{d}(i+1) = (i+1)(i+2)/2 \approx O(d^2), \tag{11}$$

$$N_4(d) = \sum_{i=0}^{d} N_3(i) = \sum_{i=0}^{d}(i+1)(i+2)/2 \approx O(d^3), \tag{12}$$

$$\vdots \qquad \vdots \qquad \vdots$$

$$N_n(d) = \sum_{i=0}^{d} N_{n-1}(i) \approx O(d^{n-1}), \tag{13}$$

where $N_j(d)$, $j = 2,...,n$ are the number of generated cells for j-objective problems. We can see from the above equations that the number of cells can be calculated recursively. We also see that the order of the number of cells is d^{n-1} for n-objective problems. Since the number of cells is determined from the value of d, the population size of our cellular algorithm can be specified by d. In other words, we should specify the value of d according to an appropriate population size.

4.2 Definition of Neighborhood

We can arbitrary define a neighborhood structure among cells. That is, we can utilize any distance between cells in the n-dimensional space in which cells are structured. For example, the Euclid distance can be used for measuring the distance between cells. In this paper, we use the Manhattan distance. That is, we define the distance between a cell with the weight vector $\mathbf{w} = (w_1, w_2, ..., w_n)$ and another cell with $\mathbf{v} = (v_1, v_2, ..., v_n)$ as follows:

$$Distance(\mathbf{w}, \mathbf{v}) = \sum_{i=1}^{n} |w_i - v_i|. \tag{14}$$

We define the neighborhood of the weight vector $\mathbf{w}$ as

$$N(\mathbf{w}) = \{\mathbf{v} \,|\, Distance(\mathbf{w}, \mathbf{v}) \leq D\}. \tag{15}$$

For example, when $D = 2$ in Fig. 5, the cell with the weight vector (2,1,1) has six neighboring cells (i.e., shaded cells, (1,2,1), (1,1,2), (2,0,2), (3,0,1), (3,1,0), and (2,2,0) in Fig. 4) and that cell itself in its neighborhood. As shown in this example, the neighborhood of each cell is defined by its nearby cells within the distance D including that cell itself.

4.3 Selection

Two parents for generating a new individual in a cell are selected from its neighborhood. When $D = 2$ in Fig. 5, the parent solutions for the cell on (2,1,1) can be selected from that cell and its six neighbors. It is noted that the fitness value of each neighbor is recalculated based on the weight vector assigned to the cell for which a new individual is generated. That is, each individual is differently evaluated by this recalculation procedure of the fitness function in the selection for each cell. This corresponds to the selection procedure of our original MOGA where the selection of each pair of parents was governed by a different weight vector (see Step 2 in Subsection 3.3).

It is noted that the modification of the normalization condition from (5) to (8) has no effect on the selection procedure (Step 2 (a) in Subsection 3.3). Let $\mathbf{w}'$, which satisfies the normalization condition (8), be a weight vector generated by the proposed weight specification method. A weight vector $\mathbf{w} = (w_1, w_2, ..., w_n)$ satisfying the normalization condition (5) can be easily obtained from the relation $\mathbf{w}' = d \cdot \mathbf{w} = (dw_1, dw_2, ..., dw_n)$. Let us denote our scalar fitness function with the weight vectors $\mathbf{w}$ and $\mathbf{w}'$ by $f(\mathbf{x}, \mathbf{w})$ and $f(\mathbf{x}, \mathbf{w}')$, respectively, for a solution $\mathbf{x}$ in the current population. Since our scalar fitness function in linear with respect to weight values, we have $f(\mathbf{x}, \mathbf{w}') = d \cdot f(\mathbf{x}, \mathbf{w})$ and $f_{\min}(\Psi, \mathbf{w}') = d \cdot f_{\min}(\Psi, \mathbf{w})$ in the selection procedure defined by (7). Thus the same selection probability is obtained from $\mathbf{w}$ and $\mathbf{w}'$ for each solution $\mathbf{x}$ in the current population Ψ. This means that the modification of the normalization condition from (5) to (8) has no effect on the selection procedure.

4.4 Other Genetic Opearations

In the previous Subsections 4.1 to 4.3, we show the characteristic features in our C-MOGA. As for other genetic operations such as crossover, mutation, and elite preserve strategy, we can employ the same operations which can be used in the MOGA. That is, the same crossover and mutation operations can be employed for the C-MOGA. As shown in Fig. 2, some solutions are selected from the tentative set of nondominated solutions, and add them as elite solutions to the current population randomly.

5 Computer Simulations

5.1 Test Problems

We applied the C-MOGA with the proposed weight specification method to flowshop scheduling problems. Flowshop scheduling is one of the most well-known scheduling problems. Since Johnson's work [21], various scheduling criteria have been considered. Among them are makespan, maximum tardiness, total tardiness, and total flowtime. Several researchers extended single-objective flowshop scheduling problems to multi-objective problems (see, for example, Daniels & Chambers [22]).

In this paper, we use the makespan and the total tardiness as two scheduling criteria in our two-objective flowshop scheduling problems. The makespan is the maximum completion time of all jobs to be processed. The total tardiness is the total overdue of all jobs. We also employ the total flowtime together with these two criteria in our three-objective flowshop scheduling problems. The total flowtime is the total completion time of all jobs. Let $g_1(\mathbf{x})$, $g_2(\mathbf{x})$, and $g_3(\mathbf{x})$ be the makespan, the total tardiness, and the total flowtime, respectively. Since these scheduling criteria are to be minimized, we specify the three objectives $f_1(\mathbf{x})$, $f_2(\mathbf{x})$ and $f_3(\mathbf{x})$ of our flowshop scheduling as $f_1(\mathbf{x}) = -g_1(\mathbf{x})$, $f_2(\mathbf{x}) = -g_2(\mathbf{x})$ and $f_3(\mathbf{x}) = -g_3(\mathbf{x})$.

Since flowshop scheduling is to find a job permutation that optimizes the given objectives, a sequence of jobs is handled as an individual (i.e., as a string) in our algorithm. As test problems, we generated ten 20-job and 10-machine flowshop scheduling problems with two and three objectives. The processing time of each job on each machine was specified as a random integer in the interval [1, 99], and the duedate of each job was defined randomly. Our task is to find a set of Pareto-optimal solutions of each test problem.

5.2 Quality Measures of Solution Sets

Since multi-objective algorithms find a set of nondominated solutions with respect to multiple objectives (not a single final solution with respect to a single objective), the comparison between different multi-objective algorithms is not easy. For this purpose, we use the following measures for evaluating the quality of a solution set obtained by each algorithm.

1) The number of obtained nondominated solutions

The number of nondominated solutions obtained by each algorithm is a measure to evaluate the variety of the solution set.

2) The number of solutions that are not dominated by other solution sets

For comparing different solution sets with one another, we examine whether each solution is dominated by any other solutions in other sets. If a solution is dominated by another solution, we remove that solution. In this manner, we remove solutions dominated by other solution sets. The number of remaining solutions in each solution set is a measure for evaluating its relative quality with respect to the other solution sets.

3) Set quality measure proposed by Esbensen

Esbensen [23] proposed an evaluation method of the quality of a solution set. Let us denote a solution set by Ω. The best solution $\mathbf{x}^*$ for a given weight vector $\mathbf{w} = (w_1, w_2, ..., w_n)$ can be chosen from Ω for the n-objective optimization problem as follows:

$$f(\mathbf{x}^*) = w_1 f_1(\mathbf{x}^*) + w_2 f_2(\mathbf{x}^*) + \cdots + w_n f_n(\mathbf{x}^*)$$
$$= \max\{w_1 f_1(\mathbf{x}) + w_2 f_2(\mathbf{x}) + \cdots + w_n f_n(\mathbf{x}) \,|\, \mathbf{x} \in \Omega\}. \quad (16)$$

Esbensen [23] proposed an idea of measuring the quality of the solution set Ω by calculating the expected value of $f(\mathbf{x}^*)$ over possible weight vectors. In this paper, we calculate the expected value of $f(\mathbf{x}^*)$ by randomly generating 10,000 weight vectors by (6). That is, the quality of the solution set Ω is calculated as follows:

$$q(\Omega) = \frac{1}{10000} \sum_{i=1}^{10000} \max\{w_1^i f_1(\mathbf{x}) + w_2^i f_2(\mathbf{x}) + \cdots + w_n^i f_n(\mathbf{x}) \,|\, \mathbf{x} \in \Omega\}, \quad (17)$$

where $q(\Omega)$ is the quality of the solution set Ω and $\mathbf{w}^i = (w_1^i, w_2^i, ..., w_n^i)$, $i = 1, 2, ..., 10000$ are randomly specified weight vectors.

5.3 Simulation Results on Two-Objective Flowshop Scheduling Problems

In our computer simulations, we applied three algorithms (i.e., the MOGA with random weights, the MOGA with weights generated by the proposed method, and the C-MOGA) to test problems in order to show the effectiveness of the weight specification method and compare the search ability of the C-MOGA with that of the MOGAs. We employed the following parameter specifications:

Crossover: Two-point order crossover (crossover rate: 0.8),
Mutation: Shift mutation (mutation rate: 0.3),
Number of elite solutions: $N_{\text{elite}} = 3$,
Neighborhood structure for the local search: Shift,
Stopping condition: Examination of 50,000 solutions.

We used the above stopping condition in order to compare the three algorithms under the same computation load. In a single trial of each algorithm, 50,000 solutions were examined. The parameter d, which determines the population size, was specified as $d = 100$ for two-objective problems. This means that the population size was 101 (see the equation (10)). The weight vectors of 101 cells were $(w_1, w_2) = (100, 0), (99, 1), ..., (0, 100)$. For the C-MOGA, we specified the value of D as $D = 20$. Therefore parent solutions for each cell are selected from neighboring cells within the Manhattan distance 20.

We examined the effect of the introduction of the weight specification method and the cellular structure (i.e., the locally restricted genetic operations) in this experiment. We obtained a set of nondominated solutions by each algorithm. In Table 1, we summarize the average results over 100 trials for each algorithm (i.e. 10 trials for each of 10 test problems). In this table, "A" is the number of nondominated solutions obtained by each algorithm, and "B" is the number of solutions that are not dominated by other solutions obtained by the other algorithm. The ratio of these two numbers is shown in the column of B/A. "Quality" is the set quality measure of Esbensen, and "SD of Q" shows the standard deviation of the value of Quality. In the calculation of "SD of Q", we averaged the standard deviation for each of ten test problems.

From Table 1, we can see that most solutions obtained by the MOGA with random weights are dominated by solutions obtained by the MOGA with the proposed method or the C-MOGA. Thus we can conclude that the weight specification method proposed in this paper is effective in the MOGA and the C-MOGA. We can also observe that the average and the standard deviation of the Quality value for the C-MOGA are better than those for the MOGAs.

Next, we examined the specification of the parameter D in the C-MOGA. Table 2 shows the average results over 100 trials for each specification of the neighborhood structure (i.e., each specification of D) in the C-MOGA. In this table, the C-MOGA with $D = 200$ is the MOGA with the proposed weight specification method. When $D = 200$, all solutions in the current population are considered as neighbors of every cell in the selection procedure. From Table 2, we can observe that the restriction of the genetic operations within some neighboring cells is effective for improving the ability to find good sets of nondominated solutions.

Table 1. Comparison of MOGA with C-MOGA (Two-Objective)

	A	B	B/A	Quality	SD of Q
MOGA (random)	14.6	2.8	0.200	-1063.4	48.4
MOGA (proposed)	17.1	10.8	0.648	-967.4	15.2
C-MOGA ($D = 20$)	17.5	9.3	0.536	-963.6	10.6

A: The number of nondominated solutions of the method.
B: The number of nondominated solutions that are not dominated by those obtained by the other method.
Quality: Set quality measure of Esbensen.
SD of Q: Standard deviation of Quality.

Table 2. Effect of the choice of D in C-MOGA (Two-Objective)

D	4	10	20	50	200
Quality	-976.4	-968.6	-963.6	-966.0	-967.4
SD of Q	17.9	14.3	10.6	12.9	15.2

Table 3. Comparison of MOGA with C-MOGA (Three-Objective)

	A	B	B/A	Quality	SD of Q
MOGA	44.8	8.4	0.190	-8004.7	65.54
C-MOGA	61.2	63.8	0.966	-7850.9	42.13

Table 4. Effect of the choice of d in C-MOGA (Three-Objective)

D	10	11	12	13	14
Poplation Size	66	78	81	105	120
# of Generations	758	642	618	477	417
Quality	-7871.0	-7866.7	-7852.9	-7850.9	-7864.7

5.4 Simulation Results on Three-Objective Flowshop Scheduling Problems

We also applied the C-MOGA and the MOGA with random weights to three-objective test problems. We used the same parameter specifications as in the previous subsection except for the population size. Since we defined $d = 13$ for the C-MOGA, the population size was 105 from the equation (11). In order to compare the two algorithms under the same computation load, we specified the population size in the MOGA as 105. Simulation results are summarized in Table 3 and Table 4. From Table 3, we can see that the performance of the C-MOGA is better than that of the MOGA. Table 4 shows the effect of the choice of a value of d on the performance of the C-MOGA. The second row shows the population size calculated from the value of d. It is noted that each algorithm with a different value of d was terminated when the number of examined solutions exceeded the termination condition. The number of generations is shown in the third row of Table 4 for each specification of d. From Table 4, we can see that the best quality value was obtained by the C-MOGA with $d = 13$.

6 Conclusion

In this paper, we proposed a weight specification method for the cellular multi-objective genetic algorithm (C-MOGA), which is an extension of a multi-objective genetic algorithm (MOGA) in our former study (Murata & Ishibuchi [6]). In the proposed C-MOGA, each individual is located in a cell with a different weight vector. This weight vector governs the selection operation. The selection is performed in the neighborhood of each cell. The effectiveness of the C-MOGA with the proposed weight specification method was demonstrated by computer simulations on two- and three-objective flowshop scheduling problems.

References

1. Goldberg, D.E.: *Genetic Algorithms in Search, Optimization, and Machine Learning*. Reading, MA: Addison-Wesley (1989).
2. Schaffer, J.D.: Multi-objective optimization with vector evaluated genetic algorithms. *Proc. of 1st Int'l Conf. on Genetic Algorithms* (1985) 93-100.
3. Kursawe, F.: A variant of evolution strategies for vector optimization. In H.-P.Schwefel and R.Männer (Eds.), *Parallel Problem Solving from Nature*, Springer-Verlag, Berlin (1991) 193-197.
4. Horn, J., Nafpliotis, N. and Goldberg, D.E.: A niched Pareto genetic algorithm for multi-objective optimization. *Proc. of 1st IEEE Int'l Conf. on Evolutionary Computation* (1994) 82-87.
5. Fonseca, C. M. and Fleming, P. J.: An overview of evolutionary algorithms in multiobjective optimization, *Evolutionary Computation* **3** (1995) 1-16.
6. Murata, T. and Ishibuchi, H.: Multi-objective genetic algorithm and its applications to flowshop scheduling. *International Journal of Computers and Engineering* **30**, 4 (1996) 957-968.
7. Zitzler, E. and Thiele, L.: Multiobjective evolutionary algorithms: A comparative case study and the strength Pareto Approach. *IEEE Trans. on Evolutionary Computation* **3** (1999) 257-271.
8. Ishibuchi, H. and Murata, T.: A multi-objective genetic local search algorithms and its application to flowshop scheduling. *IEEE Trans. on System, Man, and Cybernetics, Part C* **28** (1998) 392-403.
9. Murata, T. and Gen, M.: Cellular genetic algorithm for multi-objective optimization. *Proc. of 4th Asian Fuzzy System Symposium* (2000) 538-542.
10. Murata, T., Ishibuchi, H., and Gen, M.: Cellular genetic local search for multi-objective optimization. *Proc. of the Genetic and Evolutionary Computation Conference 2000* (2000) 307-314.
11. Whitley, D.: Cellular Genetic Algorithms. *Proc. of 5th Int'l Conf. on Genetic Algorithms* (1993) 658.
12. Manderick, B. and Spiessens, P.: Fine-grained parallel genetic algorithms. *Proc. of 3rd Int'l Conf. on Genetic Algorithms* (1989) 428-433.
13. Wilson, D. S.: Structured demes and the evolution of group-advantageous traits. *The American Naturalist* **111** (1977) 157-185.
14. Dugatkin, L. A. and Mesterton-Gibbons, M.: Cooperation among unrelated individuals: Reciprocal altruism, by-product mutualism and group selection in fishes. *BioSystems* **37** (1996) 19-30.

15. Nowak, M. A. and May, M.: Evolutionary games and spatial chaos. *Nature* **359** (1992) 826-859.
16. Wilson, D. S., Pollock, G. B., and Dugatkin, L. A.: Can altruism evolve in purely viscous populations? *Evolutionary Ecology* **6** (1992) 331-341.
17. Oliphant, M.: Evolving cooperation in the non-iterated Prisoner's Dilemma: The importance of spatial organization. in R. A. Brooks and P. Maes (Eds.), *Artificial Life IV*, MIT Press, Cambridge (1994) 349-352.
18. Grim, P.: Spatialization and greater generosity in the stochastic Prisoner's Dilemma. *BioSystems* **37** (1996) 3-17.
19. Ishibuchi, H., Nakari, T., and Nakashima T.: Evolution of Strategies in Spatial IPD Games with Structured Demes, *Proc. of the Genetic and Evolutionary Computation Conference 2000* (2000).
20. Knowles, J.D., and Corne, D.W.: Approximating the nondominated front using the Pareto Archived Evolution Strategy, *Evolutionary Computation* (MIT Press), **8**, 2 (2000) 149-172.
21. Johnson, S.M.: Optimal two- and three-stage production schedules with setup times included. *Naval Research Logistics Quarterly* **1** (1954) 61-68.
22. Daniels, R.L. and Chambers, R.J.: Multiobjective flow-shop scheduling. *Naval Research Logistics* **37** (1990) 981-995.
23. Esbensen, H.: Defining solution set quality. *Memorandum* (No.UCB/ERL M96/1, Electric Research Laboratory, College of Engineering, Univ. of California, Berkeley, USA, Jan., 1996).

Adapting Weighted Aggregation for Multiobjective Evolution Strategies

Yaochu Jin, Tatsuya Okabe, and Bernhard Sendhoff

Future Technology Research
Honda R&D Europe (D) GmbH
63073 Offenbach/Main, Germany
yaochu.jin@hre-ftr.f.rd.honda.co.jp

Abstract. The conventional weighted aggregation method is extended to realize multi-objective optimization. The basic idea is that systematically changing the weights during evolution will lead the population to the Pareto front. Two possible methods are investigated. One method is to assign a uniformly distributed random weight to each individual in the population in each generation. The other method is to change the weight periodically with the process of the evolution. We found in both cases that the population is able to approach the Pareto front, although it will not keep all the found Pareto solutions in the population. Therefore, an archive of non-dominated solutions is maintained. Case studies are carried out on some of the test functions used in [1] and [2]. Simulation results show that the proposed approaches are simple and effective.

1 Introduction

A large number of evolutionary multiobjective algorithms (EMOA) have been proposed [3,4]. So far, there are three main approaches to evolutionary multi-objective optimization, namely, aggregation approaches, population-based non-Pareto approaches and Pareto-based approaches [4]. In the recent years, the Pareto-based approaches have gained increasing attention in the evolutionary computation community and several successful algorithms have been proposed [5].

Despite their weaknesses, the aggregation approaches are very easy to implement and computationally efficient. Usually, aggregation approaches can provide only one Pareto solution if the weights are fixed using problem-specific prior knowledge. However, it is also possible to find more than one Pareto solution using this method by changing the weights during optimization. In [6], the weights of the different objectives are encoded in the chromosome to obtain more than one Pareto solution. Phenotypic fitness sharing is used to keep the diversity of the weight combinations and mating restrictions are required so that the algorithm can work properly.

Most of the EMOAs are based on Genetic Algorithms and relatively little attention has been paid to evolution strategies. Some exceptions are [2,7,8,9]. In [7], average ranking is used to dictate the deletion of a fraction of the population.

A predator-prey-model is proposed in [9]. A selection method that is similar to the VEGA approach [10] is adopted in [8]. An algorithm called Pareto Archived Evolution Strategy (PAES) is suggested in [2], in which a non-Pareto approach together with an archive of the found Pareto solutions are used.

This paper investigates two methods using the aggregation-based approach. To approximate the Pareto front instead of a certain Pareto solution, the weight for each objective should be changed systematically. One method is to distribute the weights uniformly among the individuals in the population. The other method is to periodically change the weights with the process of the evolution. Although these methods seem to be very simple, we will show that they work effectively for two objective optimization problems. Simulations are carried out on different test functions studied in [1,2]. Different evolution strategies, including the standard evolution strategy [11], the Evolution Strategy with Rotation Matrix Adaptation [11] and the Evolution Strategy with Covariance Matrix Adaptation [12,13] are employed.

What is quite surprising from our simulation results is that our algorithms work well even for problems with a concave Pareto front (see Section 4 for details), which is usually thought to be not obtainable by aggregation based methods [4]. Our preliminary explanation is that if the search algorithm goes through the concave region of the Pareto front (which is locally near-optimal when the objectives are aggregated into one single objective function) and if the near optimal solutions are archived, then the Pareto solutions within the concave region can also be found using the aggregation method. Further results on this issue will be reported elsewhere.

2 The Aggregation Based Multiobjective Algorithms

2.1 The Evolution Strategies

In the standard Evolution Strategy (ES), the mutation of the objective parameters is carried out by adding an $N(0, \sigma_i^2)$ distributed random number. The step sizes σ_i are also encoded in the genotype and subject to mutations. A standard Evolution Strategy can be described as follows:

$$\mathbf{x}(t) = \mathbf{x}(t-1) + \tilde{\mathbf{z}} \tag{1}$$

$$\sigma_i(t) = \sigma_i(t-1)\exp(\tau' z)\exp(\tau z_i); i = 1, ..., n \tag{2}$$

where $\mathbf{x}$ is an n-dimensional parameter vector to be optimized, $\tilde{\mathbf{z}}$ is an n-dimensional random number vector with $\tilde{\mathbf{z}} \sim N(\mathbf{0}, \boldsymbol{\sigma}(t)^2)$, z and z_i are normally distributed random numbers with $z, z_i \sim N(0,1)$. Parameters τ, τ' and σ_i are the strategy parameters, where σ_i is mutated as in equation(2) and τ, τ' are constants as follows:

$$\tau = \left(\sqrt{2\sqrt{n}}\right)^{-1}; \ \tau' = \left(\sqrt{2n}\right)^{-1} \tag{3}$$

There are several extensions to the above standard ES. In our simulations, an ES with Rotation Matrix Adaptation and an ES with Covariance Matrix Adaptation as well as the standard ES are used to investigate the effectiveness of the proposed multiobjective algorithms using different search strategies. For the detailed description of the evolution strategies, please refer to [11,12] respectively.

Two main different selection schemes are used in evolution strategies. Suppose there are μ and λ individuals in the parent and offspring population, usually $\mu \leq \lambda$. One method is to select the μ parent individuals only from the λ offspring, which is usually noted as (μ,λ)-ES. If the μ parent individuals are selected from a combination of the μ parent individuals and the λ offspring individuals, the algorithm is noted as $(\mu + \lambda)$-ES. In our study, the (μ, λ)-ES is adopted.

2.2 Random Distribution of Weights within a Population

For the sake of clarity, we consider the two objective problems in the current discussion; the extension to problems with more than two objectives is straightforward. For a conventional aggregation method, the fitness function is the weighted sum of the two different objectives f_1 and f_2:

$$Fitness = w_1 f_1 + w_2 f_2, \tag{4}$$

where w_1 and w_2 are two constants determined using a prior knowledge about the problem. It is clear that by using a pair of fixed weights, only one Pareto solution can be obtained.

Imagine that we run the algorithm so many times that every weight combination has been used. In this way, we can obtain all Pareto solutions that the Pareto front consists of. Notice, that it has been argued that the Pareto solutions locating in the concave region of the Pareto front cannot be obtained by aggregation methods. However, in the experiments in Section 4, we found that our algorithms are successful in obtaining a very complete concave Pareto front for low dimensional problems (e.g. $n = 2$), and a quite complete concave Pareto front with a dimension as high as 10.

Of course, it is unpractical, if not impossible to run the evolutionary algorithm so many times to exhaust all the weight combinations. Since we are using evolutionary optimization, it is natural to take advantage of the population for this purpose. If the different weight combinations can be distributed among the individuals, the population may be able to approach the Pareto front during the process of evolution. Suppose we use the (μ, λ)-ES, then the weight combinations can be distributed uniformly among the λ individuals in the offspring population. Let

$$w_1^i(t) = random(\lambda)/\lambda, \tag{5}$$
$$w_2^i(t) = 1.0 - w_1^i(t), \tag{6}$$

where $i = 1, 2, ..., \lambda$ and t is the index for generation number. The function $random(\lambda)$ generates a uniformly distributed random number between 0 and λ.

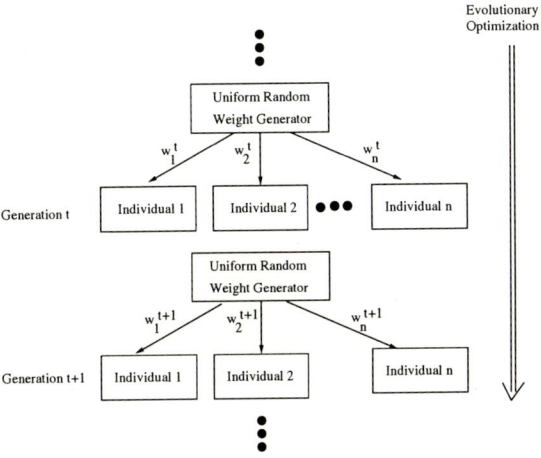

Fig. 1. Determination of the weights: Random distribution.

In this way, we can get a uniformly distributed random weight combination (w_1^i, w_2^i) among the individuals, where $0 \leq w_1^i, w_2^i \leq 1$ and $w_1^i + w_2^i = 1$, refer also to Fig.1, in which $U(0,1)$ denotes a uniform distribution. Notice that the weight combinations are regenerated in every generation.

2.3 Generation-Based Periodical Variation of the Weights

The idea of a uniformly distributed weight combination can straightforwardly be extended to a generation based approach. However, if we still use a random weight combination, convergence of the evolutionary algorithm will be in question. Therefore, instead of using a randomly distributed weight combination, we use a weight combination that is changed gradually and periodically with the process of the evolution. In this work, this is realized as follows:

$$w_1(t) = |\sin(2\pi t)/F|, \qquad (7)$$
$$w_2(t) = 1.0 - w_1(t), \qquad (8)$$

where t is the number of generation and $|\cdot|$ gives the absolute value. We can see from equation (7) that $w_1(t)$ changes from 0 to 1 periodically. The change frequency can be adjusted by F. In our study, we set $F = 400$, which means that in every 400 generations, w_1 will change from 0 to 1 and then from 1 to 0 four times. Fig.2 shows an example of how the weights change during evolution within 200 generations. We found that the results of the algorithm are not very sensitive to F, although it seems reasonable to let the weight change from 0 to 1 twice. Notice in this case, all the individuals have the same weight combination in the same generation.

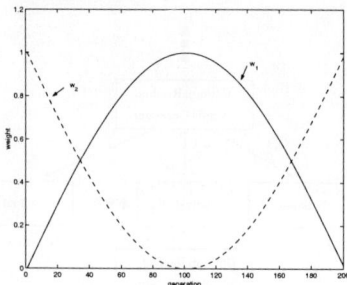

Fig. 2. Determination of weights: Generation-based periodical variation, all individuals have the same weight combination within a generation.

2.4 An Archive of Pareto Solutions

In our algorithm, the population is not able to keep all the found Pareto solutions, although it is able to approach the Pareto front dynamically. Therefore, it is necessary to record the Pareto solutions that have been found so far. The idea of building a Pareto archive is inspired from [2], although we use quite a different method to aggregate and maintain the archive. The pseudo-code for building the archive is listed in Algorithm 1. The similarity is measured by the Euclidean distance in the fitness space. It should be noticed that it is possible for one solution in the archive is dominated by another.

3 Test Functions

To evaluate the effectiveness of the proposed algorithms, simulations are carried out on four test functions used in [1,2].

- The first test function (F_1) used here is the second function in [2] and we extend it to an n-dimensional function:

$$f_1 = \frac{1}{n} \sum_{i=1}^{n} x_i^2 \tag{9}$$

$$f_2 = \frac{1}{n} \sum_{i=1}^{n} (x_i - 2.0)^2 \tag{10}$$

- The second test function (F_2) is the first function in [1], which has a convex Pareto front:

$$f_1 = x_1 \tag{11}$$

$$g(x_2, ..., x_n) = 1.0 + \frac{9}{n-1} \sum_{i=2}^{n} x_i \tag{12}$$

$$f_2 = g \times (1.0 - \sqrt{f_1/g}) \tag{13}$$

where $x_i \in [0, 1]$.

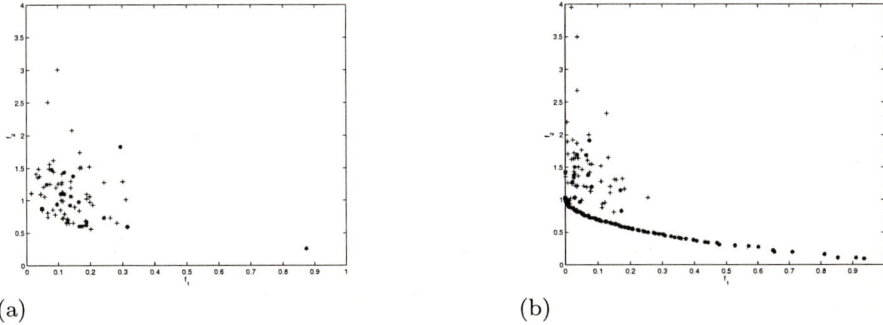

Fig. 4. Method 1 for F_2 (n=2) using the standard ES: (a) generation 10, (b) generation 500.

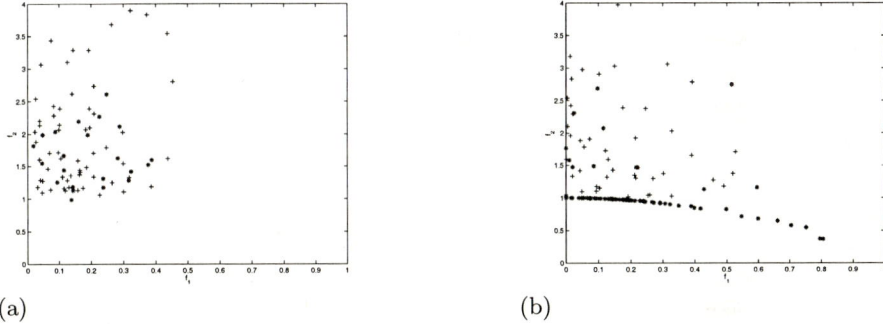

Fig. 5. Method 1 for F_3 (n=2) using the standard ES: (a) generation 10, (b) generation 500.

we find that Method 1 is working effectively for all different test functions, which shows that the idea of using a uniformly random weight distribution among the individuals of the population is for multi-objective optimization feasible.

Next, we run simulations on all four test functions with $n = 2$ using Method 2. The results are shown in Figures 7, 8, 9 and 10.

From these figures, it is demonstrated that Method 2 has been successful in obtaining a very complete Pareto front for all the four test functions. The difference between Method 1 and Method 2 is that the individuals in the population in Method 2 converged more completely to the Pareto front than the individuals in Method 1 at the end of the evolution.

The most interesting fact is that both methods have obtained very complete Pareto solutions for F_3, which has a concave Pareto front. Our empirical results show that in principle, concave Pareto solutions can be obtained by aggregation methods if the search algorithm is able to go through the concave region and if an archive is used to store the found Pareto solutions.

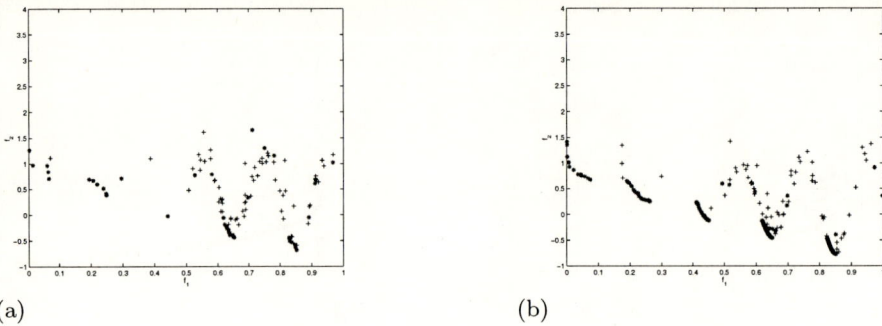

Fig. 6. Method 1 for F_4 (n=2) using the standard ES: (a) generation 10, (b) generation 500.

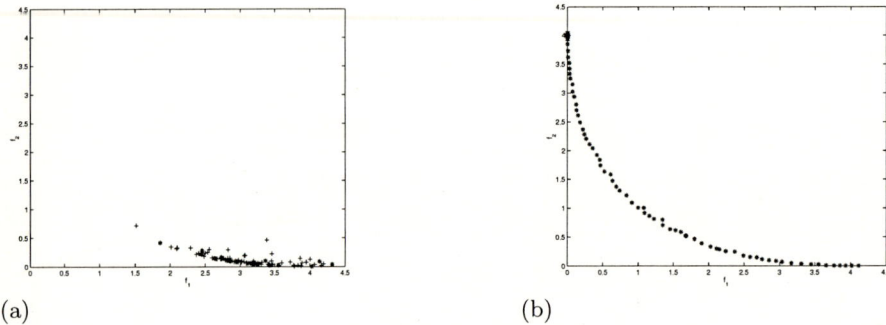

Fig. 7. Method 2 for F_1 (n=2) using the standard ES: (a) generation 10, (b) generation 500.

4.2 Comparison of Both Methods for High Dimensional Problems

The performance of Method 1 and Method 2 is compared in this part of the simulations. The purpose of the simulations is to investigate the efficiency of Method 1 and Method 2 for more complex problems. In the results presented in the following, the standard ES is used for the four test functions with a dimension of 10 and 500 generations are run. Figures 11 and 12 give the results using Method 1. It can be seen that the performance of Method 1 on 10-dimensional problems becomes worse compared to those obtained in the two dimensional problems. In contrast, Method 2 is still able to provide very good results on the same problems, as shown in Figures 13 and 14. Notice that Method 1 shows particularly bad performance on test functions F_3 and F_4, which have a concave or discontinuous Pareto front. However, Method 2 shows quite good performance on all the four test functions.

4.3 Comparison of Different Evolution Strategies for Method 2

This part of the simulation aims at comparing the performance of different ES algorithms for more difficult problems (with higher dimension in this context).

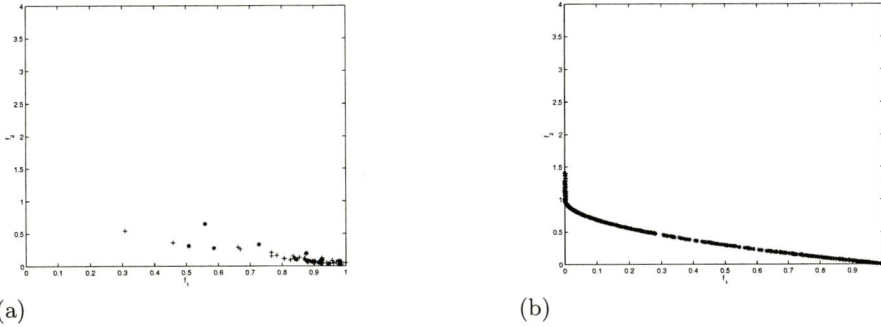

Fig. 8. Method 2 for F_2 (n=2) using the standard ES: (a) generation 10, (b) generation 500.

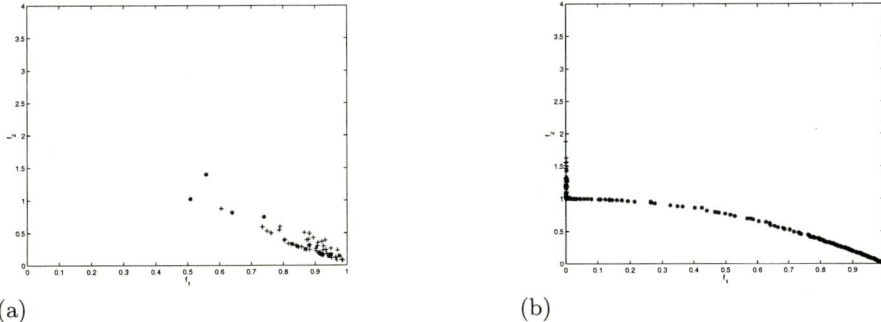

Fig. 9. Method 2 for F_3 (n=2) using the standard ES: (a) generation 10, (b) generation 500.

Since Method 2 exhibits much better performance in the above part of experiments, only Method 2 will be used in the following simulations. The algorithms considered in this work are the standard ES, the ES with Rotation Matrix Adaptation and the ES with Covariance Matrix Adaptation. The problems studied in this part of the simulation are the four test functions with $n = 30$. Results using the standard ES on the four functions are shown in Figures 15 and 16. The results using the ES with Rotation Matrix Adaptation are given in Figures 17 and 18. Finally, the ES with Covariance Matrix Adaptation is tested and the results are presented in Figures 19 and 20. In all the simulations, 500 generations are run.

As it is shown in the above figures, the standard ES together with Method 2 can always provide quite a complete Pareto front, but unfortunately, the accuracy of the solutions is not satisfactory. On the other hand, the ES with Rotation Matrix Adaptation gives consistently good results on all the four problems, which are comparable to or even better than those of the Pareto-based algorithms described in [1]. Interestingly, the ES with CMA produced very good results on F_1, but failed on F_3 and F_4. This may be ascribed to the fact that the ES with CMA

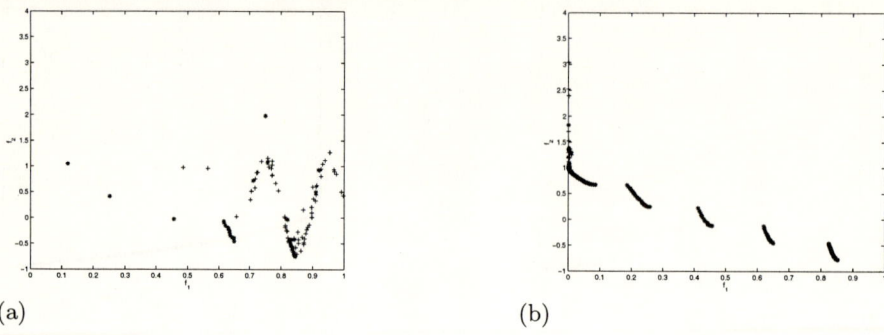

Fig. 10. Method 2 for F_4 (n=2) using the standard ES: (a) generation 10, (b) Generation 500.

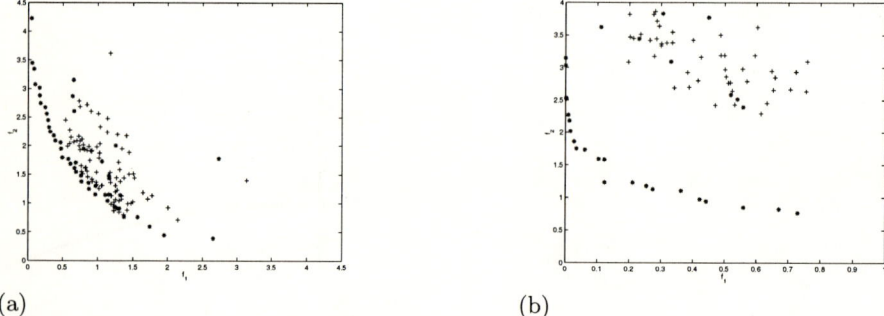

Fig. 11. Method 1 using the standard ES for (a) F_1 (n=10) and (b) F_2 (n=10).

is a more deterministic algorithm that converges quickly before it can explore a wider region of the search space.

At the same time, as for the low dimensional problems studied in Section 4.1 of this section, the ES with RMA is able to obtain a quite complete concave Pareto front with the dimension of 30. This was surprising taking into account the fact that it is a simple, dynamic aggregation based approach.

5 Conclusion

Two aggregation based methods for multiobjective optimization are proposed in this paper. The idea is to use dynamic weights instead of fixed weights to achieve the Pareto solutions. We found that both methods work well on low-dimensional problems. However, for high-dimensional problems, the second method outperforms the first one. Furthermore, the simulation results also depend on the type of evolution strategy that is employed. On the other hand, it also depends on the performance of the evolution strategy. In our experiment, the Evolution Strategy with Rotation Matrix Adaptation gives better performance than the standard Evolution Strategy. At the same time, the Evolution Strategy with Covariance

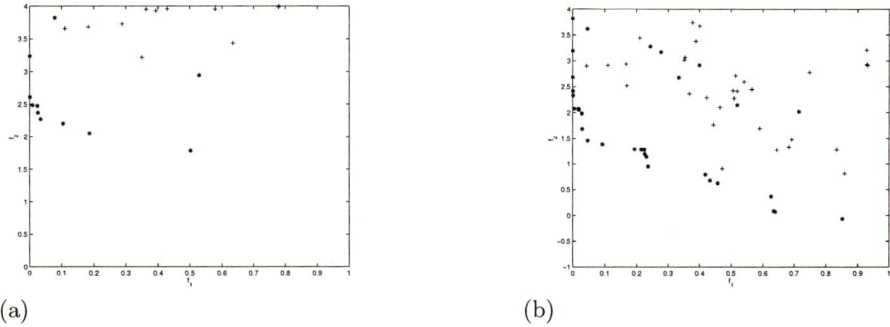

Fig. 12. Method 1 using the standard ES for (a) F_3 (n=10) and (b) F_4 (n=10).

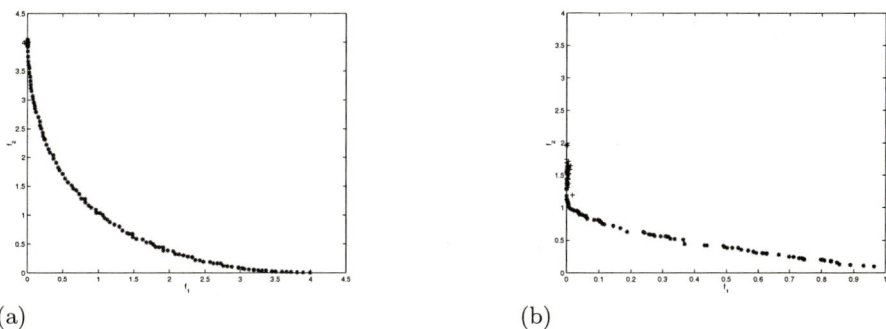

Fig. 13. Method 2 using the standard ES for (a) F_1 (n=10) (b) F_2 (n=10).

Matrix Adaptation provides very good results on smooth, high-dimensional problems, but its performance degrades seriously on problems with a discontinuous and non-convex Pareto-optimal front.

In our approach, no changes have to be made to the conventional evolutionary algorithm except for the dynamic weights and an archive of the found Pareto solutions. Therefore, the methods can straightforwardly be applied to all existing evolutionary algorithms with only minor modifications.

Another interesting phenomenon is that the proposed algorithms are able to find the Pareto solutions with a concave Pareto front. This is a very encouraging point when we are applying aggregation-based methods to multi-objective optimization. Further investigation of this issue will be part of our future work.

The problems studied in this paper are all two-objective ones. Theoretically, the proposed methods can be extended to problems with more than two objectives. Expected problems are the increasing complexity and the decreasing efficiency, which, however, is also true for the Pareto-based approaches.

Acknowledgments. The authors would like to thank E. Körner and W. von Seelen for their support and T. Arima for his insightful comments.

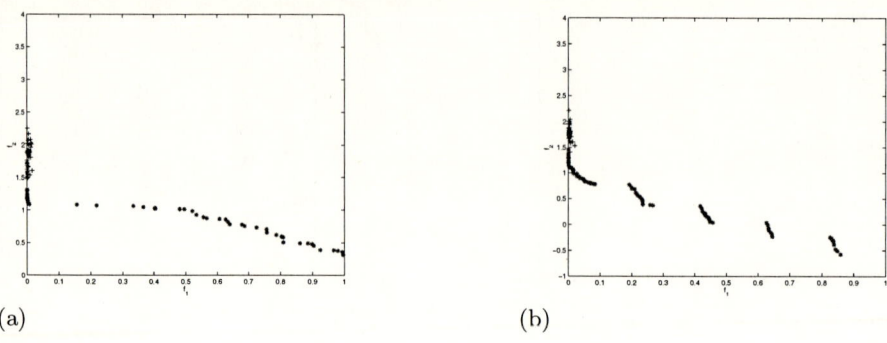

Fig. 14. Method 2 using the standard ES for (a) F_3 (n=10) and (b) F_4 (n=10).

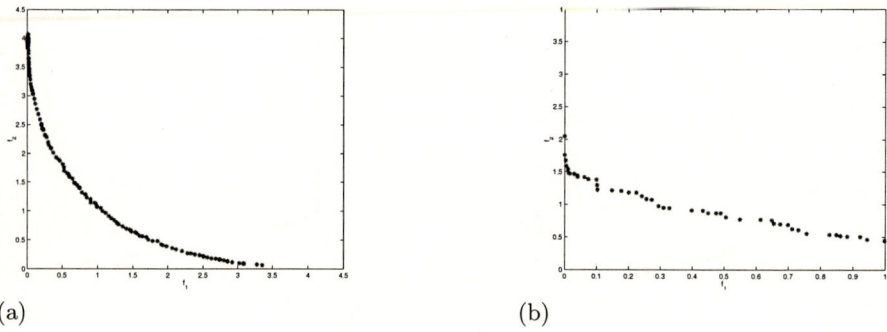

Fig. 15. Method 2 using the standard ES for (a) F_1 (n=30) and (b) F_2 (n=30).

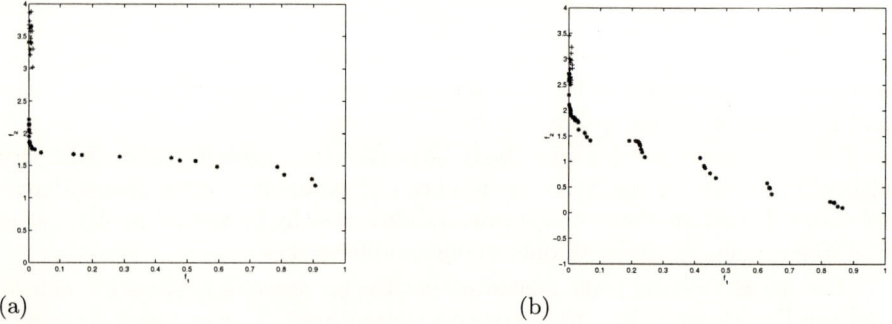

Fig. 16. Method 2 using the standard ES for (a) F_3 (n=30) and (b) F_4 (n=30).

References

1. E. Zitzler, K. Deb, and L. Thiele. Comparison of multiobjective evolution algorithms: empirical results. *Evolutionary Computation*, 8(2):173–195, 2000.
2. J. D. Knowles and D. W. Corne. Approximating the nondominated front using the Pareto archived evolution strategies. *Evolutionary Computation*, 8(2):149–172, 2000.

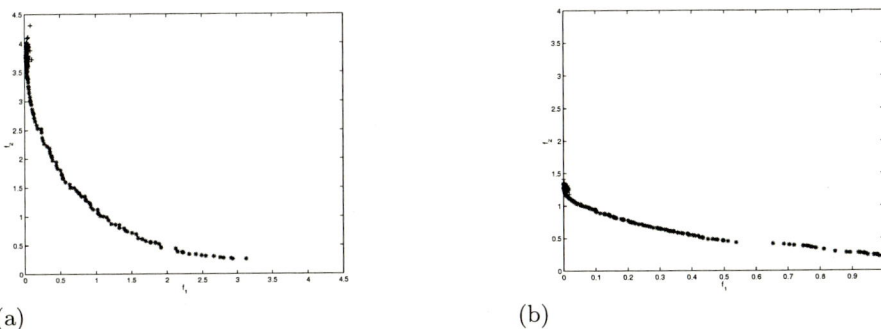

Fig. 17. Method 2 using the ES with RMA for (a) F_1 (n=30) and (b) F_2 (n=30).

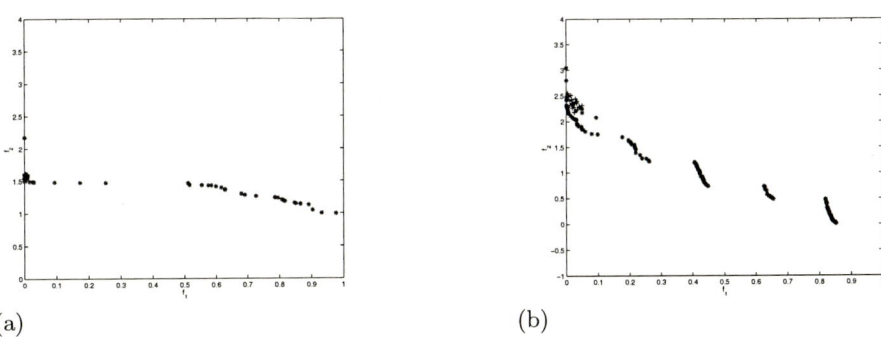

Fig. 18. Method 2 using the ES with RMA for (a) F_3 (n=30) and (b) F_4 (n=30).

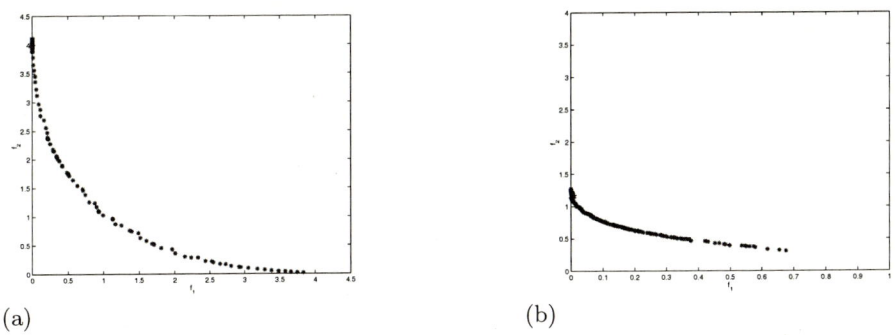

Fig. 19. Method 2 using the ES with CMA for (a) F_1 (n=30) and (b) F_2 (n=30).

3. C.A.C. Coello. A comprehensive survey of evolutionary-based multiobjective optimization techniques. *Knowledge and Information Systems*, 1(3):269–308, 1999.
4. C. M. Fonseca and P. J. Fleming. Multiobjective optimization. In Th. Bäck, D. B. Fogel, and Z. Michalewicz, editors, *Evolutionary Computation*, volume 2, pages 25–37. Institute of Physics Publishing, Bristol, 2000.
5. D. A. Van Veldhuizen and G. B. Lamont. Multiobjective evolutionary algorithms: Analyzing the state-of-art. *Evolutionary Computation*, 8(2):125–147, 2000.

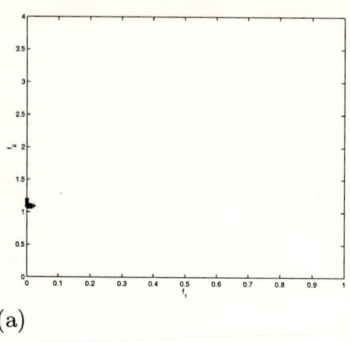
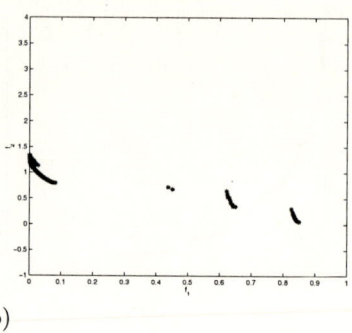

(a)   (b)

Fig. 20. Method 2 using the ES with CMA for (a) F_3 (n=30) and (b) F_4 (n=30).

6. P. Hajela and C. Y. Lin. Genetic search strategies in multicriteria optimal design. *Structural Optimization*, 4:99–107, 1992.
7. F. Kursawe. A variant of evolution strategies for vector optimization. In H.-P. Schwefel and R. Männer, editors, *Parallel Problem Solving from Nature*, volume I, pages 193–197, 1991.
8. T. Binh and U. Korn. Multiobjective evolution strategy with linear and nonlinear constraints. In *Proceedings of the 15th IMACS World Congress on Scientific Conputation, Modeling and Applied Mathematics*, pages 357–362, 1997.
9. M. Laumanns, G. Rudolph, and H.-P. Schwefel. A spatial predator-prey approach to multi-objective optimization. In A.E. Eiben, Th. Bäck, M. Schoenauer, and H.-P. Schwefel, editors, *Parallel Problem Solving from Nature*, volume V, pages 241–249, 1998.
10. J. D. Schaffer. Multiple objective optimization with vector evaluated genetic algorithms. In *Proceedins of an International Conference on Genetic Algorithms and Their Applications*, pages 93–100, 1985.
11. H.-P. Schwefel. *Evolution and Optimum Seeking*. Sixth-Generation Computer Technologies Series. John Wiley & Sons, Inc., 1994.
12. N. Hansen and A. Ostermeier. Adapting arbitrary normal mutation distributions in evolution strategies: The covariance matrix adaption. In *Proc. 1996 IEEE Int. Conf. on Evolutionary Computation*, pages 312–317. IEEE Press, 1996.
13. N. Hansen and A. Ostermeier. Completely derandomized self-adaptation in evolution strategies. *Evolutionary Computation*, 2000. To appear.

Incrementing Multi-objective Evolutionary Algorithms: Performance Studies and Comparisons

K. C. Tan[1], T. H. Lee, and E. F. Khor

Department of Electrical and Computer Engineering
National University of Singapore
10 Kent Ridge Crescent Singapore 119260
[1]Email: eletankc@nus.edu.sg

Abstract. This paper addresses the issue by presenting a novel "incrementing" multi-objective evolutionary algorithm (IMOEA) with dynamic population size that is adaptively computed according to the on-line discovered trade-off surface and its desired population distribution density. It incorporates the method of fuzzy boundary local perturbation with interactive local fine-tuning for broader neighborhood exploration to achieve better convergence as well as discovering any gaps or missing trade-off regions at each generation. Comparative studies with other multi-objective (MO) optimization are performed on benchmark problem. The new suggested quantitative measures together with other well-known measures are employed to access and compare their performances statistically.

1. Introduction

Evolutionary techniques for MO optimization are currently gaining significant attention from researchers in various fields [1-7]. The methods, however, need to 'guess' for an optimal constant population size in order to discover and distribute the non-dominated individuals along the usually sophisticated trade-off surface. As addressed by Arabas *et al.* [8], evolutionary algorithm may suffer from premature convergence if the population size is too small. If the population is too large, undesired computational resources may be incurred and the waiting time for a fitness improvement may be too long in practice. Therefore the selection of an appropriate population size in evolutionary optimization is important and could greatly affect the effectiveness and efficiency of the optimization performance [9-13]. In the case of single objective (SO) optimization, various methods of determining an optimal population size from different perspectives have been proposed. Grefenstette [14] applied a Meta genetic algorithm to control the population size of another genetic algorithm. Smith [15] and Arabas *et al.* [8] proposed the approach of determining the population size adaptively according to the "age" of a chromosome. Zhuang *et al.* [16] proposed an adaptive population size by adapting it to the degree of improvem__ achieved at each generation.
Unlike these approaches that are only applicable to single object [12,14,15,16], this paper generalizes the work in our earlier developec

proposes an incrementing multi-objective evolutionary algorithm (IMOEA) with dynamic population size for effective MO optimization. Instead of having a constant population to explore the solution space, IMOEA adaptively computes an appropriate population size according to the on-line evolved trade-offs and its desired population distribution density. This approach reduces the computational effort due to unnecessary extra individuals and avoids the pre-mature convergence or incomplete trade-offs resulting from insufficient number of individuals. In addition, the IMOEA incorporates a fuzzy boundary local perturbation technique with dynamic number of local perturbations per parent to encourage and reproduce the "incrementing" individuals for better MO optimization, whereas any 'extra' individuals will be filtered through the method of switching preserve strategy proposed in the paper. While maintaining the global search capability, the scheme enhances the local exploration and fine-tuning of the evolution at each generation so as to fill-up any discovered gaps or discontinuities among the non-dominated individuals that are loosely located or far away from each other along the trade-off surface. Details of the IMOEA with fuzzy boundary local perturbation and other advanced features are described in Section 2. Section 3 presents extensive simulations and comprehensive quantitative/statistical comparisons of IMOEA, with other famous MO optimization algorithms, on benchmark problem. Conclusions are drawn in Section 4.

2. Incrementing Multi-objective Evolutionary Algorithm

As addressed in the Introduction, current evolutionary techniques for MO optimization face a common difficulty in determining an optimal population size in order to efficiently explore the solution space as well as to distribute along the trade-off surface with sufficient number of non-dominated individuals. Intuitively, it is hard to achieve a good evolution if the population size is too small with loosely distributed individuals leading to insufficient exchange of genetic information. If the population size is too large, however, undesirable computational effort may be incurred in practice. In single objective evolutionary optimization, the population size is often "guessed" according to the size of the search space in the parameter domain [8]. This is, however, not applicable to MO optimization where the global optimal is not a single solution but is often a set of Pareto optimal points covering the entire trade-off surface in the objective domain. Further, since the exact trade-offs is often unknown in *a-priori* to an optimization process, it is difficult to estimate an optimal number of individuals necessary for effective exploration of the solution space as well as good representation of the trade-off surface.

The issue of dynamic population in MO optimization currently remains an open problem for researchers in the field of evolutionary computation as pointed out by [4]. Extending from our earlier work of MOEA [5], an incrementing multi-objective evolutionary algorithm is proposed to deal with this problem by adaptively computing an appropriate population size at each generation. The population size in IMOEA is thus dynamic and is increased or decreased based upon the on-line discovered Pareto-front and its desired population distribution density along the trade-offs.

2.1 Dynamic Population Size

Instead of 'guessing' for an optimal population size in MO optimization, an adaptation mechanism is employed in IMOEA such that the population size is evolved based upon the on-line evolved trade-offs and the required population distribution density defined by the user according to his/her preference on how close the non-dominated individuals should be distributed apart from each other on the Pareto-front. Consider an m-dimensional objective space, the desired population size, $dps^{(n)}$, with the desired population size per unit volume, ppv, and the approximated trade-off hyper-area of $A_{to}^{(n)}$ [17] discovered by the population at generation n can be defined as.

$$lowbps \leq dps^{(n)} = ppv \times A_{to}^{(n)} \leq upbps \qquad (1)$$

where $lowbps$ and $upbps$ is the lower and upper bound for the desired population size $dps^{(n)}$, respectively, which can be treated as hard bounds that are optionally defined by the user. The trade-offs for an m-objective optimization problem is in the form of an $(m-1)$ dimensional hyper-surface as pointed out by Horn and Nafpliotis [18] (note that 1-dimensional surface is actually a curve while 0-dimensional surface is in point form), which could be approximated by the hyper-surface $A_{to}^{(n)}$ of a hyper-sphere as given by [17],

$$A_{to}^{(n)} \approx \frac{\pi^{(m-1)/2}}{\left(\frac{m-1}{2}\right)!} \times \left(\frac{d^{(n)}}{2}\right)^{m-1} \qquad (2)$$

where $d^{(n)}$ is the diameter of the hyper-sphere at generation n. The above estimation of population size is valid for both convex and concave surfaces [17]. Furthermore, the same computation procedure of diameter $d^{(n)}$ can also be easily extended to any multi-dimensional objective space [17]. Clearly, eqns. 1 and 2 provide a simple estimation of the desired population size at each generation according to the on-line discovered trade-off surface A_{to} and the desired population density ppv, which is more efficient and appropriate than the idea of guessing for an optimal population size in *a-priori* to an optimization process as adopted in existing methods.

2.2 Fuzzy Boundary Local Perturbation

In this section, a fuzzy boundary local perturbation (FBLP) scheme that perturbs the set of non-dominated individuals to produce the necessary "incrementing" individuals for the desired population size in IMOEA as given by eqns. 1 and 2 is proposed. In brief, the FBLP is implemented for the following objectives:

1. Produce additional "good" individuals in filling up the gaps or discontinuities among existing non-dominated individuals for better representation of the Pareto-front as shown in Fig. 2;

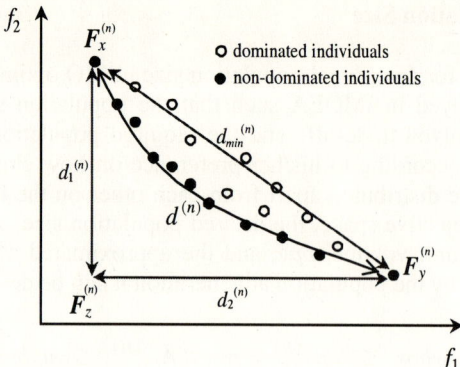

Fig. 1. The diameter $d^{(n)}$ of a trade-off curve

2. Perform interactive fine-learning to overcome weakness of local exploration in an evolutionary algorithm [19,20] and to achieve better convergence for evolutionary MO optimization;
3. Provide the possibility of perturbation beyond the neighborhood to avoid premature convergence or local saturation.

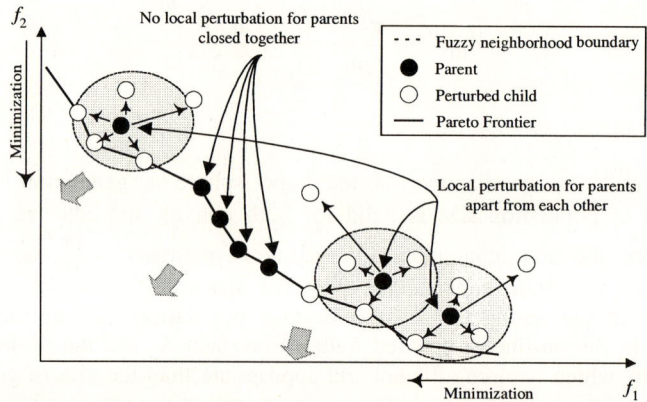

Fig. 2. FBLP for non-dominated parents with low niched cost (apart from other parents)

These additional Pareto points can be effectively obtained via the method of fuzzy boundary local perturbation at each generation. Note that only parent individuals that are being selected for reproduction from the tournament selection will be perturbed with the FBLP, and the selection criteria for the tournament is solely based upon the individuals' niched cost in the objective domain [3] instead of the cost of objective functions in order to encourage uniform distribution. Therefore parents with low niched cost (located apart from other parents) will be given higher probability to be perturbed as compared to those with a high niched cost (located close to other parents). Also, the neighborhood boundary for the parents to be perturbed is fuzzy in such a way that the probability of perturbation is higher within the neighborhood region than those outside the neighborhood [17].

In conclusion, the FBLP differs in two aspects as compared to other local search methods in single-objective evolutionary optimization [19,21]. The first is that, unlike simple mutation, the perturbation probabilities in FBLP vary according to the significance of the genes in a chromosome for ease of implementation and computation efforts. In addition, there is no hard boundary of the neighborhood size in the perturbation, which gives a non-zero chance to produce offspring that are far away from their parents in order to keep maintaining the diversity of evolution. The second is that FBLP produces a number of offspring from each selected parent without immediate cost evaluation as to determine the acceptance or rejection for each of the perturbed individual, which can be regarded as an extension of mutation with at least one (instead of one) locally perturbed offspring per parent.

2.3 Program Flowchart of IMOEA

The overall program flowchart of IMOEA is shown in Fig. 3. The dynamic sharing method for niched cost estimation from [5] is applied here to provide a simple computation of σ_{share} at each generation, which is capable of distributing the population uniformly along the Pareto-front without the need of *a-priori* knowledge in setting the σ_{share}. The detail procedures within the box of special genetic operations for IMOEA in the program flow are unveiled in Fig. 4. Instead of simple mutation, fuzzy boundary local perturbation is performed to encourage or reproduce the "incrementing" individuals needed for a better trade-off representation. While maintaining the global search capability, the scheme enhances the local exploration and fine-tuning of the evolution at each generation so as to fill-up any discovered gaps or discontinuities among the non-dominated individuals that are loosely located or far away from each other along the trade-off surface.

As shown in Fig. 3, the evolved population with the desired population size of $dps^{(n+1)}$ will be combined with the reserved non-dominated individuals at generation n to form a combined population that has a size larger or equal to $dps^{(n+1)}$. Individuals in the combined population are then selected for next generation such that a stable and well-distributed Pareto-front could be maintained at each generation. Concerning this, a switching preserved strategy (SPS) [17], as highlighted in the shaded region, that preserves the non-dominated individuals to be evolved together with the population is proposed is employed.

3. Performance Comparisons on Benchmark Problems

In this section, performance comparisons are performed among various evolutionary approaches for multi-objective optimization using both quantitative and statistical measures. Besides IMOEA, other well-known algorithms including VEGA from Schaffer [9], HLGA from Hajela Lin [22], NPGA from Horn and Nafpliotis [18], MOGA from Fonseca and Fleming [23], NSGA from Srinivas and Deb [2], SPEA from Zitzler and Thiele [7] and MOEA from Tan *et al.* [5] are included in this paper.

Also, five different performance indexes/measures for MO optimization are employed for such comparisons, apart from the usual performance measure based upon the results of final trade-offs graph. Some of these measures are taken from recent literatures while others are carefully designed and added in the paper for a more comprehensive comparison.

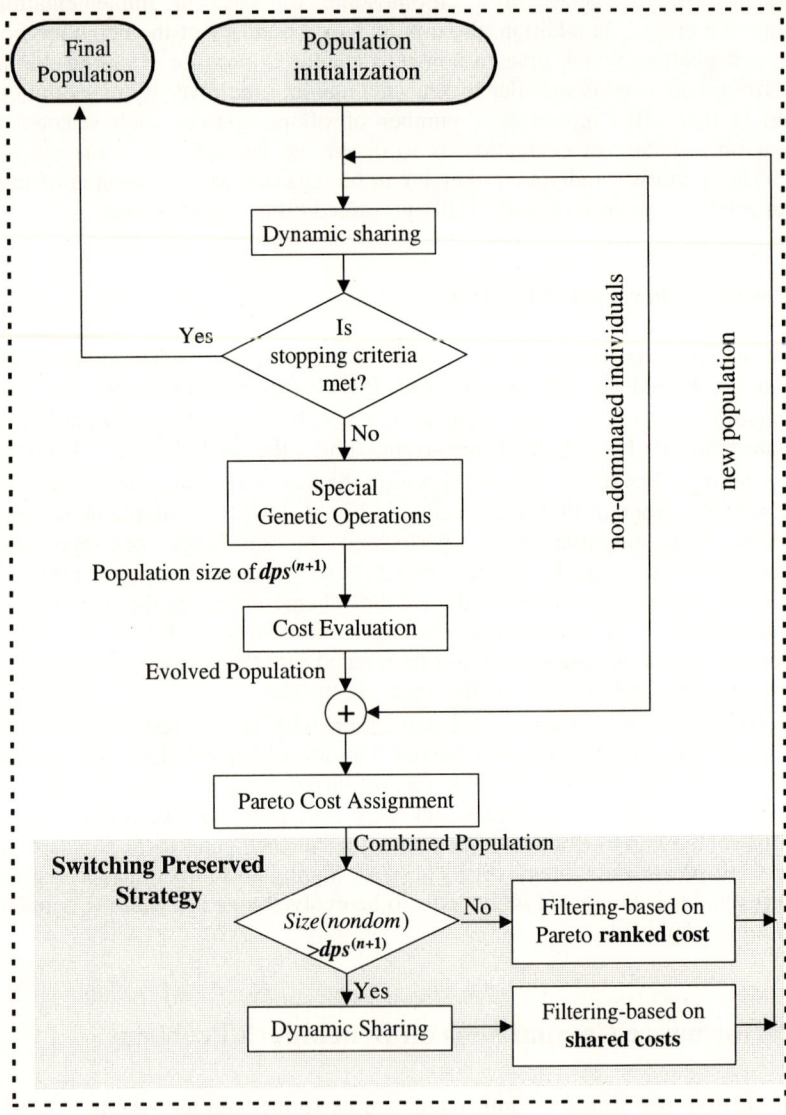

Fig. 3. Overall program flowchart with fixed (MOEA) and dynamic (IMOEA) population size

Special Genetic Operations:

Let, $dps^{(n)}$ = the size of population at current generation n
$pop^{(n)}$ = population of current generation n

Step 1) From current $pop^{(n)}$, the desired population size in next generation is computed according to the adaptation mechanism in eqn. 1:
$$dps^{(n+1)} = ppv \times A_{to}^{(n)}$$

Step 2) Perform tournament selection. The selected population is called $selpop^{(n)}$.

Step 3) Perform simple crossover with crossover probability p_c and mating restriction (Fonseca, 1995) for $selpop^{(n)}$. The resulted population is called $crosspop^{(n)}$.

Step 4) Perform FBLP. The resulted population is called $evolpop^{(n)}$.

Fig. 4. Detail procedures within the box of "Special Genetic Operations" in Fig. 3

The Fonseca's two-objective minimization problem is applied as a benchmark problem for the performance comparisons. This problem is chosen as it has been widely studied by others [5,6,23] and it has a large and non-linear trade-off curve that should challenge the MO evolutionary algorithm's ability to find and maintain the entire Pareto-front uniformly. Besides, this problem can be easily used for visualization and comparison as adopted by Fonseca and Fleming [23]. The two-objective functions, f_1 and f_2, to be minimized are given as

$$f_{1,1}(x_1,...,x_8) = 1 - \exp\left(-\sum_{i=1}^{8}\left(x_i - \frac{1}{\sqrt{8}}\right)^2\right)$$

$$f_{1,2}(x_1,...,x_8) = 1 - \exp\left(-\sum_{i=1}^{8}\left(x_i + \frac{1}{\sqrt{8}}\right)^2\right)$$

(3)

where, $-2 \le x_i < 2, \forall i = 1,2,...,8$. According to eqn. 3, there are 8 parameters $(x_1,...,x_8)$ to be optimized so that f_1 and f_2 are minimal. The trade-off line is shown by the curve in Fig. 5, where the shaded region represents the unfeasible area in the objective domain.

Five different quantitative interpretations of statistical MO optimization performances are applied. Some of them are taken from other literatures, while others are carefully designed by the authors and added in this paper for a more comprehensive comparison. Note that these measures were chosen such that the actual Pareto sets are not needed in the computation, which is often obtained through deterministic enumeration and is not always practically implementable [6].

i) *Ratio of Non-dominated Individuals (RNI)*
This performance measure is denoted here as the ratio of non-dominated individuals (*RNI*) for a given population X and is mathematically formulated as:

$$RNI(X) = \frac{nondom\_indiv}{P} \qquad (4)$$

where *nondom_indiv* is the number of non-dominated individuals in population X while P is the size of population X. Therefore the value $RNI = 1$ means all the individuals in the population are non-dominated while the opposite, $RNI = 0$ represents the situation where none of the individuals in the population are non-dominated.

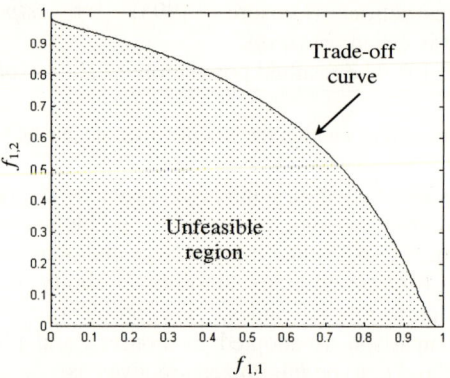

Fig. 5. Trade-off curve in the objective domain for the test problem 1

ii) *Size of Space Covered (SSC)*
This measure is originated from Zitzler and Thiele [7]. Since the original definition only accommodate maximization problem, it is generalized in this paper for both minimization and maximization process as follow. Let $x_i \in X$ be an *i*-th individual member in a population X. The function $SSC(X)$ gives the volume enclosed by the union of area in the objective domain where any point within the area is always dominated by at least one individual x_i in the population X. The higher the value of $SSC(X)$, the larger the dominated volume in the objective domain and hence the better the population is.

iii) *Coverage of Two Set (C)*
It is a measure to compare the domination of two population in a pair-wise manner, i.e., how good population *i* dominate population *j* as well as how good population *j* dominate population *i*. This measure was proposed by Zitzler and Thiele [7] and is detailed as below: Let X', $X'' \subseteq X$ be two sets of decision vectors. The function C transforms the ordered pair (X', X'') to the interval $[0, 1]$ given as,

$$C(X', X'') = \frac{\left| \{a'' \in X''; \exists a' \in X': a' \text{ dominate or nondominate } a''\} \right|}{|X''|} \qquad (5)$$

The value $C(X', X'') = 1$ indicates that all solutions in X'' are dominated by or equal to solutions in X', while $C(X', X'') = 0$ means none of the points in X'' are covered by the population X'.

iv) *Uniform Distribution (UD) of Non-dominated Population*

Most MO optimization methods attempt to spread the non-dominated individuals uniformly throughout the Pareto-front using various types of fitness sharing and niching scheme as discussed in [24]. This paper proposes an improved alternate measure of uniform distribution (*UD*) for non-dominated population. Mathematically, $UD(X')$ for a given set of non-dominated individuals X' in a population X, where $X' \subseteq X$, is defined as,

$$UD(X') = \frac{1}{1+S_{nc}} \qquad (6)$$

where S_{nc} is the standard deviation of niche count of the overall set of non-dominated individuals X', and is formulated as,

$$S_{nc} = \sqrt{\frac{\sum_{i}^{N_{x'}}\left(nc(x'_i) - \overline{nc(X')}\right)}{N_{x'} - 1}} \qquad (7)$$

where $N_{x'}$ is the size of the set X'; $nc(x'_i)$ is the niche count of i-th individual x'_i where $x'_i \in X'$; and $\overline{nc(X')}$ is the mean value of $nc(x'_i)$, $\forall\ i = 1, 2, \ldots N_{x'}$, as shown in the following equations,

$$nc(x'_i) = \sum_{j, j \neq i}^{N_{x'}} f(i,j), \quad \text{where } f(i,j) = \begin{cases} 1, & dis(i,j) < \sigma_{share} \\ 0, & \text{else} \end{cases} \qquad (8)$$

$$\overline{nc(X')} = \frac{\sum_{i}^{N_{x'}} nc(x'_i)}{N_{x'}} \qquad (9)$$

where $dis(i, j)$ is the distance between individual i and j in the objective domain. Note that the main advantage of this measure is that the actual Pareto-front is not needed in *a-priori*.

v) *Algorithm Effort (AE)*

The performance in MO optimization is often evaluated not only in terms of how good the final Pareto-front is, but also in terms of the computational time required in obtaining such optimal solutions, which includes the effort of performing genetic operations, ranking, fitness sharing and etc., Generally, the algorithm effort (*AE*) can be defined as the total number of function evaluations N_{eval} over a fixed period of simulation time as given by,

$$AE = \frac{T_{run}}{N_{eval}} \qquad (10)$$

As shown in eqn. 10, for a fixed period of T_{run}, more number of function evaluations being performed indirectly indicates that less computational effort is required by the optimization algorithm and hence resulting in a smaller AE.

For uniform comparisons, the sharing distance for MOGA, NSGA, NPGA, HLGA as well as the performance measure of *UD* for all methods are set as 0.01 in the normalized space since the population size was set at 100. Since dynamic sharing [5] was used for both MOEA and IMOEA, the sharing distance is computed dynamically at each generation, whereas no distance parameter is needed for SPEA as proposed by Zitzler and Thiele [7]. Tournament selection scheme with tournament size of 2 is used in MOGA, SPEA, MOEA and IMOEA as suggested in their original literature. The Pareto tournament selection scheme with t_{dom} = 10% of the population size is used in NPGA for a tight and complete population distribution as recommended by [1]. In order to guarantee a fair comparison, all algorithms considered are implemented with the same coding scheme, crossover and mutation. Note that each parameter is represented by 3-digit decimal and concatenated to form the chromosomes. In all cases, standard mutation with a probability of 0.01 and standard crossover with two-point crossover and a probability of 0.7 are used. For IMOEA, the parameters of *ppv* and *upbps* are set as 100 in order to distribute the non-dominated individuals with the density of 100 individuals in unit size of normalized trade-off region.

All methods under comparison were implemented with the same common sub-functions using the same programming language in Matlab [25] on an Intel Pentium II 450 MHz computer. Each of the simulation was terminated automatically when a fixed pre-specified simulation period (for each test problem) is reached, in the same platform that is free from other computation or being interrupted by other programs. The period for all algorithms being compared is 180 sec. 30 independent simulation runs have been performed for each method in each test problem so as to study the statistical performance such as consistency and robustness of the methods. Here, a random initial population was created for each of the 30 runs, except SPEA and IMOEA, with an initial population size of 100 for all methods under studied. For SPEA, as according to [7], three combinations of {P, P'}, namely {95, 5}, {70, 30} and {30, 70}, where P + P' = 100 in each case, are used. To test the ability of IMOEA in automatically adapting the population size to discover the Pareto-front, an initial population size of only 10 (instead of 100) was randomly generated for each of the 30 runs in each test problem. The indexes of compared algorithms on each test problem are in the sequence: VEGA, HLGA, NPGA, MOGA, NSGA, SPEA1, SPEA2, SPEA3, MOEA and IMOEA. SPEA 1, 2, and 3 representing the {P, P'} combinations of {95, 5}, {70, 30} and {30, 70}, respectively.

Fig. 6 summarizes the simulation results for the performances of each algorithm in respects to each performance measure. The distribution simulation data of 30 independent runs is represented in box plot format [26], which has been applied by [7] to visualize the distribution of simulation data efficiently. In the respects of the number of function evaluations (N_{eval}) and algorithm effort (*AE*) per run under a fixed CPU time, it can be seen from the first and second row of graphs in Fig. 6 that, VEGA, HLGA and NPGA have a relatively high N_{eval} and low *AE*. This indicates that these algorithms are less computational expensive as compared to others and hence more iterations were being performed per run within the fixed period of CPU time.

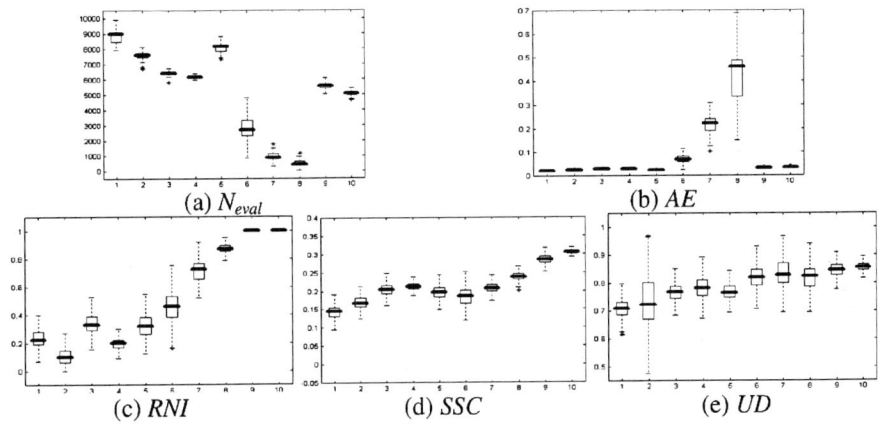

Fig. 6. Box plots based on the measures of N_{eval}, *AE*, *RNI*, *SSC* and *UD*

For the measure of ratio of non-dominated individuals (*RNI*), it is noticed that MOEA and IMOEA has relatively high mean value of *RNI*, which clearly indicates its ability to provide useful non-dominated solutions from a given size of population due to the incorporation of switching preserved strategy (SPS) and fuzzy boundary local perturbation (FBLP), where more offspring are perturbed in the neighborhood of the non-dominated individuals. The value of *RNI* in SPEA is mainly based on the setting of $\{P, P'\}$ where it is noticeable that the higher the ratio of $P'/(P + P')$, the higher the measure of *RNI* is, and as a consequence, the higher the cost of computation is needed in the clustering algorithm of SPEA in order to filter out the extra external individuals.

Concerning the measure of size of space cover (*SSC*), the performance of IMOEA is outstanding. It is also evident that MOEA is the second best in terms of measure of *SSC* for all test problems. In the context of uniform distribution (*UD*) of the non-dominated population, VEGA gives the lowest value of *UD* indicating that it has the low ability to distribute the non-dominated individuals evenly since it doesn't has any operation to take care of the population distribution. Also, IMOEA has shown to be the best in the measure of *UD*, followed by MOEA with a good performance slightly below the IMOEA. This difference is probably due to the additional operation of FBLP in IMOEA that provides the local fine-tuning to discover any gaps or missing trade-offs at each generation. Besides IMOEA and MOEA, SPEA has also shown to be performing well in the measure of *UD*.

The performance measures of $C(X_i, X_j)$ for the comparison sets between algorithms i and j where, $i, j = 1,2, ..., 10$ on test problem 1 and, $i, j = 1,2, ... 8$ on test problem 2 and 3, are shown in Fig. 7. Again, box plots are used to summarize the sample distributions of 30 independent runs per each case. The ranges of *y*- and *x*- axis of each graph are [0, 1] and [1, 10]. It is noticeable (concerning the rectangles $C(X_{10}, X_{1-10})$) that IMOEA dominate other algorithms most obviously as compared to other algorithms in the respective test problems. Also, IMOEA is dominated the least by any other algorithms in all rectangles.

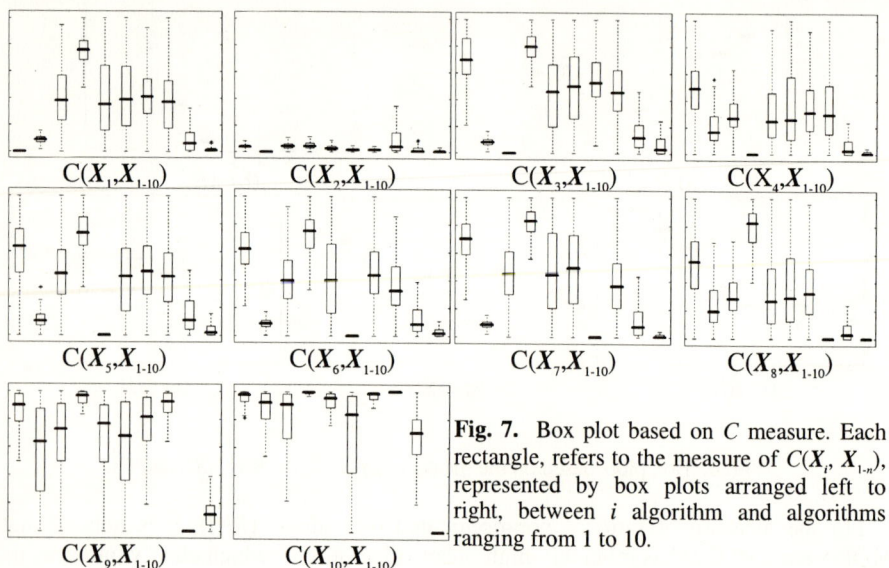

Fig. 7. Box plot based on C measure. Each rectangle, refers to the measure of $C(X_i, X_{1-n})$, represented by box plots arranged left to right, between i algorithm and algorithms ranging from 1 to 10.

Figs. 8 unveils the distribution of non-dominated individuals in the objective domain. These distributions are best selected, among the 30 independent runs, in respect to the measures of *SSC*. In general, the purpose of producing these figures are to visually inspect the performances of various algorithms in terms of the population distribution at the final generation since the quantitative performance measure alone may not provide enough information for full observation and understanding.

4. Conclusions

A novel incrementing multi-objective evolutionary algorithm that has a dynamic population size based upon the on-line discovered Pareto-front and its desired population distribution density has been proposed. The algorithm implements the concept of Pareto optimal domination with adaptive niche induction technique to evolve a set of non-dominated individuals uniformly distributing along the trade-off surface. In addition, the method of fuzzy boundary local perturbation with dynamic local fine-tuning is incorporated to achieve broader neighborhood explorations as well as to eliminate any gaps or discontinuities along the Pareto-front for better convergence and trade-offs representations. Furthermore, extensive quantitative comparisons between IMOEA and other well-known MO optimization methods on benchmark problems have been performed in this paper. Results obtained show that IMOEA has performed well in searching and maintaining the non-dominated solutions to be uniformly distributed along the global Pareto-front, with no significant computational effort needed as compared to others.

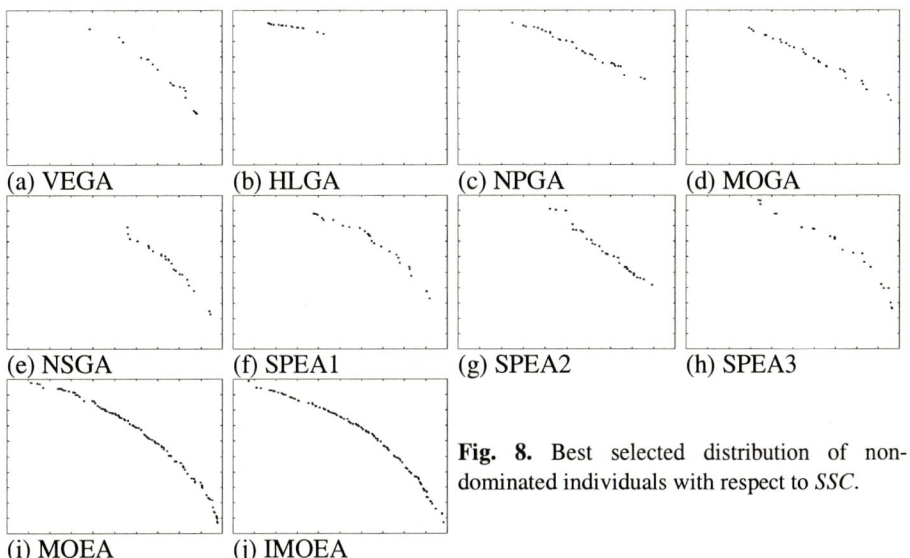

Fig. 8. Best selected distribution of non-dominated individuals with respect to *SSC*.

References

1. Horn, J., Nafpliotis, N. and Goldberg, D.E.: A Niched Pareto Genetic Algorithm for Multiobjective Optimisation. Proceeding of First IEEE Conference on Evolutionary Computation, IEEE World Congress on Computational Intelligence (1994), vol. 1, 82-87.
2. Srinivas, N. and Deb, K.: Multiobjective Optimization using Nondominated Sorting in Genetic Algorithms. Evolutionary Computation, MIT Press Journals (1994), vol. 2(3), 221-248.
3. Fonseca, C.M.: Multiobjective Genetic Algorithms with Application to Control Engineering Problems. Ph.D. Thesis, Dept. Automatic Control and Systems Eng., University of Sheffield, Sheffield, UK (1995).
4. Coello Coello, C.A.: An Updated Survey of GA-based Multiobjective Optimization Techniques. Technical Report: Lania-RD-98-08, Laboratorio Nacional de Informatica Avanzada (LANIA), Xalapa, Veracruz, Mexico (1998).
5. Tan, K.C., Lee, T.H. and Khor, E.F.: Evolutionary Algorithms with Goal and Priority Information for Multi-Objective Optimization. IEEE Proceedings of the Congress on Evolutionary Computation, Washington, D.C, USA (1999), vol. 1, 106-113.
6. Van Veldhuizen, D.A. and Lamont G.B.: Multiobjective Evolutionary Algorithm Test Suites. Symposium on Applied Computing, San Antonio, Texas (1999), 351-357.

7. Zitzler, E., and Thiele, L.: Multiobjective Evolutionary Algorithms: A Comparative Case Study and the Strength Pareto Approach. IEEE Transactions on Evolutionary Computation (1999), vol. 3(4), 257-271.
8. Arabas, J., Michalewicz, Z., and Mulawka, J.: GAVaPS-A Genetic Algorithm with Varying Population Size. Proceedings of the First Conference on Evolutionary Computation (1994), vol. 1, 73-74.
9. Schaffer, J.D.: Multiple-Objective Optimization using Genetic Algorithm. Proceedings of the First International Conference on Genetic Algorithms (1985), 93-100.
10. Goldberg, D.E.: Genetic Algorithms in Search, Optimisation and Machine Learning. Addison-Wesley, Reading, Masachusetts (1989).
11. Alander, J.T.: On Optimal Population Size of Genetic Algorithms. IEEE Proceedings on Computer Systems and Software Engineering (1992), 65-70.
12. Odetayo, M.O.: Optimal Population Size for Genetic Algorithms: An Investigation. Proceedings of the IEE Colloquium on Genetic Algorithms for Control Systems, London (1993), 2/1-2/4.
13. Sasaki, T., Hsu, C.C., Fujikawa, H., and Yamada, S.I.: A Multi-Operator Self-Tuning Genetic Algorithm for Optimization. 23rd International Conference on Industrial Electronics (1997), vol. 3, 1034-1039.
14. Grefenstette, J.J.: Optimization of Control Parameters for Genetic Algorithms. IEEE Transactions on Systems, Man and Cybernetics (1986), vol. 16(1), 122-128.
15. Smith, R.E.: Adaptively Resizing Populations: An Algorithm and Analysis. In: Forrest, S. (ed.): Proceeding of the Fifth International Conference on Genetic Algorithms, Morgan Kaufmann Publishers, Los Altos, CA (1993), 653.
16. Zhuang, N., Benten, M.S., and Cheung, P.Y.: Improved Variable Ordering of BDDS with Novel Genetic Algorithm. IEEE International Symposium on Circuits and Systems (1996), vol. 3, 414-417.
17. Khor, E.F., Tan, K.C., Wang, M.L. and Lee, T.H.: Evolutionary Algorithm with Dynamic Population Size for Multi-Objective Optimization. Accepted by Third Asia Conference on Simulated Evolution and Learning (SEAL2000) (2000), 2768-2772.
18. Horn, J. and Nafpliotis, N.: Multiobjective Optimization Using the Niche Pareto Genetic Algorithm. IlliGAL Report 93005, University of Illinois at Urbana-Champain, Urbana, Illinois, USA (1993).
19. Dengiz, B., Altiparmak, F., and Smith, A.E.: Local Search Genetic Algorithm for Optimal Design of Reliable Networks. IEEE Transactions on Evolutionary Computation (1997), vol. 1(3), 179-188.
20. Liong, S.Y., Khu, S.T., and Chan, W.T.: Novel Application of Genetic Algorithm and Neural Network in Water Resources: Development of Pareto Front. Eleventh Congress of the IAHR-APD, Yogyakarta, Indonesia (1998), 185-194.
21. Hagiwara, M.: Pseudo-hill Climbing Genetic Algorithm (PHGA) for Function Optimization. Proceedings of the International Conference on Neural Networks (1993), vol. 1, 713-716.
22. Hajela, P., and Lin, C.Y.: Genetic Search Strategies in Multicriterion Optimal Design. Journal of Structural Optimization (1992), vol. 4, 99-107.

23. Fonseca, C.M. and Fleming, P.J.: Genetic Algorithm for Multiobjective Optimization, Formulation, Discussion and Generalization. In: Forrest, S. (ed.): Genetic Algorithms: Proceeding of the Fifth International Conference. Morgan Kaufmann, San Mateo, CA (1993), 416-423.
24. Sareni, B., and Krähenbühl, L.: Fitness Sharing and Niching Methods Revisited. IEEE Transactions on Evolutionary Computation (1998), vol. 2(3), 97-106.
25. 25. The Math Works, Inc.: *Using MATLAB*, The Math Works Inc. (1998), version 5.
26. Chambers, J.M., Cleveland, W.S., Kleiner, B., and Turkey, P.A.: Graphical Methods for Data Analysis. Wadsworth & Brooks/Cole, Pacific CA (1983).

A Micro-Genetic Algorithm for Multiobjective Optimization

Carlos A. Coello Coello[1] and Gregorio Toscano Pulido[2]

[1] CINVESTAV-IPN
Depto. de Ingeniería Eléctrica
Sección de Computación
Av. Instituto Politécnico Nacional No. 2508
Col. San Pedro Zacatenco
México, D. F. 07300
ccoello@cs.cinvestav.mx
[2] Maestría en Inteligencia Artificial
LANIA-Universidad Veracruzana
Xalapa, Veracruz, México 91090
gtoscano@mia.uv.mx

Abstract. In this paper, we propose a multiobjective optimization approach based on a micro genetic algorithm (micro-GA) which is a genetic algorithm with a very small population (four individuals were used in our experiment) and a reinitialization process. We use three forms of elitism and a memory to generate the initial population of the micro-GA. Our approach is tested with several standard functions found in the specialized literature. The results obtained are very encouraging, since they show that this simple approach can produce an important portion of the Pareto front at a very low computational cost.

1 Introduction

A considerable number of evolutionary multiobjective optimization techniques have been proposed in the past [3,20]. However, until recently, little emphasis had been placed on efficiency issues. It is well known that the two main processes that consume most of the running time of an EMOO algorithm are: the ranking of the population (i.e., the comparisons required to determine non-dominance) and the mechanism to keep diversity (evolutionary algorithms tend to converge to a single solution because of stochastic noise; it is therefore necessary to avoid this with an additional mechanism).

Recent research has shown ways of improving the efficiency of an EMOO technique (see for example [14,6]). The main emphasis has been on using an external file that stores nondominated vectors found during the evolutionary process which are reinserted later in the population (this can be seen as a form of elitism in the context of multiobjective optimization [10,17,22]).

Following the same line of thought of this current research, we decided to develop an approach in which we would use a GA with a very small population size and a reinitialization process (the so-called micro-GA) combined with an external file to store nondominated vectors previously found. Additionally, we decided to include an efficient mechanism to keep diversity (similar to the adaptive grid method of Knowles & Corne [14]). Our motivation was to show that a micro-GA carefully designed is sufficient to generate the Pareto front of a multi-objective optimization problem. Such approach not only reduces the amount of comparisons required to generate the Pareto front (with respect to traditional EMOO approaches based on Pareto ranking), but also allows us to control the amount of points that we wish to obtain from the Pareto front (such amount is in fact a parameter of our algorithm).

2 Related Work

The term micro-genetic algorithm (micro-GA) refers to a small-population genetic algorithm with reinitialization. The idea was suggested by some theoretical results obtained by Goldberg [8], according to which a population size of 3 was sufficient to converge, regardless of the chromosomic length. The process suggested by Goldberg was to start with a small randomly generated population, then apply to it the genetic operators until reaching *nominal convergence* (e.g., when all the individuals have their genotypes either identical or very similar), and then to generate a new population by transferring the best individuals of the converged population to the new one. The remaining individuals would be randomly generated.

The first to report an implementation of a micro-GA was Krishnakumar [15], who used a population size of 5, a crossover rate of 1 and a mutation rate of zero. His approach also adopted an elitist strategy that copied the best string found in the current population to the next generation. Selection was performed by holding 4 competitions between strings that were adjacent in the population array, and declaring to the individual with the highest fitness as the winner. Krishnakumar [15] compared his micro-GA against a simple GA (with a population size of 50, a crossover rate of 0.6 and a mutation rate of 0.001). Krishnakumar [15] reported faster and better results with his micro-GA on two stationary functions and a real-world engineering control problem (a wind-shear controller task). After him, several other researchers have developed applications of micro-GAs [13, 7,12,21]. However, to the best of our knowledge, the current paper reports the first attempt to use a micro-GA for multiobjective optimization, although some may argue that the multi-membered versions of PAES can be seen as a form of micro-GA[1] [14]. However, Knowles & Corne [14] concluded that the addition of a population did not, in general, improve the performance of PAES, and increased the computational overhead in an important way. Our technique, on the other hand, uses a population and traditional genetic operators and, as we will show in a further section, it performs quite well.

---

[1] We recently became aware of the fact that Jaszkiewicz [11] proposed an approach in which a small population initialized from a large external memory and utilized it for a short period of time. However, to the best of our knowledgem this approach has been used only for multiobjective combinatorial optimization.

3 Description of the Approach

In this paper, we propose a micro-GA with two memories: the **population memory**, which is used as the source of diversity of the approach, and the **external memory**, which is used to archive members of the Pareto optimal set. Population memory is respectively divided in two parts: a **replaceable** and a **non-replaceable** portion (the percentages of each can be regulated by the user).

The way in which our technique works is illustrated in Fig. 1. First, an initial random population is generated. This population feeds the population memory, which is divided in two parts as indicated before. The non-replaceable portion of the population memory will never change during the entire run and is meant to provide the required diversity for the algorithm. The initial population of the micro-GA at the beginning of each of its cycles is taken (with a certain probability) from both portions of the population memory as to allow a greater diversity.

During each cycle, the micro-GA undergoes conventional genetic operators: tournament selection, two-point crossover, uniform mutation, and elitism (regardless of the amount of nondominated vectors in the population only one is arbitrarily selected at each generation and copied intact to the following one). After the micro-GA finishes one cycle (i.e., when nominal convergence is achieved), we choose two nondominated vectors[2] from the final population (the first and last) and compare them with the contents of the external memory (this memory is initially empty). If either of them (or both) remains as nondominated after comparing against the vectors in this external memory, then they are included there. All the dominated vectors are eliminated from the external memory. These two vectors are also compared against two elements from the replaceable portion of the population memory (this is done with a loop, so that each vector is only compared against a single position of the population memory). If either of these vectors dominates to its match in the population memory, then it replaces it. The idea is that, over time, the replaceable part of the population memory will tend to have more nondominated vectors. Some of them will be used in the initial population of the micro-GA to start new evolutionary cycles.

Our approach uses three types of elitism. The first is based on the notion that if we store the nondominated vectors produced from each cycle of the micro-GA, we will not lose any valuable information obtained from the evolutionary process. The second is based on the idea that if we replace the population memory by the nominal solutions (i.e., the best solutions found when nominal convergence is reached), we will gradually converge, since crossover and mutation will have a higher probability of reaching the true Pareto front of the problem over time. This notion was hinted by Goldberg [8]. Nominal convergence, in our case, is defined in terms of a certain (low) number of generations (typically, two to five in our case). However, similarities among the strings (either at the phenotypical

---

[2] This is assuming that we have two or more nondominated vectors. If there is only one, then this vector is the only one selected.

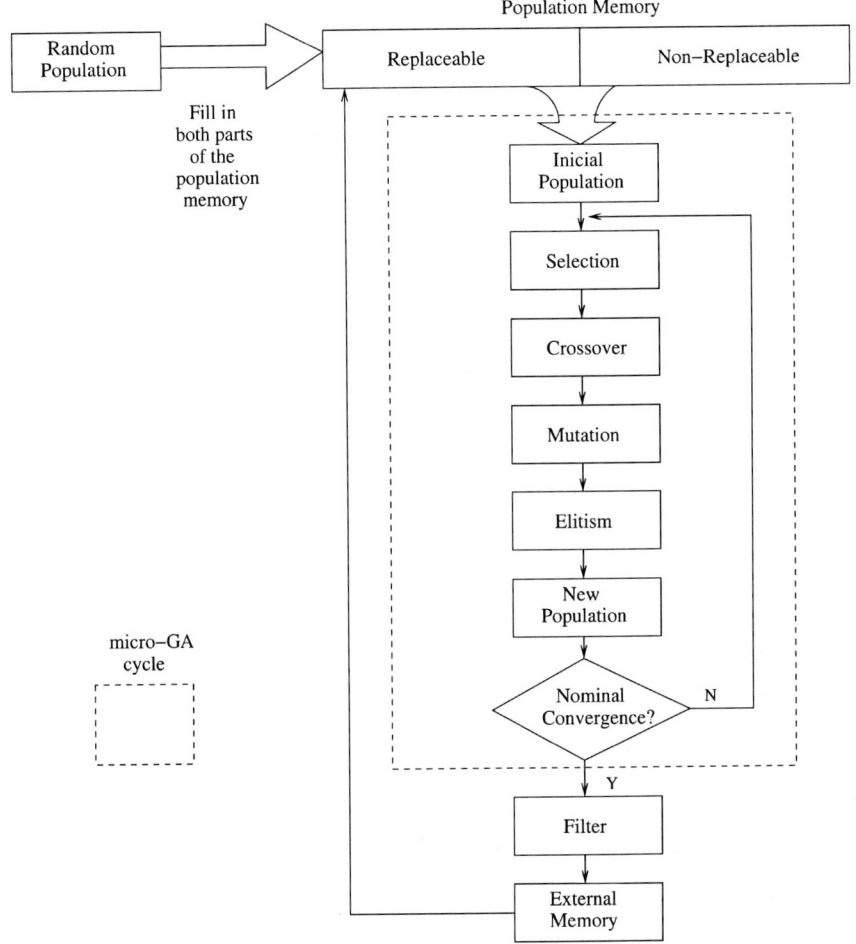

Fig. 1. Diagram that illustrates the way in which our micro-GA works.

or genotypical level) could also been used as a criterion for convergence. The third type of elitism is applied at certain intervals (defined by a parameter called "replacement cycle"). What we do is to take a certain amount of points from all the regions of the Pareto front generated so far and we use them to fill in the replaceable memory. Depending on the size of the replaceable memory, we choose as many points from the Pareto front as necessary to guarantee a uniform distribution. This process intends to use the best solutions generated so far as the starting point for the micro-GA, so that we can improve them (either by getting closer to the true Pareto front or by getting a better distribution). This also avoids that the contents of the replaceable memory becomes homogeneous.

To keep diversity in the Pareto front, we use an approach similar to the adaptive grid proposed by Knowles & Corne [14]. The idea is that once the archive that stores nondominated solutions has reached its limit, we divide the search space that this archive covers, assigning a set of coordinates to each solution. Then, each newly generated nondominated solution will be accepted only if the geographical location to where the individual belongs is less populated than the most crowded location. Alternatively, the new nondominated solution could also be accepted if the individual belongs to a location outside the previously speficied boundaries. In other words, the less crowded regions are given preference so that the spread of the individuals on the Pareto front can be more uniform.

The pseudo-code of the algorithm is the following:

function Micro-GA
  begin
    Generate starting population P of size N
    and store its contents in the population memory M
    /* Both portions of M will be filled with random solutions */
    i=0
    while i < Max **do**
      begin
        Get the initial population for the micro-GA (P_i) from M
        repeat
          begin
            Apply binary tournament selection
            based on nondominance
            Apply two-point crossover and uniform mutation
            to the selected individuals
            Apply elitism (retain only one nondominated vector)
            Produce the next generation
          end
        until nominal convergence is reached
        Copy two nondominated vectors from P_i
        to the external memory E
        if E is full when trying to insert ind_b
          then adaptive_grid(ind_b)
        Copy two nondominated vectors from P_i to M
        if i **mod** replacement_cycle
          then apply second form of elitism
        i=i+1
    end while
  end function

The adaptive grid requires two parameters: the expected size of the Pareto front and the amount of positions in which we will divide the solution space for each objective. The first parameter is defined by the size of the external memory and it is provided by the user. The second parameter (the amount of positions in which we will divide the solution space) has to be provided by the user as well, although we have found that our approach is not very sensitive to it (e.g., in most of our experiments a value of 15 or 25 provided very similar results). The process of determining the location of a certain individual has a low computational cost (it is based on the values of its objectives as indicated before). However, when the individual is out of range, we have to relocate all the positions. Nevertheless, this last situation does not occur too often, and we allocate a certain amount of extra room in the first and last locations of the grid to minimize the occurrence of this situation.

When the external memory is full, then we use the adaptive grid to decide what nondominated vectors will be eliminated. The adaptive grid will try to balance the amount of individuals representing each of the elements of the Pareto set, so that we can get a uniform spread of points along the Pareto front.

4 Comparison of Results

Several test functions were taken from the specialized literature to compare our approach. In all cases, we generated the true Pareto fronts of the problems using exhaustive enumeration (with a certain granularity) so that we could make a graphical comparison of the quality of the solutions produced by our micro-GA.

Since the main aim of this approach has been to increase efficiency, we additionally decided to compare running times of our micro-GA against those of the NSGA II [6] and PAES [14]. In the following examples, the NSGA was run using a population size of 100, a crossover rate of 0.8 (using SBX), tournament selection, and a mutation rate of 1/vars, where vars = number of decision variables of the problem. In the following examples, PAES was run using a depth of 5, a size of the archive of 100, and a mutation rate of 1/bits, where bits refers to the length of the chromosomic string that encodes the decision variables. The amount of fitness function evaluations was set such that the NSGA II, PAES and the micro-GA could reasonably cover the true Pareto front of each problem.

4.1 Test Function 1

Our first example is a two-objective optimization problem defined by Deb [5]:

$$\text{Minimize } f_1(x_1, x_2) = x_1 \qquad (1)$$

$$\text{Minimize } f_2(x_1, x_2) = \frac{g(x_2)}{x_1} \qquad (2)$$

where:

$$g(x_2) = 2.0 - \exp\left\{-\left(\frac{x_2 - 0.2}{0.004}\right)^2\right\} - 0.8\exp\left\{-\left(\frac{x_2 - 0.6}{0.4}\right)^2\right\} \quad (3)$$

and $0.1 \leq x_1 \leq 1.0$, $0.1 \leq x_2 \leq 1.0$.

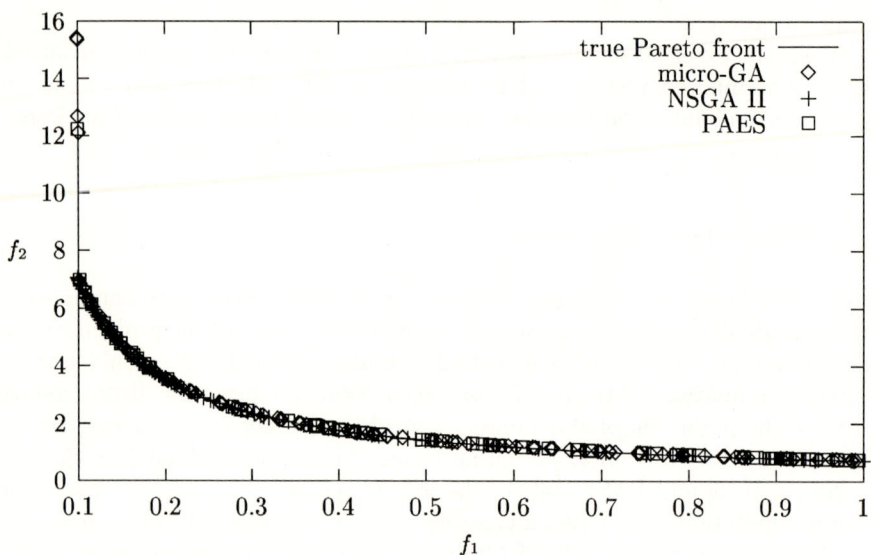

Fig. 2. Comparison of results for the first test function.

The parameters used by the micro-GA for this example were: size of the external memory = 100, size of the population memory = 50, number of iterations = 1500, number of iterations of the micro-GA (to achieve nominal convergence) = 2, number of subdivisions of the adaptive grid = 25, crossover rate = 0.7, mutation rate = 0.029, percentage of non-replaceable memory = 0.3, population size (of the micro-GA) = 4, replacement cycle at every 50 iterations.

Our first test function has a local Pareto front to which a GA can be easily attracted. Fig. 2 shows the true Pareto front for this problem with a continuous line, and the results found by the NSGA II, PAES and our micro-GA are shown with points. Similar fronts were found by the three approaches. For this example, both the NSGA II and PAES performed 12,000 evaluations of the fitness function. The average running time of each algorithm (over 20 runs) were the following: 2.601 seconds for the NSGA II (with a standard deviation of 0.33555913), 1.106 seconds for PAES (with a standard deviation of 0.25193672) and only 0.204 seconds for the micro-GA (with a standard deviation of 0.07764461).

4.2 Test Function 2

Our second example is a two-objective optimization problem proposed by Schaffer [18] that has been used by several researchers [19,1]:

$$\text{Minimize } f_1(x) = \begin{cases} -x & \text{if } x \leq 1 \\ -2 + x & \text{if } 1 < x \leq 3 \\ 4 - x & \text{if } 3 < x \leq 4 \\ -4 + x & \text{if } x > 4 \end{cases} \quad (4)$$

$$\text{Minimize } f_2(x) = (x - 5)^2 \quad (5)$$

and $-5 \leq x \leq 10$.

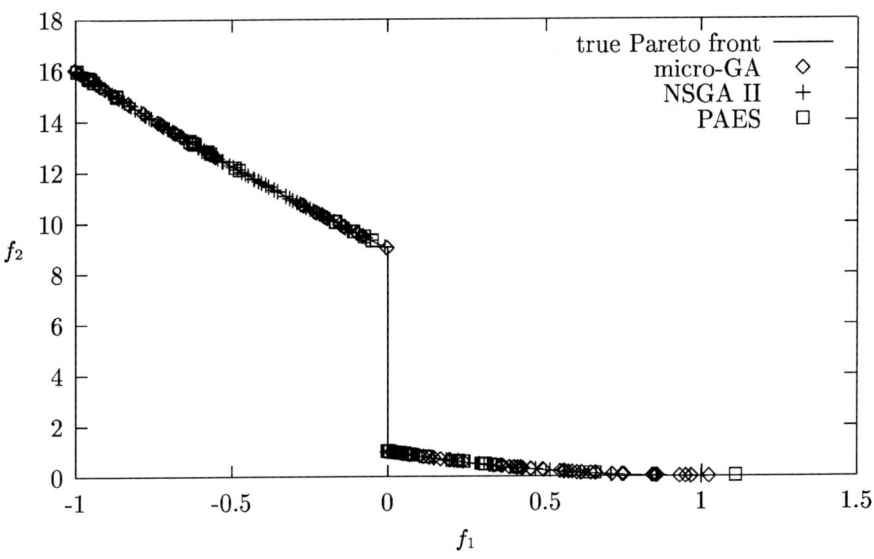

Fig. 3. Comparison of results for the second test function.

The parameters used for this example were: size of the external memory = 100, size of the population memory = 50, number of iterations = 150, number of iterations of the micro-GA (to achieve nominal convergence) = 2, number of subdivisions of the adaptive grid = 25, crossover rate = 0.7, mutation rate = 0.056 (1/L, where L=18 bits in this case), percentage of non-replaceable memory = 0.3, population size (of the micro-GA) = 4, replacement cycle at every 25 iterations.

This problem has a Pareto front that is disconnected. Fig. 3 shows the true Pareto front for this problem with a continuous line (the vertical line is obviously not part of the true Pareto front, but it appears because we used linear segments to connect every pair of nondominated points). We used points to represent the solutions found by the NSGA II, PAES and our micro-GA.

Again, similar Pareto fronts were found by the three approaches. For this example, both the NSGA II and PAES performed 1,200 evaluations of the fitness function. The average running time of each algorithm (over 20 runs) were the following: 0.282 seconds for the NSGA II (with a standard deviation of 0.00014151), 0.107 seconds for PAES (with a standard deviation of 0.13031718) and only 0.017 seconds for the micro-GA (with a standard deviation of 0.0007672).

4.3 Test Function 3

Our second example is a two-objective optimization problem defined by Deb [5]:

$$\text{Minimize } f_1(x_1, x_2) = x_1 \tag{6}$$

$$\text{Minimize } f_2(x_1, x_2) = g(x_1, x_2) \cdot h(x_1, x_2) \tag{7}$$

where:

$$g(x_1, x_2) = 11 + x_2^2 - 10 \cdot \cos(2\pi x_2) \tag{8}$$

$$h(x_1, x_2) = \begin{cases} 1 - \sqrt{\frac{f_1(x_1, x_2)}{g(x_1, x_2)}} & \text{if } f_1(x_1, x_2) \leq g(x_1, x_2) \\ 0 & \text{otherwise} \end{cases} \tag{9}$$

and $0 \leq x_1 \leq 1$, $-30 \leq x_2 \leq 30$.

This problem has 60 local Pareto fronts. Fig. 4 shows the true Pareto front for this problem with a continuous line. The results obtained by the NSGA II, PAES and our micro-GA are displayed as points.

The parameters used by the micro-GA for this example were: size of the external memory = 100, size of the population memory = 50, number of iterations = 700, number of iterations of the micro-GA (to achieve nominal convergence) = 4, number of subdivisions of the adaptive grid = 25, crossover rate = 0.7, mutation rate = 0.029, percentage of non-replaceable memory = 0.3, population size (of the micro-GA) = 4, replacement cycle at every 50 iterations.

Once again, the fronts produced by the three approaches are very similar. For this example, both the NSGA II and PAES performed 11,200 evaluations of the fitness function. The average running time of each algorithm (over 20

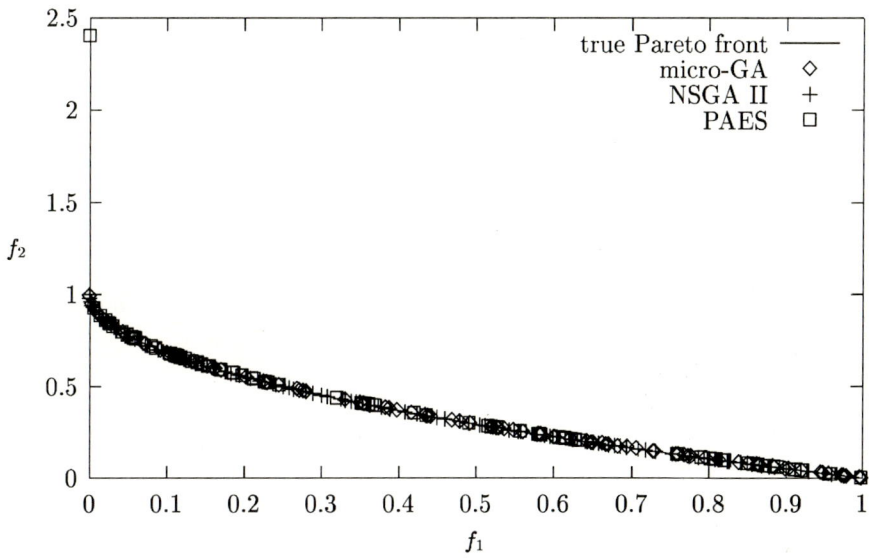

Fig. 4. Comparison of results for the third test function.

runs) were the following: 2.519 seconds for the NSGA II (with a standard deviation of 0.03648403), 2.497 seconds for PAES (with a standard deviation of 1.03348519) and only 0.107 seconds for the micro-GA (with a standard deviation of 0.00133949).

4.4 Test Function 4

Our fourth example is the so-called "unitation versus pairs" problem [9], which involves the maximization of two functions over bit strings. The first function, f_1 is the number of pairs of adjacent complementary bits found in the string, and the second function, f_2 is the numbers of ones found in the string. The Pareto front in this case is discrete. We used a string length of 28, and therefore, the true Pareto front is composed of 15 points.

The parameters used for this example were: size of the external memory = 100, size of the population memory = 15, number of iterations = 1250, number of iterations of the micro-GA (to achieve nominal convergence) = 1, number of subdivisions of the adaptive grid = 3, crossover rate = 0.5, mutation rate = 0.035, percentage of non-replaceable memory = 0.2, population size (of the micro-GA) = 4, replacement cycle at every 25 iterations.

Fig. 5 shows the results obtained by our micro-GA for the fourth test function. A total of 13 (out of 15) elements of the Pareto optimal set were found on average (only occasionally was our approach able to find the 15 target elements). PAES was also able to generate 13 elements of the Pareto optimal set on average, and the NSGA II was only able to generate 8 elements on average.

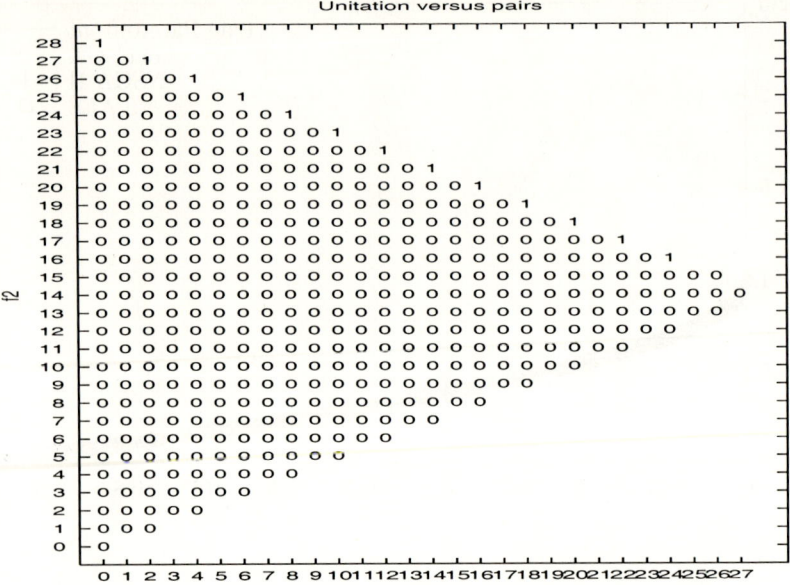

Fig. 5. Results of the micro-GA for the fourth test function.

For this example, both the NSGA II and PAES performed 5,000 evaluations of the fitness function. The average running time of each algorithm (over 20 runs) were the following: 2.207 seconds for the NSGA II, 0.134 seconds for PAES and only 0.042 seconds for the micro-GA.

Borges & Barbosa [2] reported that were able to find the 15 elements of the Pareto optimal set for this problem, using a population size of 100 and 5,000 evaluations of the fitness function, although no actual running times of their approach were reported.

4.5 Test Function 5

Our fifth example is a two-objective optimization problem defined by Kursawe [16]:

$$\text{Minimize } f_1(x) = \sum_{i=1}^{n-1} \left(-10 \exp\left(-0.2 \sqrt{x_i^2 + x_{i+1}^2} \right) \right) \quad (10)$$

$$\text{Minimize } f_2(x) = \sum_{i=1}^{n} \left(|x_i|^{0.8} + 5\sin(x_i)^3 \right) \quad (11)$$

where:

$$-5 \leq x_1, x_2, x_3 \leq 5 \quad (12)$$

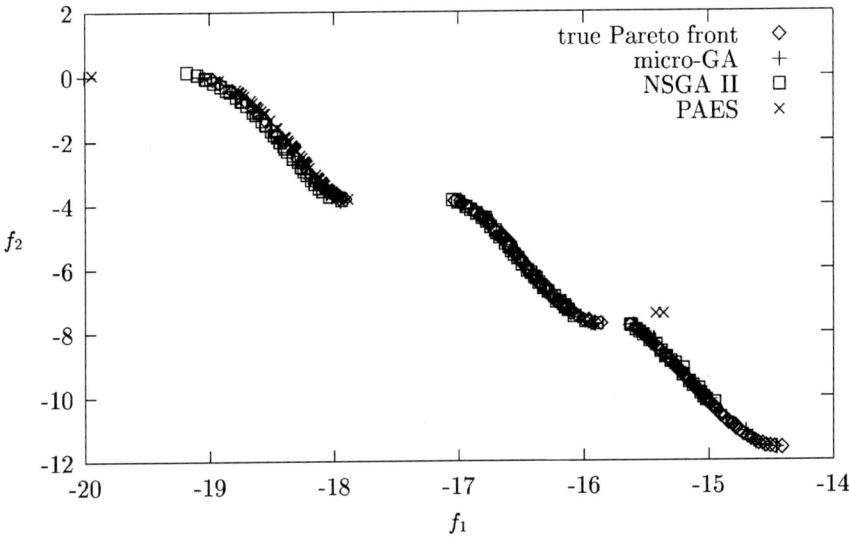

Fig. 6. Comparison of results for the fifth test function.

Fig. 6 shows the true Pareto front for this problem as points. The results obtained by the NSGA II, PAES and our micro-GA are also shown as points. It is worth mentioning that PAES could not eliminate some of the dominated points in the runs performed.

The parameters used for this example were: size of the external memory = 100, size of the population memory = 50, number of iterations = 3000, number of iterations of the micro-GA (to achieve nominal convergence) = 2, number of subdivisions of the adaptive grid = 25, crossover rate = 0.7, mutation rate = 0.019, percentage of non-replaceable memory = 0.3, population size (of the micro-GA) = 4, replacement cycle at every 50 iterations.

For this example, both the NSGA II and PAES performed 2,400 evaluations of the fitness function. The average running time of each algorithm (over 20 runs) were the following: 6.481 seconds for the NSGA II (with a standard deviation of 0.053712), 2.195 seconds for PAES (with a standard deviation of 0.25408319) and only 0.704 seconds for the micro-GA (with a standard deviation of 0.00692099).

5 Conclusions and Future Work

We have introduced an approach that uses a GA with a very small population and a reinitialization process to generate the Pareto front of a multiobjective optimization problem. The technique has a exhibited a low computational cost when compared to the NSGA II and PAES in a few test functions. The approach uses three forms of elitism, including an external file of nondominated vectors and a refilling process that allows us to approach the true Pareto front in a successive manner. Also, we use an adaptive grid (similar to the one used by PAES [14]) that maintains diversity in an efficient way.

The approach still needs more validation (particularly, with MOPs that have more decision variables and constraints), and needs to be compared with other EMOO approaches (under similar conditions) using some of the metrics that have been proposed in the literature (see for example [22]). We only provided running times produced on a PC, but a more exhaustive comparison is obviously lacking. However, the preliminary results presented in this paper, indicate the potential of the approach.

Some other future work will be to refine part of the process, so that we can eliminate some of the additional parameters that the approach needs. Since some of them are not very critical (e.g., the number of grid subdivisions, or the amount of iterations to reach critical convergence), we could probably automatically preset them to a reasonable value so that the user does not need to provide them.

Finally, we are also interested in using this approach as a basis to develop a model of incorporation of preferences from the decision maker [4].

Acknowledgements. We thank the anonymous referees for their comments that helped us improve the contents of this paper. The first author gratefully acknowledges support from CONACyT through project 34201-A. The second author acknowledges support from CONACyT through a scholarship to pursue graduate studies at the Maestría en Inteligencia Artificial of LANIA and the Universidad Veracruzana.

References

1. P. J. Bentley and J. P. Wakefield. Finding Acceptable Solutions in the Pareto-Optimal Range using Multiobjective Genetic Algorithms. In P. K. Chawdhry, R. Roy, and R. K. Pant, editors, *Soft Computing in Engineering Design and Manufacturing*, Part 5, pages 231–240, London, June 1997. Springer Verlag London Limited. (Presented at the 2nd On-line World Conference on Soft Computing in Design and Manufacturing (WSC2)).
2. Carlos C.H. Borges and Helio J.C. Barbosa. A Non-generational Genetic Algorithm for Multiobjective Optimization. In *2000 Congress on Evolutionary Computation*, volume 1, pages 172–179, San Diego, California, July 2000. IEEE Service Center.
3. Carlos A. Coello Coello. A Comprehensive Survey of Evolutionary-Based Multiobjective Optimization Techniques. *Knowledge and Information Systems. An International Journal*, 1(3):269–308, August 1999.
4. Carlos A. Coello Coello. Handling Preferences in Evolutionary Multiobjective Optimization: A Survey. In *2000 Congress on Evolutionary Computation*, volume 1, pages 30–37, Piscataway, New Jersey, July 2000. IEEE Service Center.
5. Kalyanmoy Deb. Multi-Objective Genetic Algorithms: Problem Difficulties and Construction of Test Problems. *Evolutionary Computation*, 7(3):205–230, Fall 1999.

6. Kalyanmoy Deb, Samir Agrawal, Amrit Pratab, and T. Meyarivan. A Fast Elitist Non-Dominated Sorting Genetic Algorithm for Multi-Objective Optimization: NSGA-II. KanGAL report 200001, Indian Institute of Technology, Kanpur, India, 2000.
7. G. Dozier, J. Bowen, and D. Bahler. Solving small and large scale constraint satisfaction problems using a heuristic-based microgenetic algorithm. In *Proceedings of the First IEEE Conference on Evolutionary Computation*, pages 306–311, 1994.
8. David E. Goldberg. Sizing Populations for Serial and Parallel Genetic Algorithms. In J. David Schaffer, editor, *Proceedings of the Third International Conference on Genetic Algorithms*, pages 70–79, San Mateo, California, 1989. Morgan Kaufmann Publishers.
9. Jeffrey Horn, Nicholas Nafpliotis, and David E. Goldberg. A Niched Pareto Genetic Algorithm for Multiobjective Optimization. In *Proceedings of the First IEEE Conference on Evolutionary Computation, IEEE World Congress on Computational Intelligence*, volume 1, pages 82–87, Piscataway, New Jersey, June 1994. IEEE Service Center.
10. Hisao Ishibuchi and Tadahiko Murata. Multi-Objective Genetic Local Search Algorithm. In Toshio Fukuda and Takeshi Furuhashi, editors, *Proceedings of the 1996 International Conference on Evolutionary Computation*, pages 119–124, Nagoya, Japan, 1996. IEEE.
11. Andrzej Jaszkiewicz. Genetic local search for multiple objective combinatorial optimization. Technical Report RA-014/98, Institute of Computing Science, Poznan University of Technology, 1998.
12. E.G. Johnson and M.A.G. Abushagur. Micro-Genetic Algorithm Optimization Methods Applied to Dielectric Gratings. *Journal of the Optical Society of America*, 12(5):1152–1160, 1995.
13. Charles L. Karr. Air-Injected Hydrocyclone Optimization via Genetic Algorithm. In Lawrence Davis, editor, *Handbook of Genetic Algorithms*, pages 222–236. Van Nostrand Reinhold, New York, 1991.
14. Joshua D. Knowles and David W. Corne. Approximating the Nondominated Front Using the Pareto Archived Evolution Strategy. *Evolutionary Computation*, 8(2):149–172, 2000.
15. K. Krishnakumar. Micro-genetic algorithms for stationary and non-stationary function optimization. In *SPIE Proceedings: Intelligent Control and Adaptive Systems*, pages 289–296, 1989.
16. Frank Kursawe. A variant of evolution strategies for vector optimization. In H. P. Schwefel and R. Männer, editors, *Parallel Problem Solving from Nature. 1st Workshop, PPSN I*, volume 496 of *Lecture Notes in Computer Science*, pages 193–197, Berlin, Germany, oct 1991. Springer-Verlag.
17. Geoffrey T. Parks and I. Miller. Selective Breeding in a Multiobjective Genetic Algorithm. In A. E. Eiben, M. Schoenauer, and H.-P. Schwefel, editors, *Parallel Problem Solving From Nature — PPSN V*, pages 250–259, Amsterdam, Holland, 1998. Springer-Verlag.
18. J. David Schaffer. *Multiple Objective Optimization with Vector Evaluated Genetic Algorithms*. PhD thesis, Vanderbilt University, 1984.
19. N. Srinivas and Kalyanmoy Deb. Multiobjective Optimization Using Nondominated Sorting in Genetic Algorithms. *Evolutionary Computation*, 2(3):221–248, fall 1994.
20. David A. Van Veldhuizen and Gary B. Lamont. Multiobjective Evolutionary Algorithms: Analyzing the State-of-the-Art. *Evolutionary Computation*, 8(2):125–147, 2000.

21. Fengchao Xiao and Hatsuo Yabe. Microwave Imaging of Perfectly Conducting Cylinders from Real Data by Micro Genetic Algorithm Coupled with Deterministic Method. *IEICE Transactions on Electronics*, E81-C(12):1784–1792, December 1998.
22. Eckart Zitzler, Kalyanmoy Deb, and Lothar Thiele. Comparison of Multiobjective Evolutionary Algorithms: Empirical Results. *Evolutionary Computation*, 8(2):173–195, Summer 2000.

Evolutionary Algorithms for Multicriteria Optimization with Selecting a Representative Subset of Pareto Optimal Solutions

Andrzej Osyczka[1] and Stanislaw Krenich[1]

[1] Department of Mechanical Engineering
Cracow University of Technology
Al.Jana Pawla II 37, 31-864 Krakow, Poland
{osyczka, krenich}@mech.pk.edu.pl

Abstract. In this paper the method of selecting a representative subset of Pareto optimal solutions is used to make the search of Pareto frontier more effective. Firstly, the evolutionary algorithm method for generating a set of Pareto optimal solutions is described. Then, indiscernibility interval method is applied to select representative subset of Pareto optimal solutions. The main idea of this method consists in removing from the set of Pareto optimal solutions these solutions, which are close to each other in the space of objectives, i.e., those solutions for which the values of the objective functions differ less than an indiscernibility interval. The set of Pareto optimal solutions is reduced using indiscernibility interval method after running a certain number of generations. This process can be called the filtration process in which less important Pareto optimal solutions are removed from the existing set. Finally, two design optimization problems are solved using the proposed method. From these examples it is clear that the computation time can be reduced significantly and still the real Pareto frontier obtained.

1 Introduction

While running evolutionary algorithms for multicriteria optimization a set of Pareto solutions increases with the number of generations. For some problems the final result of running the program is the Pareto set which contains hundreds or even thousands of solutions and some of them might be very close to each other considering the values of the objective functions. For the decision maker it is difficult and tiresome to analyze all these solutions. Moreover, in each generation a new generated solution is compared with all solutions from the Pareto set. If the set of Pareto solution becomes larger the calculation time increases proportionally to the size of this set. Osyczka & Montusiewicz, 1994 while using a simple random search method, noticed this problem. They proposed some methods of selecting a representative subset of Pareto optimal solutions. Montusiewicz, 1999 has developed further one of these methods. While using evolutionary algorithms these methods can be very useful considering both computation time and the decision-making problem. In this paper the method for selecting the representative set of Pareto optimal solutions is described and applied to the evolutionary algorithm to make the search more effective.

2 Method for Generating Pareto Optimal Solutions

Pareto set distribution method is used to generate the set of Pareto optimal solutions. The general idea of this is as follows: Within each new generation a set of Pareto solutions is found on the basis of two sets: the set of Pareto solutions from a previous generation and the set of solutions created by genetic algorithm operations within the considered generation. The new set of Pareto solutions, thus created is distributed randomly to the next generation for a half of the population. The remaining half of the population is bred by randomly chosen strings from the previous generation. The steps of the method are as follows:

Step 1. Let $t = 1$, where t is the index of a generation.

Step 2. Generate an initial population.

Step 3. If $t = 1$ find the set of Pareto solutions. If $t > 1$ find and create the set of Pareto solutions from the following sets: Pareto set from the $t - 1$ generation and all solutions from the t generation.

Step 4. Check if there is any improvement in the Pareto set through the last t^* generations, i.e., if any new Pareto solution was found through the last t^* generations, where t^* is the user's preassigned number. If there is an improvement go to step 5, otherwise, terminate the calculations. Such a situation means that the genetic algorithm has not been able to find a new Pareto optimal solution through the last t^* generations.

Step 5. Substitute $t = t+1$. If $t \leq T$ go to step 6, where T is the preassigned number of generations to be considered. Otherwise, terminate the calculations.

Step 6. Create a new t generation in the following way: for odd strings substitute randomly chosen strings from the Pareto set obtained in the t-1 generation whereas for even strings substitute randomly chosen strings from the $t - 1$ generation. Note that it may happen that some strings may be copied twice but in different places. This refers only to Pareto optimal strings.

Step 7. Perform the genetic algorithm operations (crossover and mutation) on the strings of the whole population, and go to step 3.

Here two types of termination criteria are used. The first, which indicates that there is absolutely no improvement in the Pareto set after running t^* number of generations, is the more important criterion. The other criterion is related to the computational time needed for generating the Pareto solutions, i.e., the method is stopped after the assumed number of generations is executed. The graphical illustration of this method is shown in Fig.1. For the continuous decision variables the method generates a large number of Pareto solutions, which are very close to each others in the space of objective functions. Thus indiscernibility interval method is used to select a representative set of Pareto optimal solutions.

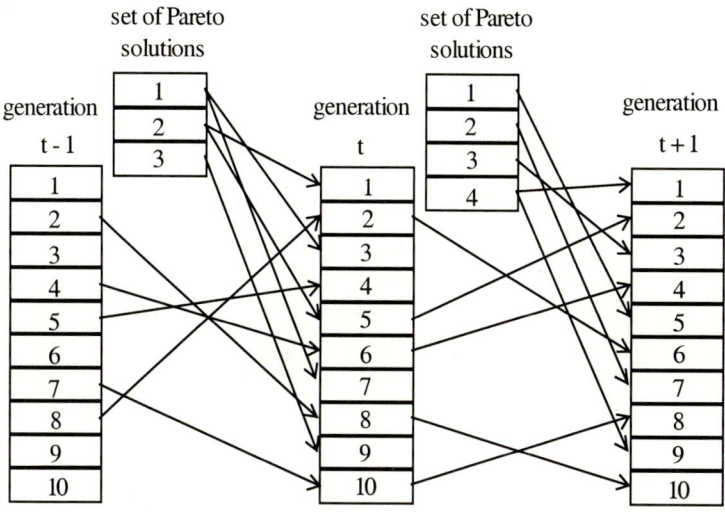

Fig. 1. Graphical illustration of the Pareto Set Distribution Method.

3 Indiscernibility Interval Method

The main idea of this method consists in removing from the set of Pareto optimal solutions these solutions, which are close to each other, i.e., those solutions for which the values of the objective functions differ less than an indiscernibility interval. The method introduces an indiscernibility interval u_i for the i-th objective function. This interval given in percentages defines differences between the values of the objective functions which qualify the solution to be removed from the Pareto set. In other words, if there is more than one solution in the interval u_i one solution remains in the Pareto set and others are discarded. The method is based on the following definition: A solution $x^\wedge \in X$ is a non-dominated solution in the indiscernibility interval sense if and only if there is no solution $x^+ \in X$ such that:

$$\text{for } f_i(x^\wedge) \geq 0: \quad f_i(x^\wedge) < f_i(x^+) \text{ and } f_i(x^\wedge) \cdot \left(1 + \frac{u_i}{100}\right) > f_i(x^+)$$

$$\text{for } f_i(x^\wedge) < 0: \quad f_i(x^\wedge) < f_i(x^+) \text{ and } f_i(x^\wedge) \cdot \left(1 - \frac{u_i}{100}\right) > f_i(x^+)$$

This definition is illustrated graphically in Fig.2 in which the indiscernibility intervals u_1 and u_2 are denoted by shaded areas. In this case for the first objective function solution 16 eliminates solutions 15 and 17 and solution 20 eliminates solution 21. For the second objective function solution 6 eliminates solutions 5 and 6. Montusiewicz, 1999, gives a more detailed description of this method.

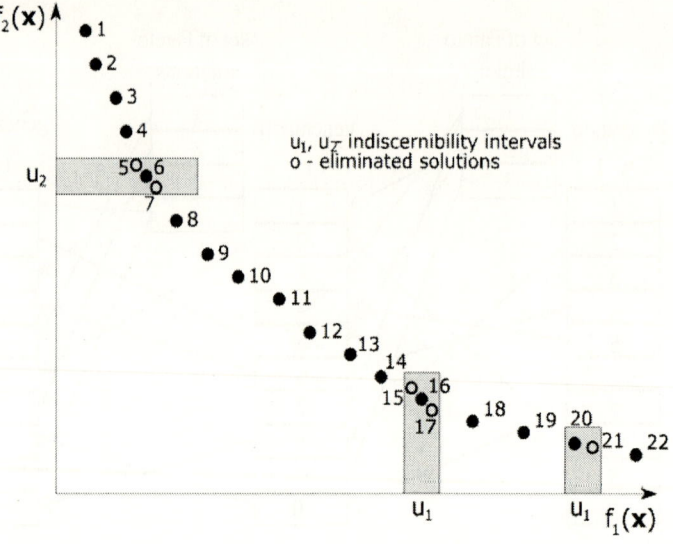

Fig. 2. Graphical illustration of the indiscernibility interval method.

Note that for selecting the representative subset of Pareto optimal solutions the minmax method as presented by Osyczka & Montusiewicz, 1994 can also be applied here. Note also that the method of selecting the representative subset of Pareto optimal solutions may be also used at the end of running any evolutionary algorithm multicriteria optimization method. In this case instead of making the decision on the basis of hundreds of Pareto optimal solutions generated by the evolutionary algorithm the decision maker chooses the most preferable solution from only several solutions, which are evenly distributed over the Pareto frontier.

4 Evolutionary Algorithm Method with Selecting the Representative Subset of Pareto Solutions

The idea of the method consists in reducing the set of Pareto optimal solutions using indiscernibility interval method after running a certain number of generations. This process can be called the filtration process in which less important Pareto optimal solutions are removed from the existing set.

The steps of the method are as follows:

Step 1. Set $t = 1$, where t is the number of the currently run generation.

Step 2. Generate the set of Pareto optimal solutions using any evolutionary algorithm method.

Step 3. Is the criterion for filtration the set of Pareto solutions satisfied? If yes, select the representative subset of Pareto solutions using the indiscernibility interval method and go to step 4. Otherwise, go straight to step 4.

Step 4. Set $t = t + 1$ and if $t \leq T$, where T is the assumed number of generations, go to step 2. Otherwise, terminate the calculations.

Note that if in the step 3 the answer is *yes* we start the process, which can be called the filtrating process since we filtrate and retain in the Pareto set only these solutions which are not close to each other in the space of objectives. Note also that in Step 3 the term *the criterion for filtration* is introduced. Three types of criteria can be used here:

Type 1. The number of solutions in the Pareto set exceeds the assumed number P, for example 100.

Type 2. The number of solutions in the Pareto set is assumed as P. The first filtration is made if the number of solutions in the Pareto set exceeds this number. The following filtration is made when P new Pareto optimal solutions is added to the set.

Type 3. The filtration is made after running the assumed number of generations P and in this case the number of the solutions in the Pareto set is not controlled.

These three types of criteria may produce slightly different results but generally all of them reduce the computation time significantly. The choice of the criterion depends on the problem to be solved. Using these three criteria the choice of P should be made with great care. If P is too small the number of Pareto solutions might not be representative for the problem and the evolutionary algorithm may not reach the real Pareto frontier. If P is too large we lose the effect of reducing the calculation time. Also the choice of the indiscernibility interval u_i is very important. If u_i is too small the number of rejected solutions is also too small and there is no effect in reducing the set of Pareto solutions, whereas too big a value of u_i may make the subset of the obtained solutions to small too be representative.

5 Numerical Examples

5.1 Beam Design Problem

Let us consider the problem of optimum design of the beam the scheme of which is presented in Fig.3. The multicriteria optimization problem is formulated as follows:
The vector of decision variables is:

$$\mathbf{x} = [x_1, x_2, ..., x_N]^T,$$

where x_n is the thickness of the n-th part of the beam.

The objective functions are:
- the volume of the beam

$$f_1(\mathbf{x}) = bl \sum_{n=1}^{N} x_n$$

- displacement under the force F:

$$f_2(\mathbf{x}) = \frac{Fl^3}{2E}\left(\frac{1}{I_1} + \sum_{n=2}^{N} \frac{i^3 - (i-1)^3}{I_i}\right)$$

where

$$I_n = \frac{bx_n^3}{12} \quad \text{for } n = 1, 2, ..., N$$

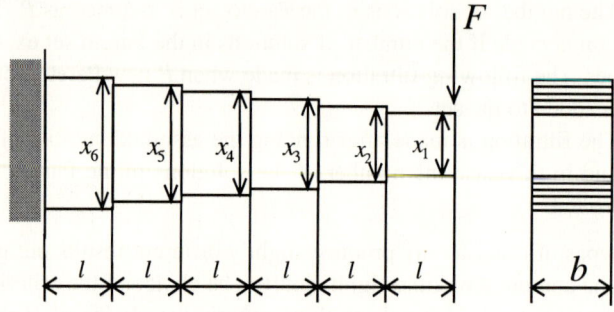

Fig. 3. Scheme of the beam.

The constraints are:

$$\frac{6F \times nl}{bx_n^2} \leq \sigma_g \quad \text{for } n = 1, 2, ..., N$$

$$0 \leq x_1, x_n \leq x_{n+1}, x_N \leq d \quad \text{for } n = 1, 2, ..., N-1$$

The problem was run for the following data:
$N = 6$, $l = 50$[mm], $b = 50$[mm], $F = 10000$[N],

$E = 2.06 \times 10^5$[N/mm^2], $\sigma_g = 360$[N/mm^2], $d = 32$[mm]

The problem was considered as a continuous programming problem with the following ranges of decision variables: $12 < x_i < 32$ [mm] for $i = 1,2...6$
The evolutionary algorithm parameters were as follows:
- Number of generation T=1000
  - Population size $J = 100$
  - Crossover rate $R_c = 0.6$
  - Mutation rate $R_m = 0.08$
  - Penalty rate $r = 10^5$

Experiments were carried out using indiscernibility interval $u_i = 5\%$ for i = 1,2,...,6 and for P = 100. The results of experiments are shown in Fig.4 in which Pareto frontiers for solutions without filtration and with filtration are compared for three types of the filtration criteria. The solutions depicted by black points are obtained while running

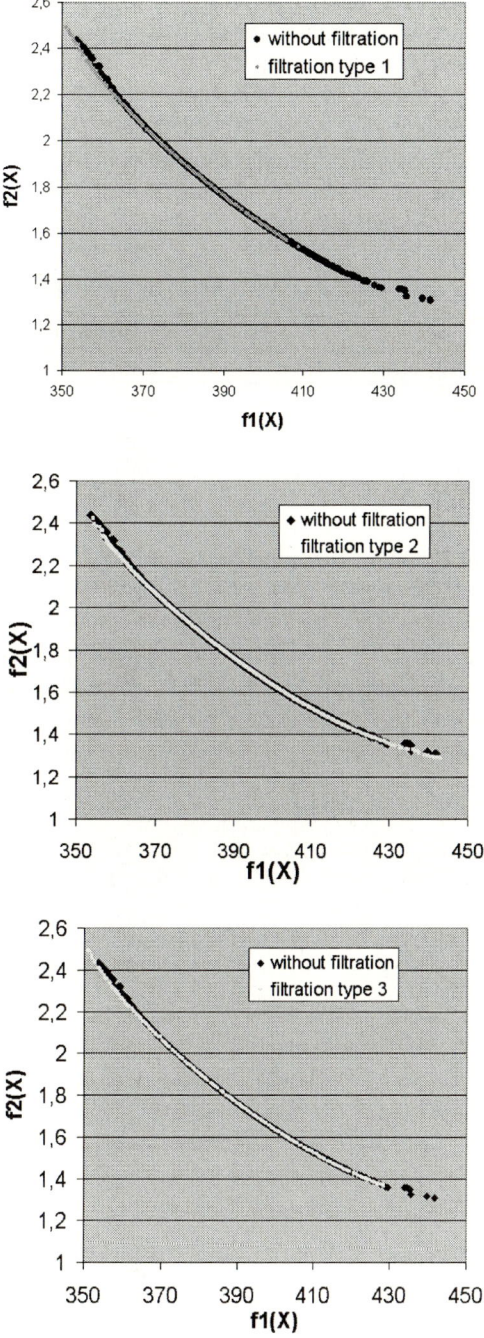

Fig. 4. Sets of Pareto optimal solutions for the beam design problem.

the evolutionary algorithm without filtration process whereas, depicted by almost white points are obtained while running the evolutionary algorithm with the assumed type of filtration. In Table 1 the number of generated Pareto optimal solutions obtained during each experiment and the computing time for each experiment are presented. From these experiments it is clear that using the indiscernibility interval method with the evolutionary algorithm the computation time may be reduced several times and still almost the same Pareto frontier is obtained, which in some parts is even better than the one obtained by the ordinary evolutionary algorithm.

Table 1. Comparison of the results for different filtration types for the beam design problem.

Method	Number of Pareto solution	Time [s]
Without filtration	2007	608
Filtration Type 1	152	29
Filtration Type 2	1102	184
Filtration Type 3	207	34

5.2 Robot Gripper Problem Design

Let us consider an example of the optimum design of a robot gripper as formulated by Krenich and Osyczka, 2000. The scheme of the gripper is presented in Fig.5. For this gripper the vector of decision variables is $\mathbf{x} = [\ a,b,c,e,f,l,\delta]^T$, where a,b,c,e,f,l, are dimensions of the gripper and δ is the angle between b and c elements of the gripper. The geometrical dependencies of the gripper mechanism are (see Fig.6):

$$g = \sqrt{(l-z)^2 + e^2}, \quad \phi = atan\left(\frac{e}{l-z}\right),$$

$$a^2 = b^2 + g^2 - 2 \cdot b \cdot g \cdot \cos(\beta + \phi), \quad b^2 = a^2 + g^2 - 2 \cdot a \cdot g \cdot \cos(\alpha - \phi),$$

$$\alpha = arc\cos\left(\frac{a^2 + g^2 - b^2}{2 \cdot a \cdot g}\right) + \phi, \quad \beta = arc\cos\left(\frac{b^2 + g^2 - a^2}{2 \cdot b \cdot g}\right) - \phi$$

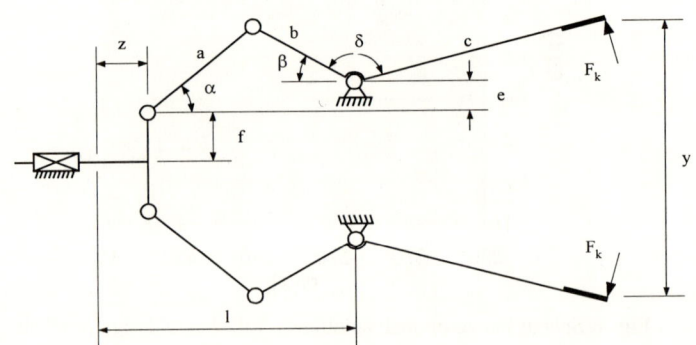

Fig. 5. Scheme of the robot gripper mechanism.

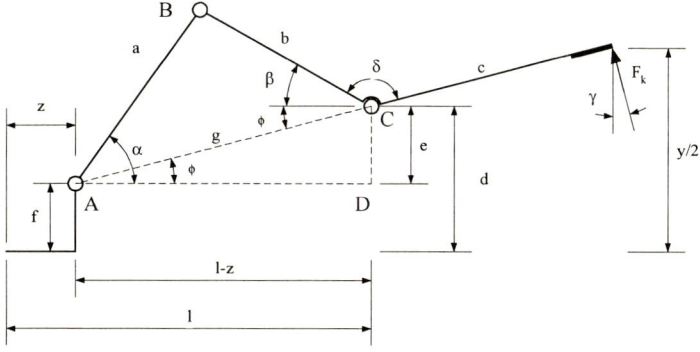

Fig. 6. Geometrical dependencies of the gripper mechanism.

The distribution of the forces is presented in Fig.7 and so we have:

$$R \cdot \sin(\alpha + \beta) = F_k \cdot c, \quad R = \frac{P}{2 \cdot \cos(\alpha)}, \quad F_k = \frac{P \cdot b}{2 \cdot c \cdot \cos(\alpha)} \cdot \sin(\alpha + \beta)$$

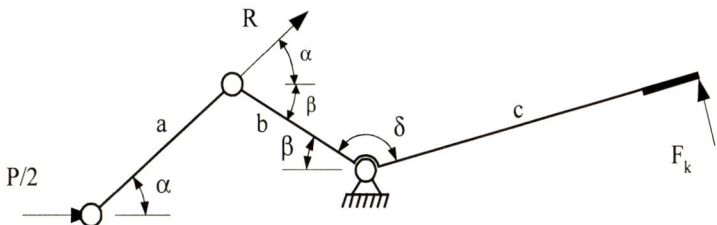

Fig. 7. Distribution of the forces in the mechanism of the griper.

Using the above formulas the objective functions can be evaluated as follows:
- The first objective function is the difference between the maximum and minimum griping forces for the assumed range of gripper ends displacement:

$$f_1(\mathbf{x}) = \max_z F_k(\mathbf{x}, z) - \min_z F_k(\mathbf{x}, z)$$

- The second objective function is the force transmission ratio:

$$f_2(\mathbf{x}) = \frac{P}{\min_z F_k(\mathbf{x}, z)}$$

Note that both objective functions depend on the vector of decision variables and on the displacement z. Thus for the given vector **x** the values of $f_1(\mathbf{x})$ and $f_2(\mathbf{x})$ can be evaluated by using a special procedure which finds the maximum of $F_k(\mathbf{x}, z)$ and the minimum of $F_k(\mathbf{x}, z)$ for different values of z. This procedure makes the objective functions computationally expensive and the problem becomes more complicated than a general nonlinear programming problem.

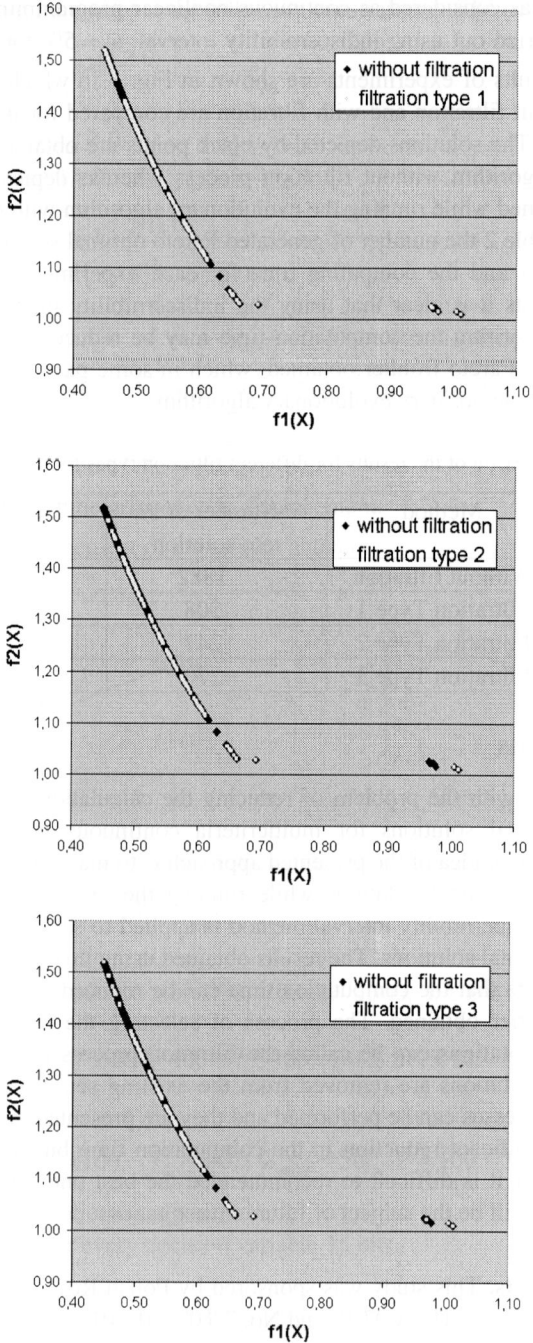

Fig. 9. Sets of Pareto optimal solutions for the gripper design problem.

References

1. Coello C. A. C. (1999) *A Comprehensive Survey of Evolutionary-Based Multiobjective Optimization Techniques*, Knowledge and Information Systems, Vol. 1, No. 3, pp. 269-308.
2. Krenich, S. & Osyczka, A. (2000): Optimization of Robot Grippers Parameters Using Genetic Algorithms. The 13th *CISM-IFToMM Symposium on the Theory and Practice of Robots and Manipulators*. July 3-6, 2000, Zakopane, Poland.
3. Montusiewicz J. (1999) *Division of the Set of Nondominated Solutions by Means of the Undifferentiation Interval Method*, In: The Technological Information Systems, pp. 65-72, Societas Scientiarum Lublinensis, Lublin.
4. Osyczka A., Montusiewicz J. (1994) *A Random Search Approach to Multicriterion Discrete Optimization*, IUTAM Symposium on Discrete Structural Optimization, Springer-Verlag, pp. 71-79.
5. Osyczka A., Tamura H., Saito Y. (1997) *Pareto Set Distribution Method for Multicriteria Design Optimization Using Genetic Algorithm*, In: Proceedings of Engineering Design and Automation Conference, Bangkok, Thailand, pp. 228-231.
6. Osyczka A., Tamura H. (1996) *Pareto Set Distribution Method for Multicriteria Optimization Using Genetic Algorithm*, In: Proceedings of Genetic Algorithms '96, (MENDEL-96)- International Conference, Brno, Czech Republic, pp.135-143.
7. Steuer R.E. (1986) Multiple Criteria Optimization – Theory, Computation and Application, Wiley, New York.

Multi-objective Optimisation Based on Relation *Favour*

Nicole Drechsler, Rolf Drechsler, and Bernd Becker

Institute of Computer Science
Albert-Ludwigs-University
79110 Freiburg im Breisgau, Germany
email: ndrechsler,drechsler,becker@informatik.uni-freiburg.de

Abstract. Many optimisation problems in circuit design, in the following also refereed to as VLSI CAD, consist of mutually dependent sub-problems, where the resulting solutions must satisfy several requirements. Recently, a new model for Multi-Objective Optimisation (MOO) for applications in Evolutionary Algorithms (EAs) has been proposed. The search space is partitioned into so-called Satisfiability Classes (SCs), where each region represents the quality of the optimisation criteria. Applying the SCs to individuals in a population a fitness can be assigned during the EA run. The model also allows the handling of infeasible regions and restrictions in the search space. Additionally, different priorities for optimisation objectives can be modelled. In this paper, the model is studied in further detail. Various properties are shown and advantages and disadvantages are discussed. The relations to other techniques are presented and experimental results are given to demonstrate the efficiency of the model.

1 Introduction

Evolutionary Algorithms (EAs) become more and more important as a tool for search and optimisation. Especially for hard combinatorial problems they often have been applied successfully (see e.g. [Mic94,Dre98]). This type of problem is often encountered in *Very Large Scale Integration CAD* (VLSI CAD), since there often problem instances of several million components have to be considered. Multiple, competing criteria have to be optimised subject to a large number of non-trivial constraints. One strategy is to artificially divide a problem into a number of sub-problems, which are then solved in sequence. Obviously, this is not a promising strategy if the objectives are conflicting. EAs are well suited for solving this kind of problems, when mutually dependent sub-problems are considered in parallel. One problem that arises when using EAs is to evaluate the solutions of a population. Thus it is necessary to determine a ranking of the elements to see which solutions are better than others.

Traditionally, the evaluation is done by an objective function which maps a solution of multiple objectives to a single value. A classical method is the *linear combination by weighted sum*, where the value of each objective is weighted by a constant coefficient. The values of the weights determine how strong the specific objective influences the value of a single fitness value. Disadvantages are that e.g. the weights have to be known in advance to find good solutions or have to be determined by experiments. Obviously, this is time consuming and not desirable, since the parameters resulting from different runs may vary, ending in "in-stable" algorithms [Esb96].

1.1 Previous Work

Advanced methods for ranking solutions with multiple objectives have been developed over the years. If priorities exist between the objectives, a simple *lexicographic* order can be used. (Information on *lexicographic* sorting of vectors can be found in standard mathematical literature.)

In [Gol89] a method is described where solutions with multiple objectives without preferences can be compared. This is realised by a relation, called *dominate*. A solution x *dominates* y, if x is equal or better for each objective than y, and x is for at least one component strongly better than y. Thus, the solutions in the search space can be ranked by the relation *dominate*. This approach of ranking solutions is the core of many EA-tools for *Multi-Objective Optimisation* (MOO) [SD95,FF95,ZT99].

In [EK96] another approach is proposed, where the search space is divided in a *satisfiable, acceptable*, and *invalid* range. This model has successfully been applied to one specific problem in the area of VLSI CAD, but it requires user interaction. The designer has to specify the limits between *satisfiable, acceptable*, and *invalid* solutions. The limits have to be adapted during the program run to obtain high quality results.

Recently, in [DDB99] a new model for ranking solutions of MOO problems has been proposed. A relation *favour* is defined analogously to *dominate* in [Gol89]. The search space is divided into several *Satisfiability Classes* (SCs). Thus the approach can be seen as a generalisation of the approach in [EK96] using a finer granularity.

1.2 Results

In this paper the model from [DDB99] is studied in further detail. The relation *favour* in comparison to other models, like *weighted sum* and *dominate* [Gol89], is shown, e.g. *favour* is able to compare solutions which are not comparable using relation *dominate*.
By this technique no user interaction is required any more. This is a very important aspect in VLSI CAD tools, since they are often so complex that it is very hard for a user to keep control. Furthermore, handling of priorities is also supported. Infeasible solutions are assigned to their own SCs depending on the objectives which are in an infeasible range. The SCs can be efficiently manipulated using operations on graphs. Finally, also experimental evaluations are given.

2 Multi-objective Optimisation Problems

In general, many optimisation problems consist of several mutually dependent sub-problems. MOO problems can be defined as follows: Let Π be the feasible range of solutions in a given search space. The objective function $f_c : \Pi \to \mathbf{R}_+^n$ assigns a cost to each objective of a solution s, where $f_c(s) = (f_{c1}(s), f_{c2}(s), \ldots, f_{cn}(s))$, with $s \in \Pi$, $n \in \mathbf{N}$ and $\mathbf{R}_+$ the positive real valued numbers[1].

To compare several multi-objective solutions *superior points* in the search space can be determined. In the following we restrict to minimisation problems. The relation *dominate* is defined as proposed in [Gol89]. Let $x, y \in \mathbf{R}_+^n$ be the costs of two different solutions. x *dominates* y ($x <_d y$), if $x \neq y$ and y is as large as x in each component. More formally:

Definition 2.1

$x <_d y : \Leftrightarrow (\exists i: f_{ci}(x) < f_{ci}(y)) \wedge (\forall j \neq i: f_{cj}(x) \leq f_{cj}(y))$

$x \leq_d y : \Leftrightarrow (\forall i: f_{ci}(x) \leq f_{ci}(y))$

A non-dominated solution is called a *Pareto-optimal* solution. $\leq_d$ defines the set of all Pareto-optimal solutions, called the *Pareto-set*, and additionally $\leq_d$ is a partial order. All elements $x \in \Pi$ in the Pareto-set are equal or *not comparable*. Usually, all points in this set are of interest for the decision maker or designer.

3 The Model

We briefly review the model from [DDB99]. The main idea of the proposed model is to extend the approach from [EK96], such that the search space is divided into more than three categories, like e.g. *superior, very good, good, satisfiable,* and *invalid*.

In the following the solutions are divided into so-called *Satisfiability Classes* (SCs) depending on their quality. The SCs are computed by a relation denoted *favour*, i.e. no limits have to be specified by user interaction. Solutions of "similar" quality belong to the same SC and thus each SC corresponds to a class of solutions of the same quality. After sorting the SCs with respect to their quality a ranking of the solutions is obtained.

Let Π be a finite set of solutions and $n \in \mathbf{N}$. The objective function $f_c : \Pi \to \mathbf{R}_+^n$ assigns a cost to each $x \in \Pi$ as defined above. To classify solutions in SCs we define the relation *favour* ($<_f$):

Definition 3.1

$x <_f y \Leftrightarrow |\{i : f_{ci}(x) < f_{ci}(y), 1 \leq i \leq n\}| > |\{j : f_{cj}(x) < f_{cj}(y), 1 \leq j \leq n\}|$

---

[1] In general MOO problems the objective functions are not necessarily restricted to $\mathbf{R}_+^n$ but it is convenient for our purposes.

4 Properties of Relation Favour

We start with some general properties (that have partially already been observed in [DDB99]). Then we study the advantages and disadvantages in more detail and also focus on several properties of the model.

4.1 Basics

Using Definition 3.1 we are able to compare elements $x, y \in \Pi$ in pairs more precisely. x is *favoured* to y ($x <_f y$) iff i ($i \leq n$) components of x are smaller than the corresponding components of y and only j ($j < i$) components of y are smaller than the corresponding components of x.

We use a graph representation to describe the relation, where each element is a node and "preferences" are given by edges. Relation $<_f$ is not a partial order, because it is not transitive, as can be seen as follows:

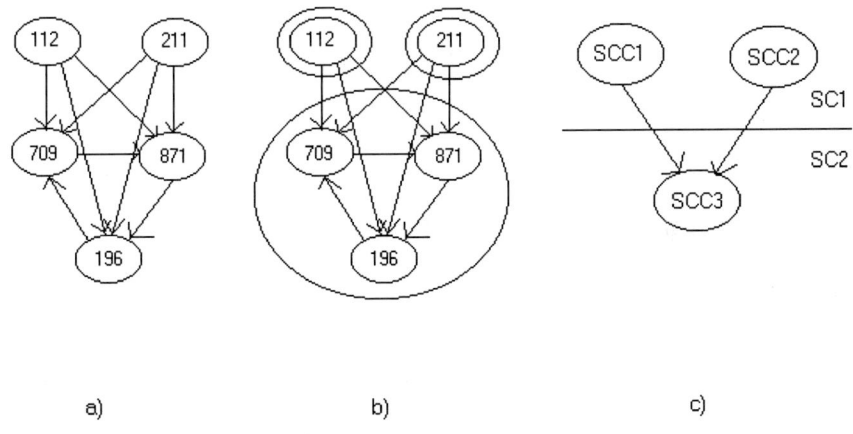

Fig. 4.1. Relation graph

Example 4.1: Consider some solution vectors from $\mathbf{R}^3$:
 (8,7,1) (1,9,6) (7,0,9) (1,1,2) (2,1,1)

The relation graph of $<_f$ is given in Figure 4.1 a). Vectors (1,1,2) and (2,1,1) are preferred to all other vectors, but they are not comparable. The remaining three vectors (8,7,1), (1,9,6), and (7,0,9) are comparable in pairs. But as can be seen in the relation graph they describe a "cycle". Thus relation $<_f$ is **not** transitive.

To get some more insight in the structure of the model we briefly focus on the meaning of the cycles in the relation graph: Elements are ranked equally when they are included in a cycle, because no element is superior to all the others. Elements that describe a cycle are denoted as not comparable. The determination of the *Strongly Connected Components* (SCC) of the relation-graph groups all elements which are not comparable in one SCC. The SCCs are computed by a DFS-based linear time graph algorithm [CLR90]. A directed graph G_{sc} is constructed by replacing each SCC in G by one node representing this SCC. Thus, all cycles in G are eliminated.

Let $G = (V,E)$ be the graph that represents relation $<_f$ and a set of solutions and let Z be the set of SCCs in G, with $Z=(Z_1,...,Z_r)$ and $Z_i = (V_i,E_i)$, $1 \le i \le r \le |V|$. Let $G_z = (V_z,E_z)$ be the graph where all SCCs Z_i of G are replaced by a node v_{zi} representing the SCC. An edge (v_{zi},v_{zj}), $1 \le i,j \le r, i \ne j$, is set in G_z, if and only if there exists an edge $(v_k,v_l) \in E$, $1 \le k,l \le |V|$, $k \ne l$, with $v_k \in Z_i$ and $v_l \in Z_j$, $Z_i \ne Z_j$. It directly follows:

Lemma 4.1: The directed graph G_z has no cycles.

The relation that is described by the relation graph G_z is denoted by $\prec_f$. Since G_z is acyclic, it is possible to determine a *level sorting* of the nodes.
For each node in G_z we define a SC. Level sorting of the nodes in G_z determines the ranking of the SCs; each level contains at least one node of G_z. Then each level corresponds to exactly one SC. Using the level sorting it is possible to group nodes (set of solutions) that are not connected by an edge in the same SC. These solutions are not comparable with respect to relation $\prec_f$ and thus they should be ranked in the same level of quality. There are two possibilities to perform the level sorting:

1. starting with nodes v that have indegree(v) = 0 or
2. starting with nodes v that have outdegree(v) = 0.

In the following the first strategy is used, since the best solutions that are not comparable should be placed in the same SC.

Example 4.2: In Figure 4.1 b) SCC 1, SCC 2, and SCC 3 are illustrated. As can be seen the elements of SCC 1 and SCC 2 are superior to the elements of SCC 3. Figure 4.1 c) shows the relation graph G_z after level sorting. Level 1 corresponds to SC 1 and level 2 corresponds to SC 2.

In the following some properties of relation *favour* are presented. For the case n = 2 it holds:

Lemma 4.2: Let $f_c : \Pi \to \mathbf{R}^2$. Relations $<_d$ and $\prec_f$ are equal, i.e. for $x, y \in \Pi$ it holds:
$$x <_d y \Leftrightarrow x \prec_f y$$
Proof: Using Definition 3.1 with n = 2 it follows: Let i (j) $\in \mathbf{N}_0$ be the number of components of x (y) that are smaller than the corresponding components of y (x). Then for $x \prec_f y$ and i and j it holds $j < i \le 2$ and $i + j \le 2$. Then for i and j it follows that i = 1 and j = 0 or i = 2 and j = 0.

"$\Rightarrow$":
If $x <_d y$, it follows $i = 2$ and $j = 0$ or $i = 1$ and $j = 0 \Rightarrow x \prec_f y$
"$\Leftarrow$":
Case 1: $i = 1, j = 0$
It follows: $(x_1 < y_1) \wedge (x_2 = y_2) \Rightarrow x <_d y$
Case 2: $i = 2, j = 0$
It follows: $(x_1 < y_1) \wedge (x_2 < y_2) \Rightarrow x <_d y$

Lemma 4.2 does not hold for $n > 2$ as can be stated by a counterexample:

Example 4.3: Consider the vectors given in Example 4.1. Applying relation *dominate* vectors (1,1,2), (2,1,1) and (7,0,9) are not dominated by any other element; thus they are ranked as the best elements of the given vector set. In comparison to relation *favour* only the two vectors (1,1,2) and (2,1,1) are the best elements.

For the computation of the execution time of the SCs an approximation can be given:

Theorem 4.1: The computation time for the SC classification is $O(|P|^2 \cdot n)$.

Proof: Each element in a population P is compared to all other elements in P and in each comparison n components are tested. The number of nodes in the relation graph G_z is $|P|$ and the operations on G_z have linear time complexity. Thus the computation time for the classification is $O(|P|^2 \cdot n)$.

First, we mention two points that should be considered when using the model. These points can be seen as disadvantages, if compared to other approaches.

1. Incomparable solutions may be placed in different SCs, even though there might be cases where the user want them to be in the same class.
2. The different optimisation goals are considered in parallel and not relative to each other (see also discussion below).

But, the model has several advantages over other approaches:
- No problem specific knowledge is needed for the choice of the weights of the fitness function, like in the case of weighted sum.
- Even if the components of a solution have different measures, a scaling of these components is not needed, since there are no distances between the solutions computed (like e.g. using a weighted sum or information about distances between solutions).
- Dependent on the population the model dynamically adapts the relation graph that performs the comparison of the fitness function. In each generation the relation graph is re-computed and the ranking is updated. (This is done totally automatic and neither user interaction is required nor limits between the SCs have to be specified.) Thus, the granularity of the SCs is dynamically adapted to present conditions.
- Relation $\prec_f$ can also handle infeasible solutions. An infeasible component of the solution is considered as the worst possible value.

- If the structure of the search space changes during the EA run these changes are directly included in the relation that is updated online.
- Experiments have shown that the model results in a finer granularity than relation *dominate*.
- Due to the efficient representation based on graphs the run times are very low.
- Handling of priorities of all (or some) objectives and of infeasible solutions is fully supported.
- No user interaction is required. This is very important in complex applications where the user cannot keep control of all parameters by hand.

In summary, there exist many advantages of the model compared to the "standard" techniques, which is demonstrated in the experiments.

In the following some further scenarios are discussed that can occur in practice.

4.2 Sorting Using Priorities

So far in the model it is not possible to handle priorities, e.g. to give a priority to the different optimisation criteria. This is not desired in most cases of complex applications, but we now outline how it can easily be added to the model as defined above. Let $pr \in [0,1]^m$ be the priority vector, where for each $x_i \in \Pi$, pr_i denotes the priority of x_i. The lower pr_i is chosen, the higher is the priority of the corresponding component. If all values of pr_i's are different, a *lexicographic sorting* can be used.

Components with same priority value are compared using relation $\prec_f$. Only for these components a ranking is computed. Then based on this sorted set of elements a final ranking is determined by *lexicographic sorting*.

4.3 Invalid Solutions

The model is very well suited to handle invalid solutions. The classification, when an element is to be seen as "invalid" has to be defined by the user. The main idea is to modify the comparison operator. The invalid elements (for one or some specific components) always lose all comparisons carried out during the evaluation. I.e. if component x_i, $1 \leq i \leq n$, of vector $x \in \mathbf{R}^n_+$ is invalid then it holds $x > y$ for all valid $y \in \Pi$. This approach has been shown to work very well in practice.

4.4 Relation to Weighted Sum

Finally, we want to briefly comment on the relation of the model to the weighted sum approach, that is still used in many applications. The new model is not a cover of the weighted sum, i.e. the weighted sum cannot be embedded. The difference is that using the presented model the solutions are ranked relatively to each other. Using the weighted sum the absolute values of each component are combined to determine a single fitness value. This is demonstrated by constructing an example:

Example 4.4: Assume that the function $f : \mathbf{R} \to \mathbf{R}^2$ with $f_1(x) = x^2$ and $f_2(x) = (x-2)^2$ is given. If $\omega = (0.5, 0.5)$ is the weight vector for the weighted sum, the minimal solution is $x = 1$ with $f(x) = (1,1)$ and the corresponding weighted sum is $0.5 \cdot 1^2 + 0.5 \cdot (1-2)^2 = 1$. It can be seen that for all other x values the sum is larger, e.g. the weighted sum of solutions $x = 0$ with $f(x) = (0,4)$ and $x = 2$ with $f(x) = (4,0)$ is $0.5 \cdot 0^2 + 0.5 \cdot (0-2)^2 = 2$ and $0.5 \cdot 2^2 + 0.5 \cdot (2-2)^2 = 2$ respectively. Using the above presented model it is not possible to perform a ranking such that $(1,1)$ is better than solution $(0,4)$ or $(4,0)$. These three solutions are *not comparable* by relation favour.

5 Case Study: Heuristic Learning

Analogously to [DDB99] for carrying out experiments we choose the application "heuristic learning" of *Binary Decision Diagram* (BDD) [Bry86] variable orderings as proposed in [DB95,DGB96]. This application has been chosen due to several reasons:

- The problem has a high practical relevance and finds many applications in VLSI CAD.
- Due to the many studies already carried out, the problem is well understood. This allows for clear interpretation of the results.
- The problem is multi-objective in nature, i.e. quality (counted in number of nodes) and run time has to be optimised.

To make the paper self-contained we review the main ideas from [DDB99].

5.1 The Learning Model

First, the basic learning model is briefly explained (for more details see [DB95,Dre00]). It is assumed that the problem to be solved has the following property:
A non empty set of optimisation procedures can be applied to a given (non-optimal) solution in order to further improve its quality. These procedures are called *Basic Optimisation Modules* (BOMs). Each heuristic is a sequence of BOMs. The length of the sequence depends on the problem instances that are considered. The goal of the approach is to determine a good (or even optimal) sequence of BOMs such that the overall results obtained by the heuristic are improved.
The set of BOMs defines the set H of all possible heuristics that are applicable to the problem to be solved in the given environment. H may include problem specific heuristics, local search operators but can also include some random operators.
The individuals of our EA make use of multi-valued strings. The sequences of different length are modelled by a variable size representation of the individuals.

To each BOM $h \in H$ we associate a *cost function* denoted by $cost : H \to \mathbf{R}$. *cost* estimates the resources that are needed for a heuristic, e.g. execution time of the heuristics. The quality fitness measures the *quality* of the heuristic that is applied to a given example.

Binary Decision Diagrams

As well-known each Boolean function $f : \mathbf{B}^n \to \mathbf{B}$ can be represented by a *Binary Decision Diagram* (BDD), i.e. a directed acyclic graph where a Shannon decomposition is carried out in each node. We make use of reduced and ordered BDDs [Bry86].

We now consider the following problem that will be solved using EAs:

> How can we develop a good heuristic to determine variable orderings for a BDD representing a given Boolean function f such that the number of nodes in the BDD is minimised?

Notice we do **not** optimise BDDs by EAs directly. Instead we optimise the heuristic that is applied to BDD minimisation.

Dynamic Variable Ordering

The algorithms that are used as BOMs in the EA are based on dynamic variable ordering. *Sifting* (S) is a local search operation for variable ordering of BDDs which allows hill climbing. *Sift light* (L) is a restricted form of sifting, where hill climbing is not allowed. The third BOM is called *inversion* (I) which inverts the variable ordering of a BDD. For more details see [DGB96,Dre00].

5.2 Evolutionary Algorithm

The configuration of the EA is described in the following sections.

Representation

In our application we use a multi-valued encoding, such that the problem can easily be formulated. Each position in a string represents an application of a BOM. Thus, a string represents a sequence of BOMs which corresponds to a heuristic. The size of the string has an upper limit of size l_{max} which is given by the designer and limits the maximum running time of the resulting heuristics. In the following we consider four-valued vectors: S (L, I) represents sifting (sift light, inversion) and value N denotes *no* operation. It is used to model the variable size of the heuristic.

Objective Function

As an *objective function* that measures the *fitness* of each element we apply the heuristics to k benchmark functions in a *training set* $T = \{b_1, \ldots, b_k\}$. The quality of an individual is calculated by constructing the BDD and counting the number of nodes for each b_i, $1 \le i \le k$. Additionally, the execution time (measured in CPU seconds) that is used for the application of the newly generated heuristic is minimised. Then, the objective function is a vector of length k+1 and is given by:

$$f_c(T) = (\#nodes(b_1), \ldots, \#nodes(b_k), time(T)),$$

where #$nodes(b_i)$, $1 \leq i \leq k$, denotes the number of BDD nodes that are needed to represent function b_i. The execution time that is needed to optimise all functions of the considered training set T is denoted by *time*.

The choice of the fitness function largely influences the optimisation procedure. It is also possible to chose a fitness function as a vector of length $2 \cdot k$ by considering the execution time for each benchmark b_i separately instead of using the sum. By our choice the EA focuses more on quality of the result than on the run time needed.

Algorithm
1. The initial population P is randomly generated.
2. Then P/2 elements are generated by the genetic operators *reproduction* and *crossover*. The newly created elements are then mutated.
3. The offspring is evaluated. Then the new population is sorted using relation *favour*.
4. If no improvement is obtained for 50 generations the algorithm stops.

For more details about the algorithm see [DGB96].

6 Experimental Results

All experiments in the following are carried out on a SUN Sparc 20 using the benchmark set from [Yan91].

First, the proposed model is compared to the weighted sum approach, because this is a "classical" method and used in many applications. Then a comparison to the "pure" Pareto-based approach from [Gol89] is performed. Notice, that comparisons to other approaches, as presented in [ZT99], are not given, because there the users' interaction is required, e.g. if distances between solutions are computed.

6.1 Comparison to Weighted Sum

In a first series of experiments we applied heuristic learning to a set of 5 benchmarks. Thus 6 objectives (5 benchmarks and the execution time) are optimised in parallel. First, only small benchmarks were studied, for which the optimal result is known. We applied an exact algorithm, an iterated greedy heuristic, learning based on weighted sum, and learning based on relation *favour*. The results are summarised in Table 6.1. Column *name* denotes the considered function, *in* (*out*) shows the number of inputs (outputs) of the function and in column optimal the minimal BDD size is shown. The applied heuristics for BDD minimisation are shown in the following columns: *sift* is an iterated greedy heuristic, *sum* denotes the heuristic resulting from the EA with weighted sum and *f1* and *f2* are the two alternative heuristics resulting from the EA using relation *favour* for ranking the individuals.

Regarding quality of course the exact algorithm determines the best results, but the run time is more then 100 CPU seconds. Iterated greedy is in some cases more than 30% worse than the exact result. Both learning approaches determine the same (optimal) quality, but the run time of the heuristic constructed by relation *favour* is nearly two times faster.

Table 6.1

name	in	out	optimal	sift	sum	f1	f2
add6	12	7	28	54	28	28	28
addm4	9	8	163	163	163	163	163
cm85a	11	3	27	35	27	27	30
m181	15	9	54	56	54	54	54
risc	8	31	65	65	65	65	65
average time			> 100	0,4	1,6	0,9	0,4

Table 6.2

name	in	out	sift	sum	favour
bc0	26	11	522	522	522
chkn	29	7	261	266	257
frg1	28	3	84	80	72
ibm	48	17	410	216	354
misj	35	14	39	39	39
x6dn	39	5	241	240	229
average time			0,9	1,0	3,5

In a second series larger benchmarks are considered for which the exact result cannot be computed. The application to larger examples with 7 objectives to be optimised showed very good results. They are summarised in Table 6.2. In this case (as can be expected) the greedy approach performs even worse. Relation *favour* outperforms the weighted sum on 50% of all examples considered.

It is important to notice that for the weighted sum approach a lot of "fine-tuning" was necessary, while relation *favour* runs without any further user interaction.

6.2 Comparison to Relation *Dominate*

We first applied the learning algorithms to the same (small) training set as above (Table 6.3). After the optimisation the learning algorithm based on *dominate* computes 14 different solutions that cannot be further compared. Relation *favour* returned two elements only (see also Table 6.1). Regarding quality, there was not a single component where one of the 14 elements outperformed one of the two. Beside this the learning time for the algorithm based on *favour* was more than four times faster, i.e. 2.5 CPU hours instead of 14.

When applying heuristic learning to construct heuristics in VLSI CAD it is very important that the number of final solutions is not too large, since finally the designer has to choose one. If the list becomes too long, it is not feasible to test them all, since the designs are too large.

Table 6. 3

name	in	out	d1	d2	d3	d4	d5	d6	d7	d8	d9	d10	d11	d12	d13	d14	f1	f2
add6	12	7	42	63	30	62	260	256	300	185	256	132	52	67	51	310	28	28
addm4	9	8	163	191	163	187	245	198	189	181	198	231	201	167	163	206	163	163
cm85a	11	3	33	39	35	30	37	37	43	39	46	49	30	35	33	37	27	30
m181	15	9	60	74	54	55	87	87	84	74	80	83	54	60	61	86	54	54
risc	8	31	70	71	65	66	94	79	84	95	97	94	65	65	65	90	65	65
average time			1,9	1,8	2,5	1,9	1,2	1,2	1,2	1,2	1,2	2,7	1,9	3,3	1,2	0,9	0,4	

To further study the selection process, we finally applied our technique to a larger training set, where the algorithm based on relation *dominate* computed 16 solutions. To this result we applied the relation *favour* and this reduced the set to one element. More details are given in [Dre00].

6.3 Further Applications

In the meantime, the MOO model based on relation *favour* has been used in many projects and has been included in the software environment GAME (Genetic Algorithm Managing Environment) – a software tool developed for applications in VLSI CAD using evolutionary techniques.
A method for test pattern generation for digital circuits and a tool for detailed routing of channels and switchboxes has been developed underlining the flexibility of the model (see [Dre98]).

7 Conclusion

Recently, in [DDB99] a new model for Multi-Objective Optimisation has been proposed that overcomes the limitations of classical EA approaches that often require several runs to determine good starting parameters. Furthermore, the model gives very robust results since the number of parameters is reduced without reducing the quality of the result. Only "non-promising" candidates (for which can be guaranteed that they are not optimal and already better individuals exist) are not considered.
In this paper, we gave a detailed description of the model. Several properties have been discussed and advantages and disadvantages are described. We especially compared from a theoretical and practical point of view the relation of the new model to the weighted sum approach and to the relation *dominate*, respectively.
As an advantage it turned out that components with different measures do not have to be scaled. This is done automatically comparing elements with relation *favour*. This may result in significant speed-ups in EAs and simplifies the handling of the algorithms.

References

[Bry86] R.E. Bryant. Graph-based algorithms for Boolean function manipulation. IEEE Trans. on Comp., 35(8):677-691, 1986.

[CLR90] T.H. Cormen, C.E. Leierson, and R.C. Rivest. Introduction to Algorithms. MIT Press, McGraw-Hill Book Company, 1990.

[Dre00] N. Drechsler. Über die Anwendung Evolutionärer Algorithmen in Schaltkreisentwurf, PhD thesis, University of Freiburg, Germany, 2000.

[Dre98] R. Drechsler. Evolutionary Algorithms for VLSI CAD. Kluwer Academic Publisher, 1998.

[DB95] R. Drechsler and B. Becker. Learning heuristics by genetic algorithms. In ASP Design Automation Conf., pages 349-352, 1995.

[DDB99] N. Drechsler, R. Drechsler, and B. Becker. Multi-objective optimization in evolutionary algorithms using satisfiability classes. In International Conference of Computational Intelligence, 6th Fuzzy Days, LNCS 1625, pages 108-117, 1999.

[DGB96] R. Drechsler, N. Göckel, and B. Becker. Learning heuristics for OBDD minimization by evolutionary algorithms. In Parallel Problem Solving from Nature, LNCS 1141, pages 730-739, 1996.

[Esb96] H. Esbensen. Defining solution set quality. Technical report, UCB/ERL M96/1, University of Berkeley, 1996.

[EK96] H. Esbensen and E.S. Kuh. EXPLORER: an interactive floorplaner for design space exploration. In European Design Automation Conf., pages 356-361, 1996.

[FF95] C.M. Fonseca and P.J. Fleming. An overview of evolutionary algorithms in multiobjective optimization. Evolutionary Computation, 3(1):1-16, 1995.

[Gol89] D.E. Goldberg. Genetic Algorithms in Search, Optimization & Machine Learning. Addision-Wesley Publisher Company, Inc., 1989.

[HNG94] J. Horn, N. Nafpliotis, and D. Goldberg. A niched pareto genetic algorithm for multiobjective optimization. In Int'l Conference on Evolutionary Computation, 1994.

[Mic94] Z. Michalewicz. Genetic Algorithms + Data Structures = Evolution Programs. Springer-Verlag, 1994.

[SD95] N. Srinivas and K. Deb. Multiobjective optimization using nondominated sorting in genetic algorithms. Evolutionary Computation , 2(3):221-248, 1995.

[Yan91] S. Yang. Logic synthesis and optimization benchmarks user guide. Technical Report 1/95, Microelectronic Center of North Carolina, Jan. 1991.

[ZT99] E. Zitzler and L. Thiele. Multiobjective evolutionary algorithms: A comparative case study and the strength pareto approach. IEEE Trans. on Evolutionary Computation, 3(4):257-271, 1999.

Comparison of Evolutionary and Deterministic Multiobjective Algorithms for Dose Optimization in Brachytherapy

Natasa Milickovic[1], Michael Lahanas[1], Dimos Baltas[1,2], and Nikolaos Zamboglou[1,2]

[1] Department of Medical Physics and Engineering, Strahlenklinik, Klinikum Offenbach, 63069 Offenbach, Germany.
[2] Institute of Communication and Computer Systems, National Technical University of Athens, 15773 Zografou, Athens, Greece.

Abstract. We compare two multiobjective evolutionary algorithms, with deterministic gradient based optimization methods for the dose optimization problem in high-dose rate (HDR) brachytherapy. The optimization considers up to 300 parameters. The objectives are expressed in terms of statistical parameters, from dose distributions. These parameters are approximated from dose values from a small number of points. For these objectives it is known that the deterministic algorithms converge to the global Pareto front. The evolutionary algorithms produce only local Pareto-optimal fronts. The performance of the multiobjective evolutionary algorithms is improved if a small part of the population is initialized with solutions from deterministic algorithms. An explanation is that only a very small part of the search space is close to the global Pareto front. We estimate the performance of the algorithms in some cases in terms of probability compared to a random optimum search method.

1 Introduction

High dose rate (HDR) brachytherapy is a treatment method for cancer where empty catheters are inserted within the cancer volume. Once the correct position of these catheters is verified, a single ^{192}Ir source is moved inside the catheters at discrete positions (dwell positions) using a computer controlled machine. The problem which we consider is the determination of the n times (dwell times) which are sometimes termed dwell weights or weights for which the source is at rest at each of the n dwell positions so that the resulting three-dimensional dose distribution will fulfill defined quality criteria. In modern brachytherapy the dose distribution has to be evaluated with reference to irradiated normal tissues and the planning target volume (PTV), which includes the cancer volume and an additional margin. For a more detailed description see [1], [2]. The number of source positions varies from 20 to 300. We consider the optimization of the dose distribution using as objectives the variance of the dose distribution on the PTV surface and in the PTV calculated at 1500 − 4000 positions (sampling points).

For variances and in general for quadratic convex objective functions $f(x)$ of the form

$$f(x) = (Ax - d)^T(Ax - d) \qquad (1)$$

it is known that a weighted sum optimization method converges to the global Pareto front [3], where A is a constant matrix and d is a constant vector of the prescribed dose values in the PTV target or surface.

We have successfully applied multiobjective evolutionary algorithm with dose-volume based objectives [1],[2]. In the past comparisons of the effectiveness of evolutionary algorithms have been made with either other evolutionary algorithms [4] or with manually optimized plans [5], [6]. We have compared the Pareto fronts obtained by multiobjective evolutionary algorithms with the Pareto fronts obtained by a weighted sum approach using deterministic optimization methods such as quasi-Newton algorithms and Powells modified conjugate gradient algorithm which does not requires derivatives of the objective function [7].

We use here only objectives where gradient based algorithms are superior. However, we must consider also critical structures partly inside the target or close to it which have to be protected by excessive radiation. Other objectives are the optimum position and the minimum number of sources. In such cases the gradient based algorithms can not be used. Therefore before applying evolutionary algorithms for the solution of these more complex problems we have compared their efficiency with deterministic methods using only as objectives the dose variance within the target and on its surface.

2 Methods

2.1 Objectives

We use as objectives the normalized variance of the dose distribution on the PTV surface f_S and in the PTV f_V:

$$f_S = \frac{1}{m_S^2 N_S} \sum_{i=1}^{N_S} (d_i - m_S)^2 \qquad (2)$$

$$f_V = \frac{1}{m_V^2 N_V} \sum_{i=1}^{N_V} (d_i - m_V)^2 \qquad (3)$$

m_S, m_V are the corresponding mean values, N_S, N_V the number of points used to estimate these parameters and d_i the dose value of the i-th point.

We use a weighted sum approach for the multiobjective optimization with the deterministic algorithms, where for a set of weights for the volume and surface variance we perform a single objective optimization of f_w:

$$f_w = w_S f_S + w_V f_V \qquad (4)$$

where $w_S, w_V \geq 0$ are the surface and volume importance factors, respectively and $w_S + w_V = 1$. We used 21 optimization runs where w_S varied from 0 to 1 in steps of 0.05 to determine the shape of the trade-off curve.

2.2 Genetic Operators

We use a real representation for the chromosomes. The following variants of genetic operators have been used in this study.

Uniform mutation. In uniform mutation if g_k is the kth element of a chromosome selected for mutation, then it is replaced by a random number from the interval $[LB, UB]$ where LB and UB are the lower and upper bounds of the kth element.

Non-uniform mutation. In non-uniform mutation if g_k is the kth element of a chromosome at generation t it will be transformed after a non-uniform mutation to g'_k

$$g'_k = \begin{cases} g_k + \Delta(t, UB - g_k) & r_1 = 0 \\ g_k + \Delta(t, g_k - UB) & \text{else} \end{cases} \quad (5)$$

$$\Delta(t, y) = y(1 - r^{(1-t/T)^b}) \quad (6)$$

where r_1 is a random bit (0 or 1), r is a random number in the range $[0, 1]$, T is the maximal generation number and b a parameter controlling the dependency of $\Delta(t, y)$ on the generation number. The function $\Delta(t, y)$ returns a value in the range $[0, y]$ such that the probability of $\Delta(t, y)$ being close to 0 increases as t increases. Initially when t is small the space is searched uniformly and very locally at later stages [8].

Flip Mutation. In the case of Flip mutation, we randomly pick elements (genes) g_k to g_j from the chromosome and then set those elements to any other of genes of this chromosome. This will work for any number of gene sets of a given chromosome.

Swap Mutation. In the case of swap mutation, we randomly swap the elements in the chromosome. The number of swapped elements depends on the mutation probability.

Gaussian Mutation. This mutation operator adds a gaussian distributed random value to the chosen gene. The new value is clipped if it falls outside the specified lower and upper bounds for that gene.

Blend Crossover. In the case of Blend crossover [9] we generate a uniform distribution based on the distance between parent values, then choose the offspring value based upon that distribution. This distance is defined for each pair of the parents corresponding genes. If g_1 and g_2 are parent chromosomes, and g'_1 and g'_2 are the offsprings then for the parents' i-th genes, this distance is given by:

$$dist = \mid g_{1i} - g_{2i} \mid \tag{7}$$

Further, lower and higher bounds, lo, hi are found as:

$$lo = \min(g_{1i}, g_{2i}) - 0.5 \cdot dist, \tag{8}$$
$$hi = \max(g_{1i}, g_{2i}) + 0.5 \cdot dist,$$

and correcting those values if necessary to retain within the wanted gene-value boundaries. Then g'_{1i} and g'_{2i} are determined as the random numbers from the range $[lo, hi]$.

Geometric Crossover. In geometric crossover if a and b are parent chromosomes then the new offspring c_1 and c_2 are:

$$c_1 = <a_1, a_2, a_3, ..., a_n, \sqrt{a_{n+1}, b_{n+1}}, ..., \sqrt{a_{2n}, b_{2n}}> \tag{9}$$
$$c_2 = <b_1, b_2, b_3, ..., a_n, \sqrt{a_{n+1}, b_{n+1}}, ..., \sqrt{a_{2n}, b_{2n}}>.$$

One of these progenies will be selected according to their fitness values.

Two Point Crossover. The two point crossover operator randomly selects two crossover points within a chromosome. Further, it interchanges the two parent chromosomes between these points (segment interchange) to produce the two new offsprings.

Arithmetic Crossover. If g_i^t and g_j^t are two chromosomes of the population at generation t, then after arithmetic crossover two new chromosomes g_i^{t+1}, g_j^{t+1} at generation $t+1$ are produced:

$$g_i^{t+1} = \alpha g_j^t + (1-\alpha) g_i^t \tag{10}$$
$$g_j^{t+1} = \alpha g_i^t + (1-\alpha) g_j^t$$

α is either a constant (uniform arithmetical crossover), or a variable depending on the generation number (non-uniform arithmetical crossover).

2.3 Estimation of the Probabilities of Random Solutions

Using random sets of decision variables (weights) we extrapolated the number of function evaluations required by a random search method to obtain points on the Pareto front. We have found that a two-dimensional normal density function

$f(x)$ in some cases can be used to describe the distribution of these random points in the objectives space known also as bi-loss map for two objectives [10].
The density function is:

$$f(x) = \frac{1}{2\pi \mid \Sigma \mid^{1/2}} \exp\left[-\frac{1}{2}(x - \mu)^T \Sigma^{-1}(x - \mu)\right] \qquad (11)$$

where

$$\Sigma = \begin{pmatrix} \sigma_{11} & \sigma_{12} \\ \sigma_{12} & \sigma_{22} \end{pmatrix} \qquad (12)$$

with $\sigma_{11} = \sigma_x^2$ and $\sigma_{12} = \sigma_{21} = cov(x,y) = E[(x - \mu_x)(y - \mu_y)] = \rho\sigma_x\sigma_y$. $E(x)$ is the expectation value of x, σ_x, σ_y are the variances in x and y respectively and ρ the correlation coefficient.

3 Results

The convergence of the Strength Pareto Evolutionary Approach algorithm (SPEA) [11] is shown in Fig. 1. For a breast implant with 250 sources the population averages of the surface and volume dose variances $<f_S>$ and $<f_V>$ are shown for 500 generations. A convergence is observed after 100 generations.

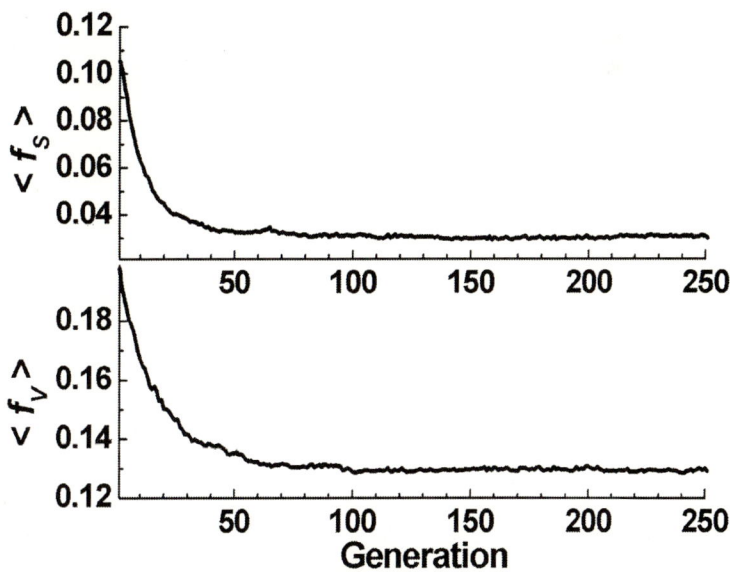

Fig. 1. An example of the convergence of $<f_S>$ and $<f_V>$ for the SPEA algorithm.

In order to compare the efficiency of SPEA and the deterministic algorithms we have evaluated the probability to obtain at random a solution presented by these algorithms. We generated 2500000 random sets of weights. The distribution of these points in search space for a cervix implant with 215 sources is shown in Fig. 2. We use Powells optimization method [7] to fit the parameters $\sigma_{ij}, i, j = 1, 2$ and μ of (12) to this distribution. The result is shown in Fig. 3 where the fit and the experimental value for bins with a non-zero number of entries is shown unfolded as a one dimensional distribution.

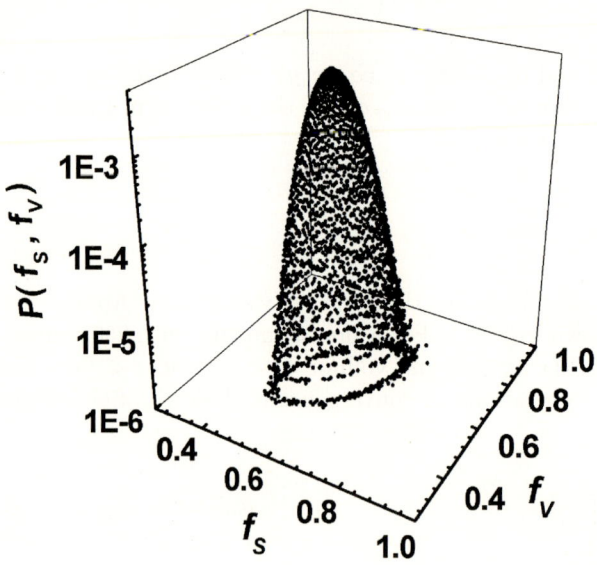

Fig. 2. Probability $P(f_S, f:_V)$ of a generation of a random point at $f_S \pm 0.05, f_V \pm 0.05$.

In Fig. 4 we show the probability distribution for the cervix implants of obtaining a solution at a point $f_S \pm 0.05, f_V \pm 0.05$. The distributions is shown for 100000 random solutions and for the non-dominated solutions from SPEA and the deterministic algorithm.

The distribution of weights obtained for the gradient based optimization for the cervix implant is shown in Fig. 5. Only a small part of the weights is significant.

Fig.6 like Fig.4 shows the result for a implant with 250 sources where the global Pareto front is closer to the majority of the random solutions in the search space. Consequently the difference between probabilities of the solutions of the deterministic and the SPEA algorithm is smaller.

In all cases which we studied the Pareto fronts of the multiobjective evolutionary algorithms MOEAs are local and between the global Pareto front found

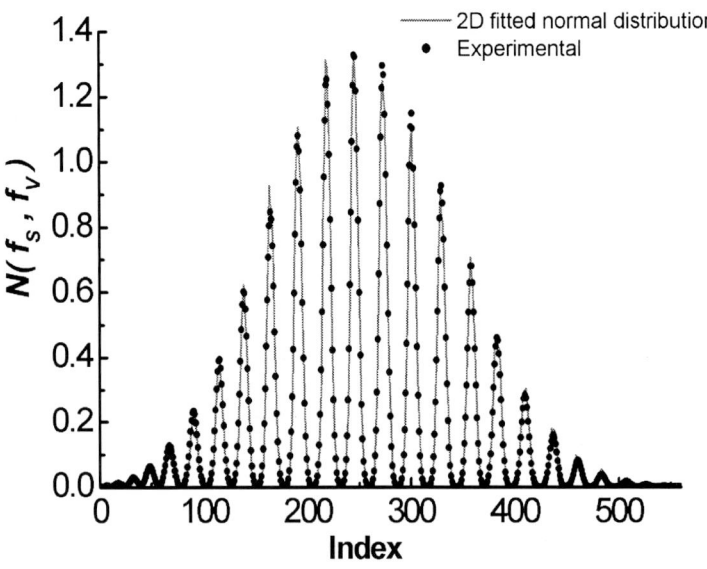

Fig. 3. A fit of a two-dimensional normal distribution.

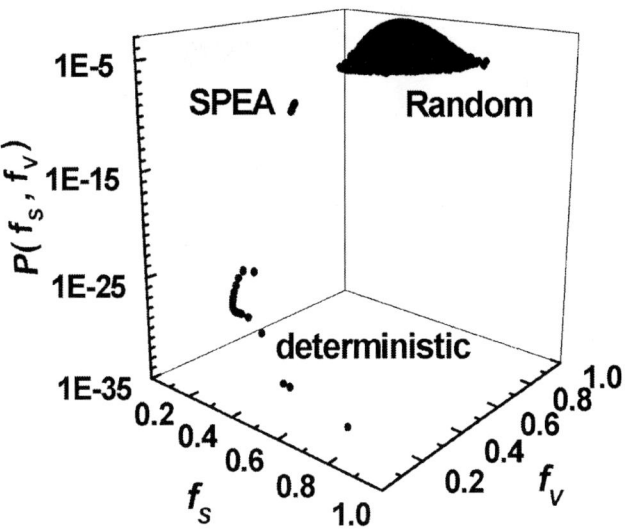

Fig. 4. Probability distribution for a cervix implant with 215 sources for 100000 random, SPEA and deterministic algorithm solutions.

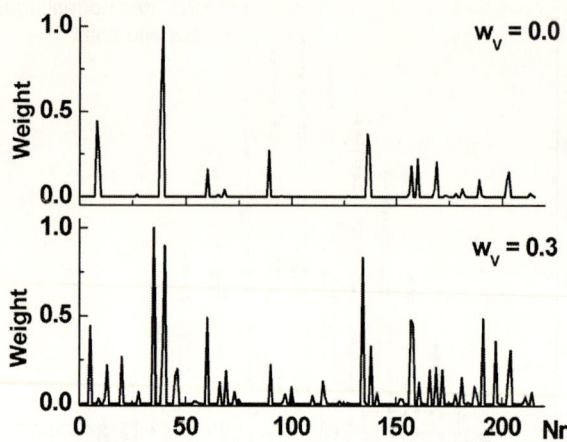

Fig. 5. Distribution of the weights for the deterministic algorithm for the importance factors $w_V = 0.0$ and 0.3, see 4.

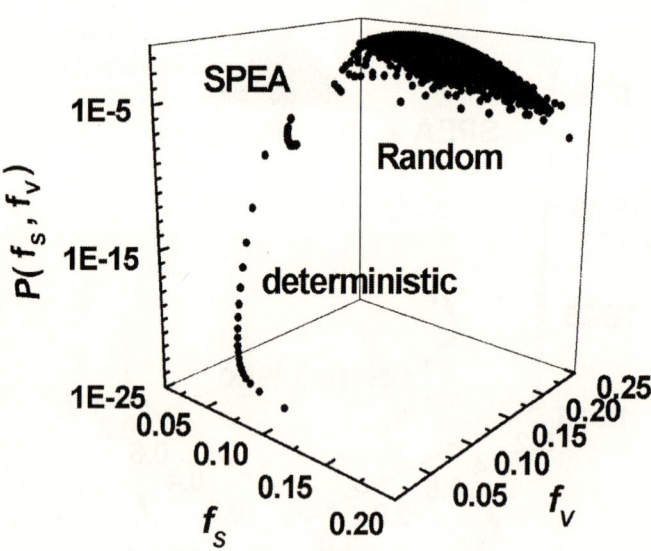

Fig. 6. Probability distribution for a breast implant with 250 sources for 100000 random, SPEA and deterministic algorithm solutions.

by the deterministic algorithms and the majority of random solutions in the search space. The shape of this set depends on the geometry of the implant and the topology of the sources. For implants with a large number of sources the majority of the random solutions is far from the the global Pareto front.

K. Deb described some reasons why convergence to the true Pareto-optimal front may not occur: Multimodality, deception, isolated optimum and collateral noise [12]. For the objectives used multimodality can be excluded. The most important reason is the isolated minimum. Since only a extreme small part of the search space is located close to the true Pareto front it is not possible for the evolutionary algorithm to acquire information about the position of the Pareto front from crossover and mutation. As described in [13] code-substrings that have above average fitness values are sampled at an exponential rate for inclusion in subsequent generations. This does however not imply that a convergence to the global Pareto-front will occur. In contrast the gradient based algorithms use very efficient the information from local gradients and converge extremely fast to the Pareto-optimal front. Collateral noise seems to be present when there is an excessive noise coming from other parts of the solution vector. Without initialization the population moves smoothly and converges to a local Pareto front. If a small part of the population is initialized with solutions from the deterministic algorithm then members of the population cover a much larger part of the search space. This shows that the solutions require a extreme fine tuning of the decision variables which the conventional genetic operators used in this study can not handle.

We analyzed the influence of different mutation and crossover operators, described previously in Methods, on the efficiency of different evolutionary algorithms. In all cases the parameters of 0.85 for the cross-over probability and 0.0065 for the mutation probability are used.

The best coverage of objective space was obtained by using the geometric crossover. The mutation operator did not influence the evolutionary algorithm efficiency as much as the crossover.

In order to compare the effectiveness of the evolutionary algorithms we generated 100000 points in the objective space from a corresponding number of random sets of weights. The efficiency of evolutionary algorithms highly depends on how far these random solutions are from the global Pareto front. In Fig. 7 the case of a prostate implant with 20 sources is presented. We can conclude that the random solutions cover the objective space very satisfactorily and approach the Pareto front. This means that the evolutionary algorithms should be able to produce Pareto sets which will converge toward the deterministic solutions. In different case, see Fig. 8, evolutionary algorithms could only theoretically approach the deterministic Pareto set after a extreme large number of generations. The Pareto set obtained by the evolutionary algorithms converges toward the global Pareto set better in the direction in which the majority of random solutions is nearer to the global Pareto set.

In the case that the majority of random solutions in objective space is far from the Pareto global front, then the evolutionary algorithms produce only local

Pareto sets. An optimal set of parameters and genetic operators does not improve the convergence to the global Pareto set as significantly as an initialization with solutions from the deterministic algorithm. With only four deterministic solutions, see Figs. 7 and 8, the evolutionary algorithms reproduce the Pareto global front. We have found that the objective functions used in deterministic algorithms for this initialization need not to be exactly the same as used by the evolutionary algorithms. For example the avoidance of high dose values inside the PTV can be satisfied by a small dose variance inside the PTV. Another objective function which can be used is for this purpose is a penalty function which penalizes solutions with dose values above a given limit. In this case the deterministic gradient based algorithms due to the presence of local minima can not be applied.

We analyzed the efficiency of different evolutionary algorithms compared with deterministic ones. The Pareto sets produced by SPEA, FFGA, FFGA with elitism [14],[15], and the niched Pareto algorithm (NPGA) [16], and the Pareto set evaluated by the Fletcher deterministic method, are compared. Pareto fronts from SPEA and FFGA with elitism are converging to the deterministic evaluated Pareto front much better than FFGA and NPGA, see Fig. 8. The main reason for this is the elitism implemented in the former two methods. In the case of multiobjective evolutionary algorithms it is important to save the nondominated solutions. Even in the case that the previously described initialization is applied, the non-elitistic algorithms do not produce better results as the algorithm does not "remember" the external nondominated set. This means that an initialization of evolutionary algorithms requires the inclusion of elitism. For NPGA and FFGA an additional problem in comparison to SPEA is that a value for the sharing radius is required which can vary from case to case.

The problem of the SPEA algorithm is that the extension of the Pareto front is not very large as FFGA. It does not cover the "ends", see Figs. 7 and 8, as the extension requires a fine tuning of the weights which can not be reached by the evolutionary algorithms, except if an additional initialization was made first. We usually run algorithms for up to 500 generations although the population converges to a local Pareto front as shown in Fig. 9 after 100 generations. The Pareto front is not significant modified even if the calculations has been extended up to 10000 generations in some cases. For implants with 200 and more sources, the computation time with 10000 generations requires up to 10 hours. We solved this problem by initialization of four members of the initial population with solutions from deterministic algorithm. In this case, the algorithm converges to the deterministic evaluated Pareto set in less than 100 generations. Even if we could use for other objectives a weighted sum approach with deterministic algorithms, this would require to use a very large number of weights to reconstruct the Pareto set, especially with increasing number of objectives. Here the evolutionary algorithms in combination with deterministic methods are more effective in generating non dominated solutions.

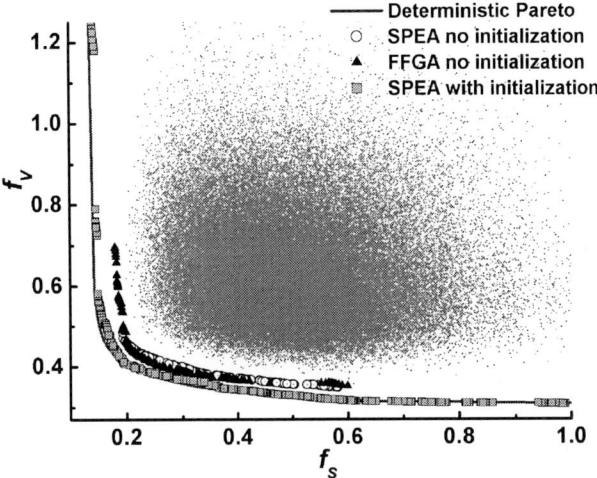

Fig. 7. An example of the distribution of 100000 random solutions in objective space for a prostate implant with 20 sources. The Pareto set obtained by the deterministic algorithm is presented by the line. The non dominated solutions of SPEA and FFGA with elitism are shown without initialization. For SPEA additional the Pareto set is show if four members of the population are initialized with solutions from a deterministic algorithm.

4 Conclusions

We have compared standard multiobjective evolutionary algorithms with deterministic optimization methods with objectives such that a weighted sum approach can be used to obtain the global Pareto fronts. This comparison was done for the dose-optimization problem in HDR brachytherapy. The number of decision variables can be as high as 300.

From the distribution of random generated weights we were able to estimate in some cases the probabilities of generating randomly a point in the objective space. This enabled us to estimate the performance of the multiobjective evolutionary and the deterministic algorithms in comparison to a random search. The evolutionary algorithms have been found to be a factor $10^5 - 10^{12}$ more effective than a random search. The deterministic are more efficient which exceeds in some cases 10^{30}. This could explain why evolutionary algorithms were not able to generate solutions close to the global Pareto front.

The Pareto front reached depends on the probability of generating a point in the objective space. The evolutionary algorithms with the standard genetic operators described in this work are not able to significantly improve the performance. An initialization from deterministic algorithms improves the performance and helps to reconstruct the Pareto front around the initial seeds of the deterministic algorithm.

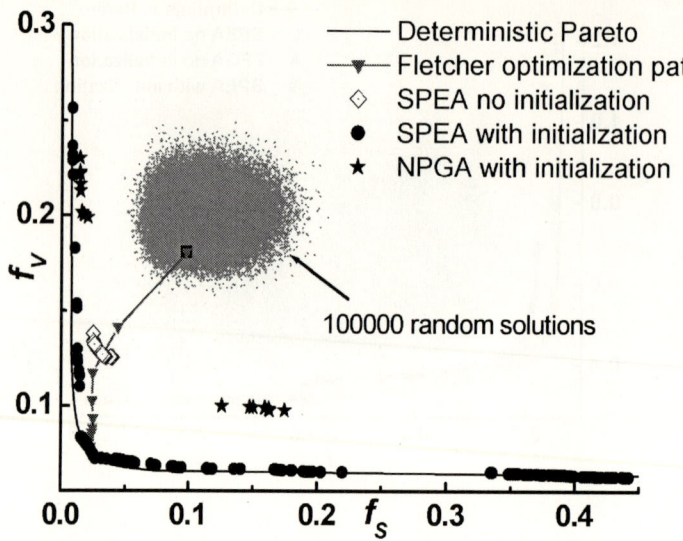

Fig. 8. An example of the SPEA algorithm for the case of a breast implant with 250 sources. The majority of random solutions in the objective space is very far from the global Pareto front. Without initialization the evolutionary algorithms do not reach the Pareto front. The path of the deterministic algorithm is shown for a set of fixed weights.

Our previous results of evolutionary algorithms with objectives where gradient based deterministic algorithms can not be used showed that the results were compatible or even better than by other phenomenological dose-optimization methods [1]. With this study we have found that if a part of the population is initialized with a good initial estimate of the Pareto front, that the results of the evolutionary algorithms improve significant more better than by any optimization of the GA parameters.

We do not know if there are special genetic operations which if applied could fill the gap between the Pareto fronts found by the evolutionary and deterministic algorithms. It seems that the global solutions requires a fine tuning of the decision variables which is far beyond what evolutionary algorithms can achieve in an acceptable number of generations. If a part of the population reaches via initialization a part of this region then the evolutionary algorithm is able to find solutions around this region.

A weakness of the MO evolutionary algorithms is the large number of function evaluations required to obtain a reasonable good local Pareto front. In the past MOEAs were compared for problems involving only very few decision variables. In such cases a random set of a few thousand sets of decision variables covers a large part of the objective space and evolutionary algorithms are able to produce fast solutions very close to the global Pareto front. The population size increases the efficiency of the MOEA algorithms by generating solutions

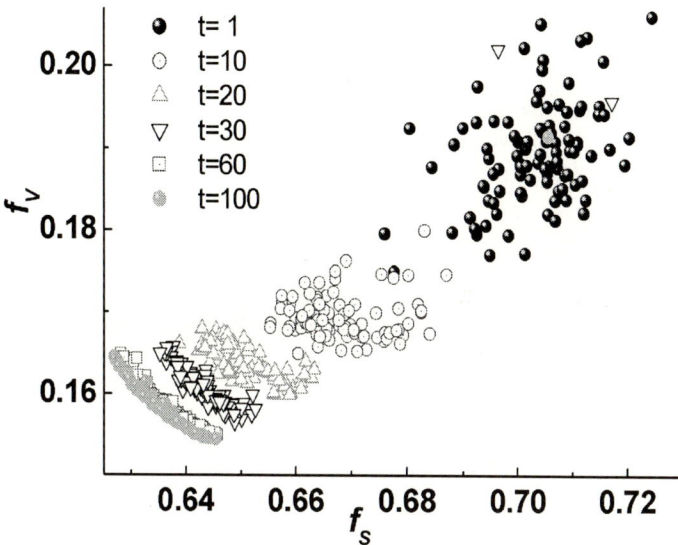

Fig. 9. An example of the SPEA algorithm for the case of an breast implant with 250 sources. The convergence in the objective space is presented at different number of generations t.

closer to the global Pareto set although population sizes of more than few hundred are not very practical and useful. In problems like in brachytherapy with up to 300 decision variables points close to the global Pareto front are in some cases of extreme low probability. In this case the evolutionary algorithms can not reach these regions whereas deterministic algorithms using information from local gradients are guided into these regions after only 10-20 iterations. In each such iteration a few times a line minimization is performed.

We have considered here only a simple coupling of the deterministic algorithm which delivers some optimal solutions as a starting point. We have found that the performance of the MOEA algorithms is significantly enhanced even if the initial solutions are obtained using different objectives for the deterministic algorithm. In the past algorithms were proposed where the evolutionary algorithms produce starting points for deterministic algorithms. Another possibility is to use a hybrid version of the evolutionary and deterministic algorithms. There are version where a hill-climbing operator is applied, with some probability which could be adapted to the performance of the algorithm.

We have estimated in some cases the performance of the MOEAs from Monte Carlo sampling experiments. We will in the future consider other approaches such as simulated multiobjective annealing where an external population is filled with the nondominated solutions found. This population is used in the optimization process by picking random members as starting points. We have to compare if such an algorithm with a single member can produce better results than MOEAs where it is assumed that the performance can be explained by mechanisms such as the implicit parallelism.

Acknowledgments. We would like to thank Dr. E. Zitzler and P. E. Sevinç for the FEMO library. This investigation was supported by a European Commission Grant (IST-1999-10618, Project: MITTUG).

References

1. Lahanas, M., Baltas, D., Zamboglou, N.: Anatomy-based three-dimensional dose optimization in brachytherapy using multiobjective genetic algorithms. Med. Phys. **26** (1999) 1904–1918
2. Lahanas, M., Milickovic, N., Baltas, Zamboglou, N.: Application of multiobjective evolutionary algorithms for dose optimization problems in brachytherapy. These proceedings
3. Bazaraa, M. S., Sherali, H. D., Shetty, C. M.: Nonlinear Programming, Theory and Algorithms. Wiley, New York. 1993
4. Zitzler, E., Deb, K., Thiele, L.: Comparison of Multiobjective Evolutionary Algorithms: Empirical Results. Evolutionary Computation. **8** (2000) 173–195
5. Yang, G., Reinstein, L. E., Pai, S., Xu, Z.: A new genetic algorithm technique in optimization of permanent ^{125}I prostate implants. Med. Phys. **25** (1998) 2308–2315
6. Yu, Y., Schell, M. C.: A genetic algorithm for the optimization of prostate implants. Med. Phys. **23** (1996) 2085–2091
7. Press,W. H., Teukolsky,S. A., Vetterling, W.T., Flannery,B. P.: Numerical Recipes in C. 2nd ed. Cambridge University Press, Cambridge, England. 1992
8. Michalewicz, Z.: Genetic Algorithms + Data Structures = Evolution Programs. Springer Verlag. 1996
9. Vicini, A., Quagliarella, Q.: Airfoil and wing design through hybrid optimization strategies. American Insitute of Aeronautics and Astronautics. Report AIAA-98-2729 (1998)
10. Das, I. Dennis, J.: A Closer Look at Drawbacks of Minimizing Weighted Sums of Objectives for Pareto Set Generation in Multicriteria Optimization Problems. Structural Optimization **14** (1997) 63–69
11. Zitzler, E., Thiele, L.: Multiobjective Evolutionary Algorithms: A Comparative Case Study and the Strength Pareto Approach. IEEE Transactions on Evolutionary Computation. **37** (1999) 257–271
12. Deb, K.: Multi-objective Genetic Algorithms: Problem Difficulties and Construction of Test Problems. Evolutionary Compuation **7** (1999) 205–230
13. Holland, J. H.: Adaptation in Natural and Artificial Systems. Ann Arbor, Unicersity o Michigan Press. 1975
14. Fonseca, M., Fleming, P. J.: Multiobjective optimization and multiple constraint handling with evolutionary algorithms I: A unified formulation Research report 564, Dept. Automatic Control and Systems Eng. University of Sheffield, Sheffield, U.K., Jan. 1995
15. Fonseca, M., Fleming, P. J.: An overview of evolutionary algorithms in multiobjective optimization. Evolutionary Computation **3** (1995) 1–16
16. Horn, J., Nafpliotis, N.: Multiobjective optimization using the niched Pareto genetic Algorithm. IlliGAL Report No.93005. Illinois Genetic Algorithms Laboratory. University of Illinois at Urbana-Champaign, 1993

On The Effects of Archiving, Elitism, and Density Based Selection in Evolutionary Multi-objective Optimization

Marco Laumanns, Eckart Zitzler, and Lothar Thiele

ETH Zürich, Institut TIK, CH–8092 Zürich, Switzerland,
{laumanns,zitzler,thiele}@tik.ee.ethz.ch,
http://www.tik.ee.ethz.ch/aroma

Abstract. This paper studies the influence of what are recognized as key issues in evolutionary multi-objective optimization: archiving (to keep track of the current non-dominated solutions), elitism (to let the archived solutions take part in the search process), and diversity maintenance (through density dependent selection). Many proposed algorithms use these concepts in different ways, but a common framework does not exist yet. Here, we extend a unified model for multi-objective evolutionary algorithms so that each specific method can be expressed as an instance of a generic operator. This model forms the basis for a new type of empirical investigation regarding the effects of certain operators and parameters on the performance of the search process. The experiments of this study indicate that interactions between operators as well as between standard parameters (like the mutation intensity) cannot be neglected. The results lead not only to better insight into the working principle of multi-objective evolutionary algorithms but also to design recommendations that can help possible users in including the essential features into their own algorithms in a modular fashion.

1 Introduction

Evolutionary algorithms have shown to be a useful auxiliary tool for approximating the Pareto set of multi-objective optimization problems. Several surveys of evolutionary multi-objective algorithms can be found in the literature, e.g., [5,17,1], which reflects the large number of different evolutionary approaches to multi-objective optimization proposed to date. While most of these algorithms were designed with regard to two common goals, fast and reliable convergence to the Pareto set and a good distribution of solutions along the front, virtually each algorithm represents a unique combination of specific techniques to achieve these goals.

As a consequence, there are no common guidelines available, how to best tailor an evolutionary algorithm to an application involving multiple objectives. Recently, some researchers have tried to address this problem by carrying out extensive comparative case studies, which can roughly be divided into two different categories. The first group compares different algorithms [17,8,19,18,20], but as the algorithms usually differ in more that just one aspect, it is very difficult to identify the features which are mainly responsible for the better performance of one algorithm over another. On the contrary, a few other studies take one algorithm and focus on a specific operator or parameter to tune, e.g. the selection method [10,12]. In this case the results are valid for the algorithm

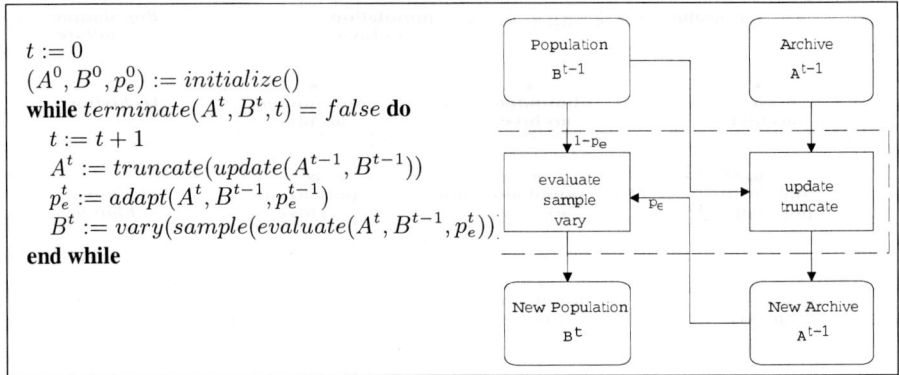

Fig. 2. Left: The general algorithm based on the unified model, A^t denotes the archive, B^t the population and p_e^t the elitism intensity at generation t. Right: A schematic view of the transition from one generation to the other.

The *evaluate* operator calculates the target sampling rates for each individual depending on the individuals in the archive and the population. For example, the primary fitness of individuals is derived from their rank with respect to their dominance level, usually referred to as 'non-dominated sorting' [15]. From these ranks, the sampling rates of an individual b in a set of individuals (population) B are calculated to simulate binary tournament selection.

According to these sampling rates, the *sample* operator then selects the required number of individuals from both the archive and the population as parents of the next generation. These parents are finally modified in the *vary* operator, where both recombination and mutation take place. In this study we use use one-point crossover and constant mutation rates σ_{mut}.

3 Experimental Design

Our aim is to investigate how the quality of the Pareto set approximation of the algorithms depends on certain algorithmic variables or parameters. This is a challenging task since the design space of the algorithm is huge and the performance indicators are noisy. In this study we pursue a novel approach which will be explained next, followed by a short descriptions of the test problem and the performance measure.

3.1 Methodology

In the design of an algorithm and specifically in the instances of our unified model, we face different types of design variables. In order to investigate the effect of these (independent) variables on the performance of the algorithm, we want to deal with all of them simultaneously in a common framework. For this purpose we distinguish

- ordinal, continuous variables (like the elitism intensity p_e)
- ordinal, discrete variables (like the population size)
- categorical variables (like different operators)

Categorical variables are mainly given by different operator instances, e. g. the archive limitation strategy in $truncate$.

For each experiment, we first decide on the variable(s) whose influence is to be assessed, and fix all other variables to some standard or optimal value from previous experiments (if those are available). The free variables are then varied randomly in their respective domain, and from each setting one observation of the performance indicator is drawn, i. e. the EA run once with with each setting.[1]

After a large number of data has been collected in this manner (usually 1000 observations per experiment), we look for a linear model that yields a good description of the data. The general form of this model for the i-th observation is

$$\mathcal{V}_i = a_0 + a_1 \cdot x_i^{(1)} + a_2 \cdot x_i^{(2)} + \ldots + a_n \cdot x_i^{(n)} + E_i \qquad (1)$$

where $x^{(1)}, \ldots, x^{(n)}$ are the explanatorial variables and $\mathcal{V}_i$ is the response (or target variable) – in our case the performance value. E_i are the random error caused by the randomness in the algorithm itself.[2]

The variables $x^{(j)}, 1 \leq j \leq n$ of the model can be any transformation or combination of original variables. Thus, the model is only linear in the coefficients, and non-linear dependencies from the variables (which are very likely for complex systems like evolutionary algorithms) can easily be traced. It furthermore allows to include the categorical variables, which have no natural value or order. This is done by indicator variables, where one variable is used for each category. If an observation falls into a specific category, the respective indicator variable is set to one, otherwise to zero. The coefficient of this indicator variable then shows the difference in the response that is caused by this category.

If only categorical variables were used, this method would be equivalent to analysis of variance (ANOVA), which has been applied to parameter studies of evolutionary algorithms by Schaffer et at. [14]. Here, we try to keep variables in their respective domain and want to use the order of the variables wherever possible. From the models we can then identify, which variables significantly effects the algorithmic performance.

3.2 Multi-objective 0/1 Knapsack Problems

To study the performance of different algorithmic configurations the multi-objective 0/1 knapsack problem is used, which has been subject to recent empirical case studies, both in the evolutionary computation community [19,8] and in the field of multiple criteria decision analysis [16].

The multi-objective 0/1 knapsack problem is a function of the form $k : \{0, 1\}^n \mapsto \mathbb{R}^m$, where n is the number of decision variables and m the number of objectives. In this work we refer to the definition in [19] and use the same parameters for the weights

---

[1] We prefer this over doing replications with identical settings for it leads to a better distribution of samples in the design space of the algorithm.

[2] In order for a linear regression to be viable, the random errors are assumed to be independent and identically distributed by a normal distribution with zero mean and equal variance. In this study we verified these assumptions using graphical diagnosis tools for the residuals $R_i = \mathcal{V}_i - \tilde{\mathcal{V}}_i$, like Normal plot and Tukey-Anscombe plot. Generally, an appropriate transformation of the target variable $\mathcal{V}$ led to the desired results.

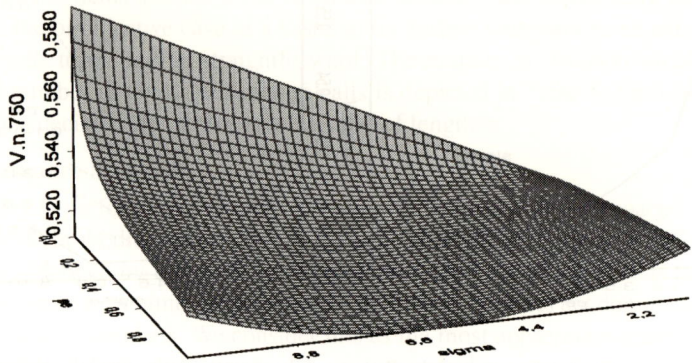

Fig. 4. Estimated response surface of the performance measure $\mathcal{V}$ to the normalized mutation rate σ and the elitism intensity p_e for $n = 750$. Two local minima arise at $(p_e = 1, \sigma \approx 6)$ and $(p_e = 0, \sigma = 1)$. A third order orthogonal polynomial was used as the model, the coefficients were estimated via linear regression.

- If the usually recommended mutation rate $\sigma = 1$ is taken, the best performance is achieved without elitism (p_e=0).
- The best overall performance can be reached with strong elitism ($p_e > 0.7$) in combination with high mutation rates ($\sigma \approx 5$).
- The combination of strong elitism with low mutation rates or weak elitism with high mutation rates shows the worst performance.

5 Bounding the Archive Size

In the previous section the influence of the mutation rate and the elitism intensity has been explored under the assumption that the archive is unbounded and can contain all non-dominated individuals found so far. In some cases, however, it may be desirable – or even mandatory – to limit its size for several reasons:

- The size of the true non-dominated set of the multi-objective problem may be exponentially large or even infinite.
- All implementations are restricted to limited resources (i. e. storage space).
- The complexity of the archive updating operator increases with the archive size.
- Genetic drift can occur since over-represented regions of the search space are favored in the (uniform) sampling process.

While the first three points mean that one has to limit the archive size for practical considerations though it would ideally be unlimited, the last point indicates that a (useful) truncation of the archive may also lead to a performance gain. In the following we set the maximum archive size $a := b$ and examine how different archive truncation methods affect the algorithmic performance.

5.1 Truncation Operators

Rudolph and Agapie [13] provide theoretical results about convergence properties of different archiving and selection strategies for multi-objective optimizers in finite search spaces. The authors state that algorithms with unlimited archive sizes do have the desired convergence properties provided that the variation operators match certain preconditions. However, they are usually not of practical relevance because of limited resources. Instead, they propose an updating operator that respects a maximum archive size via a strong elite preserving strategy, thus keeping the desired convergence properties. In our experiment this operator is named $truncate_1$, and it assures that for each dominated former archive member, at least one dominating individual must be included in the new archive. In contrast, the operator $truncate_2$ just makes a random choice which individuals to delete.

Another possibility to reduce the archive size is clustering: The possible members of the archive are grouped into distinct clusters based on some similarity measure. Then, each cluster will be represented in the archive by certain individuals. Clustering-based approaches are not strongly elite preserving, and they can be very time consuming. Many algorithms are based on iterative melioration of a given partitioning according to a predefined necessary optimality condition and can therefore lead to partitions which are only locally optimal. Clustering-based archive reduction is used in SPEA [20]: The individuals are clustered by the average linkage method into a distinct clusters, from which the centroid individual is included in the new archive. Here, we implement this as the $truncate_3$ operator.

Other approaches to limit the archive size can roughly be categorized as density dependent ranking techniques. We will discuss density estimation later in the context of biasing selection. The idea for this concept is quite intuitive: Though the archive must be truncated, one would like it to be as 'diverse' as possible. Hence, individuals in a densely populated area receive lower values and are discarded from the archive in favor of others. Different implementations of this concept are applied in [12], [7], [8], [2], or [3]. In this study we represent the method used in NSGA-II [3] by the operator $truncate_4$: For each objective coordinate the absolute difference of its predecessor and successor is aggregated for each individual, higher total values lead to better ranks.

Table 2 gives an overview of these different techniques and the implementations we use in our experiments. As a baseline, the $truncate_0$ operator is included, which represents the unlimited archive.

5.2 Experiments

At first two experiments were carried out where the normalized mutation rate was fixed at $\sigma = 1$ and $\sigma = 4$, respectively, while for each trial p_e was again varied randomly in $[0, 1]$ and one of the four $truncate$ operators was picked with equal probability.

For $\sigma = 4$ the $truncate$ operator had no significant effect on the results at all. This is not surprising since in these cases the archive size rarely exceeds a. For the small mutation rate $\sigma = 1$, where we had very large archives in the experiments before, the effect of the $truncate$ operator becomes significant. Fig. 5 (left) shows the box plots for the different operators. However, there is a strong interaction with the elitism intensity: The effect of the $truncate$ operator increases with p_e. Once again, if the archive is not

Table 2. Archive truncation methods in multi-objective evolutionary algorithms and operator instances for this study.

Method	No Reduction	Conservative	Random	Clustering	Density-based
Operator	$truncate_0$	$truncate_1$	$truncate_2$	$truncate_3$	$truncate_4$
Examples	VV [13] PR [13]	AR-1 [13] AR-2 [13]		SPEA [20]	PAES [7] M-PAES [8] PESA [2] NSGA-II [3]
Features	may grow very large, genetic drift	efficiency preserving, unreachable points	easy implementation, low complexity, genetic drift	good discrimination, adaptive metrics, high complexity	good discrimination, medium complexity

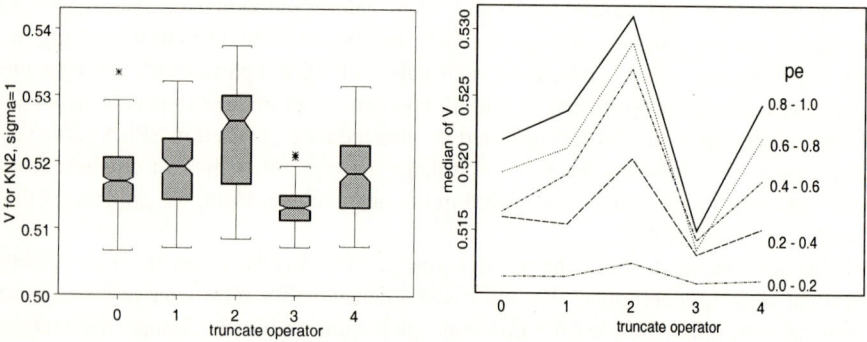

Fig. 5. Left: Box plots for $n = 500$, $\sigma = 1$ and different $truncate$ operators. The boxes range from the lower to the upper quartile of the observations, the median is marked with a bold line, outliers with an asterisk. The notches show confidence intervals for the group medians, by not overlapping they indicate a difference in location on a rough 5% significance level. Right: Interaction plots for the combined influence of the $truncate$ operator and the p_e level on $\mathcal{V}$. The median for each combination is shown.

used in the reproduction of offspring ($p_e = 0$), the effect of the $truncate$ operator is the weakest, this can be visualized by the interaction plots in Fig. 5 (right).

As a result we can claim that, if a reduction of the archive is necessary, is should be done carefully to minimize information loss. The random choice is always worst. In no case was the clustering approach significantly inferior to the unbounded archive, while it was the best for large p_e and low $\sigma = 1$. This shows that a 'good' reduction method is able to bias the search in favor of under-represented regions. Surprisingly, the density-based method does not reach significantly higher $\mathcal{V}$ values than the conservative method. The reason may be found in the specific technique, which has difficulties to deal with identical objective vectors, and not in this type of reduction method.

Nonetheless, there are other methods which directly influence the sampling rate of individuals based on their density, these will be discussed in the next section. One of the questions will be whether density dependent selection itself will lead to a higher performance gain than archive reduction.

5.3 Key Results

Depending on the choice of the truncation operator, archive reduction can either decrease or increase the performance:

- The clustering-based method leads to an increase of the average performance over the whole range of p_e and σ values. In other words: Clustering reduces the algorithm's sensitivity to suboptimal parameter settings without, however, a further improvement for the optimal settings.
- Random truncation lead to a significantly worse performance than the conservative method, which in turn was not significantly worse than the unbounded archive.

6 Density Dependent Selection

In multi-objective optimization, a uniform distribution of efficient points may be desirable in general. Unfortunately, the randomness in the evolutionary operators (genetic drift), the granularity of the search space or the topology of the objective function can make the archived solutions as well as the population itself exhibit a rather non-uniform distribution. A uniform sampling from the archive even reinforces these effects: Over-represented regions will be sampled by parents more often, and - given at least a certain locality of the variation operators - more offspring will be produced there.

One way to tackle this problem indirectly can be through a reduction of the archive, as described in the last section. A more direct approach, however, would be to bias the selection process in favor of under-represented regions. This will then be independent of the actual archive size and is much more flexible.

6.1 Density Estimation Methods

The density of individuals in a set A can serve as an a posteriori estimate of the probability density for the creation of the individuals in this set. This probability distribution is usually implicitly defined by the stochastic process which governs the evolutionary algorithm. However, it can easily be estimated (using standard probability density estimation techniques) and then be used to bias the sampling rates accordingly.

The relevance of density estimation in the context of (multi-objective) evolutionary algorithms has been put forward by [5], where the authors noted that the standard fitness sharing concept is essentially the application of a kernel density estimator. In [4], existing results from kernel density estimation were used to derive guidelines for the fitness sharing parameters.

Many advanced multi-objective evolutionary algorithms use some form of density dependent selection. Furthermore, nearly all techniques can be expressed in terms of density estimation, a classification is given in Table 3. We will make use of this as a further step towards a common framework of evolutionary multi-objective optimizers, and present the relevant enhancement of the unified model.

A straightforward density estimate is the histogram: The (multi-variate) space is divided into equally sized cuboids, and the number of points inside a cuboid is the estimate of the density for this subspace. Here we apply this technique in the $evaluate_1$ operator. Kernel density estimation is represented in $evaluate_2$ by the niche count function

Table 3. Density estimation techniques in multi-objective evolutionary algorithms and operators used in this study.

Method	None	Histogram	Kernel	Nearest Neighbor
Operator	$evaluate_0$	$evaluate_1$	$evaluate_2$	$evaluate_3$
Examples		PAES M-PAES PESA	all algorithms with fitness sharing	NSGA-II
Features: continuous smoothing control		no bin width	as kernel function window width	no local density

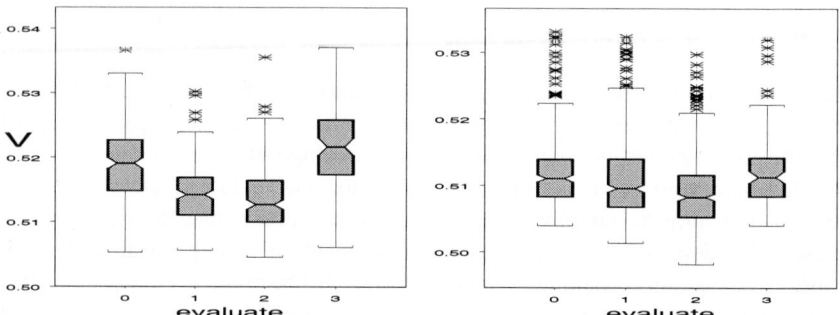

Fig. 6. Box plots for $n = 500$, $\sigma = 1$ (left) and $\sigma = 4$ (right) for different *evaluate* operators. For $n = 250$ and $n = 750$ the plots a similar.

known from standard fitness sharing. The $evaluate_3$ operator uses the technique from NSGA-II, where the distances to neighboring individuals is used to calculate a volume around each individual. The reciprocal value of this volume then serves as an estimate of the density, which is typical for nearest neighbor density estimates.

For the calculation of the target sampling rates, the normalized density estimate for each solution is added to its basic cost. If the minimal basic cost difference of differently valued solutions is at least 1 (like in our case of the dominance level), it will be assured that lower valued individuals always receive lower sampling rates regardless of their density. Thus, this method is used to bias selection *only between equally valued individuals*. Finally, a rank based assignment of target sampling rates is performed as before.

6.2 Experiments

In order to first investigate the effect of the density estimation alone, the elite preserving $truncate_1$ operator is used[3] and the maximal archive size is set to $a := b$. $\sigma \in [1, 10]$ and $p_e \in [0, 1]$ are again chosen randomly. The $evaluate_0$ operator is the reference case, where no density estimation is applied.

In contrast to the previous section, the influence of the *evaluate* operator is now significant for the whole range of $\sigma \in [1, 10]$. Figure 6 shows the box plots, again for

---

[3] We chose this as the common baseline since it did not show significant differences to the unlimited archive.

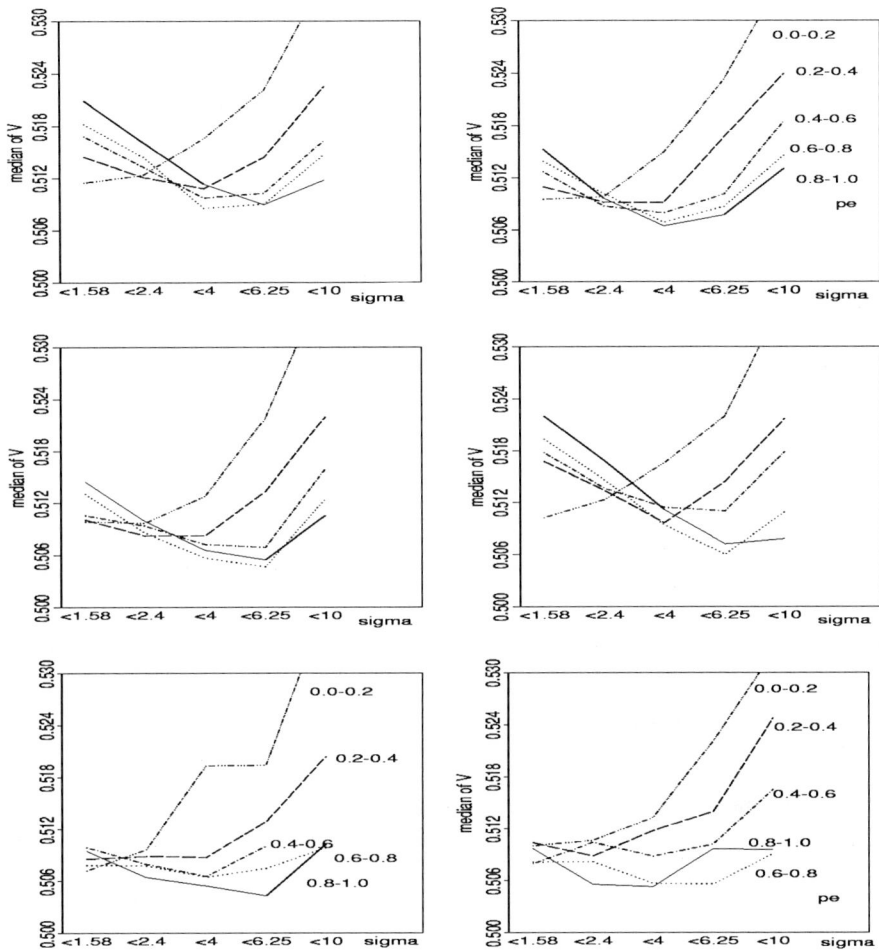

Fig. 7. Interaction plots for $n = 500$ for different *evaluate* operators. The upper row shows no density estimation (left) and the histogram method (right), the middle row the kernel density estimator (left) and the NSGA-II-based method (right), the lower row the clustering based truncation operator combined with the kernel density estimator (left) and the histogram method (right).

the medium size problem with $n = 500$. Notice that for $\sigma = 1$ the location differences of the $\mathcal{V}$ distribution between the first three operators are stronger than for $\sigma = 4$, while for $\sigma = 4$ the differences in the minimal $\mathcal{V}$ values are stronger.

Fig. 7 visualizes how the behavior of the different *evaluate* operators interacts with different p_e and σ settings. Obviously, the kernel density estimator as well as the histogram method lead to lower $\mathcal{V}$ values in all groups. The NSGA-II based method, however, does not change performance at lower mutation rates and only improves a few groups with stronger mutation. The corresponding fronts show that the performance gain of density dependent selection is only due to a broadening of the current non-dominated set, while approximation does not come closer in the middle. Comparisons to the real efficient set show that the evolutionary algorithm still finds solutions only in the central region of the real efficient set, even with the density dependent selection operators considered here.

Finally, the different density estimation techniques are combined with the clustering based *truncate* operator. Though this does not improve much on the top performance, the lower row of Fig. 7 shows that the truncation method reduces the sensitivity to suboptimal parameter settings, especially for low mutation rates.

6.3 Key Results

The performance gain of density dependent selection strongly depends on the accuracy of the applied density estimator:

- In general, the kernel density estimator as well as the histogram method improves the performance for all p_e and mutation rate settings.
- The combination of these techniques with the clustering-based archive truncation method leads to synergy effects in the sense that the algorithm becomes even less sensitive to suboptimal parameter settings than without density based selection.

7 Conclusion and Outlook

In this study we identified and analyzed the key elements of evolutionary multi-objective optimizers. From the results we can derive some design guidelines:

- Strong elitism should be used to achieve best performance, but in connection with a high mutation rate. The right combination of elitism intensity and mutation rate is the decisive factor for the performance.
- Density based selection can further improve the algorithmic performance by a broader distribution of solutions along the trade-off surface. Here, enough effort should be put in the density estimation technique since only a good estimate brings forth the desired results; fitness sharing as a simple kernel density estimator is a good choice.
- A good archive reduction method like clustering should be incorporated to make the algorithm robust concerning suboptimal parameter settings. However, the truncation operator has to be chosen carefully as inappropriate methods can decrease the performance.

As for the methodology of comparing different algorithms it should be noted that

- In comparisons of elitist against non-elitist algorithm not only the influence of the mutation rate must be considered but also the intensity of the elitism, which is often defined implicitly.
- When new algorithms are developed, the effect of all new features should be examined separately.

At present, these results are only valid for the bi-objective knapsack problem, but we believe that some fundamental characteristics have been found. It is subject to ongoing research to verify this for other problem classes and higher objective space dimensions. It can be expected, however, that performance differences increase, when more objectives and more difficult problems are considered.

References

1. Carlos A. Coello Coello. A comprehensive survey of evolutionary-based multiobjective optimization. *Knowledge and Information Systems*, 1(3):269–308, 1999.
2. D. W. Corne, J. D. Knowles, and M. J. Oates. The pareto envelope-based selection algorithm for multiobjective optimisation. In Marc Schoenauer et al., editor, *Parallel Problem Solving from Nature – PPSN VI*, Berlin. Springer.
3. K. Deb, S. Agrawal, A. Pratap, and T. Meyarivan. A fast elitist non-dominated sorting genetic algorithm for multi-objective optimization: NSGA-II. In Marc Schoenauer et al., editor, *Parallel Problem Solving from Nature – PPSN VI*, Berlin. Springer.
4. C. M. Fonseca and P. J. Fleming. Multiobjective genetic algorithms made easy: Selection, sharing and mating restrictions. In *First Int'l Conf. on Genetic Algorithms in Engineering Systems: Innovations and Applications (GALESIA 95)*, pages 45–52, London, UK, 1995. The Institution of Electrical Engineers.
5. C. M. Fonseca and P. J. Fleming. An overview of evolutionary algorithms in multiobjective optimization. *Evolutionary Computation*, 3(1):1–16, 1995.
6. D. E. Goldberg. *Genetic Algorithms in Search, Optimization, and Machine Learning*. Addison-Wesley, Reading, Massachusetts, 1989.
7. J. D. Knowles and D. W. Corne. The pareto archived evolution strategy: A new baseline algorithm for pareto multiobjective optimisation. In *Congress on Evolutionary Computation (CEC99)*, volume 1, pages 98–105, Piscataway, NJ, 1999. IEEE Press.
8. J. D. Knowles and D. W. Corne. M-PAES: A memetic algorithm for multiobjective optimization. In *Congress on Evolutionary Computation (CEC 2000)*, volume 1, pages 325–332, Piscataway, NJ, 2000. IEEE Press.
9. M. Laumanns, E. Zitzler, and L. Thiele. A unified model for multi-objective evolutionary algorithms with elitism. In *Congress on Evolutionary Computation (CEC 2000)*, volume 1, pages 46–53, Piscataway, NJ, 2000. IEEE Press.
10. S. Obayashi, S. Takahashi, and Y. Takeguchi. Niching and elitist models for MOGAs. In A. E. Eiben et al., editor, *Parallel Problem Solving from Nature – PPSN V*, pages 260–269, Berlin, 1998. Springer.
11. G. Ochoa. Consensus sequence plots and error thresholds: Tools for visualising the structure of fitness landscapes. In M. Schoenauer et al., editor, *Parallel Problem Solving from Nature – PPSN VI*, pages 129–138, Berlin, 2000. Springer.
12. G. T. Parks and I. Miller. Selective breeding in a multiobjective genetic algorithm. In A. E. Eiben et al., editor, *Parallel Problem Solving from Nature – PPSN V*, pages 250–259, Berlin, 1998. Springer.
13. G. Rudolph and A. Agapie. Convergence properties of some multi-objective evolutionary algorithms. In *Congress on Evolutionary Computation (CEC 2000)*, volume 2, pages 1010–1016, Piscataway, NJ, 2000. IEEE Press.
14. J. D. Schaffer, R. A. Caruana, L. J. Eshelman, and R. Das. A study of control parameters affecting online performance of genetic algorithms for function optimization. In J. D. Schaffer, editor, *Proceedings of the third international conference on genetic algorithms*, pages 51–60, San Mateo, CA, 1989. Morgan Kaufmann.
15. N. Srinivas and K. Deb. Multiobjective optimization using nondominated sorting in genetic algorithms. *Evolutionary Computation*, 2(3):221–248, 1994.
16. E.L. Ulungu, J. Teghem, P.H. Fortemps, and D. Tuyttens. Mosa method: A tool for solving multiobjective compinatorial optimization problems. *Journal of Multi-Criteria Decision Analysis*, 8(4):221–236, 1999.
17. D. A. Van Veldhuizen. *Multiobjective Evolutionary Algorithms: Classifications, Analyses, and New Innovations*. PhD thesis, Graduate School of Engineering of the Air Force Institute of Technology, Air University, June 1999.

For these and other reasons, the main interest has focussed on selection strategies which use the Pareto order only. Earlier analyses have shown that additional requirements are needed to prove that such algorithms converge towards efficient solutions ([11], for other results see [21]). These are elitist and conservative features of selection which may have other disadvantages. Especially, a slower introduction of new generated data is a possible drawback. Below, we show some experimental results concerning this question.

Another open problem concerns the adaptation of the parameters controlling the strength of mutations. In evolution strategies, the alternatives are encoded as floating point numbers instead of bit strings as in genetic algorithms. This allows a more appropriate representation of multiobjective optimization problems with continuous variables[1] and a reaching of higher precision results. For a fast approaching of an optimal solution, the 1/5 rule has been proposed as an adaptation procedure for the mutation strengths in scalar evolutionary algorithms. Below, we discuss why this rule is not applicable in the multiobjective case.

2 Terminology

Let us consider a multiobjective programming (MOP) problem defined as follows:

$$(MOP) \qquad min_{a \in A} f(a) \quad \text{with} \quad f: R^n \to R^q, q \geq 2,$$

$$\text{with} \quad A = \{a \in R^n : g_j(a) \leq 0, j \in \{1, ..., m\}\} \neq \emptyset$$

f is a vector-valued function consisting of components $f_k : A \to R$ which are called objectives or criteria. A is the set of feasible alternatives (or solutions) of the optimization problem. The functions $g_j : R^n \to R$ are called restriction functions. Setting $q = 1$, the (MOP) becomes a scalar optimization problem.

The most important mathematical solution concept for an MOP problem is called efficiency, or Pareto optimality (see, e.g., [7]). For its definition, a generalization of the "$\leq$" relation for vectors x, y in the objective space R^q is necessary:

$$x \leq y \quad \text{iff} \quad x_k \leq y_k \quad \text{for all} \quad k \in \{1, ..., q\} \quad \text{and} \quad x \neq y.$$

We use "$\leq$" for the component-wise generalization of the scalar "$\leq$":

$$x \leq y \quad \text{iff} \quad x_k \leq y_k \quad \text{for all} \quad k \in \{1, ..., q\}.$$

The set $E(A, f)$ of efficient (or Pareto-optimal) solutions is defined as:

$$E(A, f) := \{a \in A : \neg \exists b \in A : f(b) \leq f(a)\}.$$

$a \in E(A, f)$ is called efficient (or Pareto-optimal).

---

[1] Although, evolution strategies have so far only been scarcely developed and applied for multiobjective programming (see [17], [18], [9]). More recent approaches using floating number representations of decision variables were presented, for instance, at the PPSN V conference [3] and at the Workshop on Multiobjective Problem Solving from Nature (MPSN) [19].

The set of efficient solutions usually consists of many alternatives. For decision problems in practice it is often desired to select one alternative as 'solution' of the problem. For this purpose additional information related to the decision maker's preferences is requested. The property of efficiency is usually regarded as a minimum requirement for any such 'compromise solution'. For details on methods supporting a decision maker see, for instance, [12], [25], [27], or [28].

Let us denote the parent population of a multiobjective evolutionary algorithm (MOEA) in generation t as $M^t = \{a_1^t, ..., a_\mu^t\}$ with $a_l^t \in A \subseteq R^n$ for $l \in \{1, ..., \mu\}$. The set of efficient alternatives of population M^t is denoted as $E(M^t, f)$. The offspring population in t is $N^t = \{b_1^t, ..., b_\lambda^t\}$ with $b_l^t \in A$ for $l \in \{1, ..., \lambda\}$.

For an alternative $a \in A$ we define the *dominating set* as

$$Dom(a) := \{a' \in A : f(a') \stackrel{\leq}{=} f(a)\}.$$

For a population M^t the dominating set is defined as

$$Dom(M^t) := \bigcup_{i \in \{1,...,\mu\}, a_i^t \in E(M^t, f)} Dom(a_i^t).$$

For an alternative $a \in A$ the *dominated set* is defined as

$$Dom^-(a) := \{a' \in A : f(a) \leq f(a')\}.$$

The *dominated set* for a population M^t is then defined as

$$Dom^-(M^t) := \bigcup_{i \in \{1,...,\mu\}} Dom^-(a_i^t).$$

More details on dominating and dominated sets can be found in [11].

3 Selection Mechanisms

Compared with scalar optimization, the multiobjective nature of the MOP problem causes difficulties for the selection step of an evolutionary algorithm while other steps like mutation or recombination of alternative values are not necessarily affected by the multiobjective nature of the alternative evaluations.

In the scalar case, alternatives are judged by a single (real-valued) objective function which allows to define a linear order on the objective evaluations. With this, alternatives can be (completely) rank-ordered, a best alternative can be defined and so on. (Canonical) genetic algorithms then define probabilities of an alternative's reproduction based on its relative fitness. In evolution strategies usually an elitist selection strategy is applied which chooses the μ best of the λ children (comma strategy) or of the λ children and μ parents together (plus strategy) as parents for the next generation.

Considering a multiobjective evaluation of alternatives, these and similar concepts cannot be applied since only a partial order (in objective space) or

Pareto order (in decision space) is naturally defined on the objective evaluations. This implies that there may be alternatives which are not comparable (better or worse) considering fitness. There is no obvious way to define a ranking order, probabilities for reproduction etc. for which fitness serves in scalar optimization. In the literature (see, e.g., [4], [5], [26], and [14] for surveys) different approaches to overcome this problem have been proposed.

A quite simple approach is to define a scalarization function which maps the q objective functions to a single aggregated one such that a problem of type (SOP) can be analyzed. Approaches based on scalarization (see [16], for a discussion of theoretical aspects) are also often used in multicriteria decision making (MCDM), for instance in utility theory. Other MCDM scalarization concepts applicable in evolutionary algorithms are additive weighting and reference point approaches. Using such a scalarization function for the q objectives the familiar selection processes for (scalar) evolutionary algorithms can be used.

Several problems may be involved with such scalarization approaches: For instance, they do not work well in approximating the efficient set, they possibly have difficulties generating all efficient solutions in the concave case (see [5]), or they do not allow a 'truly' multiobjective analysis etc.

Some modifications of the scalarizing concept have been proposed to allow a generation of a diversified set of solutions approximating the efficient set. One approach uses the q different scalar objective functions. In each selection step one of these functions is chosen randomly ([17], [18]). Another approach [22] is based on a division of the population into q groups and an selection in each group k according to objective function k. Fourman [6] proposes to use pairwise selection based on one objective k selected according to pre-selected priorities (probabilities) randomly. Also for such approaches similar problems with representing the efficient set as in the scalarization approach occur.

Another, more consequent direction of transforming scalar evolutionary algorithms into multiobjective optimization methods consists in using only the Pareto order for selecting alternatives. Some of these approaches are based on pairwise comparisons, a kind of tournament selection [15] while others consider the alternative set of the population in total. For instance, an alternative is judged by the number of other alternatives which dominate it (see below). Surveys of selection mechanisms are, for instance, given by Fonseca and Fleming [5] and Tamaki, Kita, and Kobayashi [26]. Since these methods are parsimonious with additional information and usually good in representing the efficient set, such approaches are mostly preferred today.

Two of these selection mechanisms are very simple and have been implemented within LOOPS, the Learning Object-Oriented Problem Solver (see [9],[12]) which integrates different MCDM methods, evolutionary algorithms and other decision making tools (like, e.g., neural networks). One of them is based on the judging of an alternative according to the number of alternatives (plus 1) by which it is dominated (called *dominance rank* or *dominance grade*) [4]. The dominance grade works as an auxiliary fitness function.

A similar concept is the *dominance level* which is based on an iterative definition of different layers of efficiency [8]. The first layer includes the efficient alternatives of the population. The second one is defined by those alternatives which are efficient when the first layer is removed, etc. The number of the layer of an alternative defines an auxiliary fitness function to be minimized.

Although these approaches seem to overcome some of the problems caused by other selection mechanisms several possible disadvantages should be noted: The discrimination power of the dominance grade or dominance level as criteria for selection becomes low, especially in a progressed state of evolution or in high dimensional objective spaces. Below, there are some test results demonstrating such effects (for a modified dominance level approach). When at least μ alternatives of the $M^t \cup N^t$ population become efficient, they have the same dominance grade and dominance level such that a selection among them becomes arbitrary unless some additional selection rules are applied. Such selection rules are discussed in the next section. They have been introduced in [11] to ensure convergence and other interesting properties.

4 Efficiency Preservation and Negative Efficiency Preservation

The properties of efficiency preservation and negative efficiency preservation have been developed to have a multicriteria analogue to monotonicity properties as fulfilled in some evolutionary algorithms as, for instance, $(\mu+\lambda)$-evolution strategies for scalar optimization which guarantee monotonously increasing best values of the population. Such properties are useful for proving the convergence of an algorithm. Using efficiency preservation, the convergence of multicriteria evolutionary algorithms is shown for 'regular' MOP problems in [11].

A MOEA is called *efficiency preserving* iff for all $t \geq 0$ and $\emptyset \neq M^0 \subseteq A$

$$Dom(M^{t+1}) \subseteq Dom(M^t)$$

is fulfilled. It is called *negatively efficiency preserving* iff for all $t \geq 0$ and $\emptyset \neq M^0 \subseteq A$

$$Dom^-(M^{t+1}) \supseteq Dom^-(M^t)$$

holds.

Simply said, efficiency preservation is based on the idea of reducing the dominating set during the evolution while negative efficiency preservation increases the dominated set. In [11], we have shown that these properties are mutually independent.

Although efficiency preservation implies interesting convergence properties for MOEAs, this property easily brings up problems for a good representation of the efficient set because alternatives in $E(A, f) \setminus Dom(M^t)$ are unreachable.

For negative efficiency preservation, an equivalent formulation has been found [11] which is more appropriate for algorithmic application:

$$a^t \in E(M^t, f) \Rightarrow \exists a^{t+1} \in M^{t+1} : f(a^{t+1}) \leq f(a^t) \text{ for all } t \geq 0.$$

This means that an efficient parent a^t is either transferred into the next generation or replaced by an offspring alternative $a^{t+1} \in N^t$ such that a^{t+1} dominates a^t (or $f(a^{t+1}) = f(a^t)$).

Such a 'conservative' selection property can be regarded as a multiobjective generalization of the 'plus' concept ($(\mu + \lambda)$ evolution strategies) with elitist selection where parents survive as long as they are better than their offspring as discussed for scalar evolutionary algorithms.

The dominance grade and dominance level methods are modified as follows to provide negative efficiency preservation: First, dominance grades and dominance levels are calculated as usual. Then, for all parent alternatives this value is decreased by, for instance, 0.5. Because of this, parent alternatives which are efficient in $M^t \cup N^t$ are preferred to efficient offspring. Offspring replaces parents which are efficient in M^t only when they dominate them. To provide a maximum change of data while fulfilling negative efficiency preservation, we prefer inefficient offspring to inefficient parents with the same grade or level.

Let us finally remark that also many other multicriteria selection methods for evolutionary algorithms as discussed in the literature (see above) are neither efficiency preserving nor negatively efficiency preserving.

5 Some Test Results

In this Scetion, we show some experimental results with applying a selection concept as outlined above. Especially, we focus on the speed of obtaining a situation of a population consisting of efficient alternatives (with respect to the population) only for test problems of different 'complexity' and for different population sizes.

For the application runs of the MOEA the following parameters have been chosen: The mutation strengths are initialized with $\sigma_i = 0.1 \cdot \overline{x}_i^0$ where $\overline{x}^0$ is the starting point with $\overline{x}_i^0 > 0$ for $i \in \{1, ..., n\}$. Since the 1/5 rule is not applicable for MOEAs (see Section 6) we have chosen a concept of mutating the mutation strengths which is done with $\sigma = 0.1\sigma^{old}$. This means that mutation rates are changed by 10% on the average per generation. Recombination is performed between two randomly chosen alternatives with probability 0.1. The recombination is non-intermediary. Fitness sharing techniques are not applied.

To ensure negative efficiency preservation let us summarize the following necessary adaptations and parameter assessments:

1. The selection scheme is based on Pareto information only, i.e. dominance grade or dominance level. (Because both methods produce very similar results, only those for the dominance level approach are shown below.)
2. This scheme has to be modified (see above).
3. Parents and offspring are both considered during the selection (as in plus evolution strategies).
4. An elitist selection according to the evolution strategy concept is applied.

This is necessary because technically the results of the multicriteria fitness evaluation is are scalar values, e.g. the number of alternatives (plus 1) by which the considered alternative is dominated. Therefore, it is essential to consider the μ best alternatives as parents of the next generation only (unlike in standard GAs where fitness determinates selection probabilities).

We consider the following sizes of the parent generations: $\mu = 10$, $\mu = 50$, and $\mu = 250$. For the number of offspring, λ, always $\lambda = \mu$ is chosen. The number of simulated generations, #gen, is adapted to the population size because for a small population, a situation where the number efficient alternative approximately equals μ is reached faster than for larger population sizes. For $\mu = 10$, we choose #gen = 100, for $\mu = 50$, #gen = 200, and for $\mu = 250$, #gen = 1000.

Let us consider the following three MOP problems with $n = q = 2$ already analyzed using a MOEA in [10]: The first one is a convex MOP with quadratic objective functions:

(P1) $\qquad min f(x), f : R^2 \to R^2, f_i : x \mapsto x_i^2 \text{ for } i \in \{1,2\}$

with

$$x \in X = \{x \in R^2 : x_1 \geq 0, x_2 \geq 0, x_1 + x_2 \geq 5\}.$$

The second problem captures the problem of a disconnected efficient set caused by an appropriate restriction function with multiple local optima:

(P2) $\qquad min f(x), f : R^2 \to R^2 \text{ with } f_i : x \mapsto x_i \text{ for } i \in \{1,2\}$

and

$$x \in X = \{x \in R^2 : x_1 \geq 0, x_2 \geq 0, x_2 - 5 + 0.5\, x_1\, sin(4\, x_1) \geq 0\}.$$

For this problem the efficient set consists of 7 separated regions.

The third problem is even more complex because of locally (but not globally) efficient regions which are caused by non-monotonous objective functions (which are additionally not continuous):

(P3) $\qquad min f(x), f : R^2 \to R^2 \text{ with}$

$$f_1 : x \mapsto int(x_1) + 0.5 + (x_1 - int(x_1))sin(2\pi(x_2 - int(x_2))),$$
$$f_2 : x \mapsto int(x_2) + 0.5 + (x_1 - int(x_1))cos(2\pi(x_2 - int(x_2))),$$

and

$$x \in X = \{x \in R^2 : x_1 \geq 0, x_2 \geq 0, x_1 + x_2 \geq 5\}.$$

This problem is based on a partition of the decision space into squares $[i, i+1) \times [j, j+1), i, j \in N$ which are mapped onto circles with radius 1 using a 'rectangle into polar'-type of co-ordinate transformation. Because of the int operator (which calculates the integer part of its argument) the objective functions are neither continuous nor monotonous. Multiple locally efficient regions exist which are not (globally) efficient. In [10], graphical representations for these problems and additional information are provided.

In contrast to [10], these problems are now solved using modified the dominance level method which ensure negative efficiency preservation. Results of the (modified) dominance grade model are not shown because they are very similar, i.e. selection results for efficient alternatives do not differ.

Instead, we are interested in the number of efficient solutions during the simulation which mainly influences the possibility of getting new data into the population. Below there are 3 figures showing the number of efficient alternatives in $M^t \cup N^t$ for each generations number during the experiments.[2] In each graphic we have the results for the 3 different MOP problems given for a different population size.

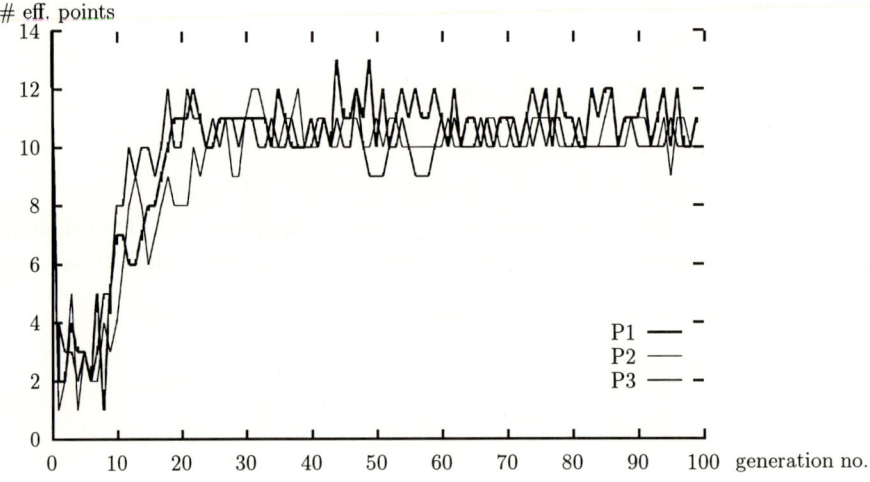

Fig. 1. Results of a MOEA for $\mu = \lambda = 10$.

In all cases we observe an increase of this number until it is approximately equal to μ. For larger populations this process obviously takes longer than for small populations. Except for the small population size with $\mu = 10$ where such an effect is not clearly visible, the different complexities of the problems influence the time to come to this type of balance: For problem (P1) μ efficient solutions are reached faster than for (P2), and for (P2) faster than for (P3). In these situations it is increasingly difficult for offspring alternatives to become parents of the next generation.

---

[2] In the Figures and Table 1, the results for an exemplary test run are shown. Other test runs with the same parameters of the evolutionary algorithm led to similar results.

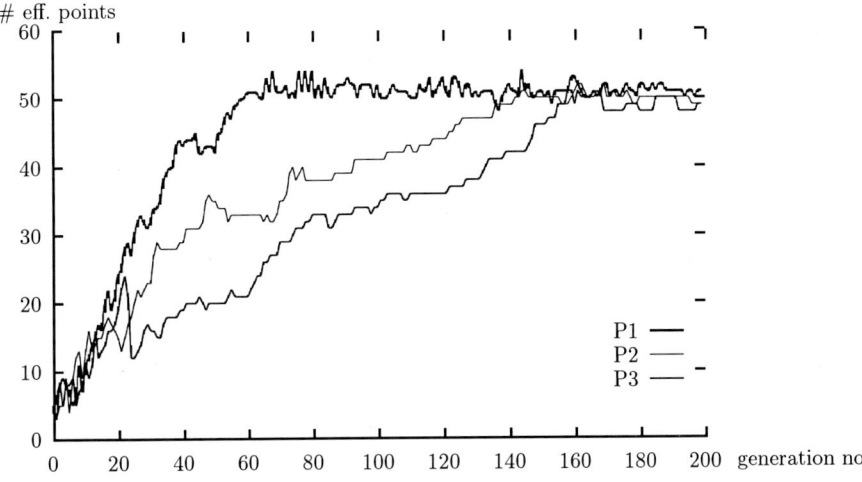

Fig. 2. Results of a MOEA for $\mu = \lambda = 50$.

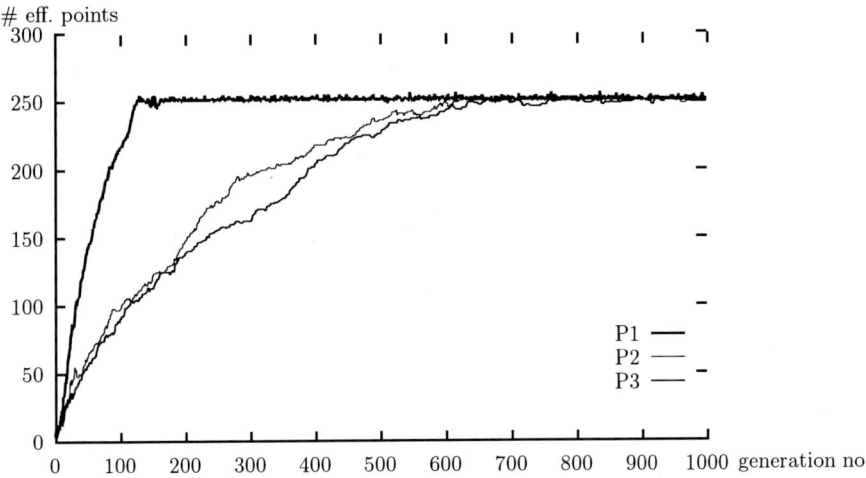

Fig. 3. Results of a MOEA for $\mu = \lambda = 250$.

Table 1. Number of generations until reaching μ efficient alternatives in $M^t \cup N^t$.

Problem	$\mu = 10$	$\mu = 50$	$\mu = 250$
(P1)	19	59	127
(P2)	23	142	601
(P3)	13	159	663

6 The Non-applicability of the 1/5 Rule

For scalar evolution strategies the 1/5 rule has been introduced by Rechenberg [20] (and further specified by Schwefel [23], p. 128-132, [24]) as a means for controlling the mutation strengths. Without an adaptation of mutation strengths, it is rather difficult to approach the optimum when the alternatives are already close to it (in relationship to the σ_i values). On the other hand, for small σ_i values it takes a long time to proceed to the optimum. Therefore, it is useful to start the optimization process with relatively large σ_i values and to reduce them while coming close to the optimum.

The 1/5 rule is based on theoretical considerations [20] which show that it is useful to have approximately 1/5 of the offspring alternatives being better than their parents. Too low values of σ_i lead to a higher percentage of successful offspring, too high values lead to a lower percentage. According to the 1/5 rule, the step sizes σ_i are increased if on average the portion of 'successful' offspring is larger than 1/5. If the portion is less than 1/5 the step sizes are decreased. Direction-specific adaptations of the σ_i (i.e. those which change the σ_i/σ_j ratios) are not supported by the 1/5 rule. Therefore, it is only possible to prescribe constant scaling factors for the co-ordinate directions because the σ_i remain in constant proportions (as long as they do not reach a minimum value > 0). From scalar optimization, it is known that the 1/5 rule fails when there are no continuous first partial derivatives of the objective function ([23], p. 136f, [24]), e.g. because of active restrictions.

For multiobjective applications additional difficulties arise for this kind of step size control. Considering multiple objectives, it is no longer possible to decide easily whether a new alternative is 'better' than an existing one. If we do not apply scalarizing functions or similar tools based on additional information we can only compare alternatives using the Pareto order. In objective space R^q an alternative a_j is better than an alternative a_i if a_j is in $a_i - (R_+^q \backslash 0)$. a_j is worse than a_i if a_j is in $a_i + (R_+^q \backslash 0)$. Otherwise, if a_i and a_j are not identical, then a_i and a_j are incomparable.

If a_j is uniformly distributed in a symmetric interval around a_i, or if a_j is normally distributed with expected value a_i then the probability of a_j being better than a_i is $1/(2q)$. The probability of a_i being better than a_j is the same. The probability of a_i and a_j being incomparable is $1-(1/(2q)+1/(2q)) = 1-1/q$. The situation of incomparability therefore increases with the dimensionality of the objective space.

In [20], the deduction of the 1/5 rule for scalar evolution strategies is based on an analysis of two optimization problems considered as representative for various other problems: the ball model and the corridor model. The ball model is an unrestricted optimization problem where the Euclidean distance to a given point $\hat{x}_1$ is minimized. This model is supposed to represent the situation of coming close to an optimal solution. In the corridor model the objective value of a point increases linearly along a tube of feasible points. Here, the main problem is to leave not the corridor. This model is assumed to be representative for optimization far away from an optimum. For both models, similar optimal values for the progress speed (average improvement in alternative space, or covered distance in the optimum direction, per generation) are obtained for a success probability (average number of offspring with better objective values than the parents) of 1/5.

The corridor model cannot easily be transformed into a multiobjective model. If we keep a one dimensional tube of feasible alternatives and linear objective functions we can come to very different situations in objective space. Let x^0 be the present point of a population of a MOEA with $\mu = 1$. Either almost all of the image of the 'positive half' of the corridor is enclosed in $f(x^0)$ minus the Pareto cone R_+^q (as in the scalar case) (unbounded problem) or only a small compact part of its image lies in the Pareto cone. The first case is very similar to the scalar model while in the second case almost all of the corridor's image in objective space is not enclosed in the Pareto cone. The points can march only towards a (compact) part of the cone's intersection with the feasible set (in objective space). This problem is rather related to close approximation than avoiding infeasibility. Therefore, we are discussing the ball model only.

The ball model for the multicriteria case can be defined as follows: Let us consider q goal vectors $\hat{x}^1, ..., \hat{x}^q \in R^n$. The objective functions $f_k, k \in \{1, ..., q\}$, are given by

$$min f_k(x) = \sum_{i=1}^{n}(x_i - \hat{x}_i^k)^2 \quad \text{for} \quad k \in \{1, ..., q\}.$$

The feasible set is $A = R^n$.

Let us consider an arbitrary point $x^0 \in R^n$ as parent. The probability density to obtain an offspring point $x \in R^n$ from x^0 is then

$$p(x^0 \to x) = (1/\sqrt{2\pi}\sigma)^n e^{-1/2\sigma^2 \sum_{i=1}^n (x_i^0 - x_i)^2}$$

The expected improvement in alternative space (progress speed) with respect to objective k, ϕ_k, is given by

$$\phi_k = (1/\sqrt{2\pi}\sigma)^n \int ... \int_{f(x) \leq f(x^0)} \left(\sqrt{\sum_{i=1}^n (x_i^0 - \hat{x}_i^k)^2} \right.$$

$$\left. - \sqrt{\sum_{i=1}^n (x_i - \hat{x}_i^k)^2} \right) e^{-\frac{1}{2\sigma^2} \sum_{i=1}^n (x_i^0 - x_i)^2} dx_1 ... dx_n.$$

Unfortunately, this integral is not analytically solvable. Additionally, it is not possible to use approximation functions as in [20] for the scalar case. Especially, one cannot simplify the n-fold integral by using a co-ordinate transformation (into ball co-ordinates). Therefore, also numerical integration using standard mathematical software (as, for instance, MAPLE V) fails since this software requires an integration area given by an interval. Here, the integration area is the intersection set of q n-balls around $\hat{x}^1, ..., \hat{x}^q$ with x^0 on the border of each

7 Conclusions

Our test results have shown some fundamental problems with applying 'conservative' selection mechanisms in multiobjective evolutionary algorithms. A population quickly reaches a situation where almost all parents are efficient in M^t. Applying efficiency preservation, such a parent can be replaced by an offspring alternative only when it is better considering the Pareto order which occurs more and more scarcely with growing q.

When negative efficiency preservation is used also offspring which is not comparable with efficient parents can become new parents. Such alternatives, however, cannot replace parents which are still efficient in $M^t \cup N^t$. When the parent population reaches the 'all efficient situation' negative efficiency preservation does not offer much advantages for offspring alternatives to survive compared with efficiency preservation. Therefore, both concepts may be rather unfavorable for quick progress.

Further test runs for other optimization problems should be done. Especially, it would be useful to compare different selection rules concerning the representation of the efficient set and the speed of computation (in relation to approximation quality). Such test runs should also help to find more heuristic information on 'good' mutation strengths and their control.

References

1. Bäck, T., Hoffmeister, F., Schwefel, H.-P.: A Survey of Evolution Strategies. In: R. K. Belew, L. B. Booker (eds.): Genetic Algorithms, Proceedings of the Fourth International Conference. Morgan Kaufmann, San Mateo (1991) 2-9
2. Calpine, H. C., Golding, A.: Some Properties of Pareto-Optimal Choices in Decision Problems. OMEGA International Journal of Management Science 4, 2 (1976) 141-147
3. Eiben, A. E. (ed.): Parallel Problem Solving from Nature - PPSN V. Proceedings of the 5th International Conference, Amsterdam, The Netherlands, September 27 - 30, 1998. Springer, Berlin (1998)
4. Fonseca, C. M., Fleming, P. J.: Genetic Algorithms for Multiobjective Optimization: Formulation, Discussion and Generalization. In: S. Forrest (ed.): Genetic Algorithms: Proceedings of the Fifth International Conference. Morgan Kaufmann, San Mateo (1993) 416-423
5. Fonseca, C. M., Fleming, P. J.: An Overview of Evolutionary Algorithms in Multiobjective Optimization. Evolutionary Computation 3, 1 (1995) 1-16
6. Fourman, M. P.: Compaction of Symbolic Layout Using Genetic Algorithms. In: Grefenstette, J. J. (Ed.): Genetic Algorithms and Their Applications. Proceedings of the First International Conference, July 24-26, 1985. Lawrence Erlbaum, Hillsdale (1988) 141-153
7. Gal, T.: On Efficient Sets in Vector Maximum Problems - A Brief Survey. European Journal of Operations Research 24 (1986) 253-264
8. Goldberg, D. E.: Genetic Algorithms in Search, Optimization, and Machine Learning. Addison-Wesley, Reading (1989)

9. Hanne, T.: Concepts of a Learning Object-Oriented Problem Solver (LOOPS). In: Fandel, G., Gal, T., Hanne, T. (eds.): Multiple Criteria Decision Making. Proceedings of the Twelfth International Conference, Hagen 1995. Springer, Berlin (1997) 330-339
10. Hanne, T.: Global Multiobjective Optimization Using Evolutionary Algorithms. Journal on Heuristics 6, 3 (Special issue on "Multiple Objective Metaheuristics" edited by X. Gandibleux, A. Jaskiewicz, A. Freville, and R. Slowinski) (2000) 347-360
11. Hanne, T.: On the Convergence of Multiobjective Evolutionary Algorithms, European Journal of Operational Research 117, 3 (1999) 553-564
12. Hanne, T.: Intelligent Strategies for Meta Multiple Criteria Decision Making. Kluwer, Boston (2001)
13. Holland, J. H.: Adaptation in Natural and Artificial Systems. University of Michigan Press, Ann Arbor (1975)
14. Horn, J.: Multicriterion Decision Making. In: T. Bäck, Fogel, D. B., Michalewicz, Z. (eds.): Handbook of Evolutionary Computation, IOP Publishing and Oxford University Press, New York and Bristol (1997) F1.9:1-F1.9:15
15. Horn, J., Nafpliotis, N., Goldberg, D. E.: A Niched Pareto Genetic Algorithm for Multiobjective Optimization. In: Proceedings of the First IEEE Conference on Evolutionary Computation, IEEE World Congress on Computational Intelligence. (1994) Vol. 1, 82-87
16. Jahn, J.: Scalarization in Vector Optimization. Mathematical Programming 29 (1984) 203-218
17. Kursawe, F.: A Variant of Evolution Strategies for Vector Optimization, In: Schwefel, H.-P., Männer, R. (eds.): Parallel Problem Solving from Nature. 1st Workshop, PPSN 1, October 1-3, 1990. Springer, Berlin (1991) 193-197
18. Kursawe, F.: Evolution Strategies for Vector Optimization. In: Proceedings of the Tenth International Conference on Multiple Criteria Decision Making. Taipei (1993) Volume III, 187-193
19. PPSN/SAB Workshop on Multiobjective Problem Solving from Nature (MPSN). Held at PPSN VI in Paris, September 2000. Papers available at http://www.rdg.ac.uk/ ssr97jdk/MPSNprogram.html
20. Rechenberg, I.: Evolutionsstrategie: Optimierung technischer Systeme nach Prinzipien der biologischen Evolution. Frommann-Holzboog, Stuttgart (1973)
21. Rudolph, G.: On a Multi-objective Evolutionary Algorithm and Its Convergence to the Pareto Set. In: Fogel, D. B., Schwefel, H.-P., Bck, T., Yao, X. (eds.): Proceedings of the Second IEEE World Congress on Computational Intelligence (WCCI'98) with Fifth IEEE Conference on Evolutionary Computation (IEEE/ICEC'98), Anchorage AK, May 4-9, 1998. IEEE Press, Piscataway NJ (1998) 511-516
22. Schaffer, J. D.: Multiple Objective Optimization with Vector Evaluated Genetic Algorithms. In: Grefenstette, J. J. (ed.): Proceedings of the First International Conference on Genetic Algorithms and Their Applications. July 24-26, 1985. Lawrence Erlbaum Associates, Hillsdale (1988) 93-100
23. Schwefel, H.-P.: Numerische Optimierung von Computer-Modellen mittels der Evolutionsstrategie. Birkhäuser, Basel (1977)
24. Schwefel, H.-P.: Numerical Optimization of Computer Models. Wiley, Chichester (1981)
25. Steuer, R. E.: Multiple Criteria Optimization: Theory, Computation, and Application. John Wiley & Sons, New York (1986)

26. Tamaki, H., Kita, H., Kobayashi, S.: Multi-objective Optimization by Genetic Algorithms: A Review. In: Proceedings of the 3rd IEEE International Conference on Evolutionary Computation. IEEE Press, Piscataway (NJ) (1996) 517-522
27. Vincke, P.: Multicriteria Decision-Aid. Wiley, Chichester (1992)
28. Zeleny, M.: Multiple Criteria Decision Making. McGraw-Hill, New York (1982)

Inferential Performance Assessment of Stochastic Optimisers and the Attainment Function

Viviane Grunert da Fonseca[1], Carlos M. Fonseca[1,2], and Andreia O. Hall[3]

[1] ADEEC, UCE
Universidade do Algarve
Faro, Portugal
[2] Instituto de Sistemas e Robótica
Coimbra, Portugal
[3] Departamento de Matemática
Universidade de Aveiro
Aveiro, Portugal
vgrunert@ualg.pt, cmfonsec@ualg.pt, andreia@mat.ua.pt

Abstract. The performance of stochastic optimisers can be assessed experimentally on given problems by performing multiple optimisation runs, and analysing the results. Since an optimiser may be viewed as an estimator for the (Pareto) minimum of a (vector) function, stochastic optimiser performance is discussed in the light of the criteria applicable to more usual statistical estimators. Multiobjective optimisers are shown to deviate considerably from standard point estimators, and to require special statistical methodology. The attainment function is formulated, and related results from random closed-set theory are presented, which cast the attainment function as a mean-like measure for the outcomes of multiobjective optimisers. Finally, a covariance-measure is defined, which should bring additional insight into the stochastic behaviour of multiobjective optimisers. Computational issues and directions for further work are discussed at the end of the paper.

1 Introduction

Stochastic optimisers, such as evolutionary algorithms, simulated annealing and tabu search, have found many successful applications in a broad range of scientific domains. However, only limited theoretical results concerning their performance are available. Typically, simple versions of the algorithms and/or objective functions must be considered in order to make the theoretical analysis possible, which limits their practical applicability. As an alternative, the performance of stochastic optimisers may be assessed experimentally on given problems by performing multiple, independent optimisation runs, and statistically analysing the results.

Two main issues are raised by such an inferential approach. Firstly, the very notion of optimiser performance must take into account the stochastic nature of the optimisers considered, as well as any other relevant optimiser characteristics, such as scale-independence, for example. As the same considerations apply to statistical estimators, optimiser performance will be discussed in that light in Section 2.

Secondly, specific statistical methodology may be needed, depending on the notion of performance adopted, in order to analyse the data produced by the optimisation runs. In particular, multiobjective optimisers such as multiobjective genetic algorithms (Fonseca and Fleming, 1995) produce sets of non-dominated objective vectors, instead of a single optimal objective value per run. Dealing with random sets introduces additional difficulties into the analysis. In Section 3, the attainment function is formally defined, and shown to relate closely to established results in random closed set theory. In particular, it is shown to be a measure analogous to the common mean, which considerably strengthens its role as a measure of multiobjective optimiser performance. Based on the same theory, variance-like and covariance-like measures are introduced which should provide additional insight into multiobjective optimiser performance.

Finally, computational issues are discussed. The paper concludes with a summary of the results, and a discussion of their implications for future work.

2 Inferential Performance Assessment

Optimiser performance can ultimately be understood in terms of the trade-off between the quality of the solutions produced and the computational effort required to produce those solutions, for a given class of optimisation problems. Experimentally, optimiser performance may be assessed in terms of:

1. The time taken to produce a solution with a given level of quality (run time),
2. The quality of the solutions produced within a given time,

where time may be measured in terms of number of iterations, number of function evaluations, CPU time, elapsed time, etc., and solution quality is defined by the problem's objective function(s). When considering *stochastic* optimisers, or deterministic optimisers under *random* initial conditions, both run time, in the first case, and solution quality, in the second case, are random, and the study of optimiser performance is reduced to the study of the corresponding distributions.

Hoos and Stützle (1998) propose the estimation and analysis of run-time distributions. It is worth noting that such time-to-event data may originate from improper distributions, since an optimiser may fail to find a solution with the desired quality in some runs. Also, the data may be subject to censoring whenever the actual run-time of the optimiser exceeds the practical time-limits of the experiment. Thus, the data may require special statistical treatment, of the kind usually encountered in statistical survival analysis. Run-time distributions are univariate distributions by definition, even if the problem considered involves multiple objectives.

Fonseca and Fleming (1996) suggested the study of solution-quality distributions. The outcome of a multiobjective optimisation run was considered to be the set of non-dominated objective vectors evaluated during that run. In the single-objective case, this reduces to a single objective value per run, corresponding to the quality of the best solution(s) found, and leads to the study of univariate distributions. In the multiple objective case, however, solution-quality distributions are either multivariate distributions, in the case where optimisers produce a single non-dominated vector per run, or set distributions, in the general case.

In this context, optimisers may be seen as estimators for the global (Pareto) optimum of a (vector) function. Therefore, optimiser performance can be viewed in the light of the performance criteria usually considered for classical statistical estimators. However, it must be noted that optimisers are actually more than simple estimators, as they must also provide the actual *solutions* corresponding to their estimates of the function's optimum.

2.1 The Single-Objective Case

As discussed above, the outcomes of single-objective optimisers consist of a single value per optimisation run, which is the objective value corresponding to the best solution(s) found. Therefore, one is interested in the stochastic behaviour of random variables X in $\mathbb{R}$, and the performance of optimisers and that of point estimators may be seen in parallel.

Good estimators should produce estimates which are *close* to the unknown estimand, both in terms of location and spread. The same applies to the outcomes of single-objective optimisers. Closeness in terms of location may be measured by the difference between the mean or the median of the corresponding distributions and the unknown estimand. This is known as the mean-bias and the median-bias, respectively. Ideally, both should be zero. Possible measures of spread are the variance and the interquartile-range, both of which should be small. Alternatively, location and spread may be combined in terms of the mean-squared-error, which should also be small.

Mean and variance are the first moment and the second centred moment of a distribution. They are efficiently estimated by the arithmetic mean $\bar{X}$ and the empirical variance s^2, respectively, when the underlying distribution is close to normal. This is the case with many statistical estimators, at least for sufficiently large sample sizes. The solution-quality distributions of optimisers, on the other hand, can (and should) be very asymmetric. Moreover, objective-scale information is ignored by some optimisers, which rely solely on order information. Thus, estimating the median and the inter-quartile range through their empirical counterparts might be preferred here, since quantiles are scale-invariant, i.e. $\tau[\gamma(X)] = \gamma[\tau(X)]$ for any quantile γ and any strictly monotonic transformation τ (Witting, 1985, p. 23).

In addition to closeness considerations, point estimates and optimisation outcomes should follow a type of distribution easy to deal with. In the case of estimators, this is usually the normal distribution. Optimisation outcomes, however, must follow a distribution which is bounded below (considering minimisation problems). Its left end-point should be as close to the unknown minimum as possible, and it should be right skewed, so that outcomes are likely to be close to the minimum. Given that the outcome of a single-objective optimisation run is the minimum of all objective values computed in the course of the run, ideal solution-quality distributions would be extreme-value distributions, the estimation of which has been vastly studied in the literature (see for instance Smith (1987), Lockhart and Stephens (1994), and Embrechts et al. (1997)), both in a parametric and in a semi/non-parametric setting.

The shape of a distribution can be assessed directly by estimating the cumulative distribution function, $F_X(\cdot)$, which completely characterises the underlying distribution. One may also wish to study specific aspects of the distribution, such as skewness (e.g. through the kurtosis) and tail behaviour (through end-point and tail-index estimation, for example). For minimisation problems, left and right-tail behaviour is related to best and worst-case performance, respectively.

2.2 The Multiobjective Case

When the optimisation problem is multiobjective, a whole front of Pareto-optimal solutions in $\mathbb{R}^d$ is to be approximated, and the outcome of an optimisation run may be a set of non-dominated objective vectors. For simplicity, the situation where the outcome of a run consists of a single objective vector shall be considered first.

Single objective vectors. The most common multivariate measure of location is possibly the arithmetic mean, which is now a vector in $\mathbb{R}^d$. If the unknown estimand is also a vector, as is the case with multivariate point estimators, this is clearly appropriate. The mean-bias of a point estimator, for example, can be written as the difference between the mean of corresponding distribution and the unknown estimand. Common measures of spread are the covariance matrix and other measures related to it (Mood et al., 1974, p. 351ff). All formulate spread in terms of deviation from the mean, which is a point.

In a multiobjective optimisation context, however, both bias and spread should be understood in terms of Pareto fronts. Note that the mean-vector of a number of non-dominated vectors could be located beyond a concave Pareto-optimal front to be approximated, outside the collection of all possible outcomes! Useful, alternative measures of bias and spread shall be given later in Section 3.

The shape of a multivariate distribution can be assessed through estimation of the cumulative multivariate distribution function, even though this is more challenging computationally than the corresponding univariate case. Again, solution-quality distributions should be skewed in the sense that outcomes should be likely to be close to the unknown Pareto front. Note that the Pareto front imposes a bound on the support of solution-quality distributions. Multivariate extreme-value theory is currently an active, but very specialised, area of research.

Multiple non-dominated objective vectors. Outcomes are represented by the random (point) sets $\mathcal{X} = \{X_j \in \mathbb{R}^d, j = 1, \ldots, M\}$ where the elements X_j are non-dominated within the set and random, and the number M of elements is random. Performance assessment oriented towards solution quality, as discussed so far, must take into account the particular set-character of the distributions involved.

Statistical estimators which produce a set of non-dominated vectors in $\mathbb{R}^d$ when applied to a data-set are not known to the authors, but curve estimators, seen as (continuous) random curve sets in $\mathbb{R}^2$, could be related. Bias measures for curve estimators $\hat{g}(\cdot)$, such as the average sum of squares

$$\frac{1}{k} \sum_{i=1}^{k} [\hat{g}(Z_i) - g(Z_i)]^2$$

or the supremum-norm

$$\sup_{i=1,\ldots,k} |\hat{g}(Z_i) - g(Z_i)|,$$

where the $Z_i \in \mathbb{R}$ are either random or deterministic, might suggest suitable analogues for the performance assessment of multiobjective optimisers. If the difference is replaced by the minimum Euclidean-distance between the random-outcomes X_j and the Pareto-optimal front to be approximated, one obtains measures similar in spirit to the *generational distance*, proposed by Van Veldhuizen and Lamont (2000).

Unlike curve-estimators, the performance of multiobjective optimisers is additionally affected by the variability of the outcomes within a set and by how uniformly the outcomes are distributed along the final trade-off surface (Zitzler, 1999; Zitzler et al., 1999; Van Veldhuizen and Lamont, 2000). Hence, taking into account the *overall* point-set character of the outcomes promises to be much more informative than just relying on summary measures such as the above. *Random closed set theory* (Matheron, 1975; Kendall, 1974) addresses precisely this issue. Note that the outcome-set $\mathcal{X}$ is closed.

The mean of a random-set distribution has been defined in various set-valued ways. One of the most popular is the *Aumann-mean*, which is defined as "the set of expected selections, where a selection is any random vector that almost surely belongs to the random set" (Cressie, 1993, p. 751). A possible estimator for this mean of some (general) random closed set $\mathcal{W}$ is formulated as

$$\bar{\mathcal{W}}_n = \frac{1}{n}(\mathcal{W}_1 \oplus \mathcal{W}_2 \oplus \ldots \oplus \mathcal{W}_n),$$

which is the Minkowski average of n independent copies $\mathcal{W}_1, \ldots, \mathcal{W}_n$ of $\mathcal{W}$ (Cressie, 1993, p. 751). Note that the Minkowski addition of two sets $\mathcal{A}_1$ and $\mathcal{A}_2$ is defined as

$$\mathcal{A}_1 \oplus \mathcal{A}_2 = \{a_1 + a_2 \mid a_1 \in \mathcal{A}_1, a_2 \in \mathcal{A}_2\}.$$

Clearly, the estimated Aumann-mean of the outcome set $\mathcal{X}$ of a multiobjective optimiser contains many more elements than the observed sets themselves (see Figure 1). In addition, the theoretical mean is typically a convex set, and

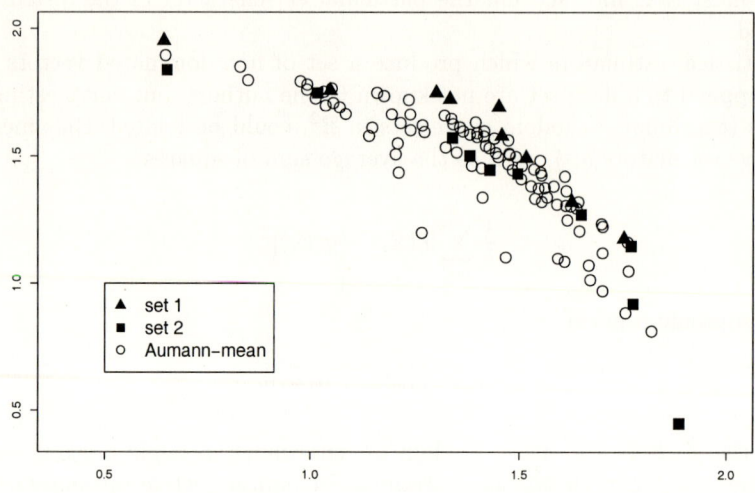

Fig. 1. The estimated Aumann-mean for two sets of non-dominated points in $\mathbb{R}^2$.

does *not* contain exclusively non-dominated elements. As for the vector-mean, some elements might even be located beyond Pareto-optimal front to be approximated, if it is concave. The Aumann-mean is therefore unsuitable as a measure of location in an optimisation context.

An alternative (empirical) mean-formula appears to be more useful. It is the *empirical covering function*, which is defined for a (general) random-set $\mathcal{W}$ as

$$p_n(z) = \frac{1}{n} \sum_{i=1}^{n} I\{z \in \mathcal{W}_i\}, \quad z \in \mathbb{R}^d. \tag{1}$$

The random sets $\mathcal{W}_1, \ldots, \mathcal{W}_n$ are independently and identically distributed like $\mathcal{W}$, and $I\{\cdot\}$ denotes the indicator function. The empirical covering function has been applied in the area of "Particle Statistics" to describe the average of possibly non-convex particles. Note that particles must be transformed into sets first, by choosing "reasonable locations and orientations" for them (Stoyan, 1998).

The attainment function and its empirical estimator (Shaw *et al.*, 1999; Fonseca and Fleming, 1996) turn out to be equivalent to the theoretical covering function $p(z) = P(z \in \mathcal{W})$ and its empirical counterpart. The definition of the attainment function and additional theoretical results are given in the following section.

3 The Attainment Function

3.1 Definition, Interpretation, and Estimation

The attainment function provides a description of the distribution of an outcome set $\mathcal{X} = \{X_j \in \mathbb{R}^d,\ j = 1, \ldots, M\}$ in a simple and elegant way, using the notion of goal-attainment. It is defined by the function $\alpha_{\mathcal{X}}(\cdot) : \mathbb{R}^d \longrightarrow [0, 1]$ with

$$\alpha_{\mathcal{X}}(z) = P(X_1 \leq z \ \vee \ X_2 \leq z \ \vee \ldots \vee \ X_M \leq z)$$
$$= P(\mathcal{X} \trianglelefteq z).$$

The symbol "$\vee$" denotes the logical "or". The expression $\alpha_{\mathcal{X}}(z)$ corresponds to the probability of at least one element of $\mathcal{X}$ being smaller than or equal to $z \in \mathbb{R}^d$, that is, the probability of an optimiser finding at least one solution which attains the goal-vector z in a single run. Clearly, the attainment function is a *generalisation of the multivariate cumulative distribution function* $F_X(z) = P(X \leq z)$. It reduces to the latter when $M = 1$, i.e. when the optimiser produces only one random objective vector per optimisation run.

The attainment function simultaneously addresses the three criteria of solution quality in the multiobjective context pointed out by Zitzler and colleagues (Zitzler, 1999; Zitzler et al., 1999), although not separately: a long tail (in the multidimensional sense) away from the true Pareto front may be due to the location of individual outcome elements in some runs (first criterion), to the lack of uniformity of the elements within runs (second criterion), or to the small extent of the outcome non-dominated sets (third criterion).

The attainment function can be estimated via its empirical counterpart

$$\alpha_n(z) = \frac{1}{n} \sum_{i=1}^{n} I\{\mathcal{X}_i \trianglelefteq z\},$$

the *empirical attainment function*, where the random sets $\mathcal{X}_1, \ldots, \mathcal{X}_n$ correspond to the outcomes of n independent runs of the optimiser. Note the similarity to the empirical covering function (1).

3.2 The Link to Random Closed Set Theory

The attainment function can be written in terms of so called "hit-or-miss probabilities", which are of fundamental importance in random closed set theory. For this, an alternative representation of the outcome set $\mathcal{X} = \{X_j \in \mathbb{R}^d,\ j = 1, \ldots, M\}$ with equivalent stochastic behaviour is chosen. It is the random (closed) set

$$\mathcal{Y} = \{y \in \mathbb{R}^d \mid X_1 \leq y \ \vee \ X_2 \leq y \ \vee \ldots \vee \ X_M \leq y\}$$
$$= \{y \in \mathbb{R}^d \mid \mathcal{X} \trianglelefteq y\} \tag{2}$$

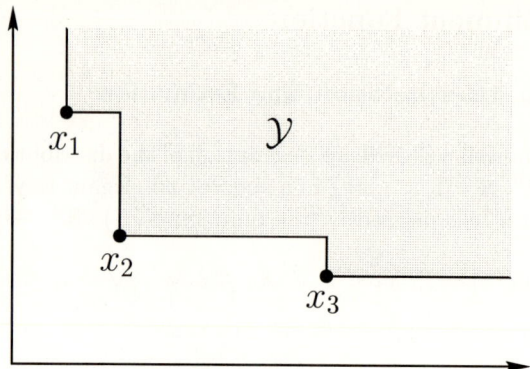

Fig. 2. Outcome set $\mathcal{X}$ with non-dominated realizations x_1, x_2, and x_3 and the set $\mathcal{Y}$ (here as a realization).

describing the region in $\mathbb{R}^d$ which is attained by $\mathcal{X}$ (see Figure 2). Using this alternative representation of $\mathcal{X}$, the attainment function may be expressed as

$$\alpha_{\mathcal{X}}(z) = P(z \in \mathcal{Y}), \quad z \in \mathbb{R}^d.$$

Hence, the attainment function of the outcome-set $\mathcal{X}$ is identical to the *covering function* of the associated random set $\mathcal{Y}$. Denoting n independent copies of the random set $\mathcal{Y}$ as $\mathcal{Y}_1, \ldots, \mathcal{Y}_n$ one can rewrite the empirical attainment function as

$$\alpha_n(z) = \frac{1}{n} \sum_{i=1}^{n} I\{z \in \mathcal{Y}_i\},$$

which shows the identity between the empirical attainment function of $\mathcal{X}$ and the empirical covering function of $\mathcal{Y}$ (compare with (1)).

Finally, the (theoretical) attainment function of $\mathcal{X}$ is identical to the *hitting function* or capacity functional (see e.g. Cressie (1993), Goutsias (1998)) of $\mathcal{Y}$ with support restricted to the collection of all one-point sets $\{z\}$ in $\mathbb{R}^d$. Hence, it can be expressed via hit-or-miss probabilities as

$$\alpha_{\mathcal{X}}(z) = P(\mathcal{Y} \cap \{z\} \neq \emptyset). \tag{3}$$

In general, the hitting function is defined over *all* compact subsets K in $\mathbb{R}^d$ (a definition for spaces more general than $\mathbb{R}^d$ is not of interest here). It fully characterises the stochastic behaviour of a random closed set in $\mathbb{R}^d$, and is of essential importance in random closed set theory. Note that the attainment function does *not* contain enough information to uniquely describe the stochastic behaviour of $\mathcal{X}$ and of $\mathcal{Y}$.

3.3 First-Order Moment Concepts

The hitting function of a (general) random closed set $\mathcal{W}$ defined over all compact subsets K in $\mathbb{R}^d$ is identical to the general first-order moment measure $C_{\mathcal{W}}^{(1)}(\cdot)$ of the same set, i.e.

$$C_{\mathcal{W}}^{(1)}(K) = P(\mathcal{W} \cap K \neq \emptyset).$$

The above definition generalises the notion of first-order moment of a binary random field $\{b(z) \mid z \in \mathbb{R}^d\}$, which is a collection of random function values $b(z)$ where $b(z)$ can be 0 or 1. Here, the first-order moment (measure) is defined as

$$P(b(z) = 1) = P(\mathcal{W} \cap \{z\} \neq \emptyset)$$

where the random closed set $\mathcal{W}$ is related to the binary random field according to $\mathcal{W} = \{z \in \mathbb{R}^d \mid b(z) = 1\}$. See Goutsias (1998).

As its formulation in (3) shows, the attainment function $\alpha_{\mathcal{X}}(\cdot)$ is the *first-order moment measure of the binary random field derived from the random set $\mathcal{Y}$* in (2) so that

$$\mathcal{Y} = \{z \in \mathbb{R}^d \mid b(z) = 1\}.$$

In other words, the attainment function $\alpha_{\mathcal{X}}(z)$ is the expected value of the binary random variable $\mathbf{I}\{\mathcal{Y} \cap \{z\} \neq \emptyset\} = \mathbf{I}\{\mathcal{X} \trianglelefteq z\}$ for all goals $z \in \mathbb{R}^d$. Hence, it makes sense to see the attainment function as a mean-measure for the set-distribution of $\mathcal{Y}$ and, in the wider sense, also of the outcome-set $\mathcal{X}$. Note again that the empirical covering function is used as a mean-measure in particle statistics.

As remarked above, the attainment function reduces to the (multivariate) distribution function $F_X(\cdot)$ for singular sets $\mathcal{X} = \{X\}$. The distribution function is a mean-measure for the distribution of the random set $\mathcal{Y} = \{y \in \mathbb{R}^d \mid X \leq y\}$ and, in a wider sense, also of $\mathcal{X} = \{X\}$. Thus, $F_X(\cdot)$ is a suitable alternative for the mean-vector of single objective vectors as a measure of location, when the optimisation problem is multiobjective and a whole Pareto-front is to be approximated (see the discussion in 2.2).

A notion of bias may be constructed in terms of the difference between the attainment function $\alpha_{\mathcal{X}}(z)$ associated with the optimisation outcomes and the ideal attainment function $\alpha_I(z) = \mathbf{I}\{z \in \mathcal{Y}^*\}$, where $\mathcal{Y}^*$ denotes the deterministic region attained by the true Pareto-optimal set of the problem. The bias, in this sense, is a function of a goal z, and indicates how far from ideal the optimiser is regarding the attainment of that goal.

3.4 Higher-Order Moment Concepts

When the first-order moment does not fully characterise a distribution, higher-order moments can contribute with additional information about the stochastic behaviour of a random variable/vector/set. Depending on the actual distribution, a finite number of higher-order moments may, or may not, be enough to fully characterise it. In statistics, this problem is known as the *problem of moments* (Mood et al., 1974, p. 81).

The attainment function, as mentioned before, does not uniquely determine the underlying set-distribution of $\mathcal{Y}$ (and of $\mathcal{X}$). In fact, it just addresses one aspect of optimiser performance, which is location-closeness. Closeness of the approximations in terms of spread (variability across runs) could be described by the variance (second centred moment). The second-order moment (measure) of $\mathcal{Y}$

$$\begin{aligned}C_{\mathcal{Y}}^{(2)}(\{z_1\},\{z_2\}) &= P\Big[(\mathcal{Y}\cap\{z_1\}\neq\emptyset) \wedge (\mathcal{Y}\cap\{z_2\}\neq\emptyset)\Big] \\ &= P[(\mathcal{X}\trianglelefteq z_1) \wedge (\mathcal{X}\trianglelefteq z_2)]\end{aligned}$$

(originally defined for the binary random field associated with $\mathcal{Y}$, see above) describes the probability of hitting two goals z_1 and z_2 simultaneously. Together with the first-order moment $C_{\mathcal{Y}}^{(1)}(\{z\})$, the attainment function, it can be used to explain the dependence structure between the two binary random variables $I\{\mathcal{Y}\cap\{z_1\}\neq\emptyset\}$ and $I\{\mathcal{Y}\cap\{z_2\}\neq\emptyset\}$. The difference

$$\begin{aligned}C_{\mathcal{Y}}^{(2)}(\{z_1\},\{z_2\}) &- C_{\mathcal{Y}}^{(1)}(\{z_1\})\cdot C_{\mathcal{Y}}^{(1)}(\{z_2\}) \\ &= P[(\mathcal{X}\trianglelefteq z_1) \wedge (\mathcal{X}\trianglelefteq z_2)] - \alpha_{\mathcal{X}}(z_1)\cdot\alpha_{\mathcal{X}}(z_2)\end{aligned}$$

can be seen as a form of covariance. If it equals zero, the two random variables are uncorrelated. On the other hand, if the event of attaining a goal z_1 is independent from the event of attaining the goal z_2 then the difference is zero (compare with Goutsias (1998)). Dependencies between more than two goals can be explored through higher-order moments of $\mathcal{Y}$. Eventually one can hope to completely characterise the distribution of the outcome-set $\mathcal{X}$ (through $\mathcal{Y}$).

Setting $z_1 = z_2 = z$, one obtains

$$C_{\mathcal{Y}}^{(2)}(\{z\},\{z\}) - C_{\mathcal{Y}}^{(1)}(\{z\})\cdot C_{\mathcal{Y}}^{(1)}(\{z\}) = \alpha_{\mathcal{X}}(z) - \alpha_{\mathcal{X}}^2(z) = \beta_{\mathcal{X}}(z)$$

which is simply the variance of the binary random variable $I\{\mathcal{Y}\cap\{z\}\neq\emptyset\} = I\{\mathcal{X}\trianglelefteq z\}$ for all $z\in\mathbb{R}^d$. The corresponding empirical estimator would be

$$\beta_n(z) = \frac{1}{n}\sum_{i=1}^{n}(\alpha_n(z) - I\{z\in\mathcal{Y}_i\})^2,$$

which is rather similar to the variance estimator defined by Stoyan (1998) for particle data.

4 Computational Issues

The practical usefulness of the attainment function as a measure of multiobjective optimiser performance is tied to the ability to estimate it from experimental data. The computation of the empirical attainment function (EAF) in arbitrary dimensions (i.e., number of objectives) is related to the computation of the multivariate empirical cumulative distribution function (ECDF), but computing the multivariate ECDF efficiently is not considered an easy task, either (see Justel et al. (1997)). In fact, whereas the univariate ECDF exhibits discontinuities at the data points only, the multivariate ECDF exhibits discontinuities at the data points *and* at other points, the coordinates of which are combinations of the coordinates of the data points. As the number of dimensions increases, the number of points needed to describe the ECDF (and the EAF) may easily become too large to store. Storing all relevant points may not always be necessary, however. The maximum difference between two EAFs, for example, can be computed without that requirement.

Similar considerations apply, to an even greater extent, to the estimation of the second-order moments. Work in this area is currently in progress.

5 Conclusions and Future Perspectives

The performance assessment of stochastic (multiobjective) optimisers was discussed in the light of existing criteria for the performance of classical statistical estimators, and theoretical foundations for the attainment function were established within the field known as random closed set theory.

The outcomes of multiobjective optimisers are random point sets in $\mathbb{R}^d$ denoted by $\mathcal{X}$. Alternatively, they can be represented by (continuous) random closed sets $\mathcal{Y}$ of a particular type with equivalent stochastic behaviour. Considering minimisation problems, the sets $\mathcal{Y}$ are unbounded towards $+\infty$ in every dimension, and are bounded below by the elements of $\mathcal{X}$.

The attainment function of an outcome set $\mathcal{X}$ is a first-order moment measure of the corresponding set $\mathcal{Y}$, defined over all possible one-point sets in $\mathbb{R}^d$ (the general moment measure is defined over all compact subsets in $\mathbb{R}^d$). Comparing the performance assessment of optimisers with that of statistical estimators showed that the attainment function is a kind of mean measure of the outcome-set $\mathcal{X}$. As such, it does indeed address a very sensible aspect of the stochastic behaviour of the optimiser, i.e. the location of the approximation. A suitable definition of bias was also suggested, which allows the location of the approximation to be seen with respect to the unknown Pareto-front.

The attainment function is a generalisation of the (multivariate) cumulative distribution function to the case of random non-dominated point sets. Thus, also the cumulative distribution function can be seen as a mean-measure for the set $\mathcal{Y}$ describing the region in $\mathbb{R}^d$ which is attained by a single objective vector X. In a wider sense, the cumulative distribution function can be seen as a mean-measure of $\{X\}$ itself. Regarding the empirical attainment function, it is hoped that it preserves some of the good properties of the empirical cumulative distribution function. Also, the attainment function makes it possible to compare the performance of multiobjective optimisers regardless of whether they produce one or many objective vectors per run!

The attainment function does not *fully* characterise the distribution of the random sets $\mathcal{X}$ or $\mathcal{Y}$. Extensions of the attainment function based on higher-order moment concepts were introduced which could contribute with additional information. They might eventually lead to the full characterisation of the distributions considered. This perspective gives the attainment function an advantage over performance measures such as the volume measure of the attained region $\mathcal{Y}$, which is related, for example, to the "size of the dominated space" in Zitzler (1999, p. 43f). In Matheron's (1975) theory, the distribution of a random closed set is characterised by hit-or-miss events (on which the attainment function is based) and "not by measures or contents" (Stoyan, 1998).

The results presented here are mainly of probabilistic nature. They are needed to support inferential methodology such as the test procedure for the comparison of optimiser performance used in Shaw et al. (1999), which is based on the maximum difference between two empirical attainment functions and on a permutation argument (see Good (2000)). Unlike the methodology proposed by Knowles and Corne (2000), such a test does not depend on auxiliary lines or suffer from multiple testing issues (see also Fonseca and Fleming (1996)). To a great extent, inferential methodology which truly exploits the attainment function and related concepts has yet to be developed.

Finally, the solution-quality view of optimiser performance could be combined with the run-time perspective by considering time an additional objective to be minimised. The outcome of an optimisation run would then be the set of non-dominated objective-vectors, augmented with time, evaluated during the run.

Acknowledgement. This work was funded by the Fundação para a Ciência e a Tecnologia under the PRAXIS XXI programme (Project PRAXIS-P-MAT-10135-1998), with the support of the European Social Fund, FEDER, and the Portuguese State. The authors wish to acknowledge the anonymous reviewers for their valuable comments on the original manuscript.

References

Cressie, N. A. C. (1993). *Statistics for Spatial Data*. Wiley Series in Probability and Mathematical Statistics. John Wiley & Sons, New York, revised edition.

Embrechts, P., Klüppelberg, C., and Mikosch, T. (1997). *Modelling Extremal Events*. Springer-Verlag, Berlin.

Fonseca, C. M. and Fleming, P. J. (1995). An overview of evolutionary algorithms in multiobjective optimization. *Evolutionary Computation*, 3(1):1–16.

Fonseca, C. M. and Fleming, P. J. (1996). On the performance assessment and comparison of stochastic multiobjective optimizers. In Voigt, H.-M., Ebeling, W., Rechenberg, I., and Schwefel, H.-P., editors, *Parallel Problem Solving from Nature – PPSN IV*, number 1141 in Lecture Notes in Computer Science, pages 584–593. Springer Verlag, Berlin.

Good, P. I. (2000). *Permutation Tests: A Practical Guide to Resampling Methods for Testing Hypotheses*. Springer Series in Statistics. Springer Verlag, New York, 2nd edition.

Goutsias, J. (1998). Modeling random shapes: An introduction to random closed set theory. Technical Report JHU/ECE 90-12, Department of Electrical and Computer Engineering, Image Analysis and Comunnications Laboratory, The John Hopkins University, Baltimore, MD 21218.

Hoos, H. and Stutzle, T. (1998). Evaluating Las Vegas algorithms — pitfalls and remedies. In *Proceedings of the 14th Conference on Uncertainty in Artificial Intelligence*, pages 238–245.

Justel, A., Peña, D., and Zamar, R. (1997). A multivariate Kolmogorov-Smirnov test of goodness of fit. *Statistics and Probability Letters*, 35:251–259.

Kendall, D. G. (1974). Foundations of a theory of random sets. In Harding, E. F. and Kendall, D. G., editors, *Stochastic Geometry. A Tribute to the Memory of Rollo Davidson*, pages 322–376. John Wiley & Sons, New York.

Knowles, J. D. and Corne, D. W. (2000). Approximating the nondominated front using the Pareto Archived Evolution Strategy. *IEEE Transactions on Evolutionary Computation*, 8(2):149–172.

Lockhart, R. and Stephens, M. (1994). Estimation and tests of fit for the three-parameter Weibull distribution. *Journal of the Royal Statistical Society, Series B*, 56(3):491–500.

Matheron, G. (1975). *Random Sets and Integral Geometry*. John Wiley & Sons, New York.

Mood, A. M., Graybill, F. A., and Boes, D. C. (1974). *Introduction to the Theory of Statistics*. McGraw-Hill Series in Probability and Statistics. McGraw-Hill Book Company, Singapore, 3rd edition.

Shaw, K. J., Fonseca, C. M., Nortcliffe, A. L., Thompson, M., Love, J., and Fleming, P. J. (1999). Assessing the performance of multiobjective genetic algorithms for optimization of a batch process scheduling problem. In *Proceedings of the Congress on Evolutionary Computation (CEC99)*, volume 1, pages 37–45, Washington DC.

Smith, R. (1987). Estimating tails of probability distributions. *The Annals of Statistics*, 15(3):1174–1207.

Stoyan, D. (1998). Random sets: Models and statistics. *International Statistical Review*, 66:1–27.

Van Veldhuizen, D. and Lamont, G. B. (2000). On measuring multiobjective evolutionary algorithm performance. In *Proceedings of the 2000 Congress on Evolutionary Computation*, pages 204–211.

Witting, H. (1985). *Mathematische Statistik I*. B. G. Teubner, Stuttgart.

Zitzler, E. (1999). *Evolutionary Algorithms for Multiobjective Optimization: Methods and Applications*. PhD thesis, Computer Engineering and Networks Laboratory, Swiss Federal Institute of Technology Zurich.

Zitzler, E., Deb, K., and Thiele, L. (1999). Comparison of multiobjective evolutionary algorithms: Empirical results (Revised version). Technical Report 70, Computer Engineering and Networks Laboratory, Swiss Federal Institute of Technology Zurich.

A Statistical Comparison of Multiobjective Evolutionary Algorithms Including the MOMGA-II

Jesse B. Zydallis[1], David A. Van Veldhuizen[2], and Gary B. Lamont[1]

[1] Dept of Electrical and Computer Engineering, Air Force Institute of Technology,
Wright-Patterson AFB, OH 45433, USA,
{Jesse.Zydallis, Gary.Lamont}@afit.af.mil
http://www.afit.af.mil

[2] Air Force Research Laboratory, Optical Radiation Branch,
Brooks AFB, TX 78235, USA
david.vanveldhuizen@brooks.af.mil

Abstract. Many real-world scientific and engineering applications involve finding innovative solutions to "hard" Multiobjective Optimization Problems (MOP). Various Multiobjective Evolutionary Algorithms (MOEA) have been developed to obtain MOP Pareto solutions. A particular exciting MOEA is the MOMGA which is an extension of the single-objective building block (BB) based messy Genetic Algorithm. The intent of this discussion is to illustrate that modifications made to the Multi-Objective messy GA (MOMGA) have further improved its efficiency resulting in the MOMGA-II. The MOMGA-II uses a probabilistic BB approach to initializing the population referred to as Probabilistically Complete Initialization. This has the effect of improving the efficiency of the MOMGA through the reduction of computational bottle-necks. Similar statistical results have been obtained using the MOMGA-II as compared to the results of the original MOMGA as well as those obtained by other MOEAs as tested with standard generic MOP test suites.

1 Introduction

We have developed an innovative Multiobjective Evolutionary Algorithm (MOEA), the Multi-Objective messy GA (MOMGA) that takes a novel approach to solving Multiobjective Optimization Problems (MOPs) [1,2,3,4]. Part of the novel approach of the MOMGA is its extension of the existing single-objective building block-based messy Genetic Algorithm (mGA) to the MOP domain [1, 3]. Building Blocks (BBs) define genes of chromosomes and contain the information that the EA is attempting to combine, evaluate and move towards the Pareto Front as extending the unproven *Building Block Hypothesis* [5]. The BB approach is used in the MOMGA to increase the number of "good" BBs that are present in each subsequent generation. These "good" BBs represent good material present in the current genotype population to be further exploited and used to move towards optimal solutions of a specific multiobjective problem. The associated test suite indicates that the MOMGA is as good, if not better, than other MOEA approaches for unconstrained numerical problems in a generic test suite.

As MOEAs have developed, studies of associated data structures and operators (recombination, mutation, selection) have reflected a variety of approaches yielding good results across a variety of MOPs. In general such MOEAs must 1) provide for an effective and efficient search towards the Pareto front, and 2) provide for a "uniform" or diverse distribution of points (finite) defining a calculated known pareto front [1,6]. Contemporary MOEA examples include the NSGA, NSGA-II, MOGA, NPGA, PAES, and the SPEA, which indirectly manipulate BBs, and the MOMGA and the MOMGA-II which explicitly and directly manipulate building blocks [1,6,7,8] In general, the successful evolution of all these MOEAs required the employment of more complex selection operators, such as niching, crowding, sharing, elitist, ranking, tournament or thresholding to achieve the above MOP Pareto front criteria. Since most implicit BB MOEAs are flowing to similar algorithmic architectures (operators, data structures), one should observe that they should have similar performance for appropriate parameter values. Of course, the NFL theorem implies that development of a generic robust MOEA will be quite difficult.

Depending on the particular MOP, various MOEAs have different statistical performance as expected across a variety of parameter settings and evaluation metrics [1,2]. Numerical unconstrained and constrained MOP test suites are considered here [1,2]. It is the intent of this paper then to indicate that the MOMGA-II is an excellent MOEA for solving MOPs with various characteristics and should be considered for a multitude of real-world applications. The basics of the MOMGA and MOMGA-II, appropriate MOP test suites, and evaluation metrics are discussed. Statistical analysis of the MOMGA, MOMGA-II, and other MOEAs using specific test suite numerical functions is presented with various conclusions indicated.

2 Multiobjective Problems

Although single-objective optimization problems may have a unique optimal solution, MOPs usually have a possibly uncountable set of solutions, which when evaluated produce vectors whose components represent trade-offs in decision space.

Pareto optimal solutions are those solutions within the search space whose corresponding objective vector components cannot be all simultaneously improved. These solutions are also termed *non-inferior*, *admissible*, or *efficient* solutions. Their corresponding vectors are termed *nondominated*; selecting a vector(s) from this vector set implicitly indicates acceptable Pareto optimal solutions (**genotypes**). These solutions may have no clearly apparent relationship besides their membership in the Pareto optimal set. It is simply the set of all solutions whose associated vectors are nondominated; we stress that these solutions are classified as such based on their *phenotypical* expression. Their expression (the nondominated vectors), when plotted in criterion (**phenotype**) space, is known as the *Pareto front*.

A MOEA's structure can easily lead to confusion when identifying Pareto components. During MOEA execution, a "local" set of Pareto optimal solutions (with respect to the *current* MOEA generational population) is determined at each EA generation and termed $P_{current}(t)$, where t represents the generation number. Many MOEA implementations also use a secondary population, storing all/some Pareto optimal solutions found through the generations [1,9]. Because a solution's classification as Pareto optimal depends upon the context within which it is evaluated (i.e., the given set of which it's a member), corresponding vectors of this set must be (periodically) tested, removing solutions whose associated vectors are dominated.

We term this secondary population $P_{known}(t)$. This term is also annotated with t (representing completion of t generations) to reflect its possible changes in membership during MOEA execution. $P_{known}(0)$ is defined as $\emptyset$ and P_{known} alone as the *final*, overall set of Pareto optimal solutions returned by an MOEA.

Different secondary population storage strategies exist; the simplest is when $P_{current}(t)$ is added at each generation (i.e., $P_{current}(t) \bigcup P_{known}(t-1)$). At any given time, $P_{known}(t)$ is thus the set of Pareto optimal solutions *yet found by the MOEA through generation t*. Of course, the *true* Pareto optimal solution set (termed P_{true}) is not explicitly known for problems of any difficulty. P_{true} is defined by the functions composing an MOP; it is fixed and does not change. Because of the manner in which Pareto optimality is defined $P_{current}(t)$ is always a non-empty solution set [1].

$P_{current}(t)$, P_{known}, and P_{true} are sets of MOEA genotypes where each set's phenotypes form a Pareto front. We term the associated Pareto front for each of these solution sets as $PF_{current}(t)$, PF_{known}, and PF_{true}. Thus, when using an MOEA to solve MOPs, the implicit assumption is that one of the following holds: $P_{known} = P_{true}$, $P_{known} \subset P_{true}$, or $PF_{known} \in [PF_{true}, PF_{true} + \epsilon]$ over some norm (Euclidean, RMS, etc.).

3 MOMGA

The MOMGA implements a deterministic process to produce the enumeration of all possible BBs, of a specified size, for the initial population. This process is referred to as Partially Enumerative Initialization (PEI). Thus, the MOMGA explicitly utilizes these building blocks in combination to attempt to solve for the optimal solutions in multiobjective problems. While finding the optimal solution is never guaranteed, the MOMGA statistically finds optimal or near optimal solutions to the functions presented in our standard MOP test suite[1,2,3]. The pseudocode for the MOMGA is presented in Figure 1.

The original messy GA consists of three distinct phases: *Initialization Phase, Primordial Phase, Juxtapositional Phase*. In the initialization phase, the messy GA begins by producing all building blocks of a specified size through a deterministic process referred to as Partially Enumerative Initialization (PEI). The

```
For n = 1 to k
    Perform Partially Enumerative Initialization
    Evaluate Each Population Member's Fitness (w.r.t. k Templates)
    // Primordial Phase
    For i = 1 to Maximum Number of Primordial Generations
        Perform Tournament Thresholding Selection
        If (Appropriate Number of Generations Accomplished)
            Then Reduce Population Size
        Endif
    End Loop
    // Juxtapositional Phase
    For i = 1 to Maximum Number of Juxtapositional Generations
        Cut-and-Splice
        Evaluate Each Population Member's Fitness (w.r.t. k Templates)
        Perform Tournament Thresholding Selection and Fitness Sharing
        P_known(t) = P_current(t) ∪ P_known(t − 1)
    End Loop
    Update k Competitive Templates (Using Best Value Known in Each Objective)
End Loop
```

Fig. 1. MOMGA Pseudocode

population grows exponentially as the BB size, k, is increased [1]. The MOMGA-II's population size initially is smaller and grows at a smaller rate. This was one of the factors in developing the MOMGA-II. To evaluate the fitness of a BB, a competitive template is used to fill in the unspecified bits prior to evaluation. This ensures that each evaluation is of a fully specified string, through the BB alone or in conjunction with the competitive template if necessary.

The primordial phase performs tournament selection on the population and reduces the population size if necessary. In the juxtapositional phase, the messy GA proceeds by building up the population through the use of the cut and splice recombination operator. The cut and splice recombination operator is used with the tournament thresholding selection operator. The process continues for a number of generations to yield strings with high fitness values for each fitness function. The combination of these three phases produces one *era* [5,10]. The algorithm continues for the user specified number of eras. Observe that an *epoch* refers to the number of times that the phases are executed for the same set of eras.

The MOMGA is an extension of the mGA to the multiobjective arena. The MOMGA proceeds through the three phases of the mGA to constitute an era. The MOMGA incorporates competitive templates for each objective function. The templates are initially created randomly and following each era, the best found individual in the population, for each objective function, becomes the new competitive template for that objective function, and is used in the following era. More detailed discussion of the MOMGA architecture is presented in [1,3].

4 MOMGA-II

We modified the MOMGA to use a probabilistic approach in initializing the population. This is referred to as Probabilistically Complete Initialization (PCI) [5]. The probabilistic BB approach initializes the population by creating a controlled number of BB clones of a specified size. These BBs then are filtered, through a Building Block Filtering (BBF) phase, to probabilistically ensure that all of the desired BBs are in the initial population. This approach should effectively reduce the computational bottlenecks encountered with PEI by reducing the size of the initial population required to obtain "good" statistical results when compared with the mGA. The pseudocode for the MOMGA-II is presented in Figure 2.

```
For n = 1 to k
    Perform Probabilistically Complete Initialization
    Evaluate Each Population Member's Fitness (w.r.t. k Templates)
// Building Block Filtering Phase
    For i = 1 to Maximum Number of Building Block Filtering Generations
        If (Building Block Filtering Required Based Off of Input Schedule)
            Then Perform Building Block Filtering
        Else
            Perform Tournament Thresholding Selection
        Endif
    End Loop
// Juxtapositional Phase
    For i = 1 to Maximum Number of Juxtapositional Generations
        Cut-and-Splice
        Evaluate Each Population Member's Fitness (w.r.t. k Templates)
        Perform Tournament Thresholding Selection and Fitness Sharing
        P_known(t) = P_current(t) ∪ P_known(t − 1)
    End Loop
    Update k Competitive Templates (Using Best Value Known in Each Objective)
End Loop
```

Fig. 2. MOMGA-II Pseudocode

The fast-messy GA consists of the following phases: *Initialization Phase, Building Block Filtering, Juxtapositional Phase*. The fmGA differs from the mGA in the Initialization and Primordial phase, which is referred to as the Building Block Filtering phase. The initialization phase utilizes PCI instead of the PEI implementation used in the mGA. The initial population is created randomly.

The BBF phase reduces the number of building blocks and stores the best building blocks found. This filtering is accomplished through a schedule consisting of the random deletion of bits in each of the chromosomes throughout the input schedule. The schedule specifies the generations to conduct the random deletion, the number of specified bits to delete from the chromosomes and the metric bars with average, median, standard deviation, max and min or tables listing explicit values of these terms, compact scatter diagrams, or Pareto fronts and Pareto solutions. In any case, the use of explicit hypothesis testing should reflect the qualitative evaluation of such comparative presentations [1].

Since MOEA observations are not normally distributed, parametric mean comparisons are not possible and therefore, non-parametric statistical techniques are employed. The Kruskal-Wallis H-Test is thus appropriate to use to determine if the hypothesis that the set of MOEA probability distribution results for a specific MOP differ with an observed significance level of 0.1. This technique then leads to comparing only two MOEAs; the hypothesis that the probability distributions of two MOEA results applied to a MOP differ uses the Wilcoxon rank sum test with a significance level of 0.2. Details of assumptions and requirements for this type of hypothesis testing can be found in [1,2]. For the various metrics, MOEA comparisons are presented using this hypothesis testing technique.

MOMGA-II effectiveness is compared with the results of the original MOMGA, as well as some other well know MOEAs, for the numerical unconstrained test function suite and one of our constrained test suite functions. In our previous research on the MOMGA, these MOEA test suites were utilized to allow for absolute comparisons of different MOEA approaches [1,2,4].

The results presented here reflect the output of each MOEA using the default values that have been specified in [1]. All of the MOMGA and MOMGA-II results are taken over 10 data runs in order to provide enough samples for statistical comparison. Each of these runs consist of 3 eras, 20 generations in each of the first 2 eras and 1 generation in the last era. The MOMGA and MOMGA-II utilize a string length of 24 bits and use a binary representation for the chromosomes. The MOMGA-II differs from the MOMGA by the BBF schedule that dictates the number of generations that BBF must take place. Where the MOMGA used the mGA calculation to determine the number of primordial generations to execute, the MOMGA-II always executes 9 generations of BBF and tournament selection, which leaves 11 generations of the juxtapositional phase. The MOMGA on the other hand may execute as many as 19 generations of the juxtapositional phase. This is an important fact to note since the creation of the solution strings takes place through the cut and splice operation of the juxtapositional phase. We show that the MOMGA-II obtains statistically similar results to the MOMGA with much fewer juxtapositional generations. This leads us to state that the MOMGA-II is finding "good" building blocks that the MOMGA finds, with the MOMGA-II finding them in less execution time.

Table 1 presents the execution time of the MOMGA and MOMGA-II across the various MOPs tested. As can be seen, the MOMGA-II is much more efficient than the MOMGA on the tested MOPs, with a 97% or higher efficiency improvement on all but one MOP over the MOMGA. The reason is that as the size of the BBs increase, the MOMGA must spend more time than the MOMGA-II in the creation and evaluation of the BBs. The MOMGA-II does not create every BB but probabilistically creates "good" BBs through the randomly deletion of bits from the population members until the correct BB size (string length) is

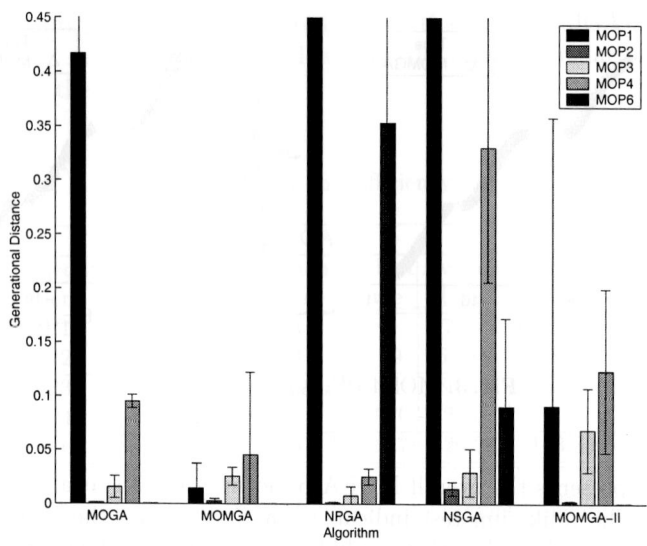

Fig. 4. Overall Generational Distance Performance

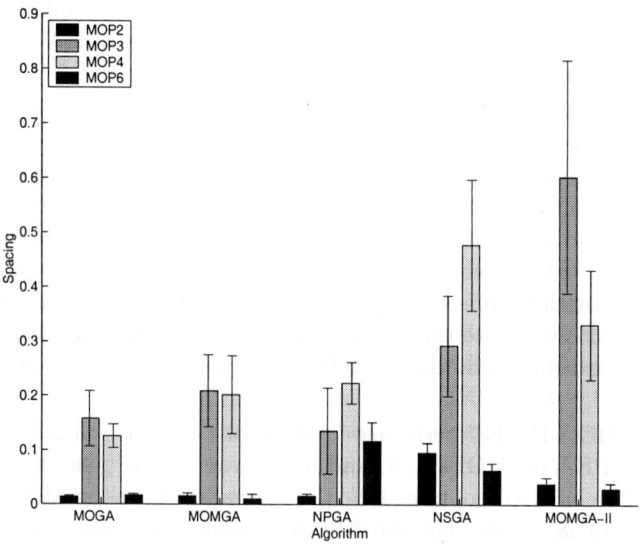

Fig. 5. Overall Spacing Performance

8   Conclusions

Through initial testing, the MOMGA-II has shown to be more efficient and has minor statistical differences when compared to its predecessor using our test suite of sufficiently complex MOPs [1]. This is as expected since we have

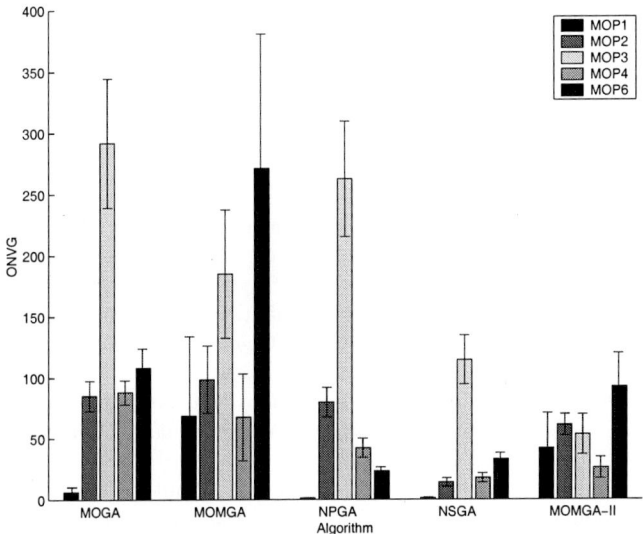

Fig. 6. Overall ONVG Performance

Table 2. Constrained MOP

Cardinality					
MOEA	Maximum	Minimum	Median	Average	Stand Dev
MOMGA	1823.00	805.00	1529.00	1418.90	346.60
MOMGA-II	443.00	347.00	396.50	398.00	32.73
Spacing					
MOEA	Maximum	Minimum	Median	Average	Stand Dev
MOMGA	0.2825	0.1150	0.1509	0.1717	0.0600
MOMGA-II	0.8666	0.5392	0.7182	0.7007	0.1048
Generational Distance					
MOEA	Maximum	Minimum	Median	Average	Stand Dev
MOMGA	0.0232	0.0096	0.0140	0.0149	0.0042
MOMGA-II	0.0250	0.0138	0.0190	0.0193	0.0035

moved from a PEI implementation to a PCI initialization of the population and have reduced the computation bottlenecks imposed by PEI. Additional focus on input parameters may yield better performance results as this testing was completed utilizing generic test parameter values. MOMGA-II results indicate that our modifications have increased the efficiency of the explicit BB MOMGA. Moreover, MOMGA and MOMGA-II are statistically different, but more effective than some other MOEAs across the test suites. Thus MOMGA-II should be considered as an efficient and effective MOEA for real-world applications.

Future work will include more extensive MOP test suite statistical evaluation with other genotype and phenotype metrics, more detailed comparison to other evolving MOEAs, and inclusion of other unconstrained, constrained, generated,

NP Complete (knapsack, TSP, etc.) and real world MOPs in larger test suites. In addition to MOMGA-II parametric studies, variations of competitive template structures will be analyzed along with parallelization of the MOMGA-II.

References

1. David A. Van Veldhuizen. *Multiobjective Evolutionary Algorithms: Classifications, Analyses, and New Innovations*. PhD thesis, Air Force Institute of Technology, Wright-Patterson AFB, 1999.
2. David A. Van Veldhuizen and Gary B. Lamont. On measuring multiobjective evolutionary algorithm performance. In *2000 Congress on Evolutionary Computation*, pages 204–211, 2000.
3. David A. Van Veldhuizen and Gary B. Lamont. Multiobjective optimization with messy genetic algorithms. In *2000 Symposium on Applied Computing*, pages 470–476, 2000.
4. David A. Van Veldhuizen and Gary B. Lamont. Multiobjective evolutionary algorithms: Analyzing state of the art. In *Journal of Evolutionary Computation*, volume 8:2, pages 125–147. MIT Press, 2000.
5. David E. Goldberg, Kalyanmoy Deb, H. Kargupta, and G. Harik. Rapid, accurate optimization of difficult problems using fast messy genetic algorithms. Technical Report 93004, University of Illinois at Urbana-Champaign, 1993.
6. Eckart Zitzler, K. Deb, and Lothar Thiele. Comparison of multiobjective evolutionary algorithms: Empirical results. In *Journal of Evolutionary Computation*, volume 8:2, pages 173–195. MIT Press, 2000.
7. Eckart Zitzler and Lothar Thiele. An evolutionary algorithm for multiobjective optimization: The strength pareto approach. Technical report, Computer Eng. and Comm. Networks Lab, Swiss Federal Institute of Technology (ETH), 1998.
8. Kalyanmoy Deb, Samir Agrawal, Amrit Pratap, and T Meyarivan. Fast elitist non-dominated sorting genetic algorithm for multi-objective optimization: Nsga-ii. In *Parallel Problem Solving from Nature - PPSN VI*, 2000.
9. David A. Van Veldhuizen and Gary B. Lamont. Multiobjective evolutionary algorithm research: A history and analysis. Technical Report TR-98-03, Air Force Institute of Technology, Wright-Patterson AFB, OH, 1998.
10. David E. Goldberg, Bradley Korb, and Kalyanmoy Deb. Messy genetic algorithms: Motivation, analysis, and first results. In *Complex Systems*, pages 493–530, 1989.
11. Hillol Kargupta. *SEARCH: Polynomial Complexity, And The Fast Messy Genetic Algorithm*. PhD thesis, University of Illinois at Urbana-Champaign, 1995.

Performance of Multiple Objective Evolutionary Algorithms on a Distribution System Design Problem - Computational Experiment

Andrzej Jaszkiewicz, Maciej Hapke, and Paweł Kominek[*]

Institute of Computing Science, Poznań University of Technology
ul. Piotrowo 3a, 60-965 Poznań, Poland
{Jaszkiewicz|Hapke|Kominek}@cs.put.poznan.pl

Abstract. The paper presents a comparative experiment with four multiple objective evolutionary algorithms on a real life combinatorial optimization problem. The test problem corresponds to the design of a distribution system. The experiment compares performance of a Pareto ranking based multiple objective genetic algorithm (Pareto GA), multiple objective multiple start local search (MOMSLS), multiple objective genetic local search (MOGLS) and an extension of Pareto GA involving local search (Pareto GLS). The results of the experiment clearly indicate that the methods hybridizing recombination and local search operators by far outperform methods that use one of the operators alone. Furthermore, MOGLS outperforms Pareto GLS.

1 Introduction

Multiple objective metaheuristics, e.g. multiple objective versions of genetic algorithms, simulated annealing or tabu search constitute one of the most active directions of research in multiple objective optimization [17]. Methods of this kind are claimed to be able to solve both continuous nonlinear optimization problems and large scale combinatorial optimization problems. A significant number of real life applications of multiple objective metaheuristics to continuous nonlinear problems (e.g. in engineering design) have been reported in the literature. At the same time the number of successful applications to real life multiple objective combinatorial optimization problems seems to be rather limited [17]. In our opinion, this is not due to the lack of potential applications but due to the difficulty of the problems.

In this paper we describe a comparative experiment on a real life combinatorial optimization problem corresponding to a distribution system design. The problem was analyzed in a project in which the authors were involved. The problem shares some characteristic features with many other real life problems:
- it involves objectives having different mathematical formulation (many test problems involve objectives that differ by parameters' values but have the same mathematical form, see e.g.[15]),
- although it has some similarities with the classic combinatorial optimization problems, e.g. location problems, it has a number of specific element, making the use of standard operational research methods unfeasible.

---
[*] This work has been supported by KBN grant No. 8T11F00619.

Four multiple objective metaheuristics have been tested on this problem. We used Pareto ranking based multiple objective genetic algorithm proposed by Fonseca and Flemming [3] (Pareto GA), multiple objective multiple start local search (MOMSLS) with random weight vectors, multiple objective genetic local search (MOGLS) [7] and an extension of Pareto GA involving local search (Pareto GLS). Two different quality measures were used in this experiment - coverage of two approximations to the nondominated set and estimation of the expected value of weighted Chebycheff scalarizing function on an approximation. In addition running time of the methods was taken into account. All the methods used the same solution encoding, the recombination and local search operators and the same way of constructing initial solutions. All the methods were implemented with MOEALIB++ library [16] and shared most of the C++ code.

The paper is organized in the following way. In the next section some basic definitions are given. The distribution system design problem is described in the third section. In the fourth section the four methods used in the experiment are briefly described. The fifth section describes customization of the methods to the problem. Design and the results of the experiment are characterized in the sixth section. The last section contains conclusions.

2 Problem Statement and Basic Definitions

The general multiple objective optimization (MOO) problem is formulated as:
$$\min\{f_1(\mathbf{x}) = z_1, ..., f_J(\mathbf{x}) = z_J\} \quad \text{(P1)}$$
s.t. $\quad \mathbf{x} \in D$,

where: *solution* $\mathbf{x} = [x_1, ..., x_I]$ is a vector of *decision variables*, D is the set of feasible solutions. If each decision variable takes discrete values from a finite set the problem is combinatorial.

The image of a solution $\mathbf{x}$ in the objective space is a *point* $\mathbf{z}^{\mathbf{x}} = [z_1^{\mathbf{x}}, ..., z_J^{\mathbf{x}}] = \mathbf{f}(\mathbf{x})$, such that $z_j^{\mathbf{x}} = f_j(\mathbf{x}), j=1,...,J$.

Point $\mathbf{z}^1$ *dominates* $\mathbf{z}^2$, $\mathbf{z}^1 \succ \mathbf{z}^2$, if $\forall j \; z_j^1 \leq z_j^2$ and $z_j^1 < z_j^2$ for at least one j. Solution $\mathbf{x}^1$ dominates $\mathbf{x}^2$ if the image $\mathbf{x}^1$ of dominates the image of $\mathbf{x}^2$.

A solution $\mathbf{x} \in D$ is *efficient (Pareto-optimal)* if there is no $\mathbf{x}' \in D$ that dominates $\mathbf{x}$. Point being an image of an efficient solution is called *nondominated*. The set of all efficient solutions is called the *efficient set*. The image of the efficient set in the objective space is called the *nondominated set* or *Pareto front*.

An *approximation to the nondominated set* is a set A of points (and corresponding solutions) such that $\neg \exists \mathbf{z}^1, \mathbf{z}^2 \in A$ such that $\mathbf{z}^1 \succ \mathbf{z}^2$, i.e. set A is composed of mutually nondominated points.

The point $\mathbf{z}^*$ composed of the best attainable objective function values is called the *ideal point*:
$$z_j^* = \min\{f_j(\mathbf{x}) | \mathbf{x} \in D\} \quad j = 1,...,J.$$

The point $\mathbf{z}_*$ composed of the worst attainable objective function values in the nondominated set is called the *nadir point*.

Range equalization factors [11, ch. 8.4.2] are defined in the following way:

$$\pi_j = \frac{1}{R_j}, j=1, \ldots, J,$$

where R_j is the (approximate) range of objective j in the set N, D or A. Objective function values multiplied by range equalization factors are called *normalized objective function values*.

Weighted Chebycheff scalarizing functions are defined in the following way:

$$s_\infty(\mathbf{z}, \mathbf{z}^0, \Lambda) = \max_j \{\lambda_j (z_j - z_j^0)\} = \max_j \{\lambda_j (f_j(\mathbf{x}) - z_j^0)\},$$

where $\mathbf{z}^0$ is a reference point, $\Lambda = [\lambda_1, \ldots, \lambda_J]$ is a weight vector such that $\lambda_j \geq 0 \ \forall j$.

Each weighted Chebycheff scalarizing function of this type has at least one global optimum (minimum) belonging to the set of efficient solutions. For each efficient solutions $\mathbf{x}$ there exists a weighted Chebycheff scalarizing function s such that $\mathbf{x}$ is global optimum (minimum) of s [11, ch. 14.8].

Linear scalarizing functions are defined in the following way:

$$s(\mathbf{z}, \Lambda) = \sum_j \lambda_j z_j = \sum_j \lambda_j f_j(\mathbf{x}).$$

Each weighted Chebycheff scalarizing function of this type has at least one global optimum (minimum) belonging to the set of efficient solutions [11, ch. 14.8].

Weight vectors that meet the following conditions:

$$\forall j \ \lambda_j \geq 0, \sum_{j=1}^{J} \lambda_j = 1,$$

are called *normalized weight vectors*.

3 The Distribution System Design Problem

An international company operating in Poland is going to completely rebuild its distribution system. The company entered the Polish market at the beginning of 90's and acquired two production plants located in two large Polish cities. The company focuses on the production and sales of cosmetics, detergents and washing articles. The annual turnover of the company is roughly $100 million. 85% of the turnover is generated at the Polish local market and the remaining 15% is an export to Eastern European countries. At the Polish market the company's products are mainly sold to wholesalers (60% of sales) and the chains of large retailers (hipermarkets, supermarkets) – 20% of sales.

The distribution system is going to be composed of up to 39 distribution centers (warehouses) placed is some predefined potential locations. Each distribution center will deliver company's products to clients located in one or more of 49 regions of Poland. The regions correspond to administrative districts. The goal of the study is to select appropriate locations for distribution centers and assignment of regions to the distribution centers. The following three objectives are taken into account:
- minimization of the total annual distribution cost, including transportation cost, warehousesing cost and cost of locked-up capital,
- minimization of the worst case riding time,

vector, constructs new random solution and applies to it local search. The outcome of the method is the set of potentially Pareto-optimal solutions found among the local optima. The algorithm is denoted by MOMSLS.

Parameters: K – size of the temporary population, optional parameter S – number of initial solutions, stopping criterion

Initialization:
    The set of potentially Pareto-optimal solution $PP:=\emptyset$
    The current set of solutions $CS:=\emptyset$

Generation of the first approximation of the ideal point:
    for each objective f_j
        Construct randomly a new feasible solution **x**
        Improve locally objective f_j starting from solution **x** obtaining **x'**
        Add **x'** to the current set of solutions CS
        Update set PP with **x'**

Generation of the initial set of solutions:
    repeat
        Draw at random a weight vector Λ
        Construct randomly a new feasible solution **x**
        Improve locally the scalarizing function s with weight vector Λ starting from solution **x** obtaining **x'**
        Add **x'** to the current set of solutions CS
        Update set PP with **x'**
    until CS contains S solutions

Main loop:
    repeat
        Draw at random a weight vector Λ
        From CS select K different solutions being the best on scalarizing function s forming temporary population TP
        Draw at random with uniform probability two solutions $\mathbf{x}_1$ and $\mathbf{x}_2$ from TP.
        Recombine $\mathbf{x}_1$ and $\mathbf{x}_2$ obtaining $\mathbf{x}_3$
        Improve locally the scalarizing function s with weight vector Λ starting from solution $\mathbf{x}_3$ obtaining $\mathbf{x}_3$'
        if $\mathbf{x}_3$' is better on s than the worst solution in TP and different in the decision space to all solutions in TP **then**
            Add $\mathbf{x}_3$' to the current set of solutions CS
        Update set PP with $\mathbf{x}_3$'
    until the stopping criterion is met
Outcome: the set of potentially Pareto-optimal solutions PP.

Fig. 1. Algorithm of multiple objective genetic local search - MOGLS.

4.3 Pareto Ranking Based Multiple Objective Genetic Algorithm

At present probably the most often used multiple objective metaheuristics are multiple objective genetic algorithms based on Pareto ranking (compare [14] ch. 3.3.2.2 and 3.3.2.3 and [17]). Although algorithms of this kind differ in some details, they share the idea of assigning solutions fitness on the basis of a ranking induced by the dominance relation proposed for the first time in [4] In this paper we use the particular implementation of the idea proposed in [3]. The algorithm uses fitness sharing in the objective space [3].

The algorithm proposed in [3] was extended by the use of the set of potentially Pareto-optimal solutions *PP* (secondary population) updated with each newly generated solution. The algorithm is denoted by Pareto GA.

4.4 Pareto Ranking Based Multiple Objective Genetic Local Search Algorithm

This algorithm hybridizes Pareto GA with local search. The starting population is composed of local optima obtained in the same way as in MOGLS and MOMSLS. In addition, after each recombination local search is applied to each offspring. The scalarizing function to be optimized is drawn at random in the same way as in MOGLS. The algorithm is denoted by Pareto GLS.

5 Adaptation of the Methods to the Distribution System Design Problem

The solutions are encoded by I lists of regions. i-th list contains regions assigned to the distribution center at location i. If the list is empty, there is no distribution center at this location (no minimum safety stock is maintained).

Initial solutions are constructed by assigning each region to a distribution center at a randomly selected location.

The idea of the recombination operator is to preserve important features common to both parents, i.e.
- common selections of distribution centers at the same locations,
- common assignments of regions to distribution centers at the same locations.

The details of the recombination operator algorithm are presented in 0.

Local search is used to improve the value of the weighted linear scalarizing function. We have tested also a version of local search that improves weighted Chebycheff scalarizing function but it resulted in worse performance of all the algorithms. Note that one of the quality measures is average value of the weighted Chebycheff scalarizing function. The value of the measure was, however, always better if the linear function was used in local search.

Local search uses neighborhood move that shifts a region from a distribution center at a location to a distribution center at a different location.

Greedy version of local search is used. It tests the neighborhood moves in random order and performs the first improving move. The local search algorithm is stopped when no improving move is found after testing all the possible neighborhood moves, which means that a local optimum is achieved.

The mutation operator shifts a randomly selected region to a distribution center at a randomly selected potential location.

> **for each** potential location of distribution center
>   **if** the distribution center at this location has non-empty list of assigned regions in both parents
>   Mark the location as *preliminary selected*
>   **else if** the distribution center at this location has non-empty list of assigned regions in one of the parents
>   Mark the location as *preliminary selected* with probability 0.5
> **for each** region
>   **if** the region is assigned in both parents to a distribution center at the same location
>   Assign the region to the distribution center at the same location
>   **else** /* The region is assigned to distribution centers at different locations */
>   **if** both the locations are marked as *preliminary selected*
>   Select at random one of the locations and assign the region
>   **else if** one of the locations is marked as *preliminary selected*
>   Assign the region to the distribution center at this location
>   **else**
>   Mark a randomly selected location as *preliminary selected* and assign the region to the distribution center at this location

Fig. 2. The algorithm of the recombination operator.

6 Computational Experiment

6.1 Quality Measures

One of the quality measures used in the experiment is the coverage of two approximations to the nondominated set. Let A and B be two approximations. The coverage measure is defined as:

$$C(A,B) = \frac{|\{z'' \in B\} | \exists z' \in A : z' \succ z''|}{|B|}.$$

The value $C(A, B) = 1$ means that all points in B are dominated by or equal to (covered by) some points in A. The value $C(A, B) = 0$ means that none point in B is covered by any point in A. Note that in general $C(A, B) \neq C(B, A)$. This quality measure was used in [15] and [8].

As the second quality measure we use the estimation of the expected value of weighted Chebycheff scalarizing function on approximation A over the set of normalized weight vectors Λ [5]:

$$R(A) = E\left(s_\infty^*(z^0, A, \Lambda)\right) = \int_{\lambda \in \Psi} s_\infty^*(z^0, A, \Lambda) p(\Lambda) d\Lambda,$$

where $s_\infty^*(\mathbf{z}^0, A, \Lambda) = \min_{\mathbf{z} \in A}\{s_\infty(\mathbf{z}, \mathbf{z}^0, \Lambda)\}$ is the best value achieved by function $s_\infty(\mathbf{z}, \mathbf{z}^0, \Lambda)$ on approximation A.

In order to estimate the expected value we use numerical integration, i.e. we calculate average value of the weighted Chebycheff scalarizing function over a large sample of systematically generated normalized weight vector. We use all normalized weight vectors in which each individual weight takes on one of the following values: $\{l/k, l = 0,...,k\}$, where k is a sampling parameter. The set of such weight vectors is denoted by Ψ_s and defined mathematically as:

$$\Psi_s = \{\Lambda = [\lambda_1,...,\lambda_J] \in \Psi \mid \lambda_j \in \{0, 1/k, 2/k, ..., k-1/k, 1\}\}.$$

With a combinatorial argument, we notice that this produces $\binom{k+J-1}{J-1}$ weight vectors. The estimation of measure R is calculated as:

$$R(A) = \frac{\sum_{\Lambda \in \Psi_s} s_\infty^*(\mathbf{z}^0, A, \Lambda)}{|\Psi_s|}$$

The parameter k was set to 50.

The C++ codes used to calculate the two quality measures are inlcuded in the MOEALIB++ library [16].

6.2 Experiment Design

Because of confidence reasons the real life data have been slightly modified. All of them, however, are in realistic ranges.

Each method has been run 10 times. MOGLS has used temporary population of size 30. The number of initial solutions has been set equal to 60. This value has been set with the approach proposed in [7]. This number of initial solutions assures that on average 30 best solutions on a randomly chosen weighted linear scalarizing function are of the same quality as 30 local optima of the same function. The sets of potentially Pareto-optimal solutions generated in each run have been saved after 60, 120,..., 600 recombinations.

Pareto GLS has used population of size 60. In other words, Pareto GLS has been starting by generating the same number of local optima as MOGLS. The number of generations has been set to 11. This has assured the same number of recombinations as in the case of MOGLS. The sets of potentially Pareto-optimal solutions have been saved after the same numbers of recombinations as in the case of MOGLS. The mutation probability has been set equal to 0.1.

MOMSLS has been performing 660 local searches, i.e. the same number of local searches as MOGLS and Pareto GLS. The sets of potentially Pareto-optimal solutions have been saved after 120, 180,..., 660 local searches, i.e. after the same number of local searches as in the case of MOGLS and Pareto GLS.

Pareto GA has used a population of size 100 and has been performing 400 generations. As this method does not use local search in performs more recombinations per second than Pareto GLS. The above parameters assured running times comparable to MOGLS and Pareto GLS. The mutation probability has been set equal to 0.1.

Table 1. Average running times of the methods.

MOGLS		Pareto GLS		MOMSLS		Pareto GA	
Number of recombinations	Average running time [s]	Number of recombinations	Running time [s]	Number of local searches	Running time [s]	Number of generations	Running time [s]
60	66.1	60	67	120	116.1	40	15.7
120	72	120	76.3	180	174.6	80	34.4
180	76.9	180	84.6	240	233.2	120	53.6
240	81	240	92.6	300	291.6	160	73.2
300	85	300	100.8	360	350.8	200	92.7
360	89.2	360	108.9	420	409.7	240	111.4
420	92.7	420	117.2	480	468.3	280	130.8
480	96.3	480	125.7	540	526.9	320	150
540	100	540	134.2	600	585.8	360	168.5
600	103.9	600	142	660	644.4	400	187.1

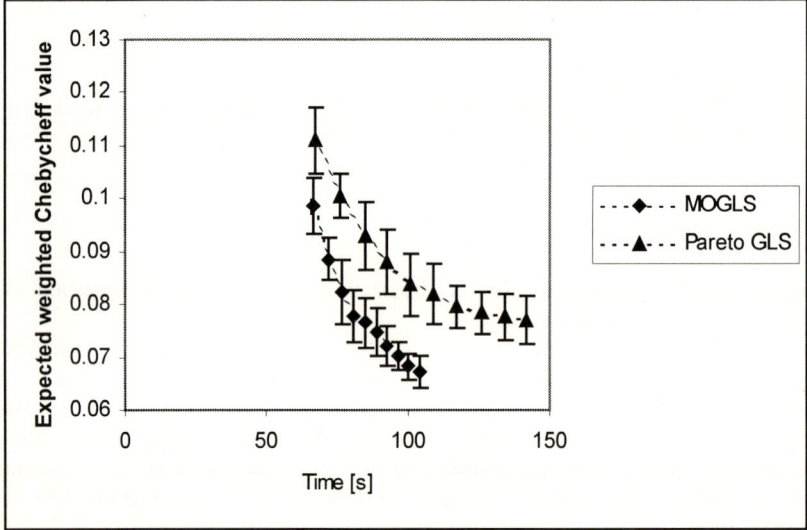

Fig. 3. Results of comparison with expected value of weighted Chebycheff scalarizing function. The chart shows ± standard deviation range of the quality measure.

6.3 Results

Table 1 presents average running times of the methods. The dispersion of the running times after a given number of iterations is not given, it has been, however, very low. Note that MOMSLS has required much more time than MOGLS and Pareto GLS to perform the same number of local searches. This is caused by the fact that solutions obtained by recombination are much closer to local optima than random solutions. Thus, local search needs less iterations to achieve a local optimum.

Table 2 presents results of evaluation with expected value of weighted Chebycheff scalarizing quality measure. The results clearly indicate that the two methods hybridizing recombination and local search operators by far outperform methods that use one of the operators alone. In particular the comparison of Pareto GLS, Pareto GA and MOMSLS is a clear example of synergy of the two operators. Pareto GLS is a simple combination of the two latter methods and does not include any element that cannot be found in Pareto GA or MOMSLS.

Table 2. Results of comparison with expected value of weighted Chebycheff scalarizing.

MOGLS			Pareto GLS		
Number of recombinations	Value of quality measure R	Standard deviation of quality measure R	Number of recombinations	Value of quality measure R	Standard deviation of quality measure R
60	0.099	0.005	60	0.111	0.006
120	0.089	0.004	120	0.101	0.004
180	0.082	0.006	180	0.093	0.007
240	0.078	0.005	240	0.088	0.006
300	0.077	0.005	300	0.084	0.006
360	0.075	0.004	360	0.082	0.006
420	0.072	0.004	420	0.080	0.004
480	0.070	0.003	480	0.078	0.004
540	0.068	0.002	540	0.078	0.004
600	0.067	0.003	600	0.077	0.005
MOMSLS			Pareto GA		
Number of local searches	Value of quality measure R	Standard deviation of quality measure R	Number of generations	Value of quality measure R	Standard deviation of quality measure R
120	0.236	0.009	40	0.402	0.023
180	0.233	0.008	80	0.389	0.028
240	0.232	0.007	120	0.387	0.027
300	0.229	0.008	160	0.383	0.025
360	0.227	0.006	200	0.378	0.024
420	0.227	0.007	240	0.371	0.022
480	0.226	0.006	280	0.367	0.023
540	0.226	0.006	320	0.363	0.024
600	0.224	0.006	360	0.357	0.020
660	0.224	0.006	400	0.355	0.021

Furthermore, MOGLS outperforms Pareto GLS. It generates better approximations to the nondominated set in shorter time. In our opinion, this is caused by the fact that MOGLS selects solutions for recombination on the basis of the same scalarizing function that is optimized by local search. Furthermore, the solutions are selected from a larger set of known solutions. From the other side, MOGLS has higher memory requirements than Pareto GLS. 0 illustrates performance of MOGLS and Pareto GLS.

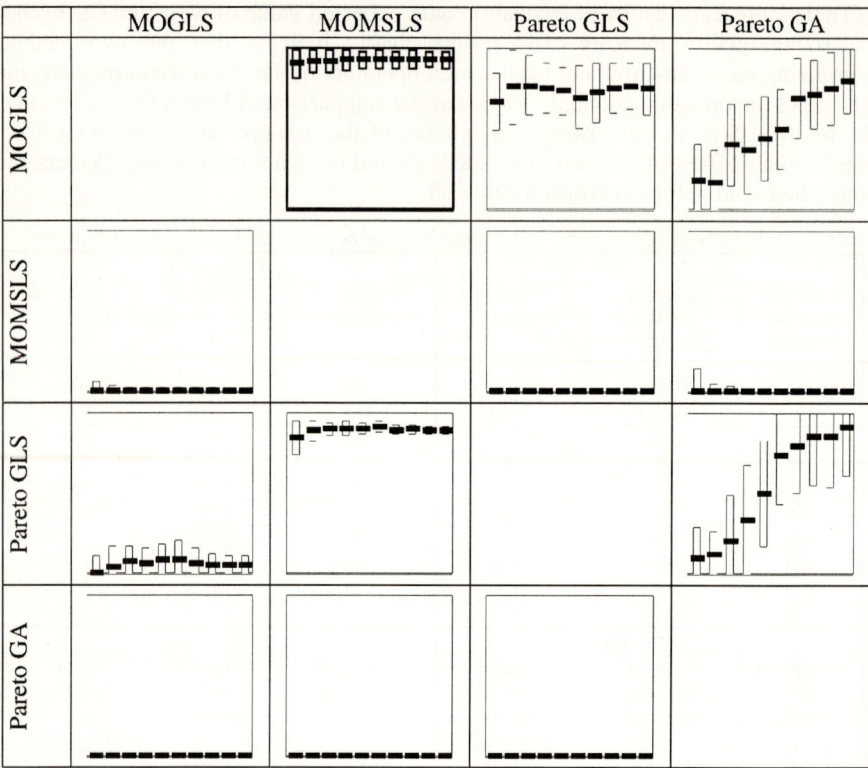

Fig. 4. Results of the comparison with coverage measure. Each chart contains 10 box plots representing the distribution of C values for a certain ordered pair of algorithms. The 10 box plots correspond to the ten sets of potentially Pareto-optimal solutions saved by each method in each run (the first set on the left). The scale is 0 at the bottom and 1 at the top for each chart. Chart in row of algorithm $A1$ and column of algorithm $A2$ presents values of coverage of approximations generated $A2$ by approximations generated by $A1$. The charts show ± standard deviation range of the quality measure.

0 presents results of the comparison with coverage measure. The results indicate the same ranking of the four methods as the evaluation with expected value of weighted Chebycheff scalarizing quality measure. Note, that no methods is able to produce approximations to nondominated set covering 100% of approximations generated by other methods. This indicates that no method is able to find all Pareto-optimal solutions.

7 Conclusions

Four multiple objective evolutionary algorithms have been tested on a real life combinatorial optimization problem corresponding to the distribution system design. The results of the computational experiment indicate that:

- hybridization of recombination and local search operators may boost performance of multiple objective evolutionary algorithms,
- the choice of solutions for recombination has crucial influence on the performance of genetic local search algorithm.

References

1. Ackley D. H. *A connectionist machine for genetic hillclimbing.* Kluwer Academic Press, Boston, 1987.
2. Galinier P., Hao J.-K. Hybrid evolutionary algorithms for graph coloring. Technical Report, Parc Scientifique Georges Besse, Nimes, 1999.
3. Fonseca C.M., Fleming P.J. (1993), Genetic algorithms for multiobjective optimization: Formulation, discussion and generalization. In S. Forrest (Ed.), Genetic Algorithms: Proceedings of 5th International Conference, San Mateo, CA, Morgan Kaufmann, 416-423.
4. Goldberg D.E. Genetic algorithms for search, optimization, and machine learning. Reading, MA, Addison-Wesley, 1989.
5. Hansen P.H, Jaszkiewicz A. Evaluating quality of approximations to the non-dominated set. Technical Report, Department of Mathematical Modelling, Technical University of Denmark, IMM-REP-1998-7, 1998.
6. Ishibuchi H. Murata T. Multi-Objective Genetic Local Search Algorithm and Its Application to Flowship Scheduling. IEEE Transactions on Systems, Man and Cybernetics, 28, 3, 392-403, 1998.
7. Jaszkiewicz A. Genetic local search for multiple objective combinatorial optimization, Technical Report RA-014/98, Institute of Computing Science, Poznan University of Technology, 1998
8. Jaszkiewicz A. On the performance of multiple objective genetic local search on the 0/1 knapsack problem. A comparative experiment. Submitted to IEEE Transactions on Evolutionary Computation.
9. Knowles J.D., Corne D.W. A Comparison of Diverse Approaches to Memetic Multiobjective Combinatorial Optimization, In Proceedings of the 2000 Genetic and Evolutionary Computation Conference Workshop Program, pages 103-108, Las Vegas, Nevada, July 2000.
10. Merz P., Freisleben B., Genetic Local Search for the TSP: New Results, In Proceedings of the 1997 IEEE International Conference on Evolutionary Computation, IEEE Press, pp. 159-164, 1997.
11. Steuer R.E. Multiple Criteria Optimization - Theory, Computation and Application, Wiley, New York, 1986.
12. Ulungu E.L., Teghem J., Fortemps Ph., Tuyttens D.. MOSA method: a tool for solving multiobjective combinatorial optimization problems. Journal of Multi-Criteria Decision Analysis, 8, 221-236, 1999.
13. Taillard É. D. Comparison of iterative searches for the quadratic assignment problem, Location science, 3, 87-105, 1995.
14. Van Veldhuizen D.A. (1999). Multiobjective Evolutionary Algorithms: Classifications, Analyses, and New Innovations. PhD thesis, Department of Electrical and Computer Engineering. Graduate School of Engineering. Air Force Institute of Technology, Wright-Patterson AFB, Ohio, May 1999.
15. Zitzler E., Thiele L. Multiobjective Evolutionary Algorithms: A Comparative Case Study and the Strength Pareto Approach, IEEE Transactions on Evolutionary Computation, Vol. 3, No. 4, pp. 257-271, November, 1999.
16. www-idss.cs.put.poznan.pl./~jaszkiewicz/MOEALib
17. www.lania.mx/~ccoello/EMOO/EMOObib.html

An Infeasibility Objective for Use in Constrained Pareto Optimization

Jonathan Wright and Heather Loosemore

Department of Civil and Building Engineering, Loughborough University,
Loughborough, Leicestershire, LE11 3TU, United Kingdom
J.A.Wright@lboro.ac.uk,
WWW home page:http://www.lboro.ac.uk

Abstract. A new method of constraint handling for multi-objective Pareto optimization is proposed. The method is compared to an approach in which each constraint function is treated as a separate objective in a Pareto optimization. The new method reduces the dimensionality of the optimization problem by representing the constraint violations by a single "infeasibility objective". The performance of the method is examined using two constrained multi-objective test problems. It is shown that the method results in solutions that are equivalent to the constrained Pareto optimal solutions for the true objective functions. It is also concluded that the reduction in dimensionality of the problem results in a more transparent set of solutions. The method retains elegance of the underlying Pareto optimization and does not preclude the representation of a constraint as an objective function where this is considered important. The method is easily implemented and has no parameters to be tuned.

1 Introduction

The aim of multi-objective optimization is to provide a set of non-dominated solutions that minimize several often conflicting criteria. Many multi-objective optimization problems are also constrained so that any optimization algorithm applied to solve them must ensure that the set of optimum solutions satisfies all constraints. Among other approaches (reviewed in [2]), two common approaches to handling constraints have been adopted for use with Pareto based ranking methods. First, the penalization of the rank of infeasible individuals, and second the transformation of the constraints to objectives.

Narayanan and Azarm [1] penalized the rank of infeasible individuals using an exterior penalty function, the penalty only being applied to the non-dominated but infeasible solutions. This approach has the disadvantage that the severity of the penalty has to be carefully chosen, especially during the early generations of a highly constrained problem (too severe a penalty may prevent any feasible solutions from being found for highly constrained problems).

Fonseca and Fleming [2] overcame the difficulty of choosing a penalty weight by treating the constraints as objectives in the Pareto ranking. The approach has the advantage that there are no additional parameters to be chosen in order to operate the algorithm with constraints. Under some circumstances, it also has the advantage that the optimum relationship between the true criteria and the constraints can be analyzed. However, for problems having many constraints, the interrelationship between the criteria and constraints is difficult, if not impossible to interpret. It is also possible that the increase in the number of objectives (and dimension of the problem), will demand an increase in population size required to give a good representation of the non-dominated solutions. As a consequence of the increased dimensionality and population size, the time required to reach convergence may also increase.

This paper examines an approach that combines the constraint violations to give a single measure of an individual's infeasibility. The infeasibility is then treated as an objective in the Pareto ranking of the solutions. The approach has the advantage that it retains the elegance of the Fonseca and Fleming approach, but that the number of objectives is reduced which makes the interrelationship between them easier to interpret. The rationale applied in this approach is that the constraints are considered to be bounds on the problem and that the interrelationship between constraints and objectives is of no interest. Where the impact of a constraint on the objectives is considered important, then the constraint may still be represented as objective (with any remaining constraints combined to produce an infeasibility objective).

1.1 The Multi-objective Genetic Algorithm

The genetic algorithm (GA) used in this study is a simple binary encoded GA with "roulette wheel" selection, single point cross-over and a non-overlapping population. The fitness assignment is based on an exponentially weighted inverse of the Pareto rank of each individual [3]. A goal attainment method [2], has also been implemented so that the search can be directed towards a particular region of the solution space (notably, the feasible region). The method penalizes the Pareto rank of an individual according to the extent of its goal violation. Finally, a phenotypic sharing function is used to ensure a reasonable spread of solutions is obtained [4] [5].

2 The Infeasibility Objective

The approach advocated here is that a single measure of a solutions infeasibility can be developed and used as an objective in a Pareto optimization. Since it is also considered that the constraints are acting as bounds on the problem, it is only the solutions that lie beyond the bounds that are of concern. The measure of infeasibility should therefore represent both the number of active constraints and the extent to which each constraint is violated. A measure of infeasibility that has these properties is the sum of the constraint values for all violated constraints. This can be evaluated in three stages. First the inequality constraints ($g_j(\mathbf{X})$), are formulated such that they are negative when feasible and the equality constraints ($h_j(\mathbf{X})$), zero when feasible (Equations 1 and 2).

Second, the feasible constraint values are reset as zero and infeasible values as positive (Equation 3). Finally, the solutions infeasibility ($č(\mathbf{X})$), is taken as the normalized sum of the reset constraint values (Equation 4). The solutions infeasibility ($č(\mathbf{X})$), is subsequently referred to as the infeasibility objective, which once minimized (to zero), ensures all constraints are satisfied.

$$g_j(\mathbf{X}) \leq 0, \quad (j = 1, ..., q) \qquad (1)$$

$$h_j(\mathbf{X}) = 0, \quad (j = q+1, ..., m) \qquad (2)$$

$$c_j(\mathbf{X}) = \begin{cases} \max(0, g_j(\mathbf{X})), & \text{if } 1 \leq j \leq q \\ |h_j(\mathbf{X})|, & \text{if } q+1 \leq j \leq m \end{cases} \qquad (3)$$

$$č(\mathbf{X}) = \sum_{j=1}^{m} \frac{c_j(\mathbf{X})}{c_{max,j}} \qquad (4)$$

Normalizing the constraint violations (by dividing by the scaling factor $c_{max,j}$), is necessary since large differences in the magnitude of the constraint values can lead to dominance of the infeasibility by the constraints having the highest values. In the procedure implemented here, the scaling factor for each constraint ($c_{max,j}$), is taken as the maximum value of the constraint violation found in the initial randomly generated population. If no infeasible solutions where found, the scaling factor is set to equal unity. This has the effect that if in subsequent generations the search moves to a region in which the constraint becomes active, its effect on the solutions infeasibility will depend on the magnitude of the constraint violation with constraint values much greater than unity having the most effect (since the normalization of the constraints tends give values in the order of unity). This may become a problem for weakly constrained problems with a high degree of feasibility since a randomly generated population may well result in a number of unity scaling factors. However, the focus of this approach is on the solution of highly constrained problems that are unlikely to yield any feasible solutions from a randomly generated population. The scaling factor is static and has been taken from the initial population so that for given constraint violations, the magnitude of infeasibility objective is consistent in every generation. This allows solutions from each generation to be included in the Pareto ranking of subsequent generations without the need to re-evaluate the infeasibility objective.

Although the infeasibility measure can be used directly as an objective function in a Pareto optimization, it is necessary to use the goal attainment method [2], to direct the optimization towards the feasible solutions (this is also the case were each constraint is treated as an individual objective) . Since all feasible solutions have the same infeasibility objective value (zero), the infeasibility objective is excluded from the Pareto ranking when if it has a value of zero. This results in the ranking for feasible solutions being a function of only the true objectives, which in turn has the effect of reducing the dimensionality of the problem and thus makes it easier to interpret the solutions.

3 Example Constrained Pareto Optimization

The infeasibility objective approach is investigated through two test problems. The first test problem is easily visualized and is an adaptation of an established multi-objective test problem. The second test problem is new and includes both equality and inequality constraints. The infeasibility objective approach is compared to the approach which treats all the constraints as individual objectives [2].

Several methods have been proposed for the performance evaluation of multi-objective genetic algorithms. For instance Srinivas and Deb [6], describe an approach to evaluating the distribution of solutions across the range of the Pareto solution space. Narayanan and Azarm [1], proposed using the Euclidean distance of the non-dominated points to zero as a indication of the rate of convergence. Such measures are best used to compare the performance of different multi-objective genetic algorithms (MOGA), in solving the same test problem. However, here we are in effect using the same MOGA to solve two different problems, one where all constraints are represented as objectives, and a transformed problem in which the constraint violations are represented by a single infeasibility objective. As such, a qualitative judgement is made here as to the effectiveness of the infeasibility objective approach. The assessment is based on the ease with which the results can be interpreted, and the extent to which the infeasibility objective approach produces Pareto optimum solutions.

3.1 A Four Function Test Problem

The four function test problem presented here is an adaptation of an existing two objective test problem (Equations 5 and 6) [2]. A third function (Equation 10), has been added to provide two inequality constraint functions (Equations 8 and 9), giving a total of four test functions ($f_1(\mathbf{X})$, $f_2(\mathbf{X})$, $g_1(\mathbf{X})$, $g_2(\mathbf{X})$). The test problem can be written as:

Minimize:

$$f_1(\mathbf{X}) = 1 - \exp\left(-\sum_{i=1}^{n}(x_i - 1/\sqrt{n})^2\right) \tag{5}$$

$$f_2(\mathbf{X}) = 1 - \exp\left(-\sum_{i=1}^{n}(x_i + 1/\sqrt{n})^2\right) \tag{6}$$

Subject to:

$$g_j(\mathbf{X}) \leq 0.0, \ \forall j \tag{7}$$

Where:

$$g_1(\mathbf{X}) = 0.4 - f_3(\mathbf{X}) \tag{8}$$

$$g_2(\mathbf{X}) = f_3(\mathbf{X}) - 0.8 \tag{9}$$

and:

$$f_3(\mathbf{X}) = 1 - \exp\left(-\sum_{i=1}^{n}\begin{cases}(x_i - 1/\sqrt{n})^2, & \text{if } i = 1,3,5...\\ (x_i + 1/\sqrt{n})^2, & \text{if } i = 2,4,6...\end{cases}\right) \quad (10)$$

In this example, the number of variables (n), has been fixed at 2. A discrete increment of 0.05 between the variable values has been chosen, and the variable range set at -2.0 to 2.0. This results in the function values being in the range 0.0 to 1.0. Figure 1 illustrates the test problem surface, with the shaded area representing the constrained Pareto solution space.

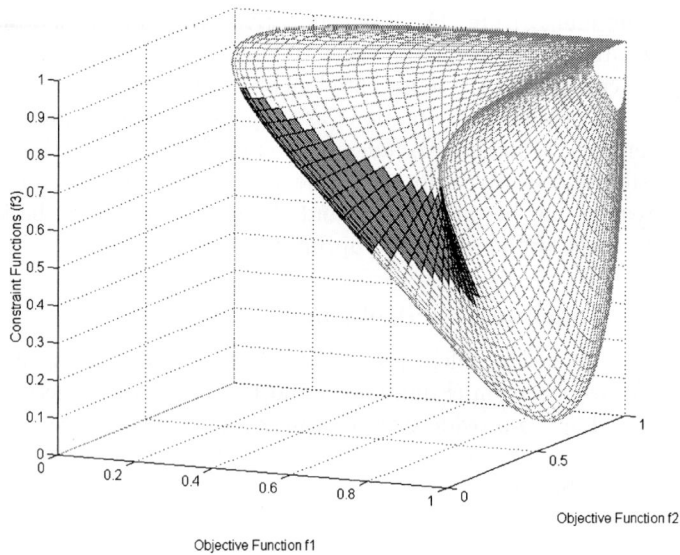

Fig. 1. The Four Function Test Problem

It would be expected that treating both constraints as separate objectives would result in solutions covering the whole of the Pareto surface. Using a population size of 100 and optimizing for 100 generations, Figure 2 illustrates that this is the case (with the solutions indicated by a "box").

However, since the infeasibility objective is only included in the Pareto ranking for infeasible solutions, it would be expected that this approach would produce a set of solutions that represent the constrained Pareto optimum solutions for only $f_1(\mathbf{X})$ and $f_2(\mathbf{X})$. This is represented by a line following the upper limit of the Pareto surface in Figure 1 (the "Pareto front"). Figure 3, illustrates that this is the case (the solution being indicated by a "box" and the remainder of the Pareto surface by "circles").

An Infeasibility Objective for Use in Constrained Pareto Optimization   261

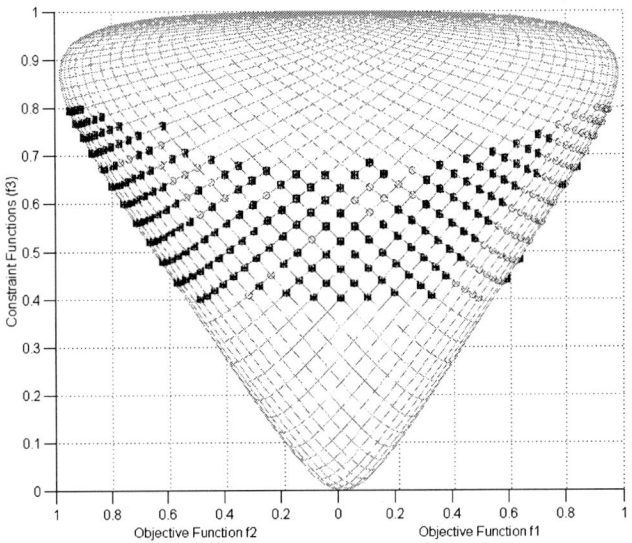

Fig. 2. Pareto Solutions for Constraint Function Optimization

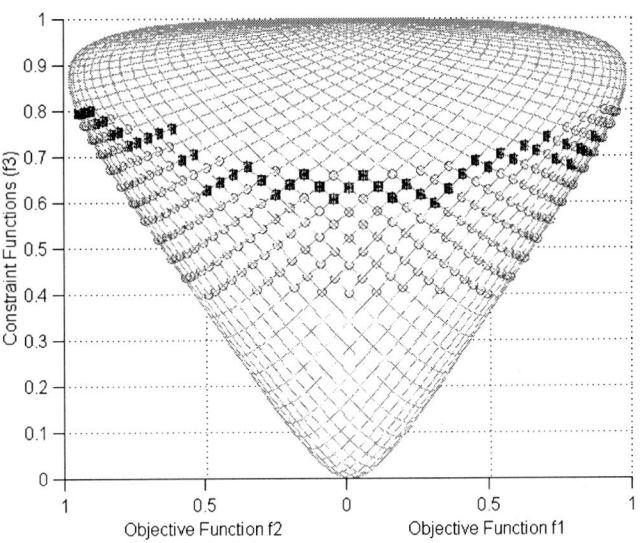

Fig. 3. Pareto Solutions for the Infeasibility Objective Approach

It can be seen from Figures 2 and 3 that the solutions from the infeasibility objective approach are a subset of the solutions obtained when the constraints are treated as separate objectives. However, for more complex problems, it may not be so easy to visualize the results and determine which set of solutions are equivalent to the constrained Pareto optimum solutions for the true problem objectives. The infeasibility objective approach and its performance is therefore illustrated further through a more complex test problem.

3.2 A Six Function Test Problem

A six function test problem has been developed that has some of the characteristics of thermal system optimization problems. For such problems, there is often a choice to be made between the capital cost and running cost of the system (Equations 11 and 12). The design of the system is normally restricted by limits on the fluid velocities (Equation 15), yet the system must also have sufficient capacity to meet the dynamic load acting on it (Equations 16 and 17). Finally, the components of the system must be physically connectable (Equation 18).

The test problem provides two true objectives, three inequality constraints and one equality constraint. The first objective function (Equation 11), is discontinuous and is only weakly coupled to the second objective function (Equation 12). The discontinuity in $f_1(\mathbf{X})$ and the weak coupling between the two objectives means that the search may have difficulty in obtaining an even distribution of solutions across both objective functions (with the continuous objective, $f_2(\mathbf{X})$ being biased). It would also be expected that the constraint functions are most active for the lower values of the first objective ($f_1(\mathbf{X})$), again adding weight to the solutions for $f_2(\mathbf{X})$.

Minimize:

$$f_1(\mathbf{X}) = \begin{cases} \left(83.0 + 14.0x_1 + 119.0\frac{x_2}{2} + 50.0x_1\frac{x_2}{2}\right) 2.0x_3 & \text{if } x_2 \leq 1.4 \\ \left(83.0 + 14.0x_1 + 119.0x_2 + 50.0x_1x_2\right) x_3 & \text{else} \end{cases} \quad (11)$$

$$f_2(\mathbf{X}) = \left(2.05(x_4 + 5) + \frac{7.03}{x_1 x_2} + \frac{5.87(x_3 - 1)}{x_1 x_2}\right) x_5 \quad (12)$$

Subject to:

$$g_j(\mathbf{X}) \leq 0.0, \ \forall j \quad (13)$$

$$h_j(\mathbf{X}) = 0.0, \ \forall j \quad (14)$$

Where:

$$g_1(\mathbf{X}) = \frac{1.66}{x_1 x_2} - 3.0 \quad (15)$$

$$g_2(\mathbf{X}) = \left(\frac{x_4 + 5.0}{85.0} - \frac{x_1 x_2 x_3}{1.0 + x_1 x_2 x_3}\right) \quad (16)$$

$$g_3(\mathbf{X}) = 15.91 - 0.58x_4 + \exp\left(-\frac{x_5}{552.0}\right)(12.0 - 0.58x_4) \tag{17}$$

$$h_1(\mathbf{X}) = \mathrm{mod}(100.0x_2x_3, 2) \tag{18}$$

and:

$0.5 \leq x_1 \leq 2.0,$ in increments of 0.05
$0.5 \leq x_2 \leq 2.0,$ in increments of 0.05
$1.0 \leq x_3 \leq 3.0,$ in increments of 1.0
$30.0 \leq x_4 \leq 40.0,$ in increments of 0.01
$600.0 \leq x_5 \leq 3600.0,$ in increments of 5.0

Both the method of including constraints as objectives and infeasibility objective approach were evaluated using five different randomly generated populations. The population size in each case was fixed at 100 individuals and the optimization continued for 100 generations. There was very little variation in the results between the different initial generations for either of the two approaches. Figure 4, illustrates the optimum results for the different initial populations for the infeasibility objective approach. Considering the consistency of the results, only one set of solutions is presented in the remainder of this paper.

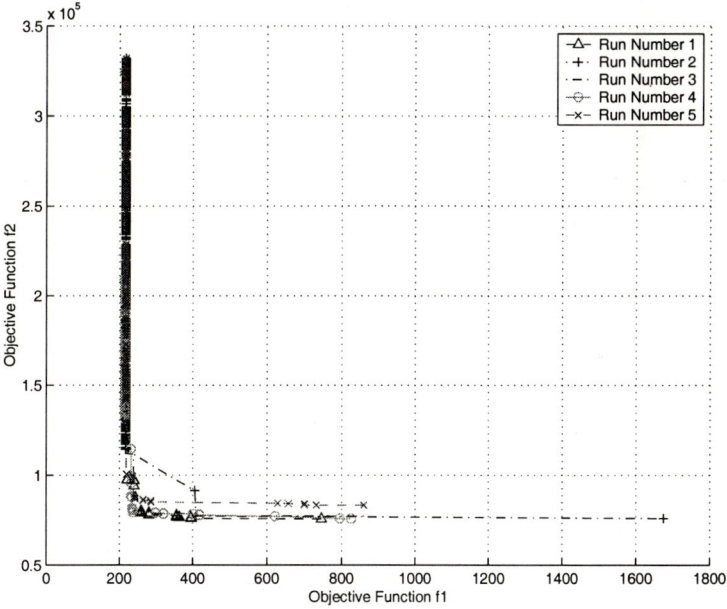

Fig. 4. Pareto Solutions for Infeasibility Objective Approach

It is clear that for two true objective functions the infeasibility objective approach produces a pay-off curve rather than surface. Treating the constraints as objectives necessarily increases the dimensions of the problem and results in a pay-off surface between the objective functions. Figure 5 illustrates this for the approaches, the "dots" being the solutions due to the constraints being treated as objectives, and the lines with "cross" points being the solutions from the infeasibility objective approach. Figure 5 shows that the infeasibility objective approach produces the Pareto optimal solutions for the two true objectives. The banding in solutions is in part due to the equality constraint function but mainly due to the discontinuity in the $f_1(\mathbf{X})$ objective function. This is most pronounced around an $f_1(\mathbf{X})$ value of approximately 800. The discontinuity here has prevented the infeasibility objective approach from producing solutions that extend further across the range of $f_1(\mathbf{X})$. For both optimization approaches, the bias of results is towards objective function $f_2(\mathbf{X})$, and is probably due to the effect of the discontinuity in $f_1(\mathbf{X})$. Similar results are also evident by comparing the pay-off between $f_1(\mathbf{X})$ and the other functions (Figures 6 and 7).

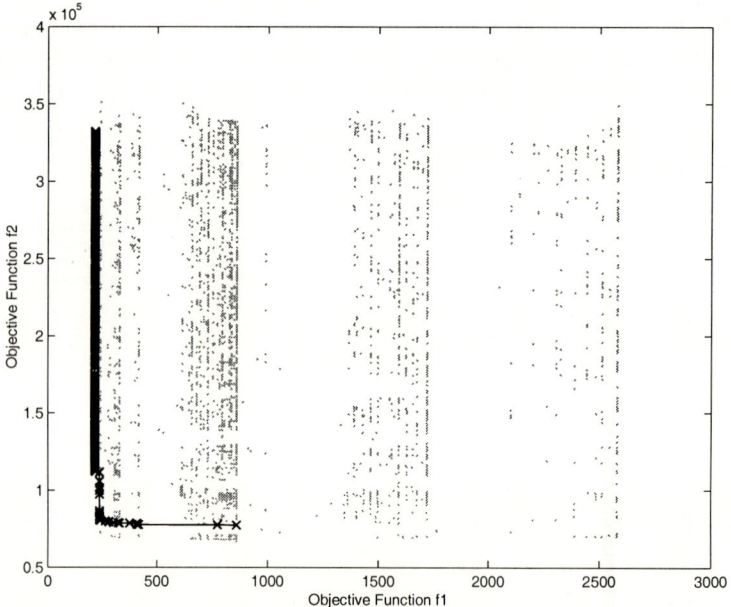

Fig. 5. Pareto Solutions for $f_1(\mathbf{X})$ and $f_2(\mathbf{X})$

Reducing the dimensionality of the problem also aids analysis of the variation in the variables for the Pareto set of solutions. Again, since for this example, the problem has been reduced to two objectives by the infeasibility approach, behaviour of the variables is much clearer than for the solutions from when all the constraints are represented as objectives (Figure 8 and 9).

The ease with which the solutions from the infeasibility objective approach can be interpreted is due, in part, to the problem being reduced to only two ob-

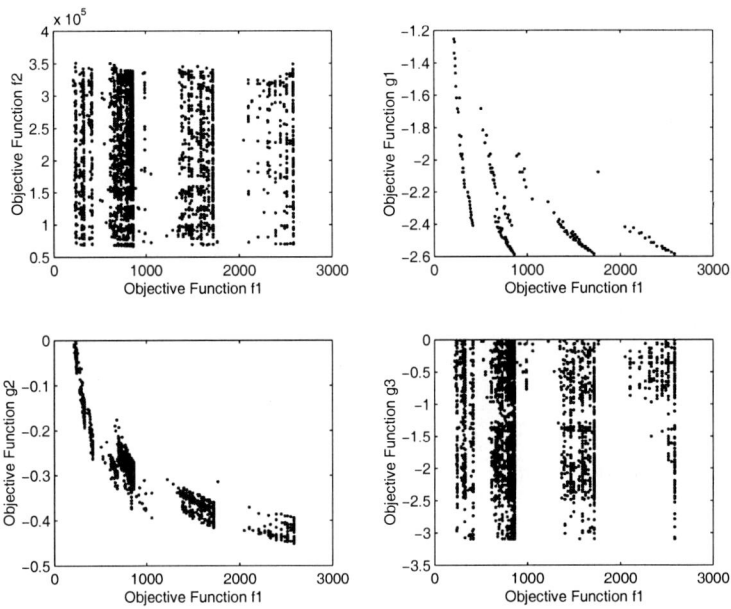

Fig. 6. Pareto Solutions for the Constraint Objective Approach

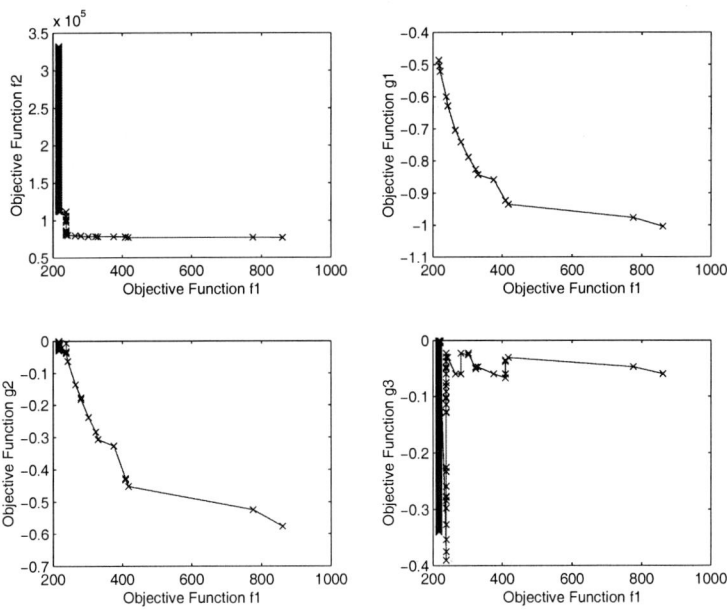

Fig. 7. Pareto Solutions for the Infeasibility Objective Approach

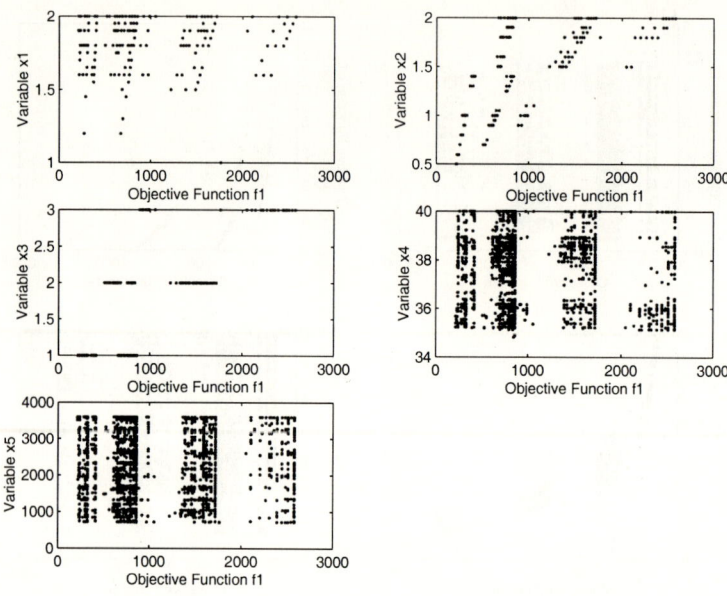

Fig. 8. Pareto Variables for the Constraint Objective Approach

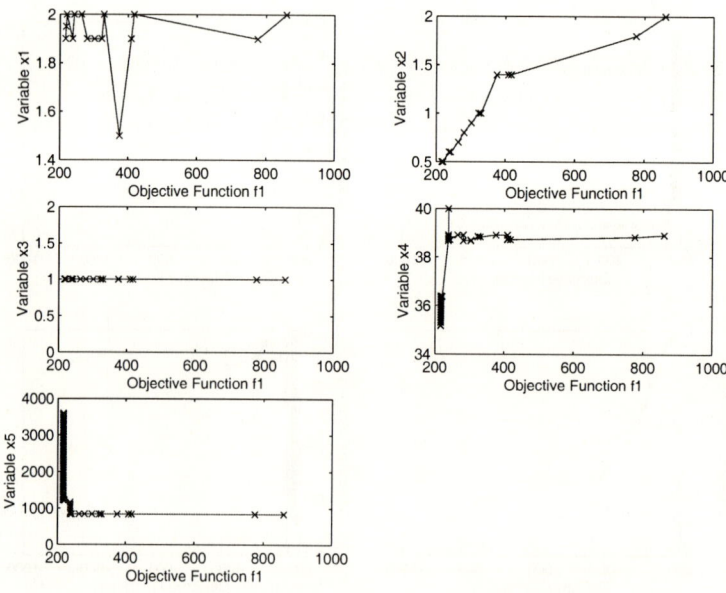

Fig. 9. Pareto Variables for the Infeasibility Objective Approach

jective functions; a higher number of objective functions necessarily resulting in a multi-dimensional surface of solutions that is more difficult to interpret. However, many real world multi-objective optimization problems are only concerned with optimizing a few objectives, the number of constraints often far out numbering the objectives. It is also clear that the infeasibility objective approach is an effective means of obtaining the Pareto optimal solutions for the true objectives and can be used to reduce the dimensionality of the optimization problem where detailed analysis of the constraint behaviour is not required.

4 Conclusions

This paper introduces a means by which the constraint violations may be combined to form an infeasibility objective for use in multi-objective Pareto optimization. The infeasibility objective has the properties that it is zero for feasible solutions, but is positive for infeasible solutions, and increases in value with both the number of active constraints and the magnitude of each constraint violation. Both inequality and equality constraints can be included in the infeasibility objective.

It is shown, that in comparison to an approach in which each constraint is treated as a separate objective, transforming the violated constraints into a single infeasibility objective results in Pareto solutions that are more transparent and easier to interpret. Using the infeasibility objective in a Pareto ranking optimization results in solutions that are equivalent to the constrained Pareto optimal solutions for only the true objective functions. This is in contrast to the solutions obtained when each constraint is treated as separate objective function which increases the dimension of the problem and limits the extent to which the interrelationship between objectives and constraints can be interpreted.

The use of the infeasibility objective does not preclude the treatment of a constraint function as an objective function if knowledge of the constraints effect is considered important. The approach is intended to allow treatment of constraints as bounds on the problem when the only concern is that the solutions are feasible, the constraint values across the Pareto optimal solutions are still available for analysis.

It can be concluded that the infeasibility objective approach retains the elegance of the existing Pareto ranking constrained optimization, but provides a means by which the dimensionality of the problem can be reduced and consequently, that the Pareto optimal solutions become more transparent. The method also has the advantage over approaches based on the use of penalty functions in that the it does not require any parameters, which overcomes the need to tune it to a particular problem.

Acknowledgements. The authors would like to acknowledge Professor Peter Fleming and Dr Andrew Chipperfield of the University of Sheffield for the advice they have given on multi-objective genetic algorithm optimization. The authors also acknowledge the UK Engineering and Physical Science Research Council for funding this work as part of an investigation into the optimization of building thermal systems.

References

[1] Narayanan S., Azarm S.: On improving multiobjective genetic algorithms for design optimization. Structural Optimization, **18 No.2-3**, (1999), 146-155.
[2] Fonseca C.M., Fleming P.J.: Multiobjective Optimization and Multiple Constraint Handling with Evolutionary Algorithms - Part 1: A Unified Formulation. IEEE Transactions on systems and Cybernetics - Part A: Systems and Humans **28 No.1**, (1998), 26-37.
[3] Fonesca C.M., Fleming P.J.: Multiobjective Genetic Algorithms Made Easy: Selection, Sharing and Mating Restriction. Genetic Algorithms in Engineering Systems: Innovations and Applications Conference Publication No.414, IEE, (1995), 12-14.
[4] Goldberg D.E., Richardson J.: Genetic Algorithms with sharing for multi-modal function optimization. Proceedings of the 7th International Conference on Genetic Algorithms', (1987), 41-49.
[5] Deb K., Goldberg D.E.: An investigation of Niche and Species Formation in Genetic Function Optimization. Proceedings Of The Third International Conference On Genetic Algorithms, (1989), 42-50.
[6] Srinivas N., Deb K .: Multiobjective optimization using nondominated sorting in genetic algorithms. Evolutionary Computation, **2**, (1995) 221-248.

Reducing Local Optima in Single-Objective Problems by Multi-objectivization

Joshua D. Knowles[1*], Richard A. Watson[2], and David W. Corne[1]

[1] School of Computer Science, Cybernetics and Electronic Engineering,
University of Reading, Reading RG6 6AY, UK
{j.d.knowles,d.w.corne}@reading.ac.uk
[2] Dynamic and Evolutionary Machine Organization,
Volen Center for Complex Systems, MS018, Brandeis University,
Waltham, MA 02454, USA
richardw@cs.brandeis.edu

Abstract. One common characterization of how simple hill-climbing optimization methods can fail is that they become trapped in local optima - a state where no small modification of the current best solution will produce a solution that is better. This measure of 'better' depends on the performance of the solution with respect to the single objective being optimized. In contrast, multi-objective optimization (MOO) involves the simultaneous optimization of a number of objectives. Accordingly, the multi-objective notion of 'better' permits consideration of solutions that may be superior in one objective but not in another. Intuitively, we may say that this gives a hill-climber in multi-objective space more freedom to explore and less likelihood of becoming trapped. In this paper, we investigate this intuition by comparing the performance of simple hill-climber-style algorithms on single-objective problems and multi-objective versions of those same problems. Using an abstract building-block problem we illustrate how 'multi-objectivizing' a single-objective optimization (SOO) problem can remove local optima. Then we investigate small instances of the travelling salesman problem where additional objectives are defined using arbitrary sub-tours. Results indicate that multi-objectivization can reduce local optima and facilitate improved optimization in some cases. These results enlighten our intuitions about the nature of search in multi-objective optimization and sources of difficulty in single-objective optimization.

1 Introduction

One of the most general heuristics used in optimization techniques is the idea that the value of solutions is to some extent correlated with how similar the solutions are; crudely, that a good solution is more likely to be found nearby to other good solutions than it is to be found nearby an arbitrary solution. Naturally, 'nearby' or 'similar' needs to be qualified. The simplest notion of similarity of solutions is their proximity as measured in the problem parameters given. But alternatively, we may define proximity in terms of the variation operators used

---
* http://www.reading.ac.uk/~ssr97jdk/

by the search algorithm [7]. In any case, the simplest way to use this heuristic is a hill-climbing algorithm: start with some random solution, try variations of this solution until a better solution (or at least, non-worse solution) is found, move to this new solution and try variations of this, and so on. But the actual success of a hill-climber requires a stronger assumption to be true: that from any point in the solution space there is a path through neighbouring points to a global optimum that is monotonically increasing in value[1]. If this is true then a hill-climber can find a global optimum - and, although a hill-climber can do better than random guessing on almost all practical problems we encounter, it usually does not find a global optimum. More likely, it gets stuck in a local optimum - a sub-optimal point or plateau that has no superior neighbouring points.

There are several approaches that can be taken to overcome the limitations of a simple hill-climber. Broadly, many approaches can be seen as one of the following: changing the neighbourhood structure of the solution space so that the strong assumption is true; or relaxing the strong assumption and, one way or another, utilizing solutions which are inferior to some extent. Changing the neighbourhood structure can be done by something as simple as increasing the neighbourhood 'radius' by increasing mutation, or by a complete redesign of how solutions are represented and new variants are created, or perhaps by adding redundancy so that a hill-climber can travel along 'neutral networks' [5] to find superior points without having to go through inferior points. Relaxing the strong assumption can be done by, amongst other things, probabilistically accepting inferior solutions, as in simulated annealing, or by the use of multi-point searchers, or multi-restart searchers, where although one searcher may become stuck another, at a different location, may continue.

In this paper, we investigate a different approach, similar to one previously proposed in [10]. Rather than changing the neighbourhood structure so that we can always find a superior point, or accepting search paths through inferior points, we use a different definition of superior and inferior. Specifically, we use a method of comparing two solution that is common in multi-objective optimization (MOO) techniques where more than one measure of a solution is provided. Briefly, under Pareto optimization, a solution **x** is superior (said to Pareto dominate) another solution **x'** if and only if it is at least as good as **x'** in all measures and better in at least one measure. Put another way, if **x'** is better than **x** in at least one measure then it is not "inferior" to **x**. Our intuition is this: if we can add other objectives to a problem to make it multi-objective, and use this relaxed notion of inferiority, then we may open up monotonically increasing paths to the global optimum that are not available under the original single-objective optimization (SOO) problem. We call this approach "multi-objectivization".

Naturally, the effect of this transformation will depend, in part, on exactly how we 'objectivize' the problem, and the particulars of the algorithm that uses the new multi-objective problem space. To begin with, we illustrate the principle on a test function that has an obvious decomposition. We explain why decomposing this problem naturally leads to the removal of all local optima in the search space, and demonstrate this fact empirically with results showing that a Pareto hillclimber (PAES) can solve the problem much more efficiently than a

---

[1] 'Efficient' success also requires that this path is not exponentially long [4].

hillclimber. We also compare the performance of PAES with that of simulated annealing on this problem, and show that increasing the freedom of movement for a hill climber, by decomposing a problem into multiple objectives, is more effective than relaxing the strong assumption for a hill-climber, as in simulated annealing. This illustrates an idealized multi-objectivization.

Following this, we take the well-known travelling salesperson problem (TSP) as an exemplar real-world problem with a single objective (to minimize the tour length) and show how it may be decomposed into sub-objectives. We then perform a number of experiments to measure the effectiveness of decomposing the problem, by comparing various single-point and multi-point hill-climbers on the original landscape and on the multi-objective landscape. Some comparison with simulated annealing is also provided. Several instances of the problem are considered, and we attempt to establish the effect the choice of different sub-objectives has on the efficacy of the resultant decomposition.

The remainder of the paper is structured as follows: Section 2 defines the concepts of single and multi-objective optimization, and introduces the technique of multi-objectivization. Section 3 defines the algorithms that we use in our experiments, and Section 4 defines the test problems we use both in their SOO and MOO versions. Section 5 describes the results of the experiments. Section 6 discusses implications and related research, and Section 7 concludes.

2 Single-Objective and Multi-objective Optimization

The general (unconstrained) single-objective combinatorial optimization problem can be expressed as:

$$\text{maximize } f(\mathbf{x}) \\ \text{subject to } \mathbf{x} \in X \quad (1)$$

where $\mathbf{x}$ is a discrete solution vector, and X is a finite set of feasible solutions, and $f(\mathbf{x})$ maps X into $\Re$.

Similarly, the multi-objective combinatorial optimization (MOCO) problem can be expressed as:

$$\text{"maximize" } \mathbf{f}(\mathbf{x}) = (f_1(\mathbf{x}), \ldots, f_K(\mathbf{x})) \\ \text{subject to } \mathbf{x} \in X \quad (2)$$

where the vector objective function $\mathbf{f}(\mathbf{x})$ maps X into $\Re^K$, where $K \geq 2$ is the number of objectives. The term 'maximize' appears in quotation marks because, in general, there does not exist a single solution that is maximal on all objectives. Instead, one may seek to find a set of solutions $X^* \subseteq X$, called the Pareto optimal set, with the property that:

$$\forall \mathbf{x}^* \in X^* \bullet \not\exists \mathbf{x} \in X \bullet \mathbf{x} \succ \mathbf{x}^* \quad (3)$$

where $\mathbf{x} \succ \mathbf{x}^* \iff ((\forall i \in 1..K \bullet (f_i(\mathbf{x}) \geq f_i(\mathbf{x}^*)) \wedge (\exists i \in 1..K \bullet f_i(\mathbf{x}) > f_i(\mathbf{x}^*)))$. The expression $\mathbf{x} \succ \mathbf{x}^*$ is read as $\mathbf{x}$ *dominates* $\mathbf{x}^*$, and solutions in the Pareto optimal set are also known as efficient or admissible solutions. In addition, for two solutions $\mathbf{x}$ and $\mathbf{x}'$, we say $\mathbf{x} \sim \mathbf{x}'$ if and only if $\exists i \in 1..K \bullet f_i(\mathbf{x}) > f_i(\mathbf{x}') \wedge \exists j \in 1..K \bullet j \neq i \bullet f_j(\mathbf{x}') > f_j(\mathbf{x})$. Such a pair of solutions are said to be *incomparable*, and each is *nondominated* with respect to the other. Since any member of a set of mutually nondominated solutions is not worse (dominated) than any other, a hillclimbing algorithm would be free to move from any one of them to any other if the variation operator allows. This is the notion that is important to allow the increased freedom of a Pareto hill-climber, and which may reduce the problem of local optima.

2.1 Multi-objectivization

To perform multi-objectivization we must either replace the original single objective of a problem with a set of new objectives, or add new objectives in addition to the original function. In either case, we want to be sure that the global optimum of the original problem is one of the points that is Pareto optimal in the multi-objective version of the problem. Specifically, the problem must be restated so as to maximize $K \geq 2$ objective functions such that the following relation between solutions in the two formulations holds:

$$\forall \mathbf{x}^{opt} \in X^{opt} \, . \, \exists \mathbf{x}^* \in X^* \, . \, \mathbf{x}^* = \mathbf{x}^{opt} \tag{4}$$

where $\mathbf{x}^{opt}$ is an optimal solution to the SOO problem, and X^{opt} is the set of all such solutions, and $\mathbf{x}^*$ and X^* relate to the MOO formulation of the problem, and have the meanings attributed above.

Part of the intuition that motivates multi-objectivization comes from notions of problem decomposition - dividing a problem into a number of smaller subproblems and solving each and all of them. Accordingly, it may be appropriate to define an objective as a function similar to the original function but over a subset of the problem parameters. For example, in our TSP example we add objectives corresponding to the length of parts of the tour. Defining functions over subsets of the problem parameters has clear connections with divide and conquer techniques, especially dynamic programming [1]. An alternative approach to multi-objectivization is to define different functions over the same (entire) set of problem parameters. For example, to take a different domain, if we desire a locomotion controller for a legged robot we might suppose that it needs to both lift its body off the floor, and swing it legs forward, both of which may depend on all of the parameters.

The skill of the researcher in either approach (similar functions over subsets, or different functions over the entire set) is to separate out the conflicting aspects of the problem - to find objectives that are as independent as possible. This is not always easy or possible. However, we suppose that in some circumstances the multi-objectivization approach may have useful tolerance of 'lazy' decompositions, where sub-problems are not completely independent. Perhaps this is because a solution that is better in all objectives is preferred if available, but a solution that exploits one objective at the expense of another objective is still valuable in the case that a dominant solution cannot be found.

In the two examples introduced later, the first uses different functions over the entire parameter set, and the second, TSP, uses similar functions over subsets of the solution. The examples serve to illustrate the different approaches to multi-objectivization, the issues involved, and show some cases where multi-objectivization can be successful.

3 Algorithms

In order to test the hypothesis that multi-objectivizing a problem can reduce the number of local optima in a way that is useful for performing local search, we employ a number of simple neighbourhood search algorithms described below. The first pair of algorithms, consisting of a simple hillclimber (SHC), and a multi-objective hillclimber similar in operation to the Pareto archived evolution strategy (PAES) [8,9], are both single-point hill-climbers. PAES represents the multi-objective analogue of SHC: both algorithms accept a neighbour of the current solution if it is not worse than any solution found so far. In the context of PAES, however, 'worse' means dominated.

The second pair of algorithms, which are a mutation-only genetic algorithm with deterministic crowding [11] (DCGA), and the Pareto envelope-based selection algorithm (PESA) [2], are both multi-point hill-climbers (neither uses recombination here). Once again, they are supposed to be analogues of each other, subject to the differences forced upon them by the different requirements of single and multiple objective optimization. The analogy between them is not as clear as between SHC and PAES because the selection and crowding methods used by them is more complicated, but they have sufficiently similar operation to warrant comparison for our purposes here.

The performance of the two pairs of algorithms above are also compared with a simulated annealing (SA) algorithm. This comparison is intended to place the effect of reducing the number of local optima achieved by multi-objectivization into a context which is familiar to the reader. SA incrementally adjusts the strictness with which it rejects moves to worse solutions, and does so very effectively; it thus serves as a useful comparison to the effect of multi-objectivization.

3.1 Single-Point Hill-Climbers

Initialization: $B \leftarrow \emptyset$
                      $\mathbf{x} \in X \leftarrow \mathsf{Init}(\mathbf{x})$
                      $B \leftarrow B \cup \mathbf{x}$
Main Loop: $\mathbf{x}' \in X \leftarrow \mathsf{Mutate}(\mathbf{x})$
                      if ($\mathsf{Inferior}(\mathbf{x}', B) \neq$ TRUE) {
                         $\mathbf{x} \leftarrow \mathbf{x}'$
                         $B \leftarrow \mathsf{Reduce}(\mathbf{x}' \cup B)$ }
Termination: return $\mathsf{Best}(B)$

Fig. 1. Generic pseudocode for hill-climbing algorithms

Pseudocode for a generic version of a hill-climbing algorithm that may be single or multi-objective is given in Figure 1. The current solution vector is denoted by $\mathbf{x}$, and B is the minimal representation of the best solutions encountered so far. The function $\mathsf{Mutate}(\mathbf{x})$ returns a new solution $\mathbf{x}'$ made by variation of $\mathbf{x}$. For the simple, single-objective hillclimber, the functions $\mathsf{Inferior}()$, $\mathsf{Reduce}()$ and $\mathsf{Best}()$ are very simple. The function $\mathsf{Inferior}(\mathbf{x}', B)$ returns true iff there is any element of the set B whose evaluation is greater than $\mathbf{x}'$, Reduce(B) returns the set of equally maximum value elements of the set B, and Best(B) returns any member of B, because they are equally good. For the PAES-based multi-objective hill-climber, these functions have Pareto versions that follow the same semantics: Inferior($\mathbf{x}', B$) returns true iff there is any element of the set B that dominates $\mathbf{x}'$, Reduce(B) returns the set of elements from B that are not dominated by any other member of B, and Best(B) returns the element of B that is maximal in the original single objective. In our experiments here, both SHC and PAES are terminated when the number of function evaluations reaches a predetermined number, num_evals.

3.2 Multi-point Hill-Climbers

Initialization: $P \leftarrow \emptyset$
                       Init_pop(P)
Main Loop: $\mathbf{x}_1 \leftarrow$ Rand_mem(P), $\mathbf{x}_2 \leftarrow$ Rand_mem(P)
                       $\mathbf{x}'_1 \leftarrow$ Mutate($\mathbf{x}_1$), $\mathbf{x}'_2 \leftarrow$ Mutate($\mathbf{x}_2$)
                       if (H($\mathbf{x}_1, \mathbf{x}'_1$) + H($\mathbf{x}_2, \mathbf{x}'_2$) > H($\mathbf{x}_1, \mathbf{x}'_2$) + H($\mathbf{x}_2, \mathbf{x}'_1$))
                          Swap($\mathbf{x}'_1, \mathbf{x}'_2$)
                       if ($f(\mathbf{x}'_1) > f(\mathbf{x}_1)$)
                          $P \leftarrow P \cup \mathbf{x}'_1 \setminus \mathbf{x}_1$
                       if ($f(\mathbf{x}'_2) > f(\mathbf{x}_2)$)
                          $P \leftarrow P \cup \mathbf{x}'_2 \setminus \mathbf{x}_2$
Termination: return Best(P)

Fig. 2. A simple form of a genetic algorithm using deterministic crowding (DCGA) as used in our experiments

Pseudocode for the mutation-only, deterministic crowding genetic algorithm (DCGA) is given in Figure 2. The set P is the population of candidate solutions initialized randomly using Init_pop(P). The function H($\mathbf{x}, \mathbf{x}'$) measures the genotypic Hamming distance between two solution vectors. At each iteration, two parents are selected from the population at random and two offspring are generated by mutation, one from each parent. Parents and offspring are then paired up so as to minimize the sum of the genotypic Hamming distance between them. Each offspring then replaces the parent it is paired with if it non-worse than that parent.

The PESA algorithm used here has been described in detail in [2], and pseudocode for it is given in Figure 3. It has an internal population IP of size P_I, and an external population of nondominated solutions EP. Here it is used without crossover so that each generation consists of selecting P_I parents from EP and mutating them to produce P_I new offspring. Then, the nondominated members of IP are incorporated into EP. The selection operator, select() is based on crowding in objective space.

The PESA and DCGA algorithms are quite different in some ways, but do represent analogues of each other at a high level: both algorithms have a popula-

```
Initialization:     IP ← ∅, EP ← ∅
                    Init_pop(IP)
                    foreach (x ∈ IP)
                        EP ← Reduce(EP ∪ x)
Main Loop:          IP ← ∅
                    while (|IP| < P_I) {
                        x ← Select(EP)
                        x' ← Mutate(x)
                        IP ← IP ∪ x }
                    foreach (x ∈ IP)
                        EP ← Reduce(EP ∪ x)
Termination:        return Best(EP)
```

Fig. 3. The Pareto envelope-based selection algorithm (PESA)

tion from which parents are selected and used to produce offspring via mutation, and both have a mechanism for maintaining diversity in the population. However, in PESA, the diversity is maintained in the objective space, while with DCGA it is in the genotype space. With PESA, the initial size of the pool of random solutions is P_I, and P_I solutions are generated at each step. The prevailing size of EP is the size of the pool from which solutions can be selected. In DCGA, the population size P determines the initial pool and the size of the pool from which solutions are selected, and 2 solutions are generated at each step. These differences in detail, however, should not affect the substance of the judgments we make about the performance of these algorithms in our experiments.

3.3 Simulated Annealing

The simulated annealing (SA) algorithm used is identical to the SHC, except for changes to the acceptance function to allow it to accept moves to worse solutions. The acceptance function we employ for accepting degrading moves is the standard exponential function:

$$p\left(\text{accept } \mathbf{x}'\right) = \frac{\exp\left(f(\mathbf{x}) - f(\mathbf{x}')\right)}{T} \qquad (5)$$

where $f(\mathbf{x})$ is the evaluation of the current solution $\mathbf{x}$, $f(\mathbf{x}')$ is the evaluation of the neighbouring solution, $\mathbf{x}'$, and $p\left(\text{accept } \mathbf{x}'\right)$ denotes the probability of accepting $\mathbf{x}'$.

We choose to use a simple form of simulated annealing, employing an inhomogeneous, geometric cooling schedule where T is updated after every iteration using:

$$T \leftarrow \alpha T \qquad (6)$$

Following general procedures outlined in [14], we attempt to set the starting temperature T_0 so that between 40% and 60% of moves are accepted. The final temperature is set so that between 0.1% and 2% of moves are accepted. We can then calculate α using:

$$\alpha = \exp\left(\frac{\ln(T_f/T_0)}{num\_evals}\right) \qquad (7)$$

given that we know the total number of function evaluations num_evals required.

4 Problems

4.1 H-IFF/MH-IFF: An Abstract Illustration

The Hierarchical-if-and-only-if function, H-IFF, is a genetic algorithm test problem designed to model a problem with a building-block structure where the sub-problem that each block represents has strong interdependencies with other blocks. That is, unlike many existing building-block problems, the optimal solution to any block in H-IFF is strongly dependent on how other building-blocks have been solved [16,17].

The fitness of a string using H-IFF can be defined using the recursive function given below. This function interprets a string as a binary tree and recursively decomposes the string into left and right halves. Each resultant sub-string constitutes a building-block and confers a fitness contribution equal to its size if all the bits in the block have the same allele value - either all ones or all zeros. The fitness of the whole string is the sum of these fitness contributions for all blocks at all levels.

$$f(B) = \begin{cases} 1, & \text{if } |B| = 1, \text{ else} \\ |B| + f(B_L) + f(B_R), & \text{if } (\forall i \{b_i = 0\} \text{ or } \forall i \{b_i = 1\}), \\ f(B_L) + f(B_R), & \text{otherwise,} \end{cases} \quad (8)$$

where B is a block of bits, $\{b_1, b_2, \ldots, b_n\}$, $|B|$ is the size of the block $= n$, b_i is the ith element of B, and B_L and B_R are the left and right halves of B (i.e. $B_L = \{b_1, b_2, \ldots, b_{n/2}\}$ and $B_R = \{b_{n/2+1}, \ldots, b_n\}$). The length of the string evaluated, n, must equal 2^p where p is an integer (the number of hierarchical levels).

Each of the two competing solutions to each block (all-ones and all-zeros) give equal fitness contributions on average. But only when neighbouring blocks match do they confer a bonus fitness contribution by forming a correct block at the next level in the hierarchy. These competing solutions and their interdependencies create strong epistatic linkage in H-IFF and many local optima (see Figure 4 left). These local optima prevent any kind of mutation based hill climber from reliably reaching one of the two global optima in H-IFF (all-ones or all-zeros) in time less than exponential in N, the number of bits in the problem [17].

Figure 4 (left) shows a particular section through the H-IFF landscape. This is the section through a 64-bit landscape starting from all zeros on the left and ending with all ones. Specifically, it shows the fitness of the strings "000...0", "100...0", "110...0", ..., "111...1". This indicates the local optima in H-IFF and the two global optima at opposite corners of the space.

$$f_k(B) = \begin{cases} 0, & \text{if } |B| = 1 \text{ and } b_1 \neq k, \text{ else} \\ 1, & \text{if } |B| = 1 \text{ and } b_1 = k, \text{ else} \\ |B| + f_k(B_L) + f_k(B_R), & \text{if } (\forall i \{b_i = k\}), \\ f_k(B_L) + f_k(B_R), & \text{otherwise.} \end{cases} \quad (9)$$

where $f_0(\mathbf{x})$ is the first objective and $f_1(\mathbf{x})$ is the second.

This particular decomposition of H-IFF results in a two-objective problem in which there are no local optima for a multi-objective hill-climber. Figure 4 (right)

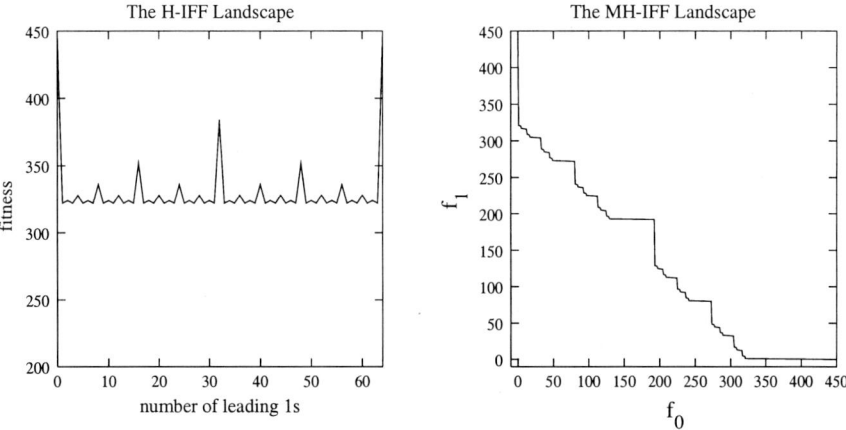

Fig. 4. H-IFF and MH-IFF

is a section through the two-objective landscape, MH-IFF, using the same strings as used in the section through H-IFF. We see that if neighbouring points on this section are worse in respect to one dimension, they are better in the other. This has transformed the problem into one that is now completely bit-climbable. That is, single bit changes can move a hill-climber from one end of the Pareto front to the other, including every point in between, without ever moving to a dominated point. (In fact, every Pareto optimal solution can be reached by bit-climbing from any start point.) This transformation is possible because we have access to the structure of the function and can thereby separate the conflicting sub-goals inherent in the SOO version. This serves to illustrate the mechanisms of multi-objectivization - but naturally, we will not have this luxury in practical applications, like the TSP.

4.2 TSP/MTSP: Decomposition via Multi-objectivization

The travelling salesperson problem (TSP) is the most well-known of all $\mathcal{NP}$-hard optimization problems. For a comprehensive review and comparison of methods used to solve it see [6], where the problem is stated as follows: We are given a set $C = \{c_1, c_2, \ldots, c_N\}$ of *cities* and for each pair $\{c_i, c_j\}$ of distinct cities there is a *distance* $d(c_i, c_j)$. Our goal is to find an ordering π of the cities that minimizes the quantity

$$\sum_{i=1}^{N-1} d(c_{\pi(i)}, c_{\pi(i+1)}) + d(c_{\pi(N)}, c_{\pi(1)}). \tag{10}$$

In order to multi-objectivize the TSP, we need to identify sub-problems to be solved. Of course, the TSP is $\mathcal{NP}$-hard for the very reason that there is no good decomposition of the problem, i.e. dependencies between components of the problem exist in most instances. However, an obvious decomposition, although by no means perfect, is simply to divide the problem into two (or more) sub-tours, each to be minimized. This can be done in a number of ways, and some may be preferable to others depending on how much is known about the problem instance, but here we use a method that is general for the TSP class. Specifically, to make a two-objective formulation of the TSP, we define two distinct sub-tours to be minimized. The sub-tours are defined by two cities, and the problem becomes:

$$\text{"minimize" } \mathbf{f}(\pi,a,b) = (f_1(\pi,a,b), f_2(\pi,a,b))$$
$$\text{where } f_1(\pi,a,b) = \sum_{i=\pi^{-1}(a)}^{\pi^{-1}(b)-1} d(c_{\pi(i)}, c_{\pi(i+1)})$$
$$\text{and } f_2(\pi,a,b) = \sum_{i=\pi^{-1}(b)}^{N-1} d(c_{\pi(i)}, c_{\pi(i+1)}) + \sum_{i=1}^{\pi^{-1}(a)-1} d(c_{\pi(i)}, c_{\pi(i+1)}) + d(c_{\pi(N)}, c_{\pi(1)}) \quad (11)$$

where a and b are the two cities specified *a priori*, and if $\pi(a) < \pi(b)$ they are swapped. It is intended that a and b are chosen arbitrarily. Notice that the sum of the two objectives is the same as the quantity to be minimized in (10). This ensures that the optimum of (10) is coincident with at least one of the Pareto optima of (11), as required by our definition of multi-objectivizing.

5 Results

5.1 H-IFF/MH-IFF

In this section, we present a comparison of hillclimbing algorithms on the H-IFF problem, and its multi-objectivized formulation, MH-IFF. All algorithms were run on a 64-bit version of the problem, for the same number of function evaluations, num_evals = 500000. Two different mutation rates $p_m = 1/n, 2/n, n = 64$ were used in each algorithm to investigate whether mutation rate choice substantially affected any of the algorithms. The choice of archive size for PAES was 100. The DCGA and PESA algorithms require population sizes to be set. For DCGA, $P = 100$, and for PESA, $P_I = 10$. For the simulated annealing algorithm, preliminary runs were performed and T_0 and T_f were adjusted until the acceptance probabilities fell into the required ranges described in Section 3.

Table 1 shows the full set of results collected on the 64-bit H-IFF problem. In each case, the results were gathered from 30 independent runs of the algorithm. For all algorithms, all best genotypically different solutions were stored off-line, and the tabulated results relate to these off-line results. The table clearly indicates that on H-IFF, removing the local optima through multi-objectivization transforms the problem from one that is very difficult for a neighbourhood searcher, to one that is much easier. The performance of the SOO search algorithms improves with increased mutation rate p_m, confirming that they require larger steps to escape the local optima in the H-IFF problem. The multi-objective algorithms are almost unaffected by the choice of p_m, indicating that they are able to reach the optima through small neighbourhood moves. These results demonstrate the multi-objectivization principle, clearly and can be understood with reference to the discussion given to the problem in Section 4. In the following section we examine how the multi-objectivization technique established here fares on a real-world problem with a much less readily decomposable structure, the TSP.

Table 1. Results of the comparison between algorithms on a 64-bit H-IFF problem. Two of the algorithms are multi-objective and use the MH-IFF decomposition of the problem, namely PAES and PESA. The other three algorithms use the H-IFF objective function directly. Essentially the results compare two single-point hill-climbers, SHC and PAES, and two multi-point hill-climbers, DCGA and PESA. In both cases, the multi-objective algorithm significantly outperforms (using any statistical test) its SOO counterpart. The results of the simulated annealing algorithm (SA) act as a benchmark, indicating the level of performance that can be achieved when escape from local optima is made possible on the original landscape. The columns, '% one' and '% both', indicate the percentage of the runs where respectively one of the optima and both optima were found over the thirty independent runs of each algorithm. Note that only PAES and PESA are able to find both optima.

Algorithm	p_m	best	mean	σ	% one	% both
SHC	$1/n$	288	242.13	22.52	0	0
	$2/n$	336	267.47	29.46	0	0
PAES	$1/n$	448	415.20	51.26	70	47
	$2/n$	448	418.13	50.68	74	43
DCGA	$1/n$	300	270.06	13.80	0	0
	$2/n$	448	323.93	26.54	3	0
PESA	$1/n$	448	448.00	0.00	100	100
	$2/n$	448	448.00	0.00	100	100
SA	$1/n$	448	435.20	26.04	80	0
	$2/n$	448	435.20	26.04	80	0

5.2 TSP/MTSP

We present results for a range of TSP instances of varying size and type. All problems are symmetric, i.e. the distance from A to B is the same as from B to A, where A and B are any two cities. The RAN-20 and RAN-50 problems have 20 and 50 cities respectively, and are non-Euclidean random weight problems where the distance between each pair of cities is a random real number in [0,1). The EUC-50 and EUC-100 are two randomly generated Euclidean problems where the cities are given co-ordinates in a 2-d plane, $x \in [0,1)$, $y \in [0,1)$ and the distance between pairs of nodes is then the Euclidean distance. The problem, kroB100, is taken from TSPLIB and is also a 100-node Euclidean TSP problem. The last problem, mnPeano-92, is a 92-node fractal problem[2] with a known optimal solution [13].

In each algorithm, the representation, initialization, and mutation operators used are identical: the representation of a tour is an N-gene permutation of the numbers $1..N$; the initialization procedure simply generates a random permutation of N cities; and the mutation operator used is the 2-change operator, originally put forward in [3]. It works by selecting two non-identical cities and reversing the order of all the cities between (and including) them. This operator preserves all but two edges in the tour.

---

[2] We conjecture that fractal TSP problems may be particularly suitable for multi-objectivization because their inherently hierarchical structure suggests they may be amenable to decomposition.

On each different TSP instance, all the algorithms are run for the same number of function evaluations, num_evals, given for each problem in Table 2. The PESA and DCGA algorithms further require population sizes to be set. As before, we use the default values of $P = 100$ for DCGA, and $P_I = 10$ for PESA. Setting the SA parameters is carried out as before using preliminary runs to derive an appropriate T_0 and T_f.

The choice of city pairs to be used in the multi-objective algorithms is investigated. First, we present results (Table 2) in which - for the multi-objective algorithms, PAES and PESA - a single, random pair of cities was selected and this pair used in all runs. Later, we report the maximum deviation from these results for other choices of city pairs. Finally, we investigate the choice of city pairs using the EUC-50 problem by selecting a pair of cities that are very close, and another pair where they are maximally distant, and repeating our experiments for these choices.

Table 2. Summary TSP results. In the 'Optimum' column, figures given in bold font represent the known optimum value; other figures are estimates. For the RAN-20 problem, the optimum is an estimate based on the fact that SA reached this value on 30 consecutive runs, and given the small size of the problem. For the RAN-50 problem, the estimated figure is based on the expected limiting value of an optimal tour [6], and similarly for EUC-50 and EUC-100, the estimates are based on the formula for expected tour length = $K\sqrt{NA}$ with N the number of cities, $A = 1.0$ the area in which the cities are placed, and $K \approx 0.7124$ [6].

Algorithm	Problem	num_evals	Optimum	Best	Mean	σ
SHC	RAN-20	500000	2.547394	2.550811	2.81	0.14
PAES	RAN-20	500000	2.547394	2.547394	2.66	0.14
SA	RAN-20	500000	2.547394	2.547394	2.55	0.00
SHC	RAN-50	500000	2.0415	2.620087	3.09	0.28
PAES	RAN-50	500000	2.0415	2.259948	2.73	0.22
DCGA	RAN-50	500000	2.0415	2.307587	2.46	0.09
PESA	RAN-50	500000	2.0415	2.189421	2.32	0.28
SA	RAN-50	500000	2.0415	2.130675	2.30	0.10
SHC	EUC-50	500000	5.0374	5.904673	6.23	0.20
PAES	EUC-50	500000	5.0374	5.801026	6.03	0.13
DCGA	EUC-50	500000	5.0374	5.707789	5.76	0.05
PESA	EUC-50	500000	5.0374	5.692169	5.78	0.08
SA	EUC-50	500000	5.0374	5.692169	5.72	0.03
SHC	EUC-100	2000000	7.124	8.143720	8.55	0.23
PAES	EUC-100	2000000	7.124	8.028227	8.35	0.24
DCGA	EUC-100	2000000	7.124	7.902731	8.16	0.14
PESA	EUC-100	2000000	7.124	7.795515	7.97	0.10
SA	EUC-100	2000000	7.124	7.853258	7.98	0.07
PESA	kroB100	2000000	**22141**	22141	22546.1	324.2
SA	kroB100	2000000	**22141**	22217	22529.2	173.0
SHC	mnPeano-92	1000000	**5697.93**	5857.47	6433.45	197.1
PAES	mnPeano-92	1000000	**5697.93**	5879.63	6255.30	197.6

From Table 2 we can see that, without exception, the results of the PAES algorithm are superior to the SHC algorithm at a statistically significant level[3], over the range of problem instances in the Table. This shows that the number of local optima in the TSP problem, using the 2-change neighbourhood, is reduced by the method of multi-objectivization we have proposed.

Although the number of local optima has successfully been reduced by multi-objectivization, this does not make PAES more effective than other methods for solving TSP. We are only using PAES to indicate the basic advantage of multi-objectivization over hill-climbing in a SOO problem. Compared with DCGA, PAES performs poorly on all problems. However, PESA, the multi-objective counterpart of DCGA, outperforms DCGA on all but one of the problems. PESA is also competitive with the SA algorithm, which has its own explicit means of escaping from local optima. This result adds further evidence that multi-objectivization can enable an algorithm to avoid local optima effectively.

To investigate the effect of the choice of city pairs, the 50-node Euclidean TSP problem was used. First, a choice of city pair in which the cities were very close to each other and relatively far from others was selected. 30 runs were performed using this pair and the results were: best = 5.804719, mean = 6.10, and $\sigma = 0.22$. Then, with cities in opposite corners of the plane the results were: best = 5.818185, mean = 6.09, $\sigma = 0.16$. So both are worse than the 'random' choice used in the results in Table 2: best = 5.801026, mean = 6.03, $\sigma = 0.13$. However, although there is some seeming dependence on city choice, all three of these results are better than the results of the SHC algorithm on this problem. Some alternative choices of city pairs were used on some of the other problems, too. On the random 20-node problem, three different choices of node pair were tried. There was a 1.5% variation in the mean over 30 runs, for the different node-pair choices. On the 50-node Euclidean problem, two different pairs of cities were chosen for the PESA algorithm. The difference in means over 30 runs was 0.4%. On no runs did the choice of city pair affect the mean TSP tour length by more than 2%.

6 Discussion

The examples in the previous sections have illustrated different ways in which additional objectives can be defined for a SOO problem and that this multi-objectivization can facilitate improved search for neighbourhood based algorithms. We suggested earlier that a successful multi-objectivization may involve decomposing the original function into a number of sub-problems and that these sub-problems should be as independent as possible. In the H-IFF/MH-IFF example, we can separate the two competing components of the problem completely, making the problem very easy for a hill-climber. In TSP this is not so easy - there is no (known) method to decompose $\mathcal{NP}$-hard problems, like TSP, that creates independent sub-problems - indeed, it is the interdependency of the problem components that puts them in this class. Nonetheless, we suggest that there may be some merit in examining further the parallels between sub-problems in SOO,

---

[3] Using a large-sample test of hypothesis for the difference in two population means [12] (which does not depend upon distribution), at the 95% confidence level.

and objectives in MOO. Specifically, if different objectives separate out different components of a problem then different points on the Pareto front correspond to solutions that are more or less specialized to different sub-problems. Our experiments support the intuition that it will be easier to discover a set of different specialists than it is to discover a generalist directly. In the experiments in this paper we used only mutation to investigate this, but if we could find some way to combine together different specialists from the front then perhaps it would facilitate discovery of generalists more directly. In a sense, this is the intuition behind recombination in the multi-objective Messy GA (MOMGA) [15].

However, the MOMGA assumes a number of objectives given *a priori*, and like other MOO methods, it is not intended as a technique for problem decomposition in SOO problems. But related research suggests that it may be possible to automatically discover objectives relating to sub-problems, and thereby apply a MOMGA-like algorithm to SOO problems. The Symbiogenic Evolutionary Adaptation Model [18] shares some features with the MOMGA but uses group evaluation of individuals to encourage them to self-organize to cover the problem space, and automatically discover sub-problems to be used as the objectives. This algorithm successfully solves the H-IFF problem without the introduction of additional objectives provided by the researcher.

7 Conclusions

In this paper, we have defined a process that we call "multi-objectivization" whereby the scalar function of a SOO problem is replaced by a vector of functions such that the resulting MOO problem has Pareto optima which coincide with the optima of the original problem. We investigated the effects of this transformation, in particular, the reduction of local optima for hill-climbing style algorithms. We illustrated the effect of the approach first on an abstract building-block problem, H-IFF, that is trivially amenable to such a decomposition. We then investigated the approach further, using several small instances of the TSP, where decomposition is inherently difficult. We defined a multi-objectivization of the problem based on minimizing two sub-tours. Our results showed that this simple multi-objectivization does seem to reduce the effect of local optima on simple hill-climbing algorithms. These preliminary results, suggest that there is a link between the presence of local optima in SOO problems and an underlying conflict of implicit objectives, and they shed some light on the processes of multi-objective search.

Acknowledgments. The authors would like to thank Anthony Bucci, Michiel de Jong, and the anonymous reviewers for their excellent comments and criticisms.

References

1. R. Bellman. Dynamic programming and multi-stage decision processes of stochastic type. In *Proceedings of the second symposium in linear programming*, volume 2, pages 229–250, Washington D.C., 1955. NBS and USAF.

2. D. W. Corne and J. D. Knowles. The Pareto-envelope based selection algorithm for multiobjective optimization. In *Proceedings of the Sixth International Conference on Parallel Problem Solving from Nature (PPSN VI)*, pages 839–848, Berlin, 2000. Springer-Verlag.
3. M. M. Flood. The travelling-salesman problem. *Operations Research*, 4:61–75, 1956.
4. J. Horn, D. Goldberg, and K. Deb. Long path problems for mutation-based algorithms. Technical Report 92011, Illinois Genetic Algorithms Laboratory, University of Illinois at Urbana-Champaign, Urbana IL, 1992.
5. M. Huynen, P. Stadler, and W. Fontana. Smoothness within ruggedness: The role of neutrality in adaptation. *Proceedings of the National Academy of Sciences (USA)*, 93:397–401, 1996.
6. D. S. Johnson and L. A. McGeoch. The travelling salesman problem: a case study. In E. Aarts and J. K. Lenstra, editors, *Local Search in Combinatorial Optimization*, pages 215–310. John Wiley and Sons, 1997.
7. T. Jones. *Evolutionary Algorithms, Fitness Landscapes and Search*. PhD thesis, University of New Mexico, Albuquerque, 1995.
8. J. D. Knowles and D. W. Corne. The Pareto archived evolution strategy: A new baseline algorithm for multiobjective optimisation. In *1999 Congress on Evolutionary Computation*, pages 98–105, Washington, D.C., July 1999. IEEE Service Center.
9. J. D. Knowles and D. W. Corne. Approximating the nondominated front using the Pareto archived evolution strategy. *Evolutionary Computation*, 8(2):149–172, 2000.
10. S. J. Louis and G. J. E. Rawlins. Pareto optimality, GA-easiness and deception. In S. Forrest, editor, *Proceedings of the Fifth International Conference on Genetic Algorithms (ICGA-5)*, pages 118–123, San Mateo, CA, 1993. Morgan Kaufmann.
11. S. W. Mahfoud. *Niching methods for genetic algorithms*. PhD thesis, University of Illinois at Urbana-Champaign, Urbana, IL, USA, 1995. IlliGAL Report 95001.
12. W. Mendenhall and R. J. Beaver. *Introduction to Probability and Statistics - 9th edition*. Duxbury Press, International Thomson Publishing, Pacific Grove, CA, 1994.
13. M. G. Norman and P. Moscato. The euclidean traveling salesman problem and a space-filling curve. *Chaos, Solitons and Fractals*, 6:389–397, 1995.
14. C. R. Reeves. Modern heuristic techniques. In V. Rayward-Smith, I. Osman, C. Reeves, and G. Smith, editors, *Modern Heuristic Search Methods*, chapter 1, pages 1–26. John Wiley and Sons Ltd., 1996.
15. D. A. Van Veldhuizen and G. B. Lamont. Multiobjective Optimization with Messy Genetic Algorithms. In *Proceedings of the 2000 ACM Symposium on Applied Computing*, pages 470–476, Villa Olmo, Como, Italy, 2000. ACM.
16. R. A. Watson, G. S. Hornby, and J. B. Pollack. Modeling building-block interdependency. In *Parallel Problem Solving from Nature - PPSN V*, pages 97–106. Springer-Verlag, 1998.
17. R. A. Watson and J. B. Pollack. Analysis of recombinative algorithms on a hierarchical building-block problem. In *Foundations of Genetic Algorithms (FOGA)*, 2000.
18. R. A. Watson and J. B. Pollack. Symbiotic combination as an alternative to sexual recombination in genetic algorithms. In *Proceedings of the Sixth International Conference of Parallel Problem Solving From Nature (PPSN VI)*, pages 425–436. Springer-Verlag, 2000.

Constrained Test Problems for Multi-objective Evolutionary Optimization

Kalyanmoy Deb, Amrit Pratap, and T. Meyarivan

Kanpur Genetic Algorithms Laboratory (KanGAL)
Indian Institute of Technology Kanpur
Kanpur, PIN 208 016, India
{deb,apratap,mary}@iitk.ac.in
http://www.iitk.ac.in/kangal

Abstract. Over the past few years, researchers have developed a number of multi-objective evolutionary algorithms (MOEAs). Although most studies concentrated on solving unconstrained optimization problems, there exists a few studies where MOEAs have been extended to solve constrained optimization problems. As the constraint handling MOEAs gets popular, there is a need for developing test problems which can evaluate the algorithms well. In this paper, we review a number of test problems used in the literature and then suggest a set of tunable test problems for constraint handling. Finally, NSGA-II with an innovative constraint handling strategy is compared with a couple of existing algorithms in solving some of the test problems.

1 Introduction

Multi-objective evolutionary algorithms (MOEAs) have amply demonstrated the advantage of using population-based search algorithms for solving multi-objective optimization problems [1,14,8,5,6,11]. In multi-objective optimization problems of varying degrees of complexities, elitist MOEAs have demonstrated their abilities in converging close to the true Pareto-optimal front and in maintaining a diverse set of solutions. Despite all these developments, there seem to be not enough studies concentrating procedures for handling constraints. Constraint handling is a crucial part of real-world problem solving and it is time that MOEA researchers focus on solving constrained multi-objective optimization problems.

To evaluate any algorithm, there is a need for well understood and tunable test problems. Although there exists a few constraint handling procedures in the literature, they all have been tried on simple problems, particularly having only a few decision variables and having inadequately nonlinear constraints. When an algorithm performs well on such problems, it becomes difficult to evaluate the true merit of it, particularly in the context of its general problem solving.

In this paper, we briefly outline a few constraint handling methods suggested in the literature. Thereafter, we review a few popular test problems for their degrees of difficulty. Next, we propose a *tunable* test problem generator having six parameters, which can be set to obtain constrained test problems of desired degree of difficulty. Simulation results with NSGA-II [1] along with the constrained-domination principle and with two other constrained handling procedures bring out the superiority of NSGA-II in solving difficult constrained optimization problems. The difficulty demonstrated by the test problems will qualify them to be used as standard constrained multi-objective test problems in the years to come.

2   Constrained Multi-objective Evolutionary Algorithms

There exists only a few studies where an MOEA is specifically designed for handling constraints. Among all methods, the usual penalty function approach [11,2] where a penalty proportional to the total constraint violation is added to all objective functions. When applying this procedure, all constraints and objective functions must be normalized.

Jimenez et al. [7] suggested a procedure which carefully compares two solutions in a binary tournament selection. If one solution is feasible and other is not, the feasible solution is chosen. If both solutions are feasible, Horn et al.'s [6] niched-Pareto GA is used. On the other hand, if both solutions are infeasible, the solution closer to the constraint boundary is more likely to be chosen.

Ray et al. [10] suggested a more elaborate constraint handling technique, where constraint violations of all constraints are not simply summed together, instead a non-domination check of constraint violations is made. Based on separate non-domination sorting of a population using objective functions alone, constraint violations alone, and objective function and constraint violation together, the algorithm demands a large computational complexity.

Recently, Deb et al. [1] defined a *constraint-domination* principle, which differentiates infeasible from feasible solutions during the non-dominated sorting procedure:

Definition 1 *A solution i is said to constrained-dominate a solution j, if any of the following conditions is true:*

1. *Solution i is feasible and solution j is not.*
2. *Solutions i and j are both infeasible, but solution i has a smaller overall constraint violation.*
3. *Solutions i and j are feasible and solution i dominates solution j.*

This requires no additional constraint-handling procedure to be used in an MOEA. Since the procedure is generic, the above constraint-domination principle can be used with any unconstrained MOEAs.

3   Past Test Problems

In the context of multi-objective optimization, constraints may cause hindrance to an multi-objective EA (MOEA) to converge to the true Pareto-optimal region

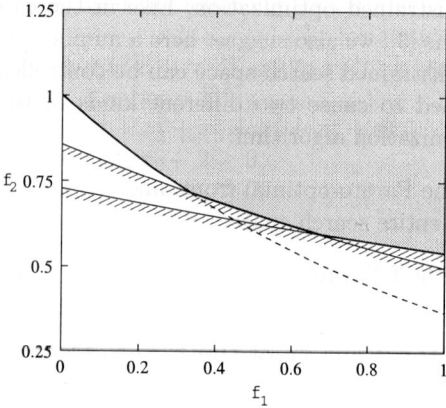

Fig. 5. Constrained test problem CTP1 with two $J = 2$ constraints.

variables (**x**) to satisfy the inequality constraints with the equality sign. Each constraint is an implicit non-linear function of decision variables. Thus, it may be difficult to find a number of solutions on a non-linear constraint boundary. The complexity of the test problem can be further increased by using more constraints and by using a multi-modal or deceptive g function.

Besides finding and maintaining correlated decision variables to fall on the several constraint boundaries, there could be other difficulties near the Pareto-optimal front. The constraint functions can be such that the unconstrained Pareto-optimal region is now infeasible and the resulting Pareto-optimal set is a collection of a number of discrete regions. At the extreme, the Pareto-optimal region can become a collection of a set of discrete solutions. Let us first present such a function mathematically and then describe the difficulties in each case.

$$\text{CTP2-CTP7:} \begin{cases} \text{Minimize } f_1(\mathbf{x}) = x_1 \\ \text{Minimize } f_2(\mathbf{x}) = g(\mathbf{x})\left(1 - \frac{f_1(\mathbf{X})}{g(\mathbf{X})}\right) \\ \text{Subject to } c(\mathbf{x}) \equiv \cos(\theta)(f_2(\mathbf{x}) - e) - \sin(\theta)f_1(\mathbf{x}) \geq \\ \qquad a \left|\sin\left(b\pi\left(\sin(\theta)(f_2(\mathbf{x}) - e) + \cos(\theta)f_1(\mathbf{x})\right)^c\right)\right|^d. \end{cases} \quad (5)$$

The decision variable x_1 is restricted in $[0, 1]$ and the bounds of other variables depend on the chosen $g(\mathbf{x})$ function. The constraint has six parameters (θ, a, b, c, d, and e). In fact, the above problem can be used as a constrained test problem generator where constrained test problems with different complexities will evolve by simply tuning these six parameters. We demonstrate this by constructing different test problems, where the effect of each parameter is also discussed.

First, we use the following parameter values:

$$\theta = -0.2\pi, \quad a = 0.2, \quad b = 10, \quad c = 1, \quad d = 6, \quad e = 1.$$

The resulting feasible objective space is shown in Figure 6. It is clear from the figure that the unconstrained Pareto-optimal region becomes infeasible in the presence of the constraint. The periodic nature of the constraint boundary makes the Pareto-optimal region discontinuous, having a number of disconnected continuous regions. The task of an optimization algorithm would be to find as many such disconnected regions as possible. The number of disconnected regions can be controlled by increasing the value of the parameter b. It is also clear that with the increase in number of disconnected regions, an algorithm will have difficulty in finding representative solutions in all disconnected regions.

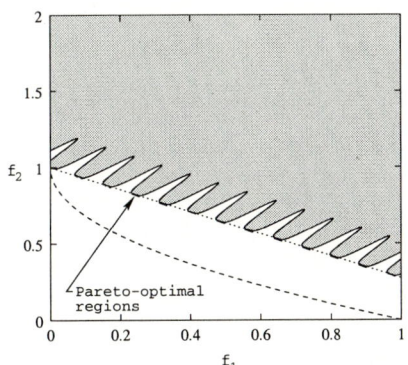
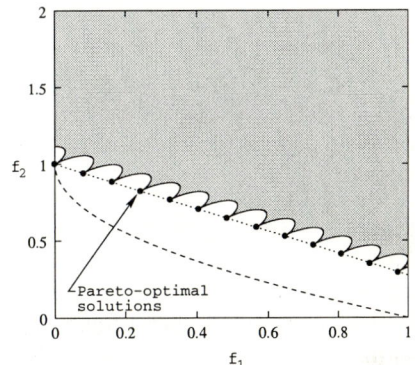

Fig. 6. Constrained test problem CTP2.   **Fig. 7.** Constrained test problem CTP3.

The above problem can be made more difficult by using a small value of d, so that in each disconnected region there exists only one Pareto-optimal solution. Figure 7 shows the feasible objective space for $d = 0.5$ and $a = 0.1$ (while other parameters are the same as that in the previous test problem). Although most of the feasible search space is continuous, near the Pareto-optimal region, the feasible search regions are disconnected, finally each subregion leading to a singular feasible Pareto-optimal solution. An algorithm will face difficulty in finding all discrete Pareto-optimal solutions because of the changing nature from continuous to discontinuous feasible search space near the Pareto-optimal region.

The problem can have a different form of complexity by increasing the value of parameter a, which has an effect of making the transition from continuous to discontinuous feasible region far away from the Pareto-optimal region. Since an algorithm now has to travel through a long narrow feasible *tunnel* in search of the lone Pareto-optimal solution at the end of tunnel, this problem will be more difficult to solve compared to the previous problem. Figure 8 shows one such problem with $a = 0.75$ and rest of the parameters same as that in the previous test problem.

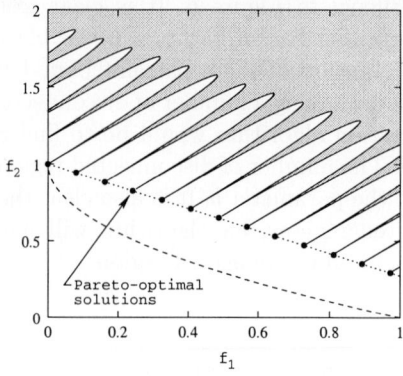 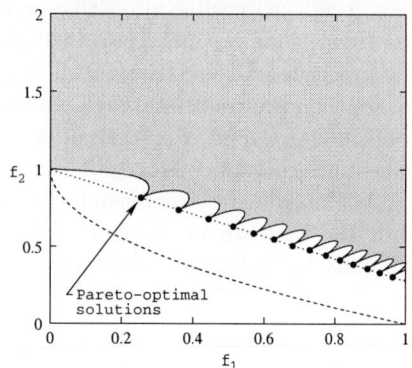

Fig. 8. Constrained test problem CTP4. **Fig. 9.** Constrained test problem CTP5.

In all the three above problems, the disconnected regions are equally distributed in the objective space. The discrete Pareto-optimal solutions can be scattered non-uniformly by using $c \neq 1$. Figure 9 shows the feasible objective space for a problem with $c = 2$. Since $c > 1$, more Pareto-optimal solutions lie towards right (higher values of f_1). If, however, $c < 1$ is used, more Pareto-optimal solutions will lie towards left. For even higher number of Pareto-optimal solutions towards right, a larger value of c can be chosen. The difficulty will arise in finding all closely packed discrete Pareto-optimal solutions.

The parameters θ and e do not have a major role to play in terms of producing a significant complexity. The parameter θ controls the slope of the Pareto-optimal region, whereas the parameter e shifts the constraints up or down in the objective space. For the above problems, the Pareto-optimal solutions lie on the following straight line:

$$(f_2(\mathbf{x}) - e) \cos \theta = f_1(\mathbf{x}) \sin \theta. \tag{6}$$

It is interesting to note that the location of this line is independent of other major parameters (a, b, c, and d). The above equation reveals that the parameter e denotes the intercept of this line on the f_2 axis. The corresponding intercept on the f_1 axis is $-e/\tan \theta$.

It is important here to mention that although the above test problems will cause difficulty in the vicinity of the Pareto-optimal region, an algorithm has to maintain an adequate diversity much before it comes closer to the Pareto-optimal region. If an algorithm approaches the Pareto-optimal region without much diversity, it may be too late to create diversity among population members, as the feasible search region in the vicinity of the Pareto-optimal region is disconnected.

4.2 Difficulty in the Entire Search Space

Above test problems provide difficulty to an algorithm in the vicinity of the Pareto-optimal region. Difficulties may also come from the infeasible search regions present in the entire search space. Fortunately, the same constrained test problem generator can also be used here.

Figure 10 shows the feasible objective search space for the following parameter values:

$$\theta = 0.1\pi, \quad a = 40, \quad b = 0.5, \quad c = 1, \quad d = 2, \quad e = -2.$$

Notice that the above parameter values are very different from that used in the previous section. Particularly, the parameter value of a is two-order magnitude larger than before.

The objective space of this function has infeasible holes of differing widths towards the Pareto-optimal region. Since an algorithm has to overcome a number of such infeasible holes before coming to the island containing the Pareto-optimal front, an algorithm may face difficulty in solving this problem. Moreover, the unconstrained Pareto-optimal region is now not feasible. The entire constrained Pareto-optimal front lies on a part of the constraint boundary. (In this particular test problem, the Pareto-optimal region corresponds to all solutions satisfying $1 \leq ((f_2 - e)\sin\theta + f_1\cos\theta) \leq 2$.) The difficulty can be increased further by widening the infeasible regions (or by using a small value of d).

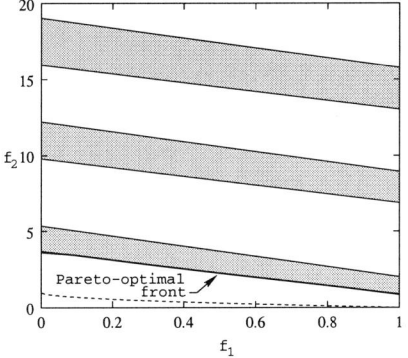
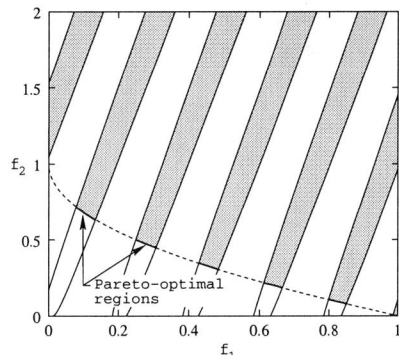

Fig. 10. Constrained test problem CTP6.   **Fig. 11.** Constrained test problem CTP7.

The infeasibility in the objective search space can also come along the Pareto-optimal region. Using the following parameter values

$$\theta = -0.05\pi, \quad a = 40, \quad b = 5, \quad c = 1, \quad d = 6, \quad e = 0,$$

we obtain the feasible objective space, as shown in Figure 11. This problem makes some portions of the unconstrained Pareto-optimal region feasible, thereby making a disconnected set of continuous regions. In order to find all such disconnected regions, an algorithm has to maintain an adequate diversity right from the beginning of a simulation run. Moreover, the algorithm also has to maintain its solutions feasible as it proceeds towards the Pareto-optimal region.

Once again, the difficulty can be increased by reducing the width of the feasible regions by simply using a smaller value of d. A careful thought will reveal that this test problem is similar to CTP4, except that a large value of the parameter a is now used. This initiates the disconnectedness in the feasible region far away from the Pareto-optimal region. If it initiates away from the search region dictated by the variable boundaries, the entire search region becomes a patch of feasible and infeasible regions.

With the above constraints, a combination of two or more effects can be achieved together in a problem by considering more than one such constraints.

4.3 More Than Two Objectives

Using the above concept, test problems having more than two objectives can also be developed. We modify equation 5 as follows. Using an M-dimensional transformation (rotational R and translational $\mathbf{e}$), we compute

$$\mathbf{f}' = R^{-1}(\mathbf{f} - \mathbf{e}). \tag{7}$$

The matrix R will involve $(M-1)$ rotation angles. Thereafter, the following constraint can be used:

$$f'_M \geq \sum_{j=1}^{M-1} a_j \left|\sin\left(b_j \pi f'^{c_j}_j\right)\right|^{d_j}. \tag{8}$$

Here, all a_j, b_j, c_j, d_j, and θ_j are parameters that must be set to get a desired effect. Like before, a combination of more than one such constraints can also be considered.

5 Simulation Results

We compare NSGA-II with the constrained-domination principle, Jimenez's algorithm and Ray et al.'s algorithms on test problems CTP1 till CTP7 with Rastrigin's function as the g functional.

In all algorithms, we have used the simulated binary crossover [4] with $\eta_c = 20$ and polynomial mutation operator [4] with $\eta_m = 20$, exactly the same as that used with NSGA-II. Identical recombination and mutation operators are used to investigate the effect of constraint handling ability of each algorithm. A crossover probability of 0.9 and mutation probability of $1/n$ (where n is the number of variables) are chosen. In all problems, five decision variables[1] are used. A population of size 100 and a maximum generation of 500 are used. For the Ray et al.'s algorithm, we have used a sharing parameter $\sigma_{share} = 0.158$. For Jimenez's algorithm, we has used $t_{dom} = 10$ and $\sigma_{share} = 0.158$. It is important to highlight that NSGA-II does not require any extra parameter setting.

In all problems, the Jimenez's algorithm performed poorly by not converging any where near the true Pareto-optimal region. Thus, we do not present the solutions using Jimenez's algorithm.

Figure 12 shows that both NSGA-II and Ray et al.'s algorithm are able to converge close to the true Pareto-optimal front on problem OSY, but NSGA-II's convergence ability as well as ability to find diverse solutions are better.

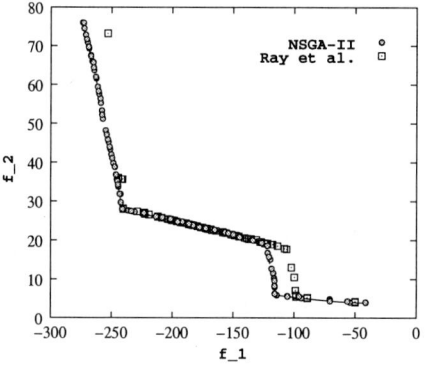

Fig. 12. Simulation results on OSY.

Figure 13 shows the population after 500 generations with NSGA-II and Ray et al.'s algorithm on CTP2. NSGA-II is able to find all disconnected Pareto-optimal solutions, whereas Ray et al.'s algorithm could not converge well on all disconnected regions.

Problem CTP3 degenerates to only one solution in each disconnected Pareto-optimal region. NSGA-II is able to find a solution very close to the true Pareto-optimal solution in each region (Figure 14). Ray et al.'s algorithm cannot quite converge near the Pareto-optimal solutions.

As predicted, problem CTP4 caused difficulty to both algorithms. Figure 15 shows that both algorithm could not get near the true Pareto-optimal solutions.

The non-uniformity in spacing of the Pareto-optimal solutions seems to be not a great difficulty to NSGA-II (Figure 16), but Ray et al's algorithm has some difficulty in converging to all Pareto-optimal solutions.

---

[1] As the simulation results show, only five variables caused enough difficulty to the three chosen MOEAs.

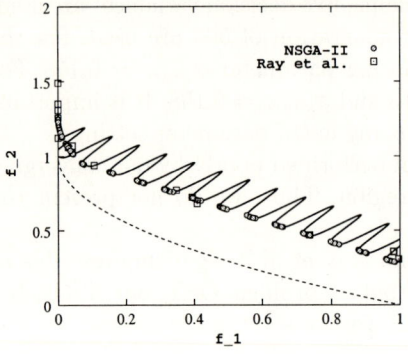
Fig. 13. Simulation results on CTP2.

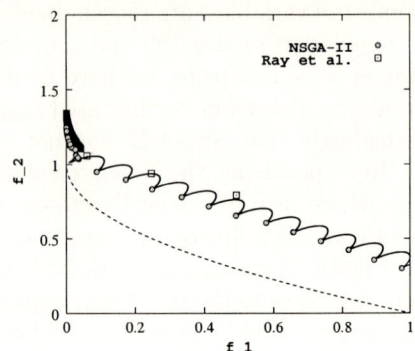
Fig. 14. Simulation results on CTP3.

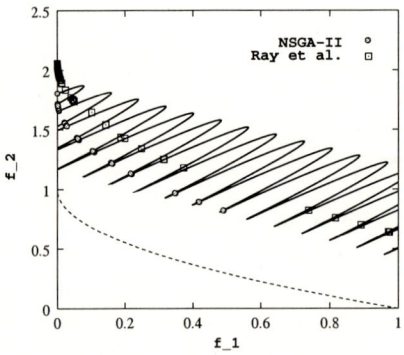
Fig. 15. Simulation results on CTP4.

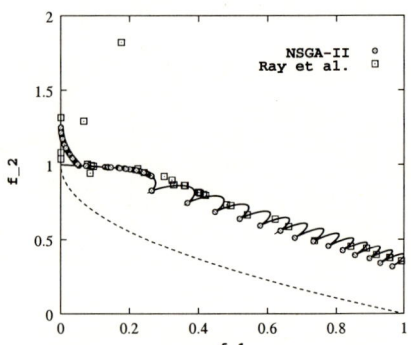
Fig. 16. Simulation results on CTP5.

When the entire search space consists of infeasible patches *parallel* to the Pareto-optimal front (CTP6 in Figure 17), both MOEAs are able to converge to the right feasible patch and finally very near to the true Pareto-optimal front. All feasible patches are marked with a 'F'. Although both algorithms performed well in this problem, NSGA-II is able to find a better convergence as well as a better spread of solutions.

However, when infeasible patches exist *perpendicular* to the Pareto-optimal front, both MOEAs had the most difficulty (Figure 18). None of them can find solutions closer to the true Pareto-optimal front. Although Ray et al's algorithm maintained a better spread of solutions, NSGA-II is able to come closer to the true front. We believe that this is one of the test problems where an algorithm must maintain a good spread of solutions from the beginning of the run and must also have a good converging ability. Algorithms which tend to converge anywhere in the Pareto-optimal front first and then work on finding a spread of

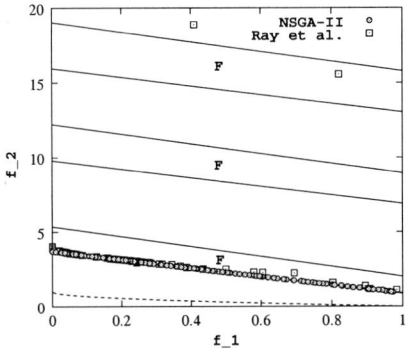
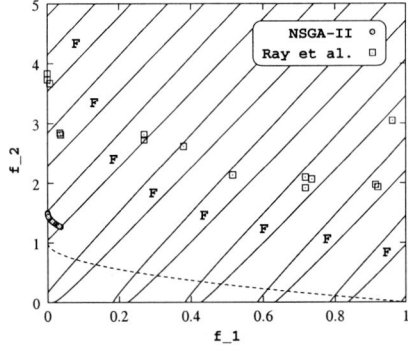

Fig. 17. Simulation results on CTP6.   **Fig. 18.** Simulation results on CTP7.

solutions will end up finding solutions in a few of the feasible patches. NSGA-II shows a similar behavior in this problem.

6   Conclusion

In this paper, we have discussed the need of having tunable test problems for constrained multi-objective optimization. Test problems commonly used in most MOEA studies are simple and the level of complexity offered by the problems cannot be changed. We have presented a test problem generator which has six controllable parameters. By setting different parameter values, we have created a number of difficult constrained test problems. The difficulty in solving these test problems has been demonstrated by applying three constraint handling MOEAs on a number of these test problems. Although all algorithms faced difficulty in solving the test problems, NSGA-II with a constrained-domination principle has outperformed the other two MOEAs. The study has also shown that certain problem features can cause more difficulty than others.

Because of the tunable feature of the proposed test problems and demonstrated difficulties on a number of constrained MOEAs, we believe that these problems would be used as standard constrained test problems in the coming years of research in evolutionary multi-objective optimization.

Acknowledgements. Authors acknowledge the support provided by Department of Science and technology (DST) for this study.

References

1. Deb, K., Agrawal, S., Pratap, A., Meyarivan, T. (2000). A Fast Elitist Non-dominated sorting genetic algorithm for multi-objective optimization: NSGA-II. *Proceedings of the Parallel Problem Solving from Nature VI Conference*, pp. 849–858.

2. Deb, K., Pratap, A., Moitra, S. (2000). Mechanical component design for multiple objectives using elitist non-dominated sorting GA. *Proceedings of the Parallel Problem Solving from Nature VI Conference*, pp. 859–868.
3. Deb, K. (1999) Multi-objective genetic algorithms: Problem difficulties and construction of test Functions. *Evolutionary Computation, 7*(3), 205–230.
4. Deb, K. and Agrawal, R. B. (1995) Simulated binary crossover for continuous search space. *Complex Systems, 9* 115–148.
5. Fonseca, C. M. and Fleming, P. J. (1993) Genetic algorithms for multi-objective optimization: Formulation, discussion and generalization. In Forrest, S., editor, *Proceedings of the Fifth International Conference on Genetic Algorithms*, pages 416–423, Morgan Kauffman, San Mateo, California.
6. Horn, J. and Nafploitis, N., and Goldberg, D. E. (1994) A niched Pareto genetic algorithm for multi-objective optimization. In Michalewicz, Z., editor, *Proceedings of the First IEEE Conference on Evolutionary Computation*, pages 82–87, IEEE Service Center, Piscataway, New Jersey.
7. Jiménez, F. and Verdegay, J. L. (1998). Constrained multiobcjctive optimization by evolutionary algorithms. *Proceedings of the International ICSC Symposium on Engineering of Intelligent Systems (EIS'98)*, pp. 266–271.
8. Knowles, J. and Corne, D. (1999) The Pareto archived evolution strategy: A new baseline algorithm for multiobjective optimisation. *Proceedings of the 1999 Congress on Evolutionary Computation*, Piscataway: New Jersey: IEEE Service Center, 98–105.
9. Osyczka, A. and Kundu, S. (1995). A new method to solve generalized multicriteria optimization problems using the simple genetic algorithm. *Structural Optimization*(10). 94–99.
10. Ray, T., Kang, T., and Chye, S. (in press). Multiobjective design optimization by an evolutionary algorithm, *Engineering Optimization*.
11. Srinivas, N. and Deb, K. (1995). Multi-Objective function optimization using non-dominated sorting genetic algorithms. *Evolutionary Computation*(2), 221–248.
12. Tanaka, M. (1995). GA-based decision support system for multi-criteria optimization. *Proceedings of the International Conference on Systems, Man and Cybernetics-2*, pp. 1556–1561.
13. Van Veldhuizen, D. (1999). Multiobjective evolutionary algorithms: Classifications, analyses, and new innovations. *PhD Dissertation and Technical Report No. AFIT/DS/ENG/99-01*, Dayton, Ohio: Air Force Institute of Technology.
14. Zitzler, E. (1999). Evolutionary algorithms for multiobjective optimization: Methods and applications. Doctoral thesis ETH NO. 13398, Zurich: Swiss Federal Institute of Technology (ETH), Aachen, Germany: Shaker Verlag.

Constraint Method-Based Evolutionary Algorithm (CMEA) for Multiobjective Optimization

S. Ranji Ranjithan[1], S. Kishan Chetan[2], and Harish K. Dakshina[3]

[1] Department of Civil Engineering, Campus Box 7908,
North Carolina State University, Raleigh, NC 27695, USA
ranji@eos.ncsu.edu

[2] Siebel Systems, 6001 Shellmound, Emeryville, CA 94608, USA
kchetan@siebel.com

[3] Department of Civil Engineering, Campus Box 7908,
North Carolina State University, Raleigh, NC 27695, USA
hkdakshi@unity.ncsu.edu

Abstract. Evolutionary algorithms are becoming increasingly valuable in solving large-scale, realistic engineering multiobjective optimization (MO) problems, which typically require consideration of conflicting and competing design issues. The new procedure, Constraint Method-Based Evolutionary Algorithm (CMEA), presented in this paper is based upon underlying concepts in the constraint method described in the mathematical programming literature. Pareto optimality is achieved implicitly via a constraint approach, and convergence is enhanced by using beneficial seeding of the initial population. CMEA is evaluated by solving two test problems reported in the multiobjective evolutionary algorithm (MOEA) literature. Performance comparisons based on quantitative metrics for accuracy, coverage, and spread are presented. CMEA is relatively simple to implement and incorporate into existing implementations of evolutionary algorithm-based optimization procedures.

1 Introduction

Tradeoff information in the form of a noninferior, or Pareto optimal set, of solutions is important in considering competing design objectives when making decisions associated with most engineering problems. The use of standard mathematical programming-based techniques, such as constraint method and weighting method (Cohon, 1978) for generating noninferior sets, in many practical engineering problems is limited because the models describing the engineering processes seldom fit into the restrictive function forms required by these mathematical programming techniques. As evolutionary algorithms offer a relatively more flexible way to analyze and solve realistic engineering design problems, their use in multi criterion decision making is becoming increasingly important. An array of multiobjective evolutionary algorithms (MOEAs) has been reported since the early eighties. Detailed summaries of the state-of-the-art in MOEA were discussed recently by Coello (1999a) and Van Veldhuizen and Lamont (2000), and are represented in the special issue of *Evolutionary Computation* (Vol. 8, No. 2, Summer 2000) on multi criterion optimization (also see Coello (1999b) for an archive of bibliography).

Cadenas (1995), Harrell and Ranjithan (1997), Coello et al. (1998), Coello and Christiansen (2000), Loughlin et al. (2000b), Obayashi et al. (2000) are some examples of use of MOEAs in multiobjective optimization of varied realistic engineering problems. VEGA-Vector Evaluated Genetic Algorithm (Schaffer, 1985), NPGA-Niched Pareto Genetic Algorithm (Horn et al., 1994), NSGA-Non-dominated Sorting Genetic Algorithm (Srinivas and Deb, 1994), and SPEA-Strength Pareto Evolutionary Algorithm (Zitzler and Thiele, 1999) are the four selected MOEAs that are used to compare the performance of CMEA presented in this paper.

3 CMEA – Constraint Method-Based Evolutionary Algorithm

Based on the concepts of the mathematical programming-based constraint approach (Cohon, 1978) for generating the noninferior set, CMEA achieves Pareto optimality in an implicit manner by ensuring that the population migrates along the noninferior surface. At each iteration, the population converges to a noninferior solution by solving the following single objective optimization problem:

$$\text{Maximize} \quad Z_h(x) \tag{6}$$

$$\text{Subject to} \quad g_i(x) \leq 0 \quad \forall i = 1,2,\ldots,m \tag{7}$$

$$Z_l(x) \geq Z_1^l \quad \forall l = 1,2,\ldots,k;\ l \neq h \tag{8}$$

$$x \in X \tag{9}$$

where, Z_h is one of the k objectives, and Z_1^l $\{l=1,2,\ldots k;\ l \neq h\}$ is the constraint value for objective l ($\neq h$) corresponding to a noninferior solution. By varying Z_1^l incrementally, the search migrates from one noninferior solution to an adjacent solution, eventually tracing the noninferior surface. A two-objective illustration is shown in Figure 1.

At each intermediate step in which one noninferior solution is obtained, the model given by Equations (6)-(9) is solved. A straightforward implementation of an algorithm that repeats this intermediate step would be similar to iterative execution of a single objective EA, which is not necessarily computationally efficient. Instead, CMEA exploits the basic concept that for some classes of problems, adjacent solutions in the decision space map to adjacent points in the objective space. Its implication is that these decision vectors (xs) (that map to adjacent noninferior points in the objective space) have solution features (i.e., values of x_js) that are only marginally different. This enables the beneficial use of the final population corresponding to the current noninferior solution to seed the search of an adjacent noninferior solution. The new search of course would have an updated constraint set (8) to represent an adjacent noninferior point in the objective space. When the new selection pressure manifesting from the updated constraint vector is applied on the previous population, the population quickly migrates to an adjacent noninferior solution. A systematic update of the constraint set (8) thus enables an efficient mechanism for incrementally tracing the noninferior set. This incremental population migration approach significantly reduces the computational burden compared to that required when solving each single objective EA as independent search problems.

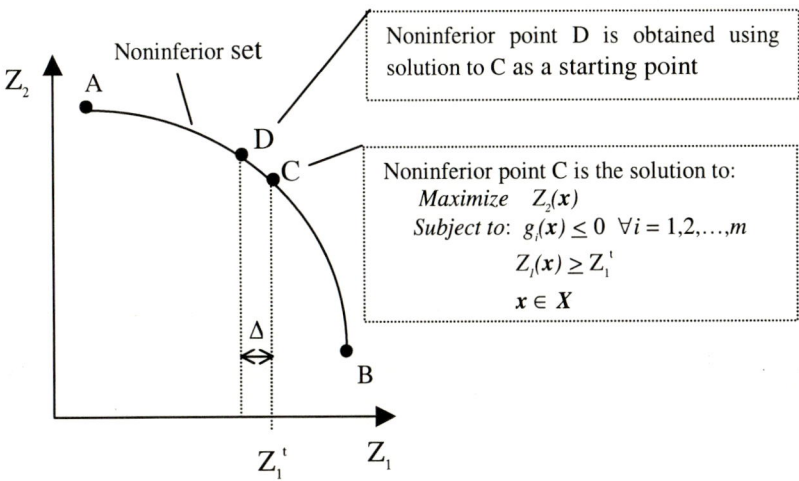

Fig. 1. Illustration of the constraint method using a two-objective example

Using a two objective problem as an illustration (Figure 1), let the current constraint value for objective 1 be Z_1^t. While at noninferior point C, the updated constraint value corresponding to the search for an adjacent noninferior solution D would be $(Z_1^t - \Delta)$, where the magnitude of Δ determines the minimum interval desired between adjacent noninferior solutions. For example, smaller values of Δ would result in a finer coverage (or better distribution) of the noninferior set, but would require execution of more intermediate steps, each of which requiring the solution of a single objective EA. At the beginning of the algorithm, the solutions to the two extreme points (A and B in Figure 1) are obtained. Using these solutions, the increment Δ is computed as $(Z_1^B - Z_1^A)/r$, where r is a predefined number of intervals. Then the algorithm starts at extreme point B, and solves the model with the constraint value of (8) set to $Z_1^B - \Delta$. Once the population converges to the adjacent noninferior solution according to some stopping criterion, the best solution is stored. Then the constraint value is incremented adaptively to $Z_1^t \leftarrow Z_1^t - \Delta$, and the current population continually undergoes the evolutionary operators and fitness pressure. To introduce higher population diversity at the beginning of each search, the mutation operator is applied in an adaptive manner during each intermediate step, starting with a higher rate and gradually reducing it (e.g., exponential decay) with generations within each step. Thus, at the beginning of each intermediate step the higher mutation rate perturbs the converged population around the previous noninferior point, introducing diversity for the new search.

This iterative process is terminated when the constraint value corresponds to the extreme point A, i.e., when $Z_1^t = Z_1^A$. Two convergence criteria are implemented to determine when to update the constraint value and when to initiate the search for the next noninferior solution. One of the criteria is to check if the number of generations, *generation,* exceeds a maximum value, *maxGenerations*. The other criterion is to track the improvement in the best solution in each intermediate step; convergence is assumed when the best solution does not improve within a certain number (N) of successive generations. If either of the above two criteria is satisfied then the

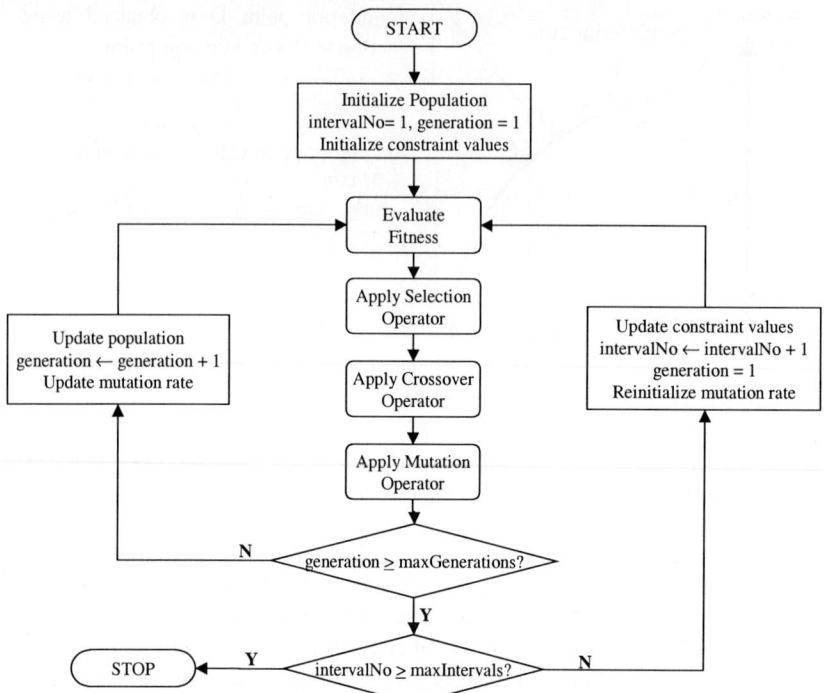

Fig. 2. Flowchart for CMEA – Constraint Method-Based Evolutionary Algorithm

constraint value is updated. The key steps of CMEA are shown as a flowchart in Figure 2.

Instead of converging the population to the noninferior set simultaneously, at each intermediate step of CMEA, a point in the noninferior set is identified through a search conducted by the whole population, and the final noninferior set is generated by storing all noninferior solutions found at the intermediate steps. The coverage of the noninferior set is achieved explicitly by traversing the noninferior surface through incremental and systematic updates of the constraint values.

4 Testing and Evaluation of CMEA

CMEA was applied to two test problems of different difficulty and characteristics. The first application uses Schaffer's F2 problem (Schaffer, 1985), which is an unconstrained, nonlinear problem. This is included since most other MOEA methods have been tested against it, providing a common basis for comparison. While this application represents a problem in a continuous search space, the second application, which uses the extended 0/1 multiobjective knapsack problem (Zitzler and Thiele, 1999), represents a problem in a combinatorial search space. This problem is a constrained, binary problem. Performance comparisons of several MOEAs in solving this problem are presented by Zitzler and Thiele (1999), and are used here to compare the performance of CMEA. In addition, a noninferior set was generated using a mathematical programming-based weighting method for the extended 0/1 knapsack problem, which was solved using a binary programming solver CPLEX®.

Several performance criteria are used to evaluate CMEA and to compare it with other approaches: 1) *accuracy*, i.e., how close are the generated noninferior solutions to the best available prediction; 2) *coverage*, i.e., how many different noninferior solutions are generated and how well are they distributed; and 3) *spread*, i.e., what is the maximum range of the noninferior surface covered by the generated solutions. Currently reported as well as newly defined quantitative measures (Chetan, 2000) are used in comparing CMEA with other MOEAs. The robustness of CMEA in solving problems with different characteristics (e.g., real vs. binary variables, constrained vs. unconstrained, continuous vs. combinatorial) is examined, in some limited manner, by applying it to these two different problems. To evaluate the robustness of CMEA in generating the noninferior set and providing good coverage, random trials were performed where the problems were solved repeatedly for different random seeds. A representative solution is used in the comparisons below

4.1 Schaffer's F2 Problem

4.1.1 Description. The F2 problem is defined as follows:

$$\text{Minimize} \quad Z_1 = x^2 \tag{10}$$

$$\text{Minimize} \quad Z_2 = (x - 2)^2 \tag{11}$$

The range for the decision variable x is [-5,7]. The Pareto optimal solutions constitute all x values varying from 0 to 2. The solution $x = 0$ is optimum with respect to Z_1, while the solution $x = 2$ is optimum with respect to Z_2. That is, objective functions Z_1 and Z_2 are in conflict in the range [0,2].

4.1.2 Results.
The F2 problem was solved using CMEA with algorithm-specific parameters as shown in Table 1. Results are compared in Figure 3 where the exact solution (obtained analytically using Equations (10) and (11)) for this problem is also shown. Although this is a relatively simple problem, the results indicate that CMEA is very accurate in generating the noninferior set for this problem. Also, it provides good coverage by generating a good distribution of noninferior solutions, and provides a full spread.

Table 1. CMEA parameters and settings for solving the test problems

Problem	Variable Type	CMEA Parameters			
		No. of intervals	Pop. size	Encoding	Crossover
F2	Real	100	100	32 bit Binary	Uniform
Knapsack	Binary	100	100	Binary	Uniform

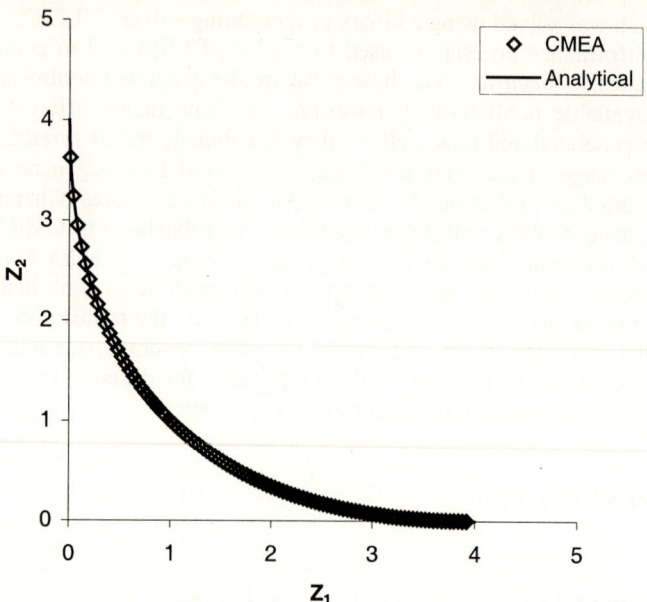

Fig. 3. The true noninferior tradeoff curve and the noninferior set determined by CMEA for Schaffer's F2 problem

4.2 Extended 0/1 Multiobjective Knapsack Problem

4.2.1 Description. Zitzler and Thiele (1999) used in their work a knapsack problem that extends the traditional single objective knapsack problem by incorporating two knapsacks that can be filled by items selected from a larger collection of items. Similar to the traditional knapsack problem, each knapsack has a limited weight capacity with different payoff when each item is included in it. The goal is to allocate a limited set of items to maximize the payoff in each knapsack without violating its weight capacity constraint. This multiobjective problem is defined mathematically as follows:

$$\text{Maximize} \quad Z_l(x) = \sum_{j=1}^{n} p_{l,j} x_j \quad \forall l = 1,2,\ldots,k \quad (12)$$

$$\text{Subject to} \quad \sum_{j=1}^{n} w_{l,j} x_j \le c_l \quad \forall l = 1,2,\ldots,k \quad (13)$$

In the formulation, $Z_l(x)$ is the total profit associated with knapsack l, $p_{l,j}$ = profit of placing item j in knapsack l, $w_{l,j}$ = weight of item j when placed in knapsack l, c_l = capacity of knapsack l, $x = (x_1, x_2, \ldots, x_n) \in \{0,1\}^n$ such that $x_j = 1$ if selected and $= 0$ otherwise, n is the number of available items and k is the number of knapsacks.

This binary MO problem was solved for the cases with two knapsacks (i.e. $k = 2$) and 250 and 500 items. The results reported here correspond to $n = 500$ and $k = 2$. The data for the problems solved were adapted from Zitzler and Thiele (1999).

4.2.2 Results. The extended knapsack problem was solved by CMEA for the parameter setting shown in Table 1. In addition, the noninferior set was generated using the constraint method for this problem by modeling it as a binary linear programming (BLP) model. This was solved using the binary linear programming solver, CPLEX®. In Figure 4, these results are shown along with the results reported by Zitzler and Thiele (1999) for the following MOEAs: VEGA, NPGA, NSGA, and SPEA. To examine the consistency of CMEA in solving this problem, ten trials with different random seeds were conducted, and the results were insensitive to the random seed, indicating robust behavior.

Accuracy of the noninferior solutions generated by CMEA should be compared with respect to the best available noninferior set, as well as with the best estimate obtained by the other MOEAs. The mathematical programming-based estimate of the noninferior set, the best available for this problem, is included in Figure 4 to make the first evaluation. Compared to this, the accuracy of noninferior solutions generated by CMEA and the other MOEAs is relatively poor. The combinatorial nature of the search can be attributed to the weak performance by all EAs. Accuracy of CMEA in comparison to other MOEA results, however, is very good. Noninferior solutions obtained by SPEA, the best performing MOEA according to Zitzler and Thiele (1999), appear to dominate the solutions generated by CMEA in the corresponding noninferior region. The spread or range covered by the CMEA generated solutions, however, is far superior to that attained by all other MOEAs, including SPEA. Further, CMEA is able to provide good coverage by identifying noninferior solutions that are better distributed over a wider range of the noninferior set.

5 Performance Metrics and Comparison of MOEAs

To compare the performance of CMEA with that of other MOEAs, the following quantitative measures are used.
- *Accuracy*: The S factor used by Zitzler and Thiele (1999) to represent the size of noninferior space covered is used to characterize and compare accuracy. In addition, the approach used by Knowles and Corne (2000) is used to characterize the degree to which a noninferior set outperforms another. An either-or criterion is used to determine if the noninferior set obtained by an MOEA dominates that obtained by another MOEA; the closeness of the two points of intersection are not differentiated statistically.
- *Spread*: Spread is quantified for each objective as the fraction of the maximum possible range of that objective in the noninferior region covered by a noninferior points A and B refer to the two extreme points, i.e., the single objective optimal solutions for objective 1 and 2, respectively, for a two objective case. The maximum range covered by the noninferior solutions represented by the ordered set $C = \{C_h, \forall\ h \in \{0,1, ..., q\}\}$ is $(Z_1^{Cq} - Z_1^{C1})$ and $(Z_2^{C1} - Z_2^{Cq})$ in Z_1 and Z_2 objective space, respectively. Therefore, the spread metrics in objective space 1 and 2 are defined as $(Z_1^{Cq} - Z_1^{C1})/(Z_1^{B} - Z_1^{A})$ and $(Z_2^{C1} - Z_2^{Cq})/(Z_2^{A} - Z_2^{B})$, respectively.

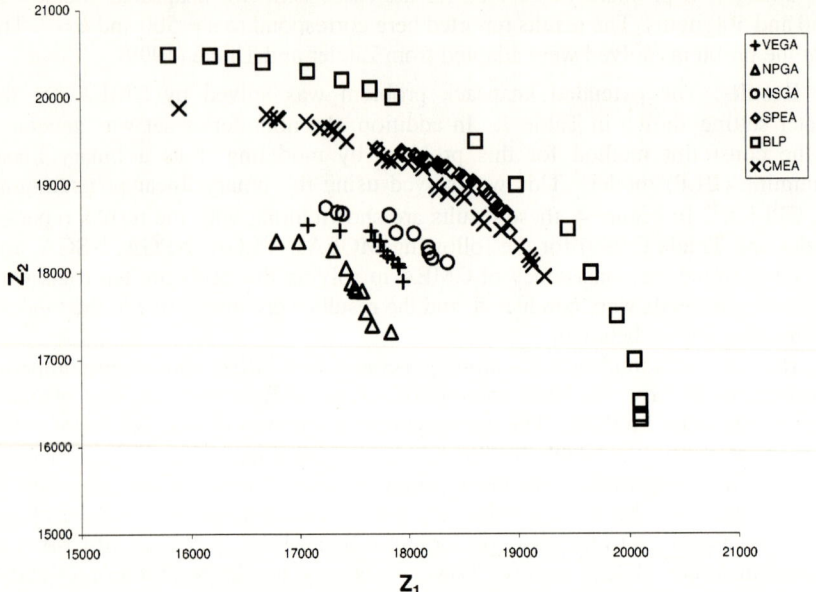

Fig. 4. A comparison of noninferior sets obtained using CMEA, VEGA, NPGA, NSGA, SPEA, and mathematical programming approach (BLP-Binary Linear Programming) for the extended 0/1 multiobjective knapsack problem

Coverage: A quantitative measure computed based on the maximum gap in coverage is defined to represent the distribution of the noninferior solutions generated by an MOEA. The Euclidean distance between adjacent noninferior points in the objective space is used to indicate the gap. A smaller value of this metric indicates better distribution of solutions in the noninferior set. This metric is defined separately as *V1* and *V2* to characterize the coverage within the range of noninferior region defined by 1) the extreme points, and 2) the solutions generated by the MOEA, respectively. Using the illustrations shown in Figure 6, *V1* is defined as $Max \{d_h, \forall h \in \{0,1, ..., q\}\}$, and *V2* is defined as $Max \{d_h, \forall h \in \{1, 2, ..., q-1\}\}$.

A summary of these metrics are compared in Tables 2a-2d for the noninferior solutions generated by all MOEAs shown in Figure 4. These results indicate that overall CMEA performs better than NPGA, NSGA, SPEA, and VEGA with respect to finding nondominated solutions with a good distribution in the noninferior region. This conclusion is specific to the 0/1 extended multiobjective knapsack problem, and similar performance comparisons for other problems are needed to make more general conclusions. Although CMEA provides the best distribution of solutions in the entire noninferior range (based on *V1* metric), SPEA provides a better distribution (based on *V2* metric) within the narrower noninferior range represented by its solutions.

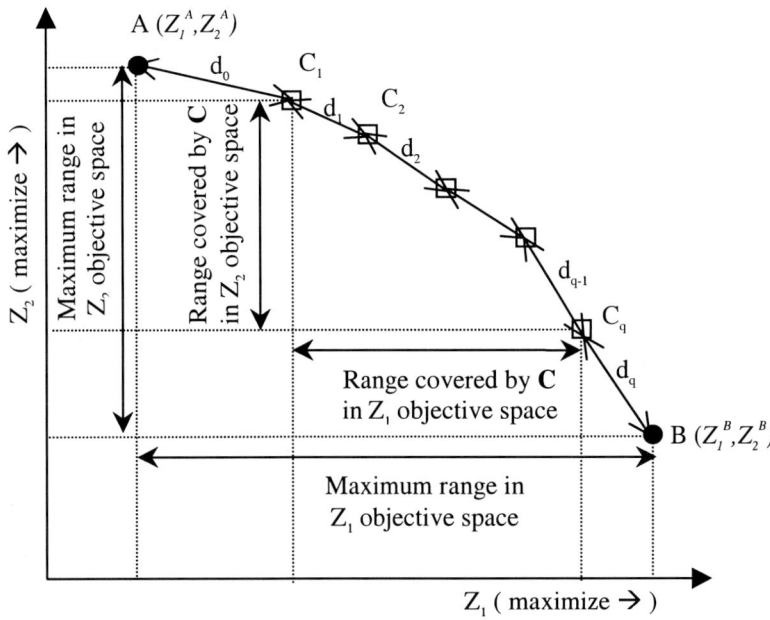

Fig. 5. An example two-objective noninferior tradeoff to illustrate the computation of: 1) *Spread* metric, and 2) *Coverage* metric

Table 2a. *Accuracy* comparison, based on the *S* factor (Zitzler and Thiele, 1999), of noninferior sets generated by different MOEAs for the extended 0/1 multiobjective knapsack problem. A larger value indicates better performance; the best is shown in bold

MOEA Method	S Factor
SPEA	0.89
NSGA	0.79
NPGA	0.83
VEGA	0.81
CMEA	**0.92**

6 Summary and Conclusions

This paper presents a new MOEA, CMEA-Constrain Method-based Evolutionary Algorithm for solving multiobjective optimization problems. CMEA is founded upon two simple, but powerful concepts borrowed from the mathematical programming literature: 1) optimization of a single objective model with target constraints on all but one objective finds a noninferior, or Pareto optimal, solution; and 2) for some classes of problems, noninferior solutions adjacent in objective space map to adjacent decision vectors with only marginal differences in the decision space. The attractive features of CMEA include: easily adaptable for use with existing implementation of evolutionary algorithms for an optimization problem since no new operators are needed; and relatively less compute intensive since Pareto optimality is ensured in an implicit manner, and therefore expensive sorting and pair-wise comparison operations that are typically required by other Pareto-based MOEAs are eliminated.

Table 2b. *Accuracy* comparison, based on the metric defined by Knowles and Corne (2000), of CMEA with different MOEAs for the extended 0/1 multiobjective knapsack problem

The MOEAs Compared ($MOEA_1$ vs. $MOEA_2$)	(P_1, P_2): (Percentage number of times $MOEA_1$ outperforms $MOEA_2$, Percentage number of times $MOEA_2$ outperforms $MOEA_1$)		
	Number of Sampling Lines		
	108	507	1083
(CMEA vs. SPEA)	(95.4, 4.6)	(95.3, 4.7)	(95.2, 4.8)
(CMEA vs. NSGA)	(100, 0)	(100, 0)	(100, 0)
(CMEA vs. NPGA)	(100, 0)	(100, 0)	(100, 0)
(CMEA vs. VEGA)	(100, 0)	(100, 0)	(100, 0)

Table 2c. Comparison of *Spread* of noninferior sets generated by different MOEAs for the extended 0/1 multiobjective knapsack problem. A larger value indicates better performance; the best is shown in bold

MOEA	Spread Metric	
	in Z_1 objective space	in Z_2 objective space
SPEA	0.28	0.23
NPGA	0.24	0.25
NSGA	0.26	0.15
VEGA	0.20	0.16
CMEA	**0.78**	**0.47**

Table 2d. Comparison of *Coverage* of noninferior sets generated by different MOEAs for the extended 0/1 multiobjective knapsack problem. A smaller value indicates better performance; the best is shown in bold

MOEA	Coverage Metric	
	V1 (includes the extreme points for each objective)	V2 (excludes the extreme points for each objective)
SPEA	0.118	**0.011**
NPGA	0.122	0.016
NSGA	0.121	0.021
VEGA	0.130	0.015
CMEA	**0.088**	0.04

To evaluate the applicability of CMEA to MO problems, it was applied to two test problems with different characteristics and levels of difficulty. Test problems covered continuous as well as combinatorial search, unconstrained as well as constrained optimization, real as well as binary variables, and as few as one variable to as high as 500 variables. This evaluation included performance comparisons with other MOEAs and with mathematical programming-based noninferior solutions. Accuracy, coverage, and spread of the noninferior solutions were used to compare the performance. To evaluate the consistency of CMEA in generating the noninferior set, several random trials were performed. Overall, CMEA performed well with respect to these criteria for both problems tested. The spread and coverage of noninferior solutions obtained using CMEA were always better than those demonstrated by other MOEAs. With respect to accuracy, SPEA did better than CMEA for a narrow range of noninferior solutions, but overall CMEA did better over a broader range of the noninferior set.

Some known limitations of CMEA include the following. The computational efficiency gain obtained in CMEA is predicated on the existence of similarities in noninferior solutions that correspond to adjacent points in the objective space. For problems where this may not hold true strongly, the search implemented by CMEA becomes analogous to solving a number of independent single objective optimization problems, and therefore, may not realize any significant computational gain. For a problem with more than two objectives, incrementally updating the constraint values to obtain an adjacent point is not necessarily as straightforward as is for the two-objective cases presented here. More investigation is needed to evaluate this issue when applying CMEA to higher dimensional problems.

In the present study, the number of functions evaluations was used as a measure to compare the computational needs of CMEA with that of a single objective-based MO analysis without seeding. CMEA required approximately 75% fewer function evaluations. This measure alone is not sufficient to compare the computational gain, if any, that may be realized by CMEA over the other MOEAs that use explicit Pareto optimality ordering. The computational performance of CMEA and other MOEAs needs to be studied further.

References

1. Chetan, S. K., (2000). *Noninferior Surface Tracing Evolutionary Algorithm (NSTEA) for Multiobjective Optimization*, MS Thesis, North Carolina State University, Raleigh, North Carolina.
2. Cieniawski, S.E., Eheart, J. W., and Ranjithan, S., (1995). Using Genetic Algorithms to Solve a Multi-Objective Groundwater Monitoring Problem, *Water Resources Research*, vol. 31, no. 2, pp. 399-409.
3. Coello, C.A.C., Christiansen, A.D., and Aguirre, A.H., (1998). Using a New GA-Based Multiobjective Optimization Technique for the Design of Robot Arms, *Robotica*, 16(4), pp. 401-414
4. Coello, C. A. C., (1999a). A comprehensive survey of evolutionary-based multiobjective optimization techniques, *Knowledge and Information System*, 1(3), pp. 269-308.
5. Coello, C. A. C., (1999b). *List of references on evolutionary multiobjective optimization*. Available: www.lania.mx/~ccoello/EMOO
6. Coello, C.A.C. and Christiansen, A.D., (2000). Multiobjective optimization of trusses using genetic algorithms, *Computers and Structures*, 75(6), pp. 647-660.

7. Cohon, J.L., (1978). *Multiobjective programming and planning*, Mathematics in Science and Engineering, Vol. 140, Academic Press, Inc.
8. Fonesca, C.M., and Fleming, P.J., (1993). Genetic Algorithms for multiobjective optimization: Formulation, Discussion and generalization, *Genetic Algorithms: Proceedings of Fifth International Conference*, pp. 416-423.
9. Fonesca, C.M, and Fleming, P.J., (1995). An overview of evolutionary algorithms in multiobjective optimization, *Evolutionary Computation*, 3(1), pp. 1-16.
10. Hajela, P. and Lin, C.-Y.,(1992). Genetic search strategies in multicriterion optimal design, *Structural Optimization*, 4, pp. 99-107.
11. Harrell, L. J. and Ranjithan, S., (1997). Generating Efficient Watershed Management Strategies Using a Genetic Algorithms-Based Method, Ed: D. H. Merritt *Proceedings of the 24th Annual Water Resources Planning and Management Conference (ASCE)*, Houston, TX, April 6-9, 1997, pp. 272-277.
12. Horn, J., and Nafpliotis, N. and Goldberg, D.E., (1994). A niched Pareto genetic algorithm for multiobjective optimization, *Proceedings of the First IEEE Conference on Evolutionary Computation, IEEE World Congress on Computational Intelligence*, 1, pp. 82-87.
13. Horn, J., (1997). *Multicriterion decision making*, In Back, T., Fogel, D., and Michalewicz, Z., editors, Handbook of Evolutionary Computation, Volume 1, pp. F1.9:1-F.1.9:15, Oxford University Press, Oxford, England.
14. Jiménez, J. and Cadenas, J.M., (1995). An evolutionary program for the multiobjective solid transportation problem with fuzzy goals, *Operations Research and Decisions*, 2, pp. 5-20.
15. Knowles, J.D., and Corne, D.W., (2000). Approximating the Nondominated Front Using the Pareto Archived Evolution Strategy, *Evolutionary Computation*, 8(2): pp. 149-172.
16. Loughlin, D.H., and Ranjithan, S., (1997). The neighborhood constraint method: a genetic algorithm-based multiobjective optimization technique, *Proceedings of the Seventh International Conference on Genetic Algorithms*, pp. 666-673.
17. Loughlin, D. H., Ranjithan, S., Brill, E. D., and Baugh, J. W., (2000a). Genetic algorithm approaches for addressing unmodeled objectives in optimization problems, to appear in *Engineering Optimization (in print)*.
18. Loughlin, D. H., Ranjithan, S., Baugh, J. W., and Brill, E. D., (2000b). Application of Genetic Algorithms for the Design of Ozone Control Strategies, *Journal of the Air and Waste Management Association*, 50, June 2000, pp. 1050-1063.
19. Menczer, F, Degeratu, M., and Street, W. N., (2000). Efficient and scalable Pareto optimization by evolutionary local selection algorithms, *Evolutionary Computation*, 8(2), pp. 223-247.
20. Obayashi, S., Sasaki, D., and Hirose, N., (2000). Multiobjective Evolutionary Computation for Supersonic Wing-Shape Optimization, *IEEE Transactions on Evolutionary Computation*, 4(2), pp. 182-187.
21. Ritzel, B. J., Eheart, J. W., and Ranjithan, S., (1994). Using Genetic Algorithms to Solve a Multi-Objective Groundwater Remediation Problem, *Water Resources Research*, vol. 30, no. 5, pp. 1589-1603.
22. Schaffer, J.D., (1984). Multiple objective optimization with vector evaluated genetic algorithms, Ph.D. Thesis, Vanderbilt University.
23. Schaffer, J.D., (1985). Multiple objective optimization with vector evaluated genetic algorithms, Genetic Algorithms and Their Applications: Proceedings of the First International Conference on Genetic Algorithms, pp. 93-100.
24. Srinivas, N., and Deb, K., (1994). Multiobjective optimization using nondominated sorting in genetic algorithms, *Evolutionary Computation*, 2(3), pp. 221-248.
25. Van Veldhuizen, D.A., and Lamont, G.B., (2000). Multiobjective evolutionary algorithms: Analyzing the state-of-the-art, *Evolutionary Computation*, 8(2), pp. 125-147.

26. Zitzler, E., and Thiele, L., (1999). Multiobjective evolutionary algorithms: a comparative case study and the strength pareto approach, *IEEE Transactions on Evolutionary Computation*, 3(4), pp. 257-271.
27. Zitzler, E., Deb, K., and Thiele, L., (2000). Comparison of multiobjective evolutionary algorithms: empirical results, *Evolutionary Computation*, 8(2), pp. 173-195.

The image of $\mathbf{X}_f$, i.e., the feasible region in the objective space, is denoted as $\mathbf{Y}_f = \boldsymbol{f}(\mathbf{X}_f) = \bigcup_{\boldsymbol{x} \in \mathbf{X}_f} \{\boldsymbol{f}(\boldsymbol{x})\}$.

With this definition, we are able to define Pareto-dominance for any two decision vectors $\boldsymbol{a}$ and $\boldsymbol{b}$ as follows:

Definition 3 (Pareto-dominance). *For any two decision vectors $\boldsymbol{a}$ and $\boldsymbol{b}$,*

$$\boldsymbol{a} \succ \boldsymbol{b} \ (\boldsymbol{a} \text{ dominates } \boldsymbol{b}) \text{ iff } \boldsymbol{f}(\boldsymbol{a}) < \boldsymbol{f}(\boldsymbol{b})$$
$$\boldsymbol{a} \succeq \boldsymbol{b} \ (\boldsymbol{a} \text{ weakly dominates } \boldsymbol{b}) \text{ iff } \boldsymbol{f}(\boldsymbol{a}) \leq \boldsymbol{f}(\boldsymbol{b})$$
$$\boldsymbol{a} \sim \boldsymbol{b} \ (\boldsymbol{a} \text{ is indifferent to } \boldsymbol{b}) \text{ iff } \boldsymbol{f}(\boldsymbol{a}) \not\geq \boldsymbol{f}(\boldsymbol{b}) \wedge \boldsymbol{f}(\boldsymbol{b}) \not\geq \boldsymbol{f}(\boldsymbol{a})$$

In this definition, the relations $=, \leq$ and $<$ on objective vectors are defined as follows:

Definition 4. *For any two objective vectors $\boldsymbol{u}$ and $\boldsymbol{v}$,*

$$\begin{aligned} \boldsymbol{u} = \boldsymbol{v} &\text{ iff } \forall i = 1, \cdots, m : u_i = v_i \\ \boldsymbol{u} \leq \boldsymbol{v} &\text{ iff } \forall i = 1, \cdots, m : u_i \leq v_i \\ \boldsymbol{u} < \boldsymbol{v} &\text{ iff } \boldsymbol{u} \leq \boldsymbol{v} \wedge \boldsymbol{u} \neq \boldsymbol{v} \end{aligned} \qquad (2)$$

The relations $\geq$ and $>$ are defined analogously.

Definition 5 (Pareto-optimality). *A decision vector $\boldsymbol{x} \in \mathbf{X}_f$ is said to be non-dominated regarding a set $\mathbf{A} \subseteq \mathbf{X}_f$ iff*

$$\not\exists \boldsymbol{a} \in \mathbf{A} : \boldsymbol{a} \succ \boldsymbol{x}$$

If it is clear from the context which set $\mathbf{A}$ is meant, is will be simply omitted in the following. Moreover, $\boldsymbol{x}$ is said to be Pareto-optimal *iff $\boldsymbol{x}$ is non-dominated regarding $\mathbf{X}_f$.*

The entirety of all Pareto-optimal points is called the *Pareto-optimal set*; the corresponding objective vectors form the *Pareto-optimal front* or *surface*.

Definition 6 (Non-dominated sets and fronts). *Let $\mathbf{A} \subseteq \mathbf{X}_f$. The function $g(\mathbf{A})$ gives the set of non-dominated decision vectors in $\mathbf{A}$:*

$$g(\mathbf{A}) = \{\boldsymbol{a} \in \mathbf{A} \mid \boldsymbol{a} \text{ is nondominated regarding } \mathbf{A}\} \qquad (3)$$

The set $g(\mathbf{A})$ is the non-dominated set *regarding $\mathbf{A}$, the corresponding set of objective vectors $\boldsymbol{f}(g(\mathbf{A}))$ is the* non-dominated front *regarding $\mathbf{A}$. Furthermore, the set $\mathbf{X}_g = g(\mathbf{X}_f)$ is called* Pareto-optimal set *and the set $\mathbf{Y}_g = \boldsymbol{f}(\mathbf{X}_g)$ is denoted as the* Pareto-optimal front.

2.2 Random Objectives and Property Intervals

In the following, we define the notion of *probabilistic dominance* of decision vectors for which each objective is not a number, but a random variable with values bounded by an interval called *property interval*.

First, we assume that our optimization problem has a single objective ($m = 1$), hence $y = f(x)$ for a given n−dimensional decision vector. Later, we extend our notation to the general case.

Definition 7 (Uncertain objective; property interval). *Given an n−dimensional decision vector and a one-dimensional objective function f. We assume $f : \mathbf{R}^n \mapsto \mathbf{R}$. f is called uncertain, if for each $x \in \mathbf{X}$,*

$$y = f(x) \in [f^s(x), \cdots, f^u(x)] \tag{4}$$

where $[f^s(x), \cdots, f^u(x)]$ with $f^s(x), f^u(x) \in \mathbf{R}$, and $f^s(x) \leq f^u(x)$ is called property interval *of f at x.*[1]

Example 1. In Fig. 1, there are three decision vectors a, b, and c that have different property intervals. Let without loss of generality the objective function to be minimized, then obviously a dominates c because no matter what value the objective of a assumes in the shown interval, the value is lower than any objective value in the interval of c. The situation is different for the pair a and b: In case $f(a) = 2$ and $f(b) = 8$, a dominates b. However, in case $f(a) = 10$, then no matter what value $f(b) \in [8, 9]$ may take, b dominates a. Hence, the dominance is uncertain.

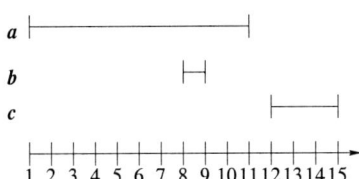

Fig. 1. Example of three decision vectors with corresponding property intervals.

From the previous example, it becomes clear that in order to make decisions concerning the dominance of decision vectors with uncertain objective functions that are given by property intervals, we need a notion of *probabilistic dominance*.

In the following, we treat $f(x)$ for each $x \in \mathbf{X}$ as a distinct random variable Y and consider *continuous uniform distributions* of $Y \in [f^s(x), \cdots, f^u(x)]$.

3 Continuous Uniform Distribution of Objective Values

In this case, we assume that the probability density function is constant over the property interval of each random variable $Y = f(x)$, $x \in \mathbf{X}$.

---

[1] W.l.o.g., we assume $f^u(x) = f^s(x) + \epsilon$ with $\epsilon \in \mathbf{R}^+$ in the following.

Definition 8 (Probability density; uniform distribution). *Let $[y^s, \cdots, y^u]$ denote the property interval of the random variable $Y = f(\boldsymbol{x})$. For a uniform distribution, its probability density function is given by*

$$p(y) = \begin{cases} 0 & \text{if } y < y^s \\ \frac{1}{y^u - y^s} & \text{if } y^s \leq y \leq y^u \\ 0 & \text{if } y > y^u \end{cases} \quad (5)$$

From the probability density function $p(y)$, we can derive the *cumulative distribution function (cdf)*, or just *distribution function*, $F(y)$, as $F(y) = P[Y \leq y]$ for all $y \in \mathbf{R}$. With $F(y) = \int_{-\infty}^{y} p(y) dy$, we obtain the *probability* of $[a < Y \leq b]$ as $P[a < Y \leq b] = F(b) - F(a) = \int_a^b p(y) dy$.

Example 2. Consider again the property intervals of the decision vectors $\boldsymbol{a}$, $\boldsymbol{b}$, and $\boldsymbol{c}$ in Fig. 1. For uniformly distributed objective values, the probability density function and the distribution function of the corresponding random variables $A = f(\boldsymbol{a})$, $B = f(\boldsymbol{b})$ and $C = f(\boldsymbol{c})$ are shown in Fig. 2 together with the probability distribution functions $P[Y \leq y]$.

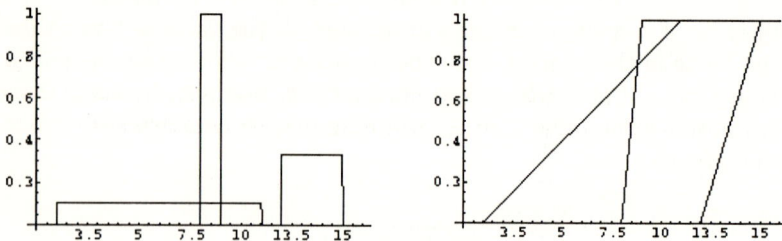

Fig. 2. Probability density and probability distribution functions (uniform distribution).

3.1 Probability of Dominance: Single-Objective Case

In the following, we elaborate the probability of a design point $\boldsymbol{a}$ with objective value $a \in [a^s, \cdots, a^u]$ to dominate a design point $\boldsymbol{b}$ with objective value $b \in [b^s, \cdots, b^u]$, again first for a single objective.

Theorem 1 (Probability of (weak) Dominance; uniform distribution). *Given two design points $\boldsymbol{a}$ and $\boldsymbol{b}$ with objective $a \in [a^s, \cdots, a^u]$ and $b \in [b^s, \cdots, b^u]$, respectively. The probability of $\boldsymbol{a}$ to dominate $\boldsymbol{b}$, written $P[\boldsymbol{a} \succeq \boldsymbol{b}]$ for uniform distribution functions is given as:*

$$P[\boldsymbol{a} \succeq \boldsymbol{b}] = \begin{cases} 0 & \text{if } b^u < a^s \\ 1 & \text{if } a^u < b^s \\ \frac{1}{a^u - a^s} \cdot \left(\int_{y=a^s}^{b^s} dy + \int_{y=\max\{a^s,b^s\}}^{\min\{a^u,b^u\}} 1 - \frac{y-b^s}{b^u-b^s} dy \right) & \text{if else} \end{cases} \quad (6)$$

Proof. The first and the second case in Eq.(6) are obvious and correspond to the case when both property intervals do not overlap. The last case is a little bit more complicated and corresponds to the case when both intervals overlap. The probability that random variable A corresponding to the objective of a is smaller than or equal to B is given as

$$P[A \leq B] = \int_{y=a^s}^{a^u} \frac{1}{a^u - a^s} \cdot P[B \geq y] dy$$

With $P[B \geq y]$ being equal to 1 for values of $y < b^s$ and being equal to $1 - P[B < y] \approx 1 - P[B \leq y] = 1 - \frac{y-b^s}{b^u-b^s}$[2] for values of y in the range from b^s to b^u, we obtain the above results.

Example 3. Consider the two decision vectors a and b introduced in Fig. 1. We want to compute the probability $P[a \succeq b]$ which is equal to $P[A \leq B]$ for the corresponding random variables of the related property intervals. Evaluating Eq.(6) with $a^s = 1, a^u = 11$, and $b_s = 8, b^u = 9$, we obtain $\frac{1}{10} \cdot (b^s - a^s) = \frac{7}{10}$ for the first term and $\frac{1}{10} \cdot \frac{1}{b^u-b^s} \int_{b^s}^{b^u} (b^u - y) dy$ for the second term resulting in $\frac{9(9-8)-(1/2(9^2-8^2))}{10} = \frac{1}{20}$, giving a total of 0.75. Hence, the probability of a (weakly) dominating b is 0.75.

Expected value

Definition 9. *Let $[y^s, \cdots, y^u]$ denote the property interval of the random variable $Y = f(x)$. For a uniform distribution, its expected value is given by*

$$E[Y] = \int_{-\infty}^{\infty} yp(y) dy = \int_{y^s}^{y^u} y \frac{1}{y^u - y^s} dy \qquad (7)$$

$$= \frac{1}{2(y^u - y^s)}((y^u)^2 - (y^s)^2) = \frac{y^s + y^u}{2}$$

Example 4. Consider again the three property intervals of decision vectors a, b and c as shown in Fig. 1. Let the corresponding random variables be A, B, and C. For an assumed uniform distribution, the expected values are computed using Eq.(7) as $E[A] = \frac{1}{2(a^u-a^s)}((a^u)^2 - (a^s)^2) = \frac{1}{2(11-1)}((11)^2 - (1)^2) = 6$. Similarly, we obtain $E[B] = \frac{1}{2(9-8)}((9)^2 - (8)^2) = 8.5$ and $E[C] = \frac{1}{2(15-12)}((15)^2 - (12)^2) = 13.5$. These values are exactly in the middle of the corresponding intervals.

3.2 Probability of Dominance: Multi-objective Case

Up to now, we have seen how to compute probabilities of dominance in case of single-objective functions. In Fig. 3, it is shown that the question of dominance becomes even more elaborated in case of multiple objectives. We extend the notion of (weak) Pareto-dominance:[3]

---

[2] To be exact, we must integrate only up to $b^u - \epsilon$ where $\epsilon \to 0$, positive, in the above equation if b^u determines the upper bound of the integration interval.

[3] We consider the notion of weak dominance here and throughout the rest of the paper.

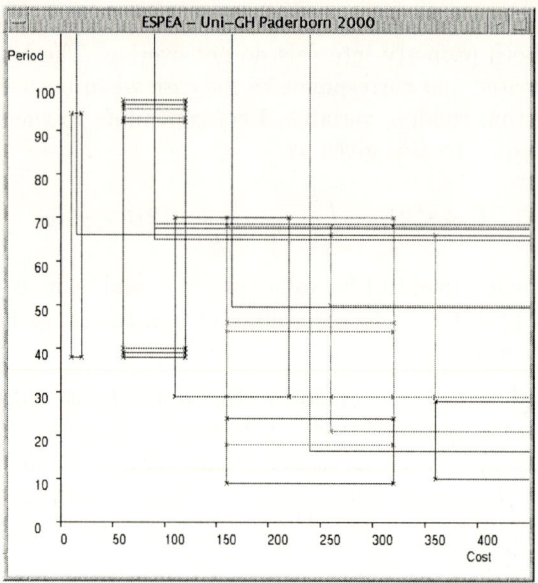

Fig. 3. Pareto-fronts with 2D property intervals. Pareto-points surrounded by dark rectangles.

Theorem 2 (Probability of (weak) Pareto-dominance). *For any two n-dimensional decision vectors a and b, and m statistically independent objective functions $f_1, f_2, \cdots, f_m$,*

$$P[a \succeq b] = \prod_{i=1}^{m} P[f_i(a) \leq f_i(b)] \qquad (8)$$

Proof. We assume that the random variables associated to each objective and each decision vector are mutually independent. Hence, the conjunction of the Pareto-conditions leads to the product of the corresponding probabilities.

Before we will make use of this probability calculus of dominance in the context of steering design space exploration, we give a few more examples.

Example 5. Consider two $n = 2$ dimensional decision vectors a and b with two objectives f_1 and f_2. Let the corresponding property intervals be given by $f_1(a) \in [2,4]$, $f_2(a) \in [2,4]$, and $f_1(b) \in [2,3]$, and $f_2(b) \in [2,3]$. We obtain $P[a \succeq b] = P[f_1(a) \leq f_1(b)] \cdot P[f_2(a) \leq f_2(b))]$ where $P[f_1(a) \leq f_1(b)] = 1/2(3(3-2) - 1/2(3^2 - 2^2)) = 0.25$, and also $P[f_2(a) \leq f_2(b)] = 1/2(3(3-2) - 1/2(3^2 - 2^2)) = 0.25$. Hence, $P[a \succeq b] = 0.0625$.

Similarly, we obtain $P[b \succeq a] = P[f_1(b) \leq f_1(a)] \cdot P[f_2(b) \leq f_2(a))]$ where $P[f_1(b) \leq f_1(a)]$ equals $P[f_2(b) \leq f_2(a)] = 2(3-2) - 1/4(3^2 - 2^2) = 0.75$. Hence, $P[b \succeq a] = 0.5625$.

The general case of not necessarily independent objective functions is dealt with in the following theorem:

Theorem 3 (Probability of (weak) Pareto-dominance). *For any two n–dimensional decision vectors $\mathbf{a}$ and $\mathbf{b}$, and m objective functions $f_1, f_2, \cdots, f_m$, let $p_a(\mathbf{X}) = p_a(X_1, X_2, \cdots, X_m)$ and $p_b(\mathbf{Y}) = p_b(Y_1, Y_2, \cdots, Y_m)$ denote the probability density functions of $\mathbf{a}$ and $\mathbf{b}$, respectively in the objective space with random variable X_i denoting the random variable for the ith objective $f_i(\mathbf{a})$ (similarly Y_i for $f_i(\mathbf{b})$). Let the property intervals for the ith objective of $\mathbf{a}$ be given by $[a_i^s, \cdots, a_i^u]$ (similar $[b_i^s, \cdots, b_i^u]$ for $\mathbf{b}$). Then*

$$P\left[\mathbf{a} \succeq \mathbf{b}\right] = P[a_1 \leq b_1, a_2 \leq b_2, \cdots, a_m \leq b_m] \qquad (9)$$

$$= \int_{X_1 = a_1^s}^{a_1^u} \cdots \int_{X_m = a_m^s}^{a_m^u} p_a(\mathbf{X}) \cdot \left(1 - \int_{Y_1 = \max\{a_1^s, b_1^s\}}^{\min\{a_1^u, b_1^u\}} \cdots \int_{Y_m = \max\{a_m^s, b_m^s\}}^{\min\{a_m^u, b_m^u\}} p_b(\mathbf{Y})\, d\mathbf{Y}\right) d\mathbf{X}$$

Proof. In analogy to Theorem 2, $P[\mathbf{a} \succeq \mathbf{b}]$ is obtained as $\int_{X_1=a_1^s}^{a_1^u} \cdots \int_{X_m=a_m^s}^{a_m^u} p_a(\mathbf{X}) \cdot P[\mathbf{X} \leq \mathbf{b}] d\mathbf{X}$. With $P[\mathbf{X} \leq \mathbf{b}] = 1 - P[\mathbf{X} > \mathbf{b}] \approx 1 - P[\mathbf{X} \geq \mathbf{b}]$,[4] we obtain the above result.

It can be seen that in case $p_a(\mathbf{X}) = p_a(X_1) \cdot p_a(X_2) \cdots p_a(X_m)$ and similarly $p_b(\mathbf{Y}) = p_b(Y_1) \cdot p_b(Y_2) \cdots p_b(Y_m)$ hold, then the result in Theorem 2 may be directly obtained as a special case.

Expected value. The definition of the expected value for the single objective case in Definition 9 directly extends to the multi-objective case as follows:

Definition 10. *For any two n–dimensional decision vectors $\mathbf{y}$ and m objective functions $f_1, f_2, \cdots, f_m$, let $\mathbf{Y} = (Y_1, Y_2, \cdots, Y_m)$ denote the vector of random variables.*

$$E[\mathbf{Y}] = (E[Y_1], E[Y_2], \cdots, E[Y_m]) \qquad (10)$$

4 Evolutionary Algorithm-Based Exploration with Property Intervals

In this chapter, we review basic ideas for design space exploration using evolutionary algorithms. For the so-called SPEA (Strength-Pareto-Approach) introduced by Zitzler and Thiele in [13] and applied to many important areas of multi-objective optimization problems in the context of embedded system design such as system-level synthesis [1] and code scheduling problems for DSPs (digital signal processors) [11,12], we provide extensions to the selection mechanisms of this population-based approach that considers objective values with property intervals.

---

[4] To be exact, we must again integrate only up to $b_i^u - \epsilon$ where $\epsilon \to 0$, positive, for all those $i = 1, \cdots, m$ in Eq. (9) for which b_i^u determines the upper bound of the integration interval.

4.1 Basic Principles of Evolutionary Algorithms

Evolutionary Algorithms (EA) denote a class of optimization methods that are characterized by a) a set of solution candidates (*individuals*) called *population* maintained during the optimization process that undergo b) a selection process and c) that are manipulated by genetic operators, usually recombination and mutation. Each individual represents a possible solution, i.e., a decision vector, to the problem under consideration. However, it is not a decision vector but rather encodes it using a particular data structure.[5] The set of all possible vectors constitutes the *individual space* **I**. A population is thus a (multi-)set of vectors $i \in \mathbf{I}$. The quality of an individual with respect to the optimization problem is represented by a scalar value, called *fitness*. Since the quality of a solution is related to the objective functions and constraints, the solution must be decoded first by a mapping function m to derive a decision vector $x = m(i)$ from i. Applying f to x yields the corresponding objective vector on the basis of which a fitness value is assigned to i. Natural evolution is simulated by an iterative process: In the selection process, low-quality individuals are removed from the population, while high-quality individuals are reproduced. The goal of this operator is to increase the average quality within the population. Recombination and mutation aim at generating new solutions within the search space by variation of existing ones. The population at iteration t (also called *generation* t) is denoted $\mathbf{M}_t$ in the following. The symbol + stands for multi-set union in conjunction with populations.

4.2 The Strength Pareto-Approach (SPEA)

In the following, we review the Strength Pareto-Approach (SPEA) as introduced by Zitzler and Thiele [13] that serves as the basis of our extension called ES-PEA. SPEA uses Pareto-Dominance based selection [4] instead of switching the objective during optimization [9] or using an aggregation function, e.g., weighted sum approach where the coefficients are changed during optimization [5]. Also, it uses elitism where the policy is to always include the B best individuals of $\mathbf{M}_t$ into $\mathbf{M}_{t+1}$ in order not to lose them during exploration. as suggested by De Jong [8]. Finally, in order to maintain a high diversity within the population, it uses a special clustering technique shown in Fig. 6 that is different from *niching* [6,7] and crowding [8] so that altogether, SPEA has shown to provide superior results compared to existing approaches, see, e.g., [13] for many problems of interest.

As it is our main goal to extend this algorithm to be able to deal with objectives bounded by property intervals, we shortly reproduce the algorithm here in Fig. 4.

---

[5] Without loss of generality, we assume that this structure is a vector, e.g., a bit vector or a real-valued vector, although other structures like trees might be considered as well.

SPEA
- IN: N : population size;
  $\overline{N}$: maximum size of external set;
  U : maximum number of generations;
  p_c: crossover probability;
  p_m: mutation rate;
- OUT: $\mathbf{A}$: non-dominated set;

BEGIN
- Step 1: Initialization:
  Set $\mathbf{M}_0 = \overline{\mathbf{M}}_0 = \emptyset$ and $t = 0$.
  FOR $i = 1$ TO N DO
      Choose i according to some probability distribution;
      $\mathbf{M}_0 := \mathbf{M}_0 + \{i\}$;
  OD
- Step 2: Update of external set:
  Set $\overline{\mathbf{M}'} = \overline{\mathbf{M}_t}$;
  $\overline{\mathbf{M}'} = \overline{\mathbf{M}'} + \{i \mid i \in \mathbf{M}_t \wedge m(i) \in g(m(\mathbf{M}_t))\}$;
  Remove individuals from $\overline{\mathbf{M}'}$ whose corresponding decision vectors are weakly dominated regarding $m(\overline{\mathbf{M}'})$;
  Reduce the number of individuals externally stored by means of clustering, i.e., call Alg. Cluster with parameters $\overline{\mathbf{M}'}$ and $\overline{N}$, and assign the resulting reduced set to $\overline{\mathbf{M}_{t+1}}$
- Step 3: Fitness assignment:
  Calculate the fitness of individuals in $\mathbf{M}_t$ and $\overline{\mathbf{M}_t}$ using Alg. Fitness;
  OD
- Step 4: Selection: Set $\mathbf{M}' = \emptyset$;
  FOR $i = 1$ TO N DO
      Select two individuals $i, j \in \mathbf{M}_t + \overline{\mathbf{M}_t}$ at random;
  IF $F(i) < F(i)$
      $\mathbf{M}' := \mathbf{M}' + \{i\}$;
  ELSE
      $\mathbf{M}' := \mathbf{M}' + \{j\}$;
  OD
- Step 5: Recombination: ...
- Step 6: Mutation: ...
- Step 7: Termination: Set $\mathbf{M}_{t+1} = \mathbf{M}'''$ and $t = t + 1$;
  IF $t \geq U$
      $\mathbf{A} = g(m(\mathbf{M}_t))$;
  ELSE GOTO Step2.

END

Fig. 4. Zitzler's Strength Pareto EA (SPEA).

4.3 ESPEA

We call our extension of SPEA to deal with estimated objective values bounded by intervals ESPEA - *Estimate Strength Pareto Evolutionary Algorithm*.

First, two variants to deal with property intervals seem to be appropriate:

- Expected value computation: A straightforward method to deal with objectives with uncertain values is to let SPEA unmodified and work directly with the vectors of expected values $E[Y]$ of the objective values. The advantage of this approach is that no changes have to be made in the code of SPEA but only code to calculate the expected values for each decision vector has to be added based on the knowledge of the property intervals and probability density functions used. As this first variant is obvious, we concentrate here on the second approach.
- ESPEA: Instead of working with expected values, we modify SPEA accordingly. Within SPEA, changes are needed in Step 2 (update of the external set, see Fig. 4) and in the two algorithms Cluster (see Fig.6) and Fitness (see Fig.5). These changes are described next.

Update of external set. We describe changes to SPEA to deal with random variables caused by property intervals. First, we consider step 2 of SPEA. Here, the external set is updated to include those members of the actual population $\mathbf{M}_t$ that are non-dominated with respect to $\mathbf{M}_t$. With probability intervals, we are only able to state the probability of a vector $m(i)$ to be dominated or not.

Given $P[m(j) \succeq m(i)]$. We define

$$R(i) = \frac{1}{N-1} \cdot \sum_{j \in \mathbf{M}_t : j \neq i} P[m(j) \succeq m(i)]$$

Each of the following three rules seems to be appropriate to decide whether a vector $m(i)$ should be taken over into the external set:

- Add $m(i)$ to $\overline{\mathbf{M}'}$ if $R(i) < \alpha$; this condition adds all vectors which are smaller in probability to be dominated then α, e.g., let $\alpha = 0.25$. Or,
- add those β % of vectors i with smallest $R(i)$. Or,
- those $\gamma \in \mathbf{N}$, $\gamma < N$ individuals with smallest $R(i)$ are copied into the external set.

After updating the external set in either of the above ways, weakly dominated individuals i must be removed regarding $m(\overline{\mathbf{M}'})$. Again, we have to deal with probabilities. We propose the following similar strategy:

Let

$$R(j) = \frac{1}{N-1} \cdot \sum_{i \in \overline{\mathbf{M}'} : j \neq i} P[m(i) \succeq m(j)]$$

Then eliminate

- those j from $\overline{\mathbf{M}'}$ where $R(j) > \alpha'$ (e.g., $\alpha' = 0.9$), or
- those $\beta'\%$ vectors j with largest $R(j)$, or
- those γ' vectors j with largest $R(j)$.

The last step concerns the clustering algorithm called in Step 2 of SPEA, see Fig. 6.

```
Fitness
    IN:   M_t : population;
          \overline{M_t}: external set;
    OUT:  F: fitness;

BEGIN
    Step 1: Strength and fitness assignment external set:
        FOREACH i ∈ \overline{M_t} DO
            S(i) = |{j | j∈M_t ∧ m(i)≽m(j)}| / (N+1);
            F(i) = S(i);
        OD
    Step 2: Fitness assignment population:
        FOREACH j ∈ M_t DO
            F(j) = 1 + Σ_{i∈\overline{M_t}, m(i)≽m(j)} S(i);
        OD
END
```

Fig. 5. Fitness assignment in SPEA.

```
Cluster
    IN:   \overline{M'} : external set;
          \overline{N}: maximum size of external set;
    OUT:  \overline{M_{t+1}}: updated external set;

BEGIN
    Step 1: Initialize each i ∈ \overline{M'} as a cluster:
        C = ∪_{i∈\overline{M'}} {{i}};
    Step 2: IF |C| ≤ \overline{N}, goto Step 5, else goto Step 3;
    Step 3: Distance calculation (objective space) of all possible pairs of clusters:
        d_c = (1/(|c_1|·|c_2|)) Σ_{i_1∈c_1, i_2∈c_2} d(i_1, i_2);
    Step 4: Amalgate two clusters with minimal distance:
        C := C \ {c_1, c_2} ∪ {c_1 ∪ c_2};
    Step 5: Select a representative for each cluster by computing the centroid
            (point with minimal average distance to all other points in the cluster)
        \overline{M'} := ∪_{c∈C} c;
END
```

Fig. 6. Clustering in SPEA.

Clustering algorithm. Depending on the above update of the external set, different numbers of clustering steps might be necessary in order to reduce the update set to its maximal size $\overline{N}$.

The only point where the objective values of two individuals are needed is for the determination of the distance between two vectors i and j $d(i,j)$ in the objective space. Here, we approximate the distance by the expected value of the distance:

$$E[d(i,j)] = E[\|f(m(i)) - f(m(j))\|] \tag{11}$$

Fitness assignment algorithm. Finally, the fitness assignment algorithm Fitness called in Step 3 of SPEA (see Fig. 5) must be changed. We consider Step 1 first. Here, the strength $S(i) = \frac{|\{j \mid j \in M_t \wedge m(i) \succeq m(j)\}|}{N+1}$ is computed. In this formula, the size of the set $\{j \mid j \in M_t \wedge m(i) \succeq m(j)\}$ is a random variable the expected value of which is computed as follows:

$$E[S(i)] = \frac{1}{N+1} E(|\{j \mid j \in M_t \wedge m(i) \succeq m(j)\}|)$$
$$= \frac{1}{N+1} \sum_{j \in M_t} P[m(i) \succeq m(j)] \tag{12}$$

Similarly, in Step 2 of the fitness assignment algorithm, we approximate $F(j) = 1 + \sum_{i \in \overline{M_t}, m(i) \succeq m(j)} S(i)$ by

$$F(j) \approx 1 + \sum_{i \in \overline{M_t} : P[m(i) \succeq m(j)] \geq \alpha} E[S(i)] \tag{13}$$

The remaining Steps of SPEA (1,4,5,6,7) remain unaffected.

5 Experiments

In this section, we describe our experiments using the example of Pareto-Front exploration for a hardware/software partitioning problem: A task graph is to be mapped onto programmable hardware and software components such that the objectives of period (execution time) and cost should be minimized simultaneously, see [1] for details. The objectives evaluated are the cost of the allocated components and the execution time (period) for the mapping (see Fig. 3 for one example population). In the following, we run a benchmark 5 runs of 200 generations each for four different variations concerning the interval sizes called ex1, ex2, ex3, and ex4 in the following for a population size of 40 and 100 individuals each. ex1 has the smallest intervals for each parameter, ex2 has larger intervals, and ex4 has the largest intervals for cost and execution time of a number of tasks. For $\alpha = 0.25$, we evaluated two different different values of α', namely A1 $\alpha' = 0.5$ and A2 $\alpha' = 0.9$. In order to compare the performance of these two algorithms for the 4 examples ex1, ex2, ex3, and ex4, we use the following definition:

Definition 11. *Let A and B be two sets of decision vectors. The function $\mathcal{C}$ maps the ordered pair (A,B) to the interval [0,1]:*

$$\mathcal{C}(A,B) := \frac{|\{b \in B; \exists a \in A : a \succeq b\}|}{|B|} \qquad (14)$$

The value $\mathcal{C}(A, B)$ gives the fraction of B that is covered by members of A. Note that both $\mathcal{C}(A, B)$ and $\mathcal{C}(B, A)$ have to be taken into account, since not necessarily $\mathcal{C}(A, B) = 1 - \mathcal{C}(B, A)$. On each example, both algorithms ran in pairs on the same initial population; then the expected values of the Pareto-points in the final population were compared for each run. The minimal, maximal, and average values are reported in Tables 1 and 2 for two different population sizes ($N = 40$ and $N = 100$).

Table 1. Comparison of algorithms A1 and A2 on four examples for population size $N = 40$.

Test case	$\mathcal{C}$(A1,A2)			$\mathcal{C}$(A2,A1)		
	mean	min	max	mean	min	max
ex1	100%	100%	100%	90%	83%	100%
ex2	50.8%	25%	75%	94%	85%	100%
ex3	64.3%	28.5%	100%	95%	85%	100%
ex4	83.4%	50%	100%	92%	85%	100%

Table 2. Comparison of algorithms A1 and A2 on four examples for population size $N = 100$.

Test case	$\mathcal{C}$(A1,A2)			$\mathcal{C}$(A2,A1)		
	mean	min	max	mean	min	max
ex1	80%	0%	100%	96.6%	83%	100%
ex2	73.4%	50%	89%	91%	80%	95%
ex3	76.2%	67%	100%	98%	95%	100%
ex4	100%	100%	100%	83%	60%	95%

First, we discovered that for each test run and test example, the run-time of algorithm A1 is about 3-5 times higher than with algorithm A2 for equal population size and equal number of populations for all test examples except ex1 (no intervals) where both algorithms had equal run-time. The reason is that in case of A1, the probability level for the selection of a design point as a point of the external set is much lower than in case of A2. We discovered also that the run-time increases about quadratically with the number of points stored in the external set. In algorithm A1, the external set almost constantly contained $\overline{N}$ points (often with many multiple copies). In case of A2, the probability level was much higher to accept a point as a Pareto-point. Here, the external set contained a considerably less number of points for all our test cases.

Concerning the quality of discovered Pareto-points, we see that no matter how large the intervals were, algorithm A2 shows a mean coverage of points explored by A1 by more than 90 %. This shows that although A1 has a much smaller acceptance level and a larger set of non-dominated points, it performed worse. Other experiments such as testing other and larger examples are necessary to support this result, see, e.g., [10].

We would finally like to mention that the concept of *probabilistic dominance* as introduced here might be used also in other Pareto-set exploring algorithms.

In order not to reject possible candidates from the Pareto-set, one should not just take the expected values of the objectives but decide on the probability of dominance if a point should be considered Pareto-point or not. For decision making in case of *probabilistic dominance*, any of the proposed styles how to update the external set may be used.

References

1. T. Blickle, J. Teich, and L. Thiele. System-level synthesis using Evolutionary Algorithms. *J. Design Automation for Embedded Systems*, 3(1):23–58, January 1998.
2. M. Dellnitz and O. Junge. An adaptive subdivision technique for the approximation of attractors and invariant measures. *Comput. Visual. Sci.*, 1:63–68, 1998.
3. Carlos M. Fonseca and Peter J. Fleming. An overview of evolutionary algorithms in multiobjective optimization. *Evolutionary Computation*, 3(1):1–16, 1995.
4. David E. Goldberg. *Genetic Algorithms in Search, Optimization and Machine Learning*. Addison-Wesley Publishing Company, Inc., Reading, Massachusetts, 1989.
5. P. Hajela and C.-Y. Lin. Genetic search strategies in multicriterion optimal design. *Structural Optimization*, 4:99–107, 1992.
6. Jeffrey Horn and Nicholas Nafpliotis. Multiobjective optimization using the niched pareto genetic algorithm. IlliGAL Report 93005, Illinois Genetic Algorithms Laboratory, University of Illinois, Urbana, Champaign, July 1993.
7. Jeffrey Horn, Nicholas Nafpliotis, and David E. Goldberg. A niched pareto genetic algorithm for multiobjective optimization. In *Proceedings of the First IEEE Conference on Evolutionary Computation, IEEE World Congress on Computational Computation*, volume 1, pages 82–87, Piscataway, NJ, 1994. IEEE Service Center.
8. K. A. De Jong. *An analysis of the behavior of a class of genetic adaptive systems*. PhD thesis, University of Michigan, 1975.
9. J. D. Schaffer. Multiple objective optimization with vector evaluated genetic algorithms. In *Proceedings of an International Conference on Genetic Algorithms and Their Applications*, pages 93–100, Pittsburgh, PA, 1985. J.J. Grefenstette (Ed.).
10. J. Teich and R. Schemann. Pareto-front exploration with uncertain objectives. Technical report, TR No. 3/00 (revised version), Computer Engineering Laboratory, University of Paderborn, December 2000.
11. E. Zitzler, J. Teich, and S. Bhattacharyya. Evolutionary algorithms for the synthesis of embedded software. *IEEE Trans. on VLSI Systems*, 8(4):425–456, August 2000.
12. E. Zitzler, J. Teich, and S. S. Bhattacharyya. Multidimensional exploration of software implementations for DSP algorithms. *J. of VLSI Signal Processing Systems*, 24:83–98, 2000.
13. E. Zitzler and L. Thiele. Multiobjective evolutionary algorithms: A comparative case study and and the strength pareto approach. *IEEE Trans. on Evolutionary Computation*, 3(4):257–271, November 1999.

Evolutionary Multi-objective Ranking with Uncertainty and Noise

Evan J. Hughes

Department of Aerospace, Power, and Sensors,
Cranfield University, Royal Military College of Science,
Shrivenham, Swindon, SN6 8LA, UK,
ejhughes@iee.org,
http://www.rmcs.cranfield.ac.uk/departments/daps/

Abstract. Real engineering optimisation problems are often subject to parameters whose values are uncertain or have noisy objective functions. Techniques such as adding small amounts of noise in order to identify robust solutions are also used. The process used in evolutionary algorithms to decide which solutions are better than others do not account for these uncertainties and rely on the inherent robustness of the evolutionary approach in order to find solutions.

In this paper, the ranking process needed to provide probabilities of selection is re-formulated to begin to account for the uncertainties and noise present in the system being optimised. Both single and multi-objective systems are considered for rank-based evolutionary algorithms.

The technique is shown to be effective in reducing the disturbances to the evolutionary algorithm caused by noise in the objective function, and provides a simple mathematical basis for describing the ranking and selection process of multi-objective and uncertain data.

1 Introduction

The use of evolutionary algorithms (EA's) in engineering is now well established and widespread. As the use of the algorithms migrates deeper into industry, and with more processing power available, the scale and characteristics of the problems being solved are changing. The objective functions are becoming more complex, nonlinear and often uncertain. Many model coefficients are derived by experiment and are therefore subject to experimental errors. In real systems, the true coefficients will not be the same as measured and are often time dependent or correlated with platform motion etc.

These errors in the modelling are unavoidable and inevitably propagate into the outputs of the objective functions, the results of which are used to classify the quality of the individual solutions to the problem. All optimisation algorithms attempt to find the problem solution that gives the most favourable output from the objective functions. With complex systems, evolutionary algorithms are a useful tool in that they can tolerate highly nonlinear and noisy system models and objective functions and still provide reasonable suggested solutions [1].

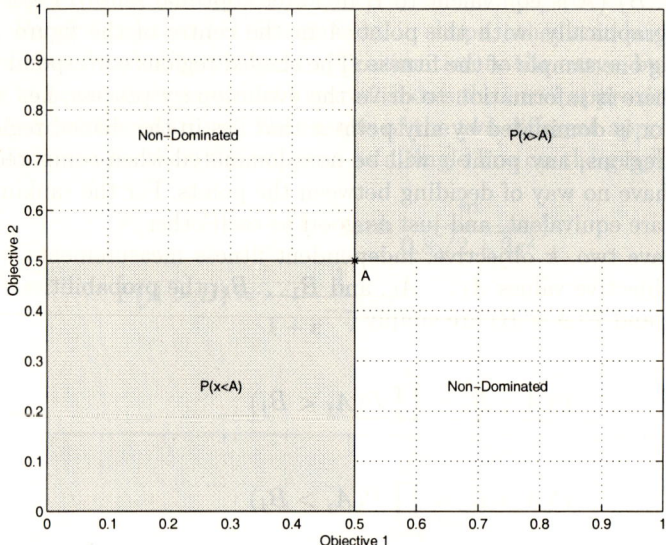

Fig. 3. Noise free non-domination map (maximisation)

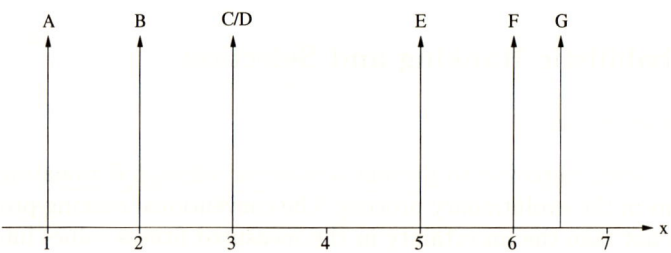

Fig. 4. Fitness values to be ranked

Table 1. Ranks of example fitness values

Value	A	B	C	D	E	F	G
Rank	0	1	2	2	4	5	6

arithmetic series zero to six, therefore the best individual will get a probability of selection of $2/n$ and the worst a probability of zero.

$$P(\text{select}_i) = \frac{(n-1) - R_i}{\sum_{j=1}^{n} R_j} = \frac{2((n-1) - R_i)}{n(n-1)} \tag{19}$$

If we use the rank values in Table 1 with both the tied fitness values being given the best 'untied' rank, we find that the sum of the ranks is no longer consistent, and in this case, $\sum_{j=1}^{n} R_j = 20$. Alternatively, as C & D are tied, it may be better to penalise them both a little and therefore take an average of the rank positions they could have shared, i.e., give them both a rank of 2.5. This would return the overall sum to be 21 and would be consistent, no matter how many fitness values share a rank. This is the method most used for ranking a vector of data.

We can view the ranking process as counting the number of fitnesses that dominate the fitness of interest [8]. If a fitness equal to the current one is encountered, then it is half dominating, and half dominated by the current fitness. Therefore we can create the rank position numbers by this simple counting process. For example, E is dominated by A, B, C & D and therefore has a rank of 4. Value C is dominated by A & B but is tied with D and so gets a rank of 2.5.

Alternatively, we could consider the dominating / not dominating decision as being the *probability* that each fitness value dominates the value of interest. For example, if we consider fitness C, the probability that A dominates C is one. The probability that G dominates C is zero. The probability that D dominates C, from (6) with $m = 0$, is $P = 0.5$. Thus we can represent the rank position as the sum of probabilities of domination as shown in (20), where $P(F_j > F_i)$ is the probability that fitness value j dominates fitness value i.

$$R_i = \sum_{j=1}^{n} P(F_j > F_i) \bigg|_{i \neq j} \tag{20}$$

In (20), we have to be sure not to compare fitness F_i with itself. If we did, we would get an extra probability of 0.5 added to the sum. We can therefore include F_i in the sum, but subtract the effect of comparing the fitness with itself. This is shown in (21).

$$R_i = \sum_{j=1}^{n} P(F_j > F_i) - 0.5 \tag{21}$$

As (21) is based on probability, if the fitness values are uncertain, we can use (6) or the approximations (13) or (14) to calculate the probability of domination. For example, if fitness values A to G have a standard deviation of $\sigma = 1$, the rank positions (using (13)) compared to the no noise case are shown in Table 2.

With $\sigma = 0$, we have conventional ranking and the probabilities will range from $2/n$ to zero. If $\sigma = \infty$, all of the fitness values will be assigned the same rank, and will have a probability of selection of $1/n$. Thus the standard deviation of the uncertainty has a similar effect to selective pressure in conventional selection processes [13].

Table 2. Ranks with uncertainty of $\sigma = 0$ and $\sigma = 1$

Value	Rank ($\sigma = 0$)	Rank ($\sigma = 1$)
A	0	0.38
B	1	1.27
C	2.5	2.31
D	2.5	2.31
E	4	4.17
F	5	5.07
G	6	5.49

4.3 Multi-objective Ranking

With multiple objectives, we now have three possible outcomes from comparing the two fitness values: A dominates B, A is dominated by B, and A and B are non-dominated. If we apply the single objective ranking equation, we find that the total of the rank positions is no longer $n(n-1)/2$ as we now have to account for the non-domination. If we have no noise, for two fitness values where A dominates B, $P(A > B) = 1$, $P(A < B) = 0$, and $P(A \equiv B) = 0$ Therefore when we sum the probabilities of domination, the contribution from this pair will be 1. If the fitness values are non-dominated, the corresponding probabilities are $P(A > B) = 0$, $P(A < B) = 0$, and $P(A \equiv B) = 1$. We have now lost the value 1 from the probability of domination calculations, therefore reducing the sum of ranks total. This state will be the same when we compare A to B and also when we compare B to A, therefore if we sum all the probabilities of non-domination, this will give us twice what was lost from the probability of domination calculations.

If we consider the ranking case for a single dimension, if A and B are identical, we cannot choose between them and so add in 0.5 to the sum. With non-domination, we also have the situation where we cannot choose between objectives and should therefore add 0.5 to the sum as required. In the case of uncertain measurements, we can multiply the value of 0.5 by the probability of non-domination, and still subtract off 0.5 to allow for comparing the individual with itself, thereby maintaining the sum of the rank positions as $n(n-1)/2$. Thus we can add the non-domination term into (21). The rank calculation for multi-objective ranking is shown in (22), where n is the number of measurements being ranked.

$$R_i = \sum_{j=1}^{n} P(F_j > F_i) + \frac{1}{2}\sum_{j=1}^{n} P(F_j \equiv F_i) - 0.5 \qquad (22)$$

This *probabilistic ranking* equation allows chromosomes to be selected based on uncertain multi-objective fitness measurements. For the objectives shown in Fig. 5, we can calculate the rankings in order to minimise the fitness values. Table 3 shows ranks (R) for no noise, and 1 standard deviation noise.

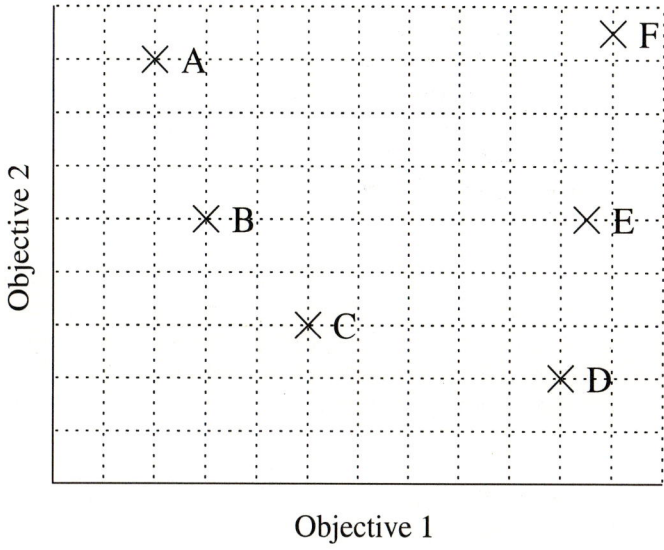

Fig. 5. Multiple fitness values to be ranked

In the example, we see that A is non-dominated with B, C, D, & E and therefore gets a rank of 2. Fitness B is non-dominated with A, C, & D but shares an objective value with E, thus being half dominating and half non-dominated with E, the rank of B is 1.5 from the three non-dominated points and 0.25 from E, giving a total of 1.75. We also see that each of the columns of Table 3 sums to 15 ($= n(n-1)/2$) as expected. The ranking process is $O(n^2)$, as are many of the other ranking methods [8,9].

Table 3. Ranks with uncertainty of $\sigma = 0$ and $\sigma = 1$

Value	R ($\sigma = 0$)	R ($\sigma = 1$)
A	2	2.27
B	1.75	1.65
C	1.5	1.42
D	1.5	1.92
E	3.25	3.22
F	5.0	4.53

In the general noisy or uncertain scenario, we see that the proximity of other fitness values, even if only close on one objective, can influence how the rank is assigned. Measurements such as C which are relatively well spaced out on all objectives are ranked more highly than other fitness values that are uncertain.

With no noise, the basic ranking by just counting how many points dominate each fitness measurement described by Fonseca and Flemming [8] is very similar, but does not allow for the non-dominated cases. The sum of the rank values will not be consistent if non-dominated solutions are present, causing a bias towards non-dominated solutions over other solutions. The ranking used by Srinivnas and Deb [9] is based on 'layers' of non-dominated solutions and has no consistency with regards to how many layers, or ranks, are produced, therefore making calculating selection probabilities awkward.

It is interesting to note that if we require an objective to be maximised, setting σ negative will cause the probabilities to be calculated for maximisation, setting σ negative has the same effect as negating the fitness values (the conventional way of converting from minimisation to maximisation). Therefore both minimisation and maximisation objectives may be handled easily by just setting the sign of the corresponding value of σ appropriately.

Limits on objectives, constraints on the chromosomes, and sharing can all be implemented easily within this ranking framework, allowing interactive decision making with uncertain or noisy systems viable. The equations for limits, constraints, and sharing are derived and discussed in [12].

5 Experiment Results

5.1 Introduction

Noise and uncertainty can be split into two broad categories relating to noise that occurs within the process (Type A) and measurement noise (Type B):

1. **Type A Noise:** Noise is applied to the chromosome *before* the objective function is calculated, i.e. $\mathcal{O} = F(\chi + N)$.
2. **Type B Noise:** Noise is applied to the objective function *after* calculation, i.e. $\mathcal{O} = F(\chi) + N$.

Both types of noise are of interest and often the observed noise will be a combination of type A and B.

Trials have been performed to assess how the noise effects the assigned rank position within a population of chromosomes. For the following results, 100 two-parameter chromosomes were generated uniformly distributed in the range [0,1]. A scaled version of the objective function MOP3, defined by Van Veldhuizen and Lamont [10] and given in (23), was used to provide input data to the ranking processes, with either type A or B noise applied as appropriate. The data were ranked and the assigned rank postion for each chromosome recorded. The process was repeated 1000 times with different values chosen for the applied noise each time. For each chromosome, the standard deviation of the rank position was calculated. The mean standard deviation of the 100 chromosome rank positions was then generated and plotted.

$$x = 6\chi(1) - 3$$
$$y = 6\chi(2) - 3$$
$$m_1 = \frac{(x^2 + y^2)}{2} + sin(x^2 + y^2)$$
$$m_2 = \frac{(3x - 2y + 4)^2}{8} + \frac{(x - y + 1)^2}{27} + 15$$
$$m_3 = \frac{1}{x^2 + y^2 + 1} - 1.1e^{-x^2 - y^2}$$
$$\mathcal{O}_1 = \frac{m_1}{8.249}$$
$$\mathcal{O}_2 = \frac{m_2 - 15}{46.940}$$
$$\mathcal{O}_3 = \frac{m_3 + 0.1}{0.296} \tag{23}$$

In (23), $\chi(1)$ and $\chi(2)$ are the two parameters of the input chromosome in the range [0,1]. The parameters x and y are scaled to lie within [-3,3] as defined by Van Veldhuizen and Lamont. The three objective functions are then calculated and scaled to give each of the objectives in the range [0,1]. Noise was then applied either to the input chromosome χ for type A noise, or to the output objectives $\mathcal{O}$ for type B noise. The applied noise was Gaussian with a standard deviation of σ.

The ranking algorithms from NSGA and MOGA were generated for comparison with the new multi-objective probabilistic selection evolutionary algorithm (MOPSEA) ranking process developed in this paper. With a different set of 100 initial chromosomes, a slightly different set of graphs will result. The differences have been found to be small however.

5.2 Results

From figures 6 & 7 it is clear that both MOGA and MOPSEA outperform the NSGA ranking process in the presence of noise for this objective function. As the uncertainty parameter σ_n is increased, it is clear that MOPSEA can out perform both alternative algorithms. The specific performance of each algorithm is dependent on the objective function though. Other objective functions are covered in [12].

6 Conclusions

The results have shown that the modified ranking process can reduce the disturbances in the rank positions caused by noisy objectives. Unlike conventional ranking processes, the rank values and therefore the corresponding selection probabilities take some account of the noise and uncertainty in the system. The theory developed in this paper forms an important first step towards addressing

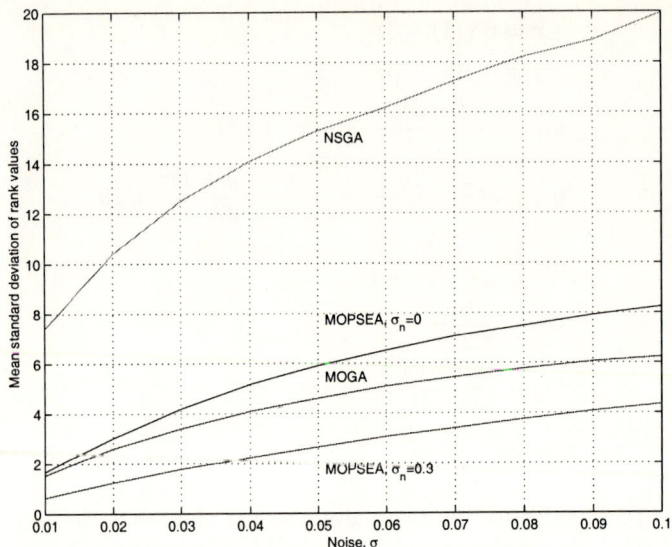

Fig. 6. Applied noise with respect to mean standard deviation of rank position for MOP3, type A noise. Performance of MOGA and NSGA ranking compared to MOPSEA with $\sigma_n = 0$ & $\sigma_n = 0.3$

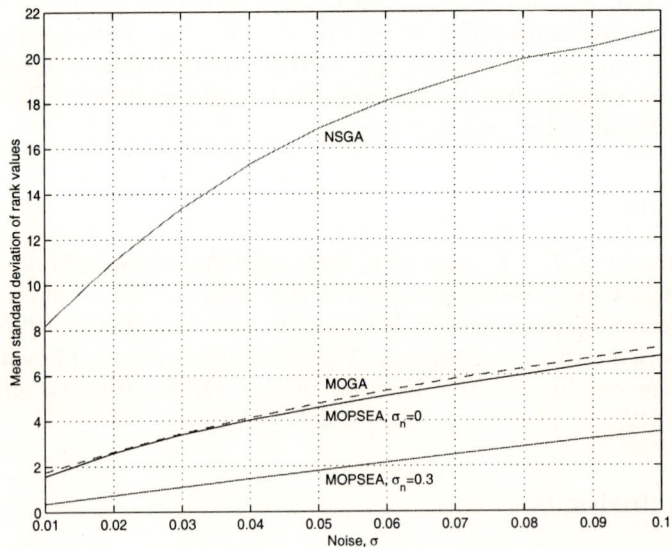

Fig. 7. Applied noise with respect to mean standard deviation of rank position for MOP3, type B noise. Performance of MOGA and NSGA ranking compared to MOPSEA with $\sigma_n = 0$ & $\sigma_n = 0.3$

directly noise and uncertainty in multi-objective problems. The simplicity of the ranking and selection equations may also provide a route to further theoretical research into the operation and performance of evolutionary algorithms.

Acknowledgements. The author would like to acknowledge the use of the Department of Aerospace, Power, and Sensors DEC Alpha Beowulf cluster for the production of the results.

References

1. T. W. Then and Edwin K. Chong. Genetic algorithms in noisy environment. In *IEEE International Symposium on Intelligent Control*, pages 225–230, Columbus, Ohio, 16-18 August 1994.
2. Adrian Thompson. Evolutionary techniques for fault tolerance. In *Control '96, UKACC International Conference on*, volume 1, pages 693–698. IEE, 2-5 September 1996. Conf. Publ. No. 427.
3. T. Bäck and U. Hammel. Evolution strategies applied to perturbed objective functions. In *IEEE World Congress on Computational Intelligence*, volume 1, pages 40–45. IEEE, 1994.
4. Shigeyoshi Tsutsui and Ashish Ghosh. Genetic algorithms with a robust searching scheme. *IEEE Transactions on Evolutionary Computation*, 1(3):201–8, September 1997.
5. Kumar Chellapilla and David B. Fogel. Anaconda defeats hoyle 6-0: A case study competing an evolved checkers program against commercially available software. In *Congress on Evolution Computation - CEC2000*, volume 1, pages 857–863, San Diego, CA, 16-19 July 2000. IEEE.
6. Carlos A. Coello Coello. List of references on evolutionary multiobjective optimization. http://www.lania.mx/~ccoello/EMOO/EMOObib.html. Last accessed 3 July 2000.
7. Anna L. Blumel, Evan J. Hughes, and Brian A. White. Fuzzy autopilot design using a multiobjective evolutionary algorithm. In *Congress on Evolution Computation - CEC2000*, volume 1, pages 54–61, San Diego, CA, 16-19 July 2000. IEEE.
8. Carlos M. Fonseca and Peter J. Flemming. Multiobjective genetic algorithms made easy: Selection, sharing and mating restriction. In *GALESIA 95*, pages 45–52, 12-14 September 1995. IEE Conf. Pub. No. 414.
9. N. Srinivas and Kalyanmoy Deb. Multiobjective optimization using nondominated sorting in genetic algorithms. *Evolutionary Computation*, 2(3):221–248, 1995.
10. David A. Van Veldhuizen and Gary B. Lamont. Multiobjective evolutionary algorithm research: A history and analysis. Technical Report TR-98-03, Air Force Institute of Technology, 1 Dec 1998.
11. William H. Press, Brian P. Flannery, Saul A. Teukolsky, and William T. Vetterling. *NUMERICAL RECIPES in C. The Art Of Scientific Computing*. Cambridge University Press, second edition, 1993.
12. Evan J. Hughes. Multi-objective probabilistic selection evolutionary algorithm. Technical Report DAPS/EJH/56/2000, Dept. Aerospace, Power, & Sensors, Cranfield University, RMCS, Shrivenham, UK, SN6 8LA, September 2000.
13. J. E. Baker. Adaptive selection methods for genetic algorithms. In *Proc. 1st Int. conf. on Genetic Algorithms*, pages 101–111, 1985.

Tabu-Based Exploratory Evolutionary Algorithm for Effective Multi-objective Optimization

E. F. Khor, K. C. Tan[1], and T. H. Lee

Department of Electrical and Computer Engineering
National University of Singapore
10 Kent Ridge Crescent Singapore 119260
[1]Email: eletankc@nus.edu.sg

Abstract. This paper proposes an exploratory multi-objective evolutionary algorithm (EMOEA) that makes use of the integrated features of tabu search and evolutionary algorithms for effective multi-objective optimization. It incorporates a tabu list and tabu constraint for individual examination and preservation to enhance the evolutionary search diversity in multi-objective optimization, which subsequently helps to avoid the search from trapping in local optima and at the same time, promotes the evolution towards the global Pareto-front. A novel method of lateral interference is also suggested, which is capable of distributing non-dominated individuals uniformly along the discovered Pareto-front at each generation. Unlike existing niching/sharing methods, lateral interference can be performed without the need of any parameter setting and can be flexibly applied in either parameter or objective domain depending on the nature of the optimization problem involved. The proposed features are experimented in order to illustrate their behavior and usefulness in the algorithm.

1. Introduction

Evolutionary algorithms have been recognized to be well suited for MO optimization problems [1,2]. Unlike conventional methods that linearly combine multiple attributes to form a composite scalar objective function, evolutionary algorithm for MO optimization incorporates the concept of Pareto's optimality or modified selection schemes to evolve a family of solutions at multiple points along the trade-off surface simultaneously. Since Schaffer's work [3], evolutionary techniques for MO optimization have been gaining significant attentions from researchers in various fields, which are reflected by the high volume of publications in this topic in the last few years (over 25 Ph.D. theses, more than 80 journal papers, and more than 300 conference papers). For more information on various techniques of handling multi-objective optimization problems via evolutionary algorithms, readers may refer to the literatures of [4-6].

This paper proposes a new exploratory multi-objective evolutionary algorithm (EMOEA), which incorporates the memory-based feature of tabu search (TS) to maintain the stability of MO optimization towards a global and uniform Pareto-front. The hybridization of TS in evolutionary optimization helps to improve the MO search performances by avoiding repeats of previously explored paths to the found peaks, i.e., local optima in the search space is avoided while good regions are being well explored. Besides, a novel method of lateral interference, which is highly efficient of distributing non-dominated individuals uniformly along the discovered Pareto-front is also proposed. It can be performed without the need of any parameter setting and can be flexibly applied in either parameter or objective domain depending on the nature of the optimization problem involved.

2. Exploratory Multi-objective Evolutionary Algorithm

In general, multi-objective (MO) optimization can be defined as the problem of optimizing a vector of non-commensurable and often competing objectives or cost functions, viz. it tends to find a parameter set P for

$$\min_{P \in \Phi} F(P) \qquad (1)$$

where $P = \{p_1, p_2, \ldots, p_n\}$ is a individual vector with n parameters and Φ defines a set of individual vectors. $\{f_1, f_2, \ldots, f_m\}$ are m objectives to be minimized and $F = \{f_1, f_2, \ldots, f_m\}$. Instead of a single optima, solution to MO optimization problem is often a family of points known as Pareto optimal set, where each objective component of any point along the Pareto-front can only be improved by degrading at least one of its other objective components [7,8]. In the total absence of information regarding the preferences of objectives, ranking scheme based upon the Pareto optimality is regarded as an appropriate approach to represent the strength of each individual in an evolutionary algorithm for MO optimization [2,8,]. A vector F_a is said to dominate another vector F_b, denoted as $F_a \prec F_b$, iff

$$f_{a,i} \leq f_{b,i} \ \forall \, i \in \{1,2,\ldots,m\} \text{ and } \exists \ j \in \{1,2,\ldots,m\} \text{ where } f_{a,j} < f_{b,j} \qquad (2)$$

The Pareto ranking scheme assigns the same smallest cost for all non-dominated individuals, while the dominated individuals are ranked according to how many individuals in the population dominating them. So, the rank of an individual x in a population can be given by $rank(x) = 1 + q_x$, where q_x is the number of individuals dominating the individual x in the objective domain [2]. They also extended the Pareto's domination scheme in their proposed multi-objective genetic algorithm (MOGA) to include goal and priority information for MO optimization. Although MOGA is a good approach, the algorithm only allows a single goal and priority vector setting, which may be difficult to define in *a-priori* to an optimization process [4].

With a modified Pareto-domination scheme, Tan *et al.* [9] proposed a unified multi-objective evolutionary algorithm (MOEA) that is capable of comparing the domination among individuals for multi-objective optimization dealing with both soft and hard optimization constraints. The scheme also allows the incorporation of multiple goals and priorities with different combinations of logical "AND" and "OR" operations for greater flexibility and higher-level decision support. Extending from the Pareto's domination and ranking schemes of Tan *et al.*, [9], this paper proposes an exploratory multi-objective evolutionary algorithm that incorporates the memory-based feature of tabu search to maintain the stability of MO optimization towards a global and uniform Pareto-front. Sections 2.1 and 2.2 detail the principle of lateral interference and tabu-based individual examination rule, respectively. The algorithm and implementation of EMOEA including the features of lateral interference and tabu-based individual examination system is then fully described in Section 2.3.

2.1 Lateral Interference (LI)

In order to evolve an equally distributed population along the Pareto-front and to distribute the population at multiple optima in the search space, many methods [1,10,11] have been proposed. Among them, the 'niche induction' technique by means of a sharing function [10] is the most popular approach for evolving an equally distributed population along the Pareto-front in MO optimization. This method creates sub-divisions in the objective domain by degrading an individual's fitness upon the existence of other individuals in its neighborhood defined by a shared distance.

To avoid the setting of sharing distance, a new population distribution method is proposed. It is capable of uniformly distributing all individuals along the Pareto-front for MO optimization without the need of any parameter setting is proposed. It can be applied in either the parameter domain or in the objective domain as needed. The method is called *Lateral Interference* which is motivated from the ecologist point of view for resources competition [12]. It works based upon the principles of exploitation competition and interference competition, which form the basis of distributing population uniformly along the Pareto-front in MO optimization without the need of any parameter setting. According to the first principle (exploitation competition), individuals with higher fitness or lower cost values will always be more likely to win when compete with individuals with lower fitness or higher cost values. The second principle (interference competition) only takes place among the individuals with same level of fitness or cost, or in other words, individuals that are equally strong in the exploitation competition.

Let us consider a sub-population P' containing N' equally-fit (or same level of fitness or cost values where the intermediate between two level, $t \geq 0$) individuals in the m-dimensional observed feature space S (can be either in objective space or parameter space), $x_1, x_2, \ldots, x_{N'}$, $x_i \in U_1 \times U_2 \times \ldots \times U_m$ $\forall$ $i = 1, 2, \ldots, N'$, $U_j = U(a_j, b_j)$, where $U(a_j, b_j)$ is any value within the range of a_j and b_j while a_j and b_j are the minimum and maximum boundary values of x_j, $\forall j = 1, 2, \ldots m$ as given by,

$$a_j = \min\{x_{i,j} \; \forall i = 1,2,\ldots,N'\} \tag{3a}$$

$$b_j = \max\{x_{i,j} \; \forall i = 1,2,\ldots,N'\} \tag{3b}$$

The metric distance between any two individuals, i and j, is defined by:

$$d(x_i, x_j) = \|x_i - x_j\|_2 \tag{4}$$

where $\|\cdot\|_2$ implies the 2-norm. To limit the distance metric to the interval of [0, 1], normalized metric distance between individual i and j, $\overline{d}(x_i, x_j)$, is used and computed by dividing the metric distance over the metric distance between a and b, where a and b are the minimum and maximum boundary vector of all the variables respectively, i.e.,

$$\overline{d}(x_i, x_j) = d(x_i, x_j) / \|a - b\|_2 \tag{5}$$

Subsequently, the nearest individual from individual i, denoted as $s(i)$, is defined as the individual that gives the smallest norm or mathematically,

$$s(i) = \{j : \overline{d}(x_i, x_j) < \overline{d}(x_i, x_k), j \neq i, \forall k = 1, 2, ..., N', k \neq i \,\&\, j\} \tag{6}$$

and the influence distance of individual i is given as,

$$\lambda_i = \overline{d}(x_i, x_{s(i)}) \tag{7}$$

The larger the value to λ_i, the stronger the individual i to interfere its nearby individuals. With the information of influence distance, the territory of each individual can be defined as given in Definition 1.

Definition 1: (Territory)
The territory of individual i, denoted as T_i, is the area where any other individuals in within will be interfered and inhibited by individual i from getting the resource. The sufficiency condition for a given point x_j to be within the territory T_i of individual i, or mathematically $x_j \in T_i$ provided that $j \neq i$, is $\overline{d}(x_j, x_{s(i)}) \leq \lambda_i$, where $s(i)$ is the index of individual that is closest to individual i as defined in eqn. 6 while the influence distance λ_i of the individual i is as defined in eqn. 7.

After determining the territory for each individual, the severity or impact of being interfered and inhibited, or simply the interfered severity, denoted as $H_s(j)$, of individual j can be measured by means of the number of times that it has been interfered and inhibited,

$$H_s(j) = \sum_{i=1, i \neq j}^{N} I_s(i, j) \tag{8a}$$

where,

$$I_s(i, j) = \begin{cases} 1, & \text{if } x_j \in T_i \\ 0, & \text{otherwise} \end{cases} \tag{8b}$$

The proposed lateral interference differs to the sharing method in the sense that, instead of defining a fix sharing distance for the whole population, it applies the concept of territory where the area of territory is dependent to the influence distance which is adaptive and dynamic from one individual to the other. This eliminates the need of prefixing the sharing distance as required by the sharing method, where good setting of sharing distance estimated upon the trade-off surface is critical and is usually unknown in many optimization problems [4].

The lateral interference can be easily incorporated with Pareto's dominance in the genetic selection process. After the cost evaluation process where every individual in a population has been evaluated based on the MO functions, these individuals will be ranked according to the Pareto's dominance and ranking scheme [9], with smaller rank representing fitter individual. The individuals are then fed to the lateral interference, in which the rank values will be interfered/modified according to the territory of each individual. As mentioned, the aim of the interference competition is to further differentiate those individuals with same rank value after the exploitation competition in order to distribute the individuals uniformly. Before the lateral interference, the non-dominated individuals in the population are classified into one category, with similar dummy cost value. To maintain the diversity of the population, these classified individuals are undergone lateral interference and the resulted severity (H_s) of being interfered for each classified individual is added to its dummy cost. This group of individuals is then ignored and another layer of non-dominated individuals (ignoring the previously classified ones) is considered, where the assigned dummy cost is set higher than the highest interfered cost of the previously classified individuals. This process continues until all individuals in the individual list have been classified. The final resulted individuals' dummy cost value after the lateral interference is referred here as interfered cost. The smaller value of the interfered cost, the better is the individual.

2.2 Tabu-Based Individual Examination Scheme

This section presents an individual examination scheme that incorporates the feature of TS (tabu search) [13] in evolutionary algorithm to avoid repetition of previous move for MO optimization. Besides maintaining the stability of evolution towards the global and uniform Pareto-front, such scheme integrates the knowledge of tabu list to enhance the MO search performance by avoiding repeats of any paths to previously discovered peaks as well as encouraging long distance exploration to discover other possible peaks. Fig. 1 depicts the heuristic reasoning for individual examination scheme. Given a tabu list and individual list, every individual is being examined with reference to the tabu list whether to accept or reject the individual from the individual list while at the same time, the tabu list is updated whenever an individual dominates any member of tabu list. Starting from the first individual in the reproduced individual list, if the examined individual dominate any member in the tabu list, the individual will replace the dominated member(s) in the tabu list. Otherwise, if the individual is dominated by any member in the tabu list, it will be kept in the individual list if any of its objective component is better than the best objective component value found in the tabu list or the individual is not a tabu. For the former case, considering a minimization problem, an individual x, represented by vector $F_x(i)\ \forall\ i \in \{1, 2,..., m\}$ in m-dimensional objective domain, is said to have any of its objective component value better than the best objective component value in tabu list if and only if $\exists\ i$ s.t. $F_x(i) < \min(\{F_l(i)\})\ \forall\ l = \in \{1, 2,..., N_t\}$ where N_t is the size of tabu list. This criterion of acceptance is used to diversify the individuals in objective domain in order to provide a wider range of choices and better information exchange while undergoes the genetic crossover operation. If both the conditions are not satisfied, the individual will be rejected from the individual list and prohibited from surviving in the next generation. In case that the individual is not dominated by any member in the tabu list and if the tabu list is not full i.e., its size does not achieve the maximum limit, the individual will be added to the tabu list. Otherwise, if the individual is able to interfere more than one member in the tabu list within its territory, it will be allowed to replace the tabu member that has the shortest distance from the individual in the space concerned. This reasoning is to promote the uniform distribution of the tabu list. If the condition is not met, the individual will be examined with the objective domain diversification test and tabu test as explained above. This process of examination is then repeated for the next individual in the individual list until all the individuals are tested. Note that although all the members in tabu list are non-dominated, not all the non-dominated individuals in the individual list are added to the tabu list when they are tabu. This is to count on the computation effort of keeping all the non-dominated individuals in tabu as well as to avoid the danger of too much emphasis on the good individuals, which may lead to premature-convergence in the evolution.

2.3 Flow Chart of EMOEA

The overall program flowchart of the proposed EMOEA algorithm, which integrates both the lateral interference for population diversity as well as the individual examination scheme with tabu list for examining evolved individuals in the individual list, is shown in Fig. 2. In general, EMOEA involves two different lists, the individual list and the tabu list, that are interacting and influencing with each other along the evolution. Solutions in individual list play the part of inductive learning on the basis of genetic evolution while the solutions in tabu list play the role of controlling the evolution through heuristic reasoning approach (deductive).

At the initial stage of evolution, a list of $N_c^{(0)}$ number of individuals is initialized randomly or biased with *a-priori* knowledge, where $N_c^{(0)}$ is the size of individual list in the evolution. The individual list is then decoded to parameter vectors for cost evaluation. Subsequently, all the evaluated individuals are ranked according to the specifications assigned. All the non-dominated individuals (those with rank = 1) are copied to the empty tabu list while the rest of the individual list are fed to lateral interference to compute for interfered cost as described in Section 2.1. If the stopping criterion is not met, genetic operations will be applied to the evaluated individuals. Here, simple genetic operations consist of tournament selection based on interfered cost, standard crossover and mutation are performed to reproduce offspring for the next generation. Note that the use of tournament selection avoids the need of chromosome sorting, interpolating and fitness averaging at every generation as required by the ranking algorithm [2], which consequently reduces the overhead global basin in the initial population and still cannot converge to the global Pareto-optimal front. Hence, this optimization problem should challenge the optimization algorithms to search for the global Pareto-optimal with the existing of high-biased local Pareto-optimal in the MO optimization.

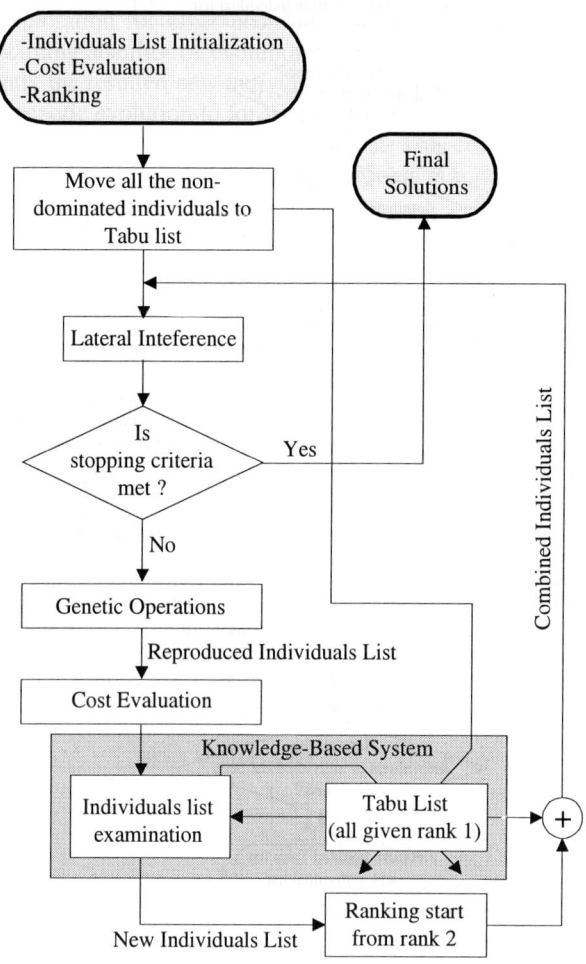

Fig. 2. Overall program flowchart of EMOEA

The simulation is implemented with the decimal coding scheme of 3-digit per parameter [9], standard two-point-crossover with a probability of 0.7, and standard mutation with a probability of 0.01. The lateral interference is performed in the objective space without the need of any parameter setting. Furthermore, a small value of $N_c = 30$ and $N_t = 10$ is employed to challenge the EMOEA. Simulation is run until it reaches the generation of 240, which is chosen with the purpose of visualizing how the EMOEA escapes itself from local optimum and finds new direction towards the global optimum, which was initially trapped at the local optimum. Fig. 5 illustrates the trend of parameters in the population versus the generation number, where each dot represents the parameter value of a particular individual in the population. Note that in Fig. 5a, as the dimension of parameter 1 is flat, the values of first parameter in the population tends to distribute themselves to uniformly cover the Pareto-front in the objective domain, as desired. For other parameters, i.e., the 2nd-4th parameters, however, there is a harmful local minimum at the value of 0.9 and a global minimum at 0.1 that is far away from the local minimum.

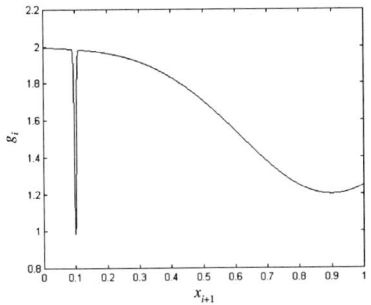 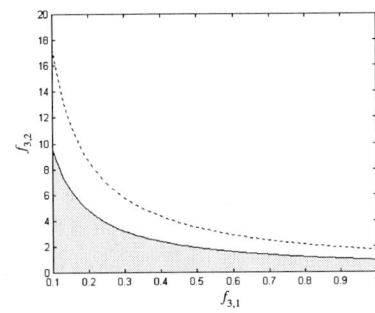

Fig. 3. global and local minimum   **Fig. 4.** Global and local Pareto-optimal curve

As can be seen in Figs. 5b-5d, the population is trapped at the local optimum of 0.9 initially, as indicated by the high population density at around 0.9 in y-axis at the initial stage of evolution. Although the EMOEA is initially trapped at the local optimum, the proposed heuristic reasoning of individual examination rule with tabu list keeps the EMOEA in exploring for other unvisited search space and subsequently discovers the global optimum of 0.1. This is illustrated in Figs. 5b-5d where the values of 2nd to 4th parameters start to converge to 0.1 as the evolution proceeds to approximately the generation of 20, 50 and 150 for the three parameters, respectively. The figures also show that EMOEA keeps exploring for other unvisited areas even after the global optimum has been found, which keeps the search diversity for other possible peaks as well as allows the diversion of evolution to other concentrated regions. This property is useful for real-world optimization problem where the final global optimum may be drifted or changed due to changes in the environment or working condition.

Fig. 6 illustrates the population distribution in the objective domain at different stage of the evolution, where the dashed curve represents the local Pareto-front while the solid line denotes the global Pareto-front. Individuals in the tabu list are represented by small circles, while those in the individual list are illustrated by solid dots. The EMOEA was trapped in local Pareto-optimum during the first stage as shown in Fig. 6a. Fig. 6b shows that besides searching for other possible optimum, the EMOEA distributes the non-dominated individuals uniformly along the currently found optimum line. Fig. 6c illustrates the situation where EMOEA has discovered the global optimum and is in the phase of migrating from the local to the global optimum. Finally, as depicted in Fig. 6d, the EMOEA is in the process of distributing the non-dominated individuals along the found global optimum, while at the same time, continues the effort of searching for other better optimum regions.

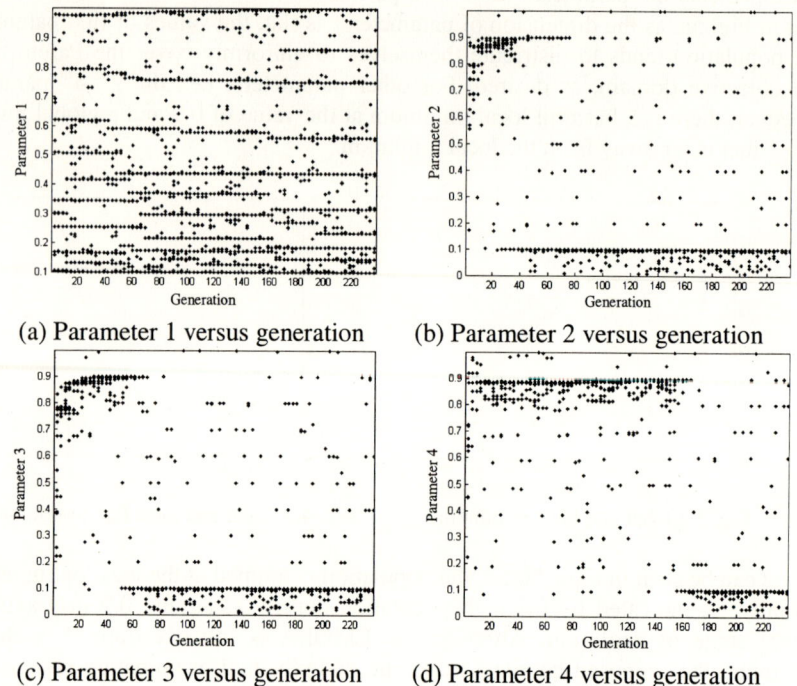

Fig. 5. Distribution of parameters along the evolution

Apart from this, EMOEA is also evaluated for noisy environment in the solution space to test their robustness in the sense that the disappearance of important individuals from the population has little effect on the global evolution behavior [16]. For this purpose, noisy version of two-objective optimization with three variables is constructed here where the function being optimized contains the elements of noise:

$$f_1 = x'_1, \qquad (10a)$$

$$f_2 = \frac{1}{x_1}\left\{1 + \left(x'^2_2 + x'^2_3\right)^{0.25}\left[\sin^2\left(50\left(x'^2_2 + x'^2_3\right)^{0.1}\right) + 1.0\right]\right\}, \qquad (10b)$$

Instead of performing the optimization on the 'real' parameters, x_i, the optimization is performed on the 'corrupted' parameters with additive noise elements:

$$x'_i = x_i + N(\sigma, \mu), \qquad (10c)$$

where $0.1 \leq x_1 \leq 1$; $-100 \leq x_i \leq 100$ $\forall i = 2,3$ and $N(\sigma,\mu)$ is a white noise. The population distribution density of the noise is given as normal distribution [17] as indicated by eqn. 11,

$$P(x\mid N(\sigma,\mu)) = \frac{1}{\sqrt{2\pi\sigma^2}}\exp\left(-\frac{(x-\mu)^2}{2\sigma^2}\right) \qquad (11)$$

where μ and σ^2 are the mean and variance of the probability distribution density. Fig. 7 illustrates the generated white noise along the generation with μ and σ^2 set as 0 and 0.1, respectively. With $N_c = 30$ and $N_t = 10$, the simulation is run for 1000 generations to acquire the steady-state performance, i.e., the optimization error in the evolution is less likely to be due to the immature individuals than the influence of noise in the observed environment at this stage.

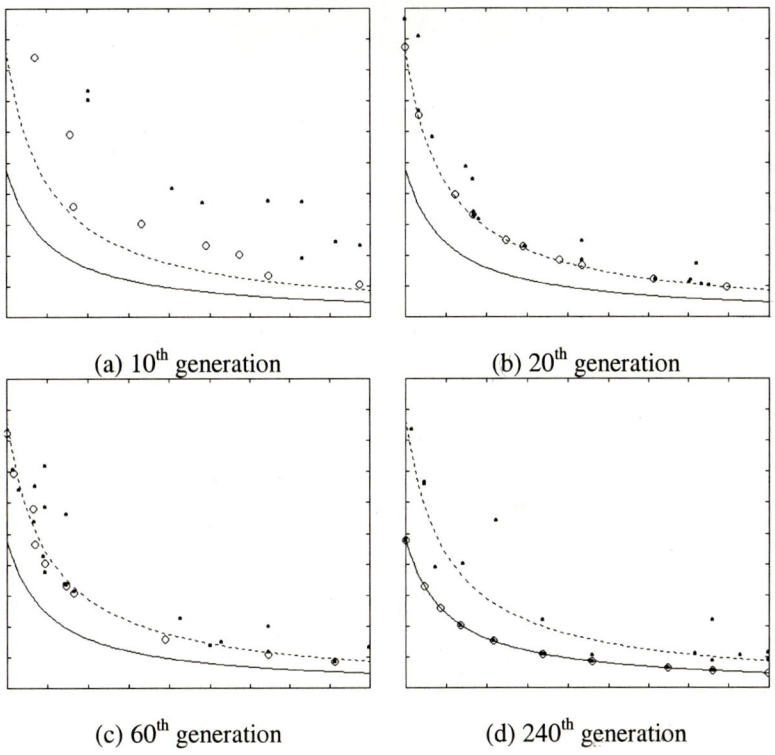

(a) 10th generation (b) 20th generation

(c) 60th generation (d) 240th generation

Fig. 6. Population distribution at different stages of evolution

Fig. 8 shows the trace of error denoting the difference between the best-found solution and the actual solution in the parameter domain for parameters 2 and 3. Parameter 1 is not considered in this observation since there is no noise element added in this parameter. As shown in Fig. 8, the optimization errors for both parameters are large initially since the best solutions at the initial stage of evolution are far away from the actual solutions. However, the error decreases significantly as the candidate solutions of EMOEA evolve towards the actual solution, which finally reaches to nearly zero at the steady-state. Note that the optimization error should not converge to zero due to the existing of white noise as shown in Fig. 7. Nevertheless, as illustrated in Fig. 8, the optimization errors for both parameters at the steady-state are much smaller than the magnitude of the noise since $|E_{op}(t\Rightarrow\infty)|_\infty$ for both parameters 2 and 3 are less than 0.05 while $|N(t\Rightarrow\infty)|_\infty$ is more than 1.5. As shown in Fig. 9, the proposed EMOEA has good robustness for global optimization in noisy environment, where it almost traces the entire actual Pareto-front in the objective domain at the end of the evolution.

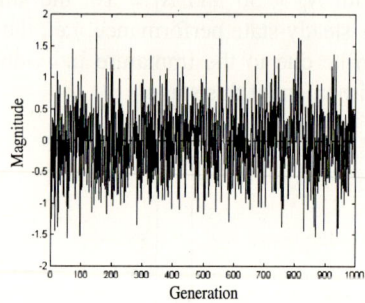

Fig. 7. Applied white noise for noise sensitivity test

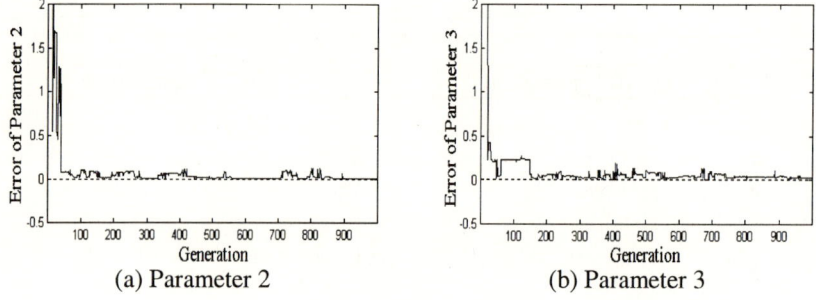

Fig. 8. E_{op} along the evolution for parameters 2 and 3

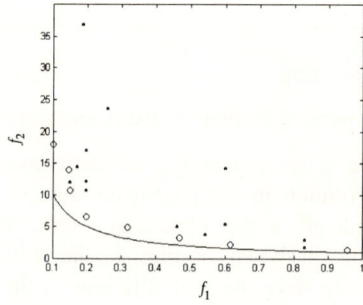

Fig. 9. Population distribution at the end of evolution under the existing of noise

4. Conclusions

This paper has proposed a high performance EMOEA, which makes use of the integrated features of tabu search and evolutionary algorithms to avoid the search from trapping in local optima as well as to promote the population diversity along the discovered Pareto-front for MO optimization. It has been shown that the proposed algorithm has look-ahead capabilities and is capable of reducing the frequency of backward drive using the concepts of tabu constraint. In addition, a novel method of lateral interference that helps to distribute non-dominated individuals uniformly along the discovered Pareto-front is proposed. Unlike existing sharing/niching methods, the lateral interference can be flexibly applied in either parameter or objective domain and allows efficient uniform population distribution without the need of any parameter setting. Validation on the proposed EMOEA have been performed and the experimental results unveiled that it has high capability to escape from local optima as well as to accurately identify the actual global optima in noisy environment.

References

1. Beasley, D., Bull, D.R., and Martin, R.R.: A Sequential Niche Technique for Multimodal Function Optimization. Evolutionary Computation, MIT Press Journals (1993), vol. 1(2), 101-125.
2. Fonseca, C.M. and Fleming, P.J.: Genetic Algorithm for Multiobjective Optimization, Formulation, Discussion and Generalization. In: Forrest, S. (ed.): Genetic Algorithms: Proceeding of the Fifth International Conference. Morgan Kaufmann, San Mateo, CA (1993), 416-423.
3. Schaffer, J.D.: Multiple-Objective Optimization using Genetic Algorithm. Proceedings of the First International Conference on Genetic Algorithms (1985), 93-100.
4. Coello Coello, C.A.: An Updated Survey of GA-based Multiobjective Optimization Techniques. Technical Report: Lania-RD-98-08, Laboratorio Nacional de Informatica Avanzada (LANIA), Xalapa, Veracruz, Mexico (1998).
5. Van Veldhuizen, D.A., and Lamont, G.B.: Evolutionary Computation and Convergence to a Pareto Front. In: Koza, J. R. (ed.): Late Breaking Paper at the Genetic Programming 1998 Conference, Stanford University Bookstore, Stanford University, California (1998), 221-228.
6. Zitzler, E., and Thiele, L.: An Evolutionary Algorithm for Multiobjective Optimization: The Strength Pareto Approach. Technical Report 43, Computer Engineering and Communication Network Lab (TIK), Swiss Federal Institute of Technology (ETH), Zurich, Switzerland (1998).
7. Richardson, J.T., Palmer, M.R., Liepins, G., and Hilliard, M.: Some Guidelines for Genetic Algorithms with Penalty Functions. In: Schaffer, J.D. (ed.): Proc. of Third Int. Conf. on Genetic Algorithms (1989), 191-197.
8. Srinivas, N. and Deb, K.: Multiobjective Optimization using Nondominated Sorting in Genetic Algorithms. Evolutionary Computation, MIT Press Journals (1994), vol. 2(3), 221-248.

9. Tan, K.C., Lee, T.H. and Khor, E.F.: Evolutionary Algorithms with Goal and Priority Information for Multi-Objective Optimization. IEEE Proceedings of the Congress on Evolutionary Computation, Washington, D.C, USA (1999), vol. 1, 106-113.
10. Deb, K., and Goldberg, D.E.: An Investigation of Niche and Species Formation in Genetic Function Optimization. In: Schaffer, J.D. (ed.): Proceedings of the Third International Conference on Genetic Algorithms (1989), 42-50.
11. Goldberg, D.E.: Genetic Algorithms in Search, Optimisation and Machine Learning. Addision-Wesley, Reading, Masachusetts (1989).
12. Encyclopaedia Britannica.: The Encyclopedia Britannica. http://www.britannica.com/bcom/eb/ article /2/0,5716,127612+29,00.html (2000).
13. Beyer, D.A., and Ogier, R.G.: Tabu Learning: a Neural Network Search Method for Solving Nonconvex Optimization Problems. IEEE International Joint Conference on Neural Networks (1991), vol. 2, 953-961.
14. Deb, K.: Multi-Objective Genetic Algorithms: Problem Difficulties and Construction of Test Problem. Journal of Evolutionary Computation, The MIT Press (1999), vol. 7(3), 205-230.
15. Whitley, D.: The Genitor Algorithm and Selection Pressure: Why Rank-Based Allocation of Reproductive Trials is Best. Proceedings of the Third International Conference on Genetic Algorithims (1989), 116-121.
16. Collard, P., and Escazut, C.: Genetic Operators in a Dual Genetic Algorithm. International Conference on Tools and Artificial Intelligence (1995), 12-19.
17. Grimm, L.G.: Statistical application for behavioral sciences. J. Wiley, New York (1993), 99.

Bi-Criterion Optimization with Multi Colony Ant Algorithms

Steffen Iredi, Daniel Merkle, and Martin Middendorf*

Institute AIFB,
University of Karlsruhe,
D-76128 Karlsruhe, Germany
iredi@informatik.uni-hannover.de
{merkle,middendorf}@aifb.uni-karlsruhe.de

Abstract. In this paper we propose a new approach to solve bi-criterion optimization problems with ant algorithms where several colonies of ants cooperate in finding good solutions. We introduce two methods for cooperation between the colonies and compare them with a multistart ant algorithm that corresponds to the case of no cooperation. Heterogeneous colonies are used in the algorithm, i.e. the ants differ in their preferences between the two criteria. Every colony uses two pheromone matrices — each suitable for one optimization criterion. As a test problem we use the Single Machine Total Tardiness problem with changeover costs.

1 Introduction

Ant Colony Optimization (ACO) is an evolutionary approach that has been applied successfully to solve various combinatorial optimization problems (for an overview see Dorigo and Di Caro [4]). In ACO ants that found a good solution mark their paths through the decision space by putting some amount of pheromone along the path. The following ants of the next generation are attracted by the pheromone so that they will search in the solution space near good solutions.

Much work has been done to apply evolutionary methods to solve multi-criterion optimization problems (see [13] for an overview). But only a few of this works used the ACO principle. Mariano and Morales [9] proposed an ant algorithm where for each objective there exists one colony of ants. In particular, they studied problems where every objective is influenced only by parts of a solution, i.e. an objective can be determined knowing only the relevant part of a solution. The objectives are assumed to be ordered by importance. In every generation ant k from colony i receives a (partial) solution from ant k of colony $i-1$. The ant then tries to improve (or extend) the (partial) solution with respect to criterion i. When the solutions have passed through all colonies those solutions that are in

---

* Corresponding author. Part of this work was done while the author stayed at the Institute of Computer Science at the University of Hannover.

the nondominated front are allowed to update the pheromone information. Gambardella et al. [7] developed an ant algorithm for a bi-criterion vehicle routing problem. They used two ant colonies — one for each criterion. The two colonies share a common global best solution which is used for pheromone update in both colonies. Criterion 1 — the number of vehicles — is considered to be more important than criterion 2 — the total travel time of the tours. Colony 1 tries to find a solution with one vehicle less than the global best solution while colony 2 tries to improve the global best solution with respect to criterion 2. Whenever colony 1 finds a new global best solution both colonies start anew (with the new global best solution). Gagné et al. [6] tested a multi-criterion approach for solving a single machine total tardiness problem with changeover costs and two additional criteria. In their approach the changeover costs were considered to be most important. The idea was to construct heuristic values for the decisions of the ants that take all criteria into account. The amount of pheromone that an ant adds to the pheromone matrix depends solely to changeover costs of the solution. All the above mentioned approaches assume that the different criteria can be ordered by importance and in the multi colony approaches there is always one colony for every objective.

In this paper we study ACO methods for multi-criterion optimization when the objectives can not be ordered by importance. The aim is to find different solutions which cover the Pareto-optimal front. A multi colony approach is proposed where the ant colonies are forced to search in different regions of the nondominated front. It should be noted, that multi colony ant algorithms have been studied before by some authors to parallelize ACO algorithms (a short overview is given in [12]). We use heterogeneous colonies where the ants in a colony weight the relative importance of the two optimization criteria differently so that they are able to find different solutions along the Pareto front. Cooperation between the colonies is done by exchanging solutions in the global nondominated front that are in regions which "belong to other colonies".

Our test problem, the Single Machine Total Tardiness Problem (SMTTP) with changeover costs is described in section 2. A short introduction to ant algorithms for solving the single objective versions of our test problem are given in Section 3. Our ACO approaches for bi-criteria optimization problems are described in Section 4. The multi colony approaches are explained in Section 5. The tests instances and parameters are described in Section 6. The Results are discussed in Section 7 and conclusions are given in Section 8.

2 The Test Problem

In this paper we use the Single Machine Total Tardiness Problem (SMTTP) with changeover costs as our bi-criterion test problem. The problem is defined as follows.

- Given: n jobs, where job j, $1 \leq j \leq n$ has a processing time p_j and a due date d_j and where for every pair of jobs i, j, $i \neq j$ there are changeover costs $c(i, j)$ that have to be paid when j is the direct successor of i in a schedule.

- Find: A non-preemptive one machine schedule that minimizes the value of $T = \sum_{j=1}^{n} \max\{0, C_j - d_j\}$ where C_j is the completion time of job j and that also minimizes the sum of the changeover costs $C = \sum_{i=1}^{n-1} c(j_i, j_{i+1})$ where $j_1, j_2, \ldots, j_n$ is the sequence of jobs in the schedule.

T is called the total tardiness of the schedule and C is the cost of the schedule. It is known that SMTTP is NP-hard in the weak sense [5] and is solvable in pseudo-polynomial time [8]. The problem to find a schedule that minimizes only the changeover costs is equivalent to the asymmetric Shortest Hamiltonian Path problem which is NP-complete in the strong sense.

3 Ant Algorithms for Single-Criteria Optimization Problems

3.1 Total Tardiness Minimization

A variant of the ACO algorithm of Merkle and Middendorf [10] for the SMTTP is described in this section (other ACO approaches for SMTTP can be found in [1,3]). In every generation each of m ants constructs one solution. An ant selects the jobs in the order in which they will appear in the schedule. For the selection of a job the ant uses heuristic information as well as pheromone information. The heuristic information, denoted by η_{ij}, and the pheromone information, denoted by τ_{ij}, are an indicator of how good it seems to have job j at place i of the schedule. The heuristic value is generated by some problem dependent heuristic whereas the pheromone information stems from former ants that have found good solutions.

The next job is chosen from the set $\mathcal{S}$ of jobs that have not been scheduled so far according to the probability distribution that is determined by

$$p_{ij} = \frac{[\tau_{ij}]^\alpha [\eta_{ij}]^\beta}{\sum_{h \in \mathcal{S}} [\tau_{ih}]^\alpha [\eta_{ih}]^\beta} \tag{1}$$

The heuristic values η_{ij} are computed according the following modified due date rule

$$\eta_{ij} = \frac{1}{\max\{\mathcal{T} + p_j, d_j\} - \mathcal{T}} \tag{2}$$

where $\mathcal{T}$ is the total processing time of all jobs already scheduled. The best solution found so far is then used to update the pheromone matrix. But before the update is done some of the old pheromone is evaporated according to

$$\tau_{ij} = (1 - \rho) \cdot \tau_{ij}$$

The reason for this is that old pheromone should not have a too strong influence on the future. Then, for every job j in the schedule of the best solution found so far some amount of pheromone Δ is added to element τ_{ij} of the pheromone

$\sum_{h\in S}\sum_{k=1}^{i}\gamma^{i-k}\tau_{kh}$ and $y_i := \sum_{h\in S}\tau_{ih}$ are factors to adjust the relative influence of local and summation evaluation. Observe, that for $c = 1$ the standard evaluation is obtained and for $c = 0$ pure summation evaluation.

Therefore it might be advantageous to use different methods for the evaluation of the two pheromone matrices. If e.g. summation evaluation is used for the first criterion the probabilities used by the ant are

$$p_{ij} = \frac{(\tau_{ij}^*)^{\lambda\alpha} \cdot \tau_{ij}^{\prime(1-\lambda)\alpha} \cdot \eta_{ij}^{\lambda\beta} \cdot \eta_{ij}^{\prime(1-\lambda)\beta}}{\sum_{h\in S}(\tau_{ih}^*)^{\lambda\alpha} \cdot \tau_{ih}^{\prime(1-\lambda)\alpha} \cdot \eta_{ih}^{\lambda\beta} \cdot \eta_{ih}^{\prime(1-\lambda)\beta}} \tag{5}$$

4.3 Pheromone Update

When all m ants of a generation have found a solution it has to be decided which of the ants are allowed to update. Here we propose that all ants in the nondominated front of the actual generation are allowed to update. An ant that updates will update both pheromone matrices M and M'. Note, that this rule makes sense only when there are not too few ants in a colony. With only very few ants in a colony the ants differ much by their λ-values. Hence, no real competition about best solutions occurs since the ants will search in different regions of the nondominated front. Thus even the ants with weak solutions will have good changes to do an update. One way to solve this problem is to allow only those ants to update that have found solutions which are good compared to the nondominated front of all solutions that have been found so far.

To give every generation of ants the same influence the amount of pheromone that is added to a pheromone matrix is the same in every generation. Therefore, every ant is allowed to update an amount of $\tau_{ij} = \tau_{ij} + 1/l$ where l is the number of ants that are allowed to update in the actual generation.

5 The Multi Colony Approach

In our multi colony ant algorithm several colonies of ants cooperate and specialize to find good solutions in different regions of the Pareto front. All p colonies have the same number of m/p ants.

5.1 Pheromone Update

In the single colony algorithm only ants in the nondominated front of the colony are allowed to update. Hence, a reasonable way for pheromone update in the multi colony algorithm is that only those ants update that found a solution which is in the local nondominated front of the colony. This corresponds to the case were there is no cooperation between the colonies. Therefore the results are the same as with a multistart approach where a single colony ant algorithm is run several times and the global nondominated front at the end is determined from the nondominated fronts of all runs. In the following we describe how to introduce collaboration between the colonies. For this, the ants in a generation put their solutions in a global solution pool that is shared by all colonies. The pool is used to determine the nondominated front of all solutions in that generation. Then, only ants that found a solution which is in the global nondominated front are allowed to update. We study two different methods to determine in which colony an ant should update the pheromone matrix:

1. Method 1 – update by origin: an ant updates only in its own colony (compare Figure 1).
2. Method 2 – update by region in the nondominated front: the sequence of solutions along the nondominated front is split into p parts of equal size. Ants that have found solutions in the ith part update in colony i, $i \in [1, p]$ (compare Figure 2). More formally, the solutions in the nondominated front are sorted with respect to the first criterion (it does not matter whether the list is sorted according to the first or the second criterion). Let L be the sorted list. The sorted list is then split into parts $L_1, L_2, \ldots, L_p$ so that their size differs by at most one. All ants that found solutions in list L_i, $i \in [1, p]$ will update the pheromone matrix of colony i.

The first method imposes a stronger selection pressure on the ants that are allowed to update. It is not enough for an ant to have a solution in the local nondominated front of its colony. Instead, the solution must be in the global nondominated front. This method might be advantageous because other colonies help to detect which of the solutions in the local nondominated front of a colony might be weak. An interesting observation is that the update by origin method might also enforce the colonies to search in different regions of the nondominated front. It is more likely that a solution from the local nondominated front of a colony might also be in the global nondominated front when only a few solutions from other colonies are in the same region. Hence, it is more likely that an ant with solutions in less dense areas of the nondominated front will be allowed to update and thereby will influence the further search process.

The aim of method 2 is to explicitly guide the ant colonies to search in different regions of the Pareto front.

5.2 Heterogeneous Colonies

As in the single colony algorithm the ants in a colony use different λ-values, i.e. when making their decisions they weight the relative importance of the two optimization criteria differently. More exactly, ant k in colony i, $i \in [1, p]$ uses $\lambda_k = \frac{k-1}{m/p-1}$, $k \in [1, m/p]$. We call this rule 1 for defining the λ-values.

An alternative could be to use different λ-values in the colonies so that λ-values of the ants in the colonies are in different subintervals of $[0, 1]$. Thus the colonies weight the optimization criteria differently.

- Rule 2 — disjoint λ-intervals: ant k, $k \in [1, m/p]$ in colony i has λ-value $(i-1) \cdot \frac{m}{p} + k$.

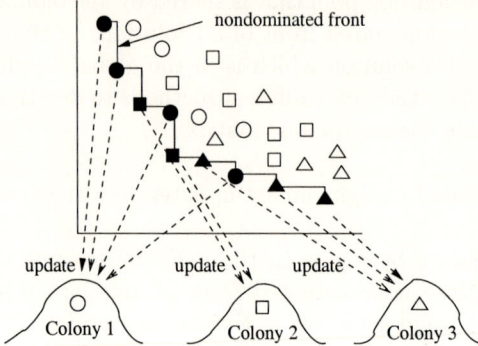

Fig. 1. Update by origin: Every ant with a solution in the nondominated front updates in its own colony.

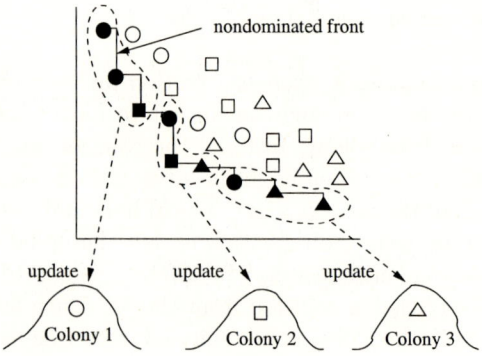

Fig. 2. Update by region in the nondominated front: Ants with a solution in the nondominated front update in the colony that are corresponds to the region of the solution.

- Rule 3 — overlapping λ-intervalls: the λ-intervall of colony i overlaps by 50% with the λ-intervall of colony $i-1$ and colony $i+1$. Formally, colony i has ants with λ-values in $[(i-1)/(p+1), (i+1)/(p+1)]$ (compare Figure 3).

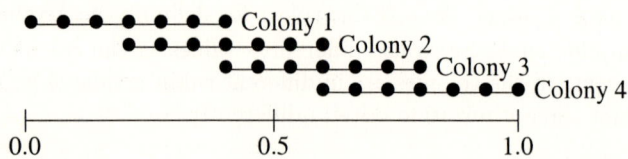

Fig. 3. λ-values when using rule 3: 4 colonies with 7 ants each.

6 Test Instances and Parameters

We tested our ant algorithms on problem instances where the jobs and their deadlines were generated after the following rule that is often used to create instances for the SMTTP [2]: for each job $j \in [1, 100]$ an integer processing time p_j is taken randomly from the interval $[1, 100]$ and an integer due date d_j is taken randomly from the interval

$$\left[\sum_{j=1}^{100} p_j \cdot (1 - TF - \frac{RDD}{2}), \sum_{j=1}^{100} p_j \cdot (1 - TF + \frac{RDD}{2}) \right]$$

The value RDD (relative range of due dates) determines the length of the interval from which the due dates were taken. TF (tardiness factor) determines the relative position of the centre of this interval between 0 and $\sum_{j=1}^{100} p_j$. We chose the values for TF from the set $\{0.4, 0.5, 0.6\}$ and RDD was set to $RDD = 0.6$. The changeover costs between the jobs were chosen randomly from one of the sets $[1, 100]$ and $[50, 100]$.

The parameters used for the test runs are: $\alpha = 1$, $\beta = 1$, $\rho = 0.02$. Pheromone evaluation was done according to formula 5 where a combination between summation evaluation and standard evaluation for matrix 1 is used. For the corresponding parameters the values $c = 0.6$ and $\gamma = 0.9$ were used (these values were shown to be suitable for the weighted SMTTP [11]). The number of ants in every generation was $m = 100$. When using several colonies the 100 ants were distributed equally to the colonies. Every element of the pheromone matrices was initialized with 1.0. Every test was performed with 11 runs. Every run was stopped after 300 generations.

7 Results

The performance of the multi colony approach was tested on 6 problem instances: three instances with changeover costs in $[1, 100]$ and three instances with changeover costs in $[50, 100]$ (and $TF \in \{0.4, 0.5, 0.6\}$). The outcome of each single run of the ant algorithm is the subset of all nondominated solutions in the set of all solutions found during the run of the algorithm. The median attainment surfaces for runs with 1 colony and 10 colonies are shown in Figure 4 (the median attainment surface is the median line of all the attainment surfaces connecting the pareto front in every of the 11 runs). For the tests with 10 colonies we used pheromone update method 2 (update by region in the pareto front) and λ-rule 2 with overlapping λ-intervalls. The figure shows that the median attainment surfaces of the runs with 10 colonies are nearly always better than those for the 1 colony runs. Only for two instances with changeover costs in $[50, 100]$ the median attainment surfaces of the 1 colony runs are slightly better in a small region. But for these instances the 10 colonies found solutions with much smaller changeover costs.

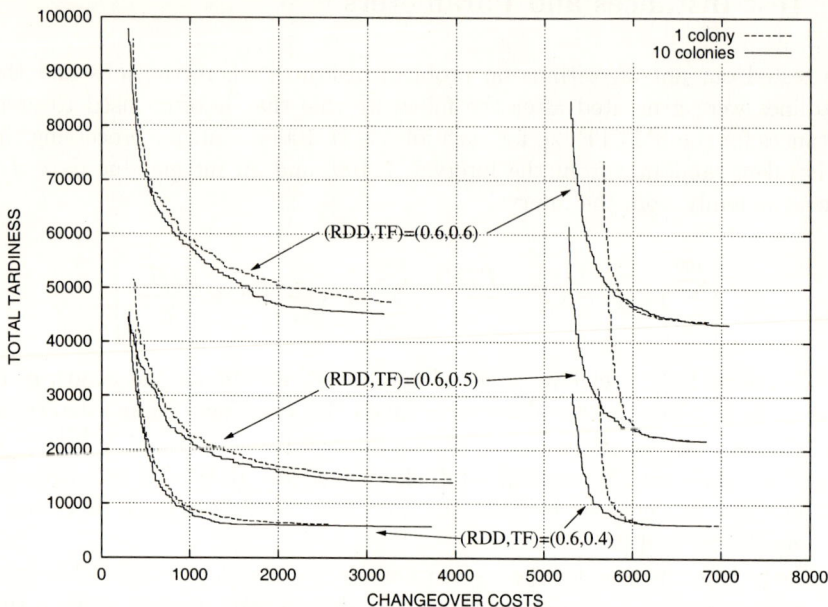

Fig. 4. Median attainment surfaces obtained with 1 colony and 10 colonies. Left part: three instances with changeover costs in [1,100]. Right part: three instances with changeover costs in [50,100].

In the following we present some results where we compare the different pheromone update methods and the λ-rules. For these tests we used the problem instance with $TF = 0.6$ and changeover costs in $[1, 100]$.

Figure 5 shows the convergence behaviour of the 10 colonies algorithm. The median attainment surfaces obtained after different numbers of generations are depicted. It can be seen that the median attainment surfaces after 200 generations and 300 generations differ not much. The results obtained for 1 colony were similar. Hence, our other results that were all obtained after 300 generations should not change with a higher number of generations.

The median attainment surfaces obtained with the different pheromone update methods are shown in Figure 6. The simple multistart strategy without cooperation between the colonies is worst along the whole surfaces. For the smaller costs values method 1 (update by origin) and method 2 (update by region in the nondominated front) show nearly the same performance. But median attainment surface when using method 2 is the best for medium and small total tardiness values.

Figure 7 shows from which colony the ants stem that found solutions which are in the final nondominated front (i.e. after 300 generations). It can be seen that the pheromone update method 2 (by region in the nondominated front) forces all colonies to specialize to those regions from where the ants come that

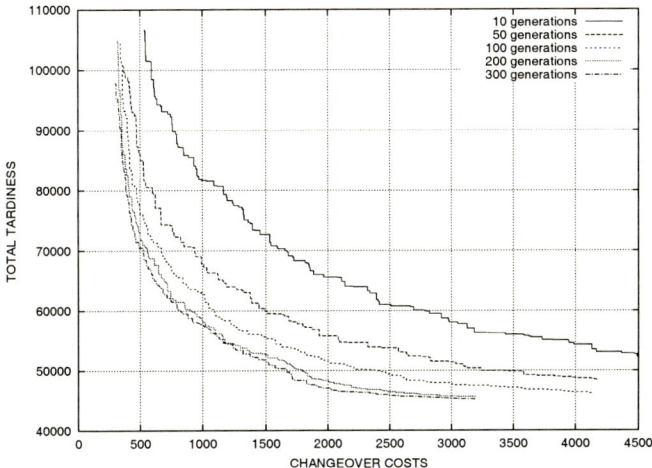

Fig. 5. Convergence behaviour with 10 colonies: median attainment surfaces are shown for generations 10, 50, 100, 200 and 300.

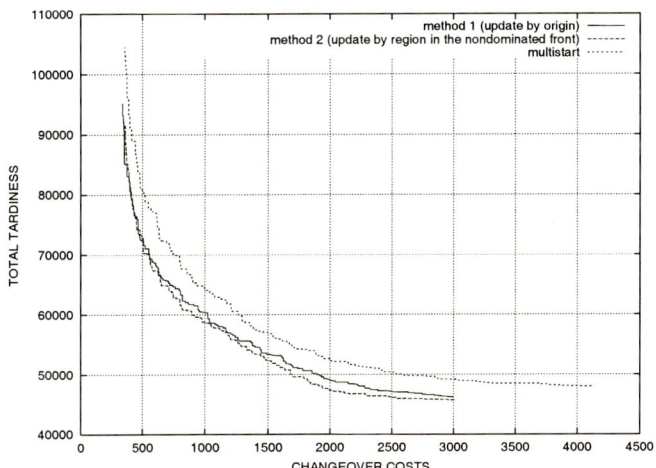

Fig. 6. Influence of the pheromone update method when using 10 colonies: method 1 — update by origin; method 2 — update by cutting the nondominated front; the multistart approach.

are allowed to update in that colony. Also, method 1 (update by origin) seems to force the colonies to specialize, though this effect is not so clear as for method 2. Clearly, for method 1 it can not be predicted to which region a colony will specialize.

The influence of the λ-rules when using pheromone update method 1 (by origin) is shown in Figure 8 for 2 and 5 colonies. When using rule 2 with disjoint λ-intervalls the case of 2 colonies clearly shows that colonies are forced to

An interesting topic for future research will be to study how the pheromone update by region in the nondominated front can be applied to optimization problems with more than two criteria. It is not obvious what regions should be used then in the nondominated front.

References

[1] A. Bauer, B. Bullnheimer, R. Hartl, and C. Strauss. An ant colony optimization approach for the single machine total tardiness problem. In *Proceedings of the 1999 Congress on Evolutionary Computation (CEC99), 6-9 July Washington D.C., USA*, pages 1445–1450, 1999.

[2] H. Crauwels, C. Potts, and L. V. Wassenhove. Local search heuristics for the single machine total weighted tardiness scheduling problem. *Informs Journal on Computing*, 10:341–350, 1998.

[3] M. den Besten, T. Stützle, and M. Dorigo. Scheduling single machines by ants. Technical Report IRIDIA/99-16, IRIDIA, Université Libre de Bruxelles, Belgium, 1999.

[4] M. Dorigo and G. Di Caro. The ant colony optimization meta-heuristic. In D. Corne, M. Dorigo, and F. Glover, editors, *New Ideas in Optimization*, pages 11–32, London, 1999. McGraw-Hill.

[5] J. Du and J.-T. Leung. Minimizing total tardiness on one machine is NP-hard. *MOR: Mathematics of Operations Research*, 15:483–496, 1990.

[6] C. Gagné, M. Gravel, and W. Price. Scheduling a single machine where setup times are sequence dependent using an ant-colony heuristic. In *Abstract Proceedings of ANTS'2000, 7.-9. September Brussels, Belgium*, pages 157–160, 2000.

[7] L. M. Gambardella, É. Taillard, and G. Agazzi. MACS-VRPTW: A multiple ant colony system for vehicle routing problems with time windows. In D. Corne, M. Dorigo, and F. Glover, editors, *New Ideas in Optimization*, pages 63–76. McGraw-Hill, London, 1999.

[8] E. Lawler. A 'pseudopolynomial' algorithm for sequencing jobs to minimize total tardiness. *Annals of Discrete Mathematics*, pages 331–342, 1977.

[9] C. E. Mariano and E. Morales. MOAQ an ant-Q algorithm for multiple objective optimization problems. In W. Banzhaf, J. Daida, A. E. Eiben, M. H. Garzon, V. Honavar, M. Jakiela, and R. E. Smith, editors, *Proceedings of the Genetic and Evolutionary Computation Conference*, volume 1, pages 894–901, Orlando, Florida, USA, 13-17 July 1999. Morgan Kaufmann.

[10] D. Merkle and M. Middendorf. An ant algorithm with a new pheromone evaluation rule for total tardiness problems. In *Proceeding of the EvoWorkshops 2000*, number 1803 in Lecture Notes in Computer Science, pages 287–296. Springer Verlag, 2000.

[11] D. Merkle, M. Middendorf, and H. Schmeck. Pheromone evaluation in ant colony optimization. IEEE Press, 2000. to appear in : *Proceeding of the Third Asia-Pacific Conference on Simulated Evolution and Learning (SEAL2000)*, Nagoya, Japan, 25-27 Oct. 2000.

[12] M. Middendorf, F. Reischle, and H. Schmeck. Information exchange in multi colony ant algorithms. In *SPDP: IEEE Symposium on Parallel and Distributed Processing*. ACM Special Interest Group on Computer Architecture (SIGARCH), and IEEE Computer Society, 2000.

[13] D. A. V. Veldhuizen and G. B. Lamont. Multiobjective Evolutionary Algorithms: Analyzing the State-of-the-Art. *Evolutionary Computation*, 8(2):125–147, 2000.

Multicriteria Evolutionary Algorithm with Tabu Search for Task Assignment

Jerzy Balicki[1] and Zygmunt Kitowski[1]

[1] Computer Science Department, The Polish Naval Academy
ul. Smidowicza 69, 81-103 Gdynia, Poland
jbalicki@gdynia.amw.pl, zkit@amw.gdynia.pl

Abstract. In this paper, three evolutionary algorithms are discussed for solving three-criteria optimisation problem of finding a set of Pareto-optimal task assignments. Finally, the algorithm with a tabu mutation is recommended for solving an established multiobjective optimisation dilemma. Some numerical results are submitted.

1 Introduction

An adaptive evolutionary algorithm and an adaptive evolution strategy are important techniques for solving multiobjective optimisation problems related with task assignment that minimize a workload of a bottleneck computer and the cost of machines [1]. From the other hand, a tabu search is the powerful meta-heuristic approach, which has been applied for crucial applications in engineering, economics and science [4]. We propose a new version of a multicriteria evolutionary algorithm with a tabu search as an advanced mutation operation.

Finding allocations of program modules is a major design problem for distributed computer systems [5]. Program component allotments may diminish the total time of a program execution by taking a benefit of the particular properties of some workstations or an advantage of the computer load. Three criteria are utilized for a quality evaluation of the module allocation: a processing load of the bottleneck machine, the cost of computers, and the total numerical performance of workstations.

The efficient network flow algorithm has been applied for the minimization of the program execution cost in a two-computer system [23]. If the number of computers is greater than 3 or the memory in a computer is constrained, then a problem of the program completion cost minimization by task dispersion is NP-hard [14]. If a tree or the parallel-sequence graph represents the structure of the intermodule communication, then efficient algorithms based on the shortest path procedure can be developed for finding an optimal solution [5]. Task assignment problems are related with scheduling questions [25], however there are several important distinguishes between them. In task assignment questions we use the other evaluation criteria, and unlike constraints are respected. Although a task assignment model is less complex, the most common used problems are NP-hard.

machine for the other task allocation in a distributed system. The workload of the bottleneck computer can be employed as an assessment quantify of an allotment quality in systems, where the minimization of a response time is required, too [6].

Let (X, F, P) be the multi-criterion optimisation question for finding the representation of Pareto-optimal solutions. It can be established, as follows:

---

1) X - an admissible solution set

$$X = \{x \in B^{I(V+J)} \mid \sum_{i=1}^{I} x_{vi}^m = 1, v = \overline{1,V}, \sum_{j=1}^{J} x_{ij}^\pi = 1, i = \overline{1,I}\}$$

where B={0, 1}

2) F - a vector superiority criterion

$$F : X \rightarrow R^3, \qquad (8)$$

where
R – the set of real numbers,

$F(x) = [Z_{max}(x), F_2(x), \tilde{F}_2(x)]^T$ for $x \in X$,

$Z_{max}(x)$ is calculated by (5),

$F_2(x)$ is calculated by (6),

$\tilde{F}_2(x)$ is calculated by (7)

3) P - the Pareto relationship [2, 7].

The relationship P is a subset of the product $Y \times Y$, where an evaluation set $Y=F(X)$. If elements $a \in Y$, $b \in Y$, and there are $a_n \leq b_n, n = \overline{1,N}$, then the pair of evaluations $(a,b) \in P$. The meaning of the Pareto relationship respects the minimization of all criteria. There is no task allocation $a \in X$ such that $(F(a), F(x^*)) \in P$ for the Pareto-optimal assignment $x^* \in X$,

5 Adaptive Genetic Algorithm

Genetic algorithms GA have been applied for solving distinguishes optimisation problems [17]. The vector evaluated genetic algorithm VEGA has been considered for solving multi-criterion optimisation questions [19].

The selection with hierarchical tournaments have been considered [10], where two randomly chosen solutions have been compared. The hierarchical alternative is chosen and it is included to a mating pool of likely parents. A selection probability is calculated for the most considerable aim. A casual choice is carried out twice according to the roulette rule.

A ranking system for non-dominated individuals has been introduced to avoid the prejudice of the interior Pareto alternatives [12]. It has been extended by Srinivas and Deb [22]. If some admissible solutions are in a population, then the Pareto-optimal individuals are determined, and after that they get the rank 1. Afterwards, they are temporary removed from the population. After that, the recent Pareto-optimal alternatives are found from the reduced population and they get the rank 2. The level is increased and the procedure is repeated until the set of admissible solutions is exhausted. All non-dominated individuals have the same reproduction fitness because of the comparable rank. The Goldberg's ranking is based on dominance layers.

The fitness for a non-feasible solution is equal to the difference between the maximal penalty P_{max} in a population and the solution penalty. If x is admissible, then the fitness function value is estimated, as below:

$$f(x) = P_{max} - r(x) + L + 1, \tag{9}$$

where $r(x)$ denotes the rank of an admissible solution.

Another ranking procedure has been introduced by Fonseca and Fleming [8]. It assigns each individual a rank based on the number of other individuals by which it is dominated. A niching procedure modifies these ranks. The surface region of the Pareto front is divided by the size of the population. The number of other member's falling within the sub-area of any individual is taken to establish the niching penalty for it [9].

Sheble and Britting have observed the quality of attained solutions increases in optimisation problems with one criterion, if the crossover probability and the mutation rate is changed in an adaptive way [21]. Let this approach be launched to a multicriteria genetic algorithm with ranking procedure (Fig. 1).

A proposed adaptive multiobjective genetic algorithm AMGA may be applied for solving a spacious class of multi-criterion optimisation problems. Binary vectors represent solutions in genetic algorithms and it is a crucial constraint.

Let it be discussed the adaptive changing of a crossover probability p_c and a mutation rate p_m. At the initial population the crossover probability is 1 and each pair of potential parents is obligatory taken for crossover operation performing. A crossover operation supports the finding of a high-quality solution area in the search space. It is important in the early search stage. If the number of generation t increases, then the crossover probability decreases, according to the formula $p_c = e^{-t/T_{max}}$, where T_{max} is a maximal number of generations.

A mutation rate is 0 at the initial generation. It is an operation that can support finding of local optimal solution in the areas determined by individuals in the population. A mutation combines with crossover procedure to focus the searching towards global optimum. The value of p_m increases with respect to the formula $p_m = e^{0.05t/T_{max}} - 1$, exponentially. In the final population 5.13% bits are chosen to a bit mutation.

In this paper, we understand that using functions for mutation and crossover makes evolutionary algorithms *adaptive*, however, according to another meaning it is related with the operators change as a result of the search process (i.e., population diversity, etc), not as a function of generation.

6 Level of Convergence to Pareto Front

The AMGA with a few modifications is able to find task assignment representation for several multiobjective optimisation problems. Simulation results corroborate that the AMGA is capable for finding the set of Pareto-suboptimal solutions for the question (8).

BEGIN
$t:=0$, set the even size of population L
generate randomly initial population $P(t)$
calculate ranks r(x) and fitness $f(x)$, $x \in P(t)$
finish:=FALSE
WHILE NOT *finish* DO
BEGIN /* new population */
  $t:= t+1$, $P(t) := \emptyset$

  calculate selection probabilities $p_S(x), x \in P(t-1)$
  FOR $L/2$ DO
  BEGIN /* reproduction cycle */
  - $p_c := e^{-t/T_{\max}}$; $p_m := e^{0.05t/T_{\max}} - 1$;
  - proportional selection of a potential parent pair (**a,b**) from the population $P(t-1)$
  - simple crossover of a parent pair (**a,b**) with the adaptive crossover probability p_c
  - bit mutation of an offspring pair (**a',b'**) with the adaptive mutation probability p_m
  - $P(t):=P(t) \cup ($**a',b'**$)$}
  END
  calculate ranks r(x) and fitness $f(x), x \in P(t)$
IF ($P(t)$ converges OR $t \geq T_{max}$) THEN *finish*:=TRUE
END
END

Fig. 1. An adaptive multicriteria genetic algorithm AMGA

Let the Pareto points $\{P_1, P_2, ..., P_U\}$ be given for any instance of the task assignment problem (8). The level of convergence to the Pareto front can measure the quality of obtained set of solutions. If the AMGA finds the efficient point (A_{u1}, P_{u2}, A_{u3}) for the cost of computers P_{u2}, then there is the uth Pareto result (P_{u1}, P_{u2}, P_{u3}) with the same cost of computers.

The distance between points (A_{u1}, P_{u2}, A_{u3}) and (P_{u1}, P_{u2}, P_{u3}) is calculated according to $\sqrt{(P_{u1} - A_{u1})^2 + (P_{u3} - A_{u3})^2}$. If the point (A_{u1}, P_{u2}, A_{u3}) is not discovered by the algorithm, then we assume the distance is $\sqrt{(P_{u1} - A_{u1}^{\max})^2 + (P_{u3} - A_{u3}^{\max})^2}$, where $A_{u1}^{\max}$ is the maximum load of the bottleneck computer for the instance of problem (8), and $A_{u3}^{\max}$ is the minimum performance of computers for the instance of question (8).

The level of convergence to the Pareto front is calculated, as follows:

$$S = \sum_{u=1}^{U} \sqrt{(P_{u1} - A_{u1})^2 + (P_{u3} - A_{u3})^2}. \tag{10}$$

An average level $\overline{S}$ is calculated for several runs of the evolutionary algorithm.

7 Tabu Search as Mutation Operation

A survey of the state-of-the-art for multi-criterion evolutionary algorithms is submitted in [7, 8, 24, 26]. Let the adaptive multicriteria evolutionary algorithm AMEA be a base for creating the AMEA with tabu mutation. A flow scheme of the AMEA is alike to the AMGA, but the preliminary population is erected in a specific manner. Individuals are constructed to satisfy constraints (1) and (2) by introducing integer representation of chromosomes, as follows:

$$X = (X_1^m, ..., X_v^m, ..., X_V^m, X_1^\pi, ..., X_i^\pi, ..., X_J^\pi), \tag{11}$$

where $X_v^m = i$ for $x_{vi}^m = 1$ and $X_i^\pi = j$ for $x_{ij}^\pi = 1$. Besides, $1 \le X_v^m \le I$ and $1 \le X_i^\pi \le J$.

The crossover point is randomly chosen between neighbour genes in the chromosome X. A bit mutation is carried out through the random swap of the integer value by another one from a feasible discrete set. If the gene X_v^m is randomly taken for mutation, then the positive integer value is taken from the set $\{1,..., I\}$. If the gene X_i^π is randomly chosen, then the value is selected from the set $\{1,..., J\}$.

The adaptive multicriteria evolutionary algorithm with a tabu mutation is a hybrid optimisation technique that combines advantages of a genetic search with a tabu search. In *a tabu search* special areas are forbidden during the seeking in a space of all possible combinations.

Tabu search algorithms have been applied for solving several optimisation problems in scheduling, computer-aided design, quadratic assignment, training and designing of neural networks [11]. Moreover, the best results have been obtained by tabu search algorithms in telecommunication call routing, volume discount acquisition in production, and vehicle routing. Its good capabilities have been confirmed during solving standard optimisation problems such as graph partitioning, graph colouring, clique partitioning. Tabu search can be treated as a general combinatorial optimisation technique for using in zero-one programming, nonconvex non-linear programming, and general mixed integer optimisation.

Tabu search uses memory structures by reference to four principal dimensions, consisting of recency, frequency, quality and influence [11]. Tabu search algorithm inherits from a simple descent method an idea of a neighbourhood $N(x^{now})$ of a current solution x^{now}. From this neighbourhood we can choose the next solution x^{next} to a search trajectory. The accepted alternative should have the best value of an objective function among the current neighbourhood. But, the descent method terminates its searching, when the chosen candidate is worse than the best one from the searching trajectory.

In the tabu search algorithm based on the short-term memory, a basic neighbourhood $N(x^{now})$ of a current solution may be reduced to a considered neighbourhood $M(x^{now})$ because of the maintaining a selective history of the states encountered during the exploration. Some solutions, which were visited during the given last term, are excluded from the basic neighbourhood according to the tabu classification of movements. If any solutions performs aspiration criterion, then it can be included to the considered neighbourhood, only.

A recency-based memory keeps track of solutions attributes that have changed during the recent past. Selected attributes that occur in solutions lately visited are branded tabu-active. Solutions with tabu-active attributes or with some combinations of these attributes became tabu, too. This prevents some solutions included to the recent part of a trajectory from belonging to a considered neighbourhood and hence from being revisited. Furthermore, other solutions with tabu-active attributes are similarly prevented being visited. While the tabu classification strictly refers to solutions that are forbidden to be visited, we also often refer to moves that lead to such solutions as being tabu [11].

In the tabu search algorithm based on the long-term memory, a considered neighbourhood may also be expanded to incorporate solutions not regular found in an essential neighbourhood. During stretched exhausting searching, there is an opportunity to count frequency measures of selected attributes. Often performed movements should be inviolable to take an ability rarely performed actions after long examination. Frequency measures of selected attributes are respected in the selecting function of a next solution from a current neighbourhood.

Hansen has proposed a multiobjective optimisation tabu search MOTS [13] to generate non-dominated alternatives. The MOTS works with a population of solutions, which, through manipulation of weights, are moved towards the Pareto front [13]. But, the MOTS do not cooperate with an evolutionary algorithm.

We propose the following approach. An integer mutation is substituted through the tabu search for randomly chosen chromosome (t_{vj}, X_i^{π}). A fitness of this chromosome can be calculated according to the formula (9). The tabu search technique increases an initial value of fitness $f(X_v^m, X_i^{\pi})$ to the final value $f^{next}(X_v^m, X_i^{\pi})$. It usually calculates better outcome in the fitness sense than initial level. A tabu search algorithm results an additional complexity $O(n^3)$.

Better outcomes from the tabu mutation are transformed into improving of solution quality obtained by the adaptive multicriteria evolutionary algorithm with tabu mutation AMEA+. This adaptive evolutionary algorithm gives better results than the AMEA and much better than the AMGA (Fig. 2). After 200 generations, an average level of Pareto set obtaining is 0.8% for the AMEA+, 1.3% for the AMEA, and 43% for the AMGA. 30 test preliminary populations were prepared, and each algorithm starts 30 times from these populations.

For integer constrained coding of chromosomes there are 12 decision variables in the test optimisation problem. The search space consists of 25 600 solutions, only.

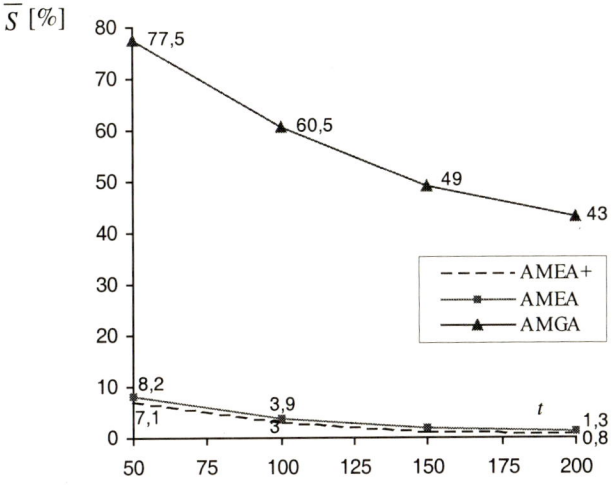

Fig. 2. Outcome convergence for the AMEA+ and the AMEA

8 Evolution Strategy

An evolution strategy [20] for a multi-criterion optimisation was introduced by Kursawe [16]. This class of optimisation techniques has been discussed and extended in [3] and [15]. We propose a chromosome in the multicriteria evolution strategy MES for problem (8) that consists of two main parts:

$$\overline{X} = (X, \sigma), \qquad (12)$$

where
X – the integer decision variable vector given by (11),
σ – the standard deviation vector for X.

The novel population is created from the μ individuals in the existing generation by 3 steps. In the step 1, λ individuals are randomly chosen from the current population to the temporary parent set. The selection rate is constant for each individual from the current population. Number of parents λ is even. So, the proposed algorithm is a version ($\mu+\lambda$) of an evolution strategy [16].

In the next step, the crossover operator is carried out by the gene recombination between randomly selected two task assignments from the temporary parent set. Each mth gene is taken as the mth gene from the parent A or as the mth gene from the parent B.

The temporary offspring set is transformed by the strategic mutation operation, in the last step. It changes a value of each decision variable X_m of every offspring by adding the random value Δx_m that represents a random variable with a normal distribution $N(0, \sigma_m)$. Δx_m is rounded to the integer number. After that, the standard devia-

15. Knowles, J., Corne, D. W.: Approximating The Nondominated Front Using The Pareto Archived Evolution Strategy. Evolutionary Computation, Vol. 8, No. 2 (2000) 149-172
16. Kursawe, E.: A Variant Of Evolution Strategies For Vector Optimisation. in H.-P. Schwefel, R. Manner (eds.), Parallel Problem Solving From Nature, 1st Workshop, Lecture Notes in Computer Science, Springer Verlag, Vol. 496 (1991) 193-197
17. Michalewicz, Z.: Genetic Algorithms + Data Structures = Evolution Programs. Springer Verlag, Berlin Heidelberg New York (1996)
18. Murthy, I., Seo, P. K.: A Dual-Descent Procedure For The File Allocation And Join Site Selection Problem On A Telecommunications Network. An International Journal Networks, Vol. 33, No. 2 (1999) 109-123
19. Schaffer, J. D.: Multiple Objective Optimisation With Vector Evaluated Genetic Algorithm. Proceedings of the First International Conference on Genetic Algorithms, Hillsdale, (1985) 93-100
20. Schwefel, H. P.: Evolution And Optimum Seeking. John Wiley and Sons, Chichester (1995)
21. Sheble, G. B., Britting, K.: Refined Genetic Algorithm – Economic Dispatch Example. IEEE Transactions on Power Systems, Vol. 10, No. 2 (1995) 117-124
22. Srinivas N., Deb K.: Multiobjective Optimisation Using Nondominated Sorting In Genetic Algorithms. Evolutionary Computation, Vol. 2, No. 3 (1994) 221-248
23. Stone, H. S.: Multiprocessor Scheduling With The Aid Of Network Flow Algorithms. IEEE Transactions on Software Engineering, Vol. SE-3, No. 1 (1977) 85-93
24. Van Veldhuizen, D. V., Lamont, G. B.: Multiobjective Evolutionary Algorithms: Analyzing The State-Of-The-Art. Evolutionary Computation, Vol. 8, No. 2 (2000) 125-147
25. Weglarz J. (ed.): Recent Advances In Project Scheduling. Kluwer Academic Publishers, Dordrecht (1998)
26. Zitzler, E., Deb, K., and Thiele, L.: Comparison Of Multiobjective Evolutionary Algorithms: Empirical Results. Evolutionary Computation, Vol. 8, No. 2 (2000) 173-195

A Hybrid Multi-objective Evolutionary Approach to Engineering Shape Design

Kalyanmoy Deb and Tushar Goel

Kanpur Genetic Algorithms Laboratory (KanGAL)
Indian Institute of Technology Kanpur
Kanpur, PIN 208 016, India
{deb,tusharg}@iitk.ac.in
http://www.iitk.ac.in/kangal

Abstract. Evolutionary optimization algorithms work with a population of solutions, instead of a single solution. Since multi-objective optimization problems give rise to a set of Pareto-optimal solutions, evolutionary optimization algorithms are ideal for handling multi-objective optimization problems. Over many years of research and application studies have produced a number of efficient multi-objective evolutionary algorithms (MOEAs), which are ready to be applied to real-world problems. In this paper, we propose a practical approach, which will enable an user to move closer to the true Pareto-optimal front and simultaneously reduce the size of the obtained non-dominated solution set. The efficacy of the proposed approach is demonstrated in solving a number of mechanical shape optimization problems, including a simply-supported plate design, a cantilever plate design, a hoister design, and a bicycle frame design. The results are interesting and suggest immediate application of the proposed technique in more complex engineering design problems.

1 Introduction

For last decade or so, a number of multi-objective optimization techniques using evolutionary algorithms are suggested [3,6,10,14,16,17]. The outcome of these studies is that different multi-objective optimization problems are possible to solve for the purpose of finding multiple Pareto-optimal solutions in one *single* simulation run. Classical means of finding one solution at a time with a weight vector or with a similar approach requires a priori knowledge of weight vector and need to be run many times, hopefully finding a different Pareto-optimal solution each time. In addition to converging close or on the true Pareto-optimal set, multi-objective evolutionary algorithms (MOEAs) are capable to finding a widely distributed set of solutions.

In this paper, we suggest a hybrid technique to take evolutionary multi-objective optimization procedures one step closer to practice. Specifically, in a real-world problem, we would like to ensure a better convergence to the true Pareto-optimal front and would also like to reduce the size of obtained non-dominated solutions to a reasonable number. The solutions obtained by an MOEA are modified using a local search method, in which a weighted objective function is minimized. The use of a local search method from the MOEA solutions will allow a better convergence to the true Pareto-optimal front.

A clustering method is suggested in general to reduce the size of the obtained set of solutions. For finite search space problems, the local search approach may itself reduce the size the the obtained set.

A specific MOEA—elitist non-dominated sorting GA or NSGA-II—and a hill-climbing local search method are used together to solve a number of engineering shape optimization problems for two objectives. Minimizing the weight of a structure and minimizing the maximum deflection of the structure have conflicting solutions. When these two objectives are considered together in a design, a number of Pareto-optimal solutions result. By representing presence and absence of small constituting elements in a binary string [1,2,5], NSGA-II uses an innovative crossover operator which seems to help in combining good partial solutions together to form bigger partial solutions. The finite element method is used to evaluate a string representing a shape. The paper shows how the proposed hybrid technique can find a number of solutions with different trade-offs between weight and deflection. On a cantilever plate design, a simply-supported plate design, a hoister plate design, and a bicycle frame design problem, the proposed technique finds interesting and well-engineered solutions. These results indicate that the proposed hybrid technique is ready to be applied to more complex engineering shape design problems.

2 Hybrid Approach

It has been established elsewhere that NSGA-II is an efficient procedure of finding a wide-spread as well as well-converged set of solutions in a multi-objective optimization problem [3,4]. NSGA-II uses (i) a faster non-dominated sorting approach, (ii) an elitist strategy, and (iii) no niching parameter. It has been shown elsewhere [3] that the above procedure has $O(MN^2)$ computational complexity. Here, we take NSGA-II a step closer to practice by

1. ensuring convergence closer to the true Pareto-optimal front, and
2. reducing the size of the obtained non-dominated set.

We illustrate both the above issues in the following subsections.

2.1 Converging Better

In a real-world problem, the knowledge of the Pareto-optimal front is usually not known. Although NSGA-II has demonstrated good convergence properties in test problems, we enhance the probability of its true convergence by using a hybrid approach. A local search strategy is suggested from each obtained solution of NSGA-II to find a better solution. Since a local search strategy requires a single objective function, a weighted objective or a Tchebyscheff metric or any other metric which will convert multiple objectives into a single objective can be used. In this study, we use a weighted objective:

$$F(\mathbf{x}) = \sum_{j=1}^{M} \bar{w}_j^{\mathbf{x}} f_j(\mathbf{x}), \tag{1}$$

where weights are calculated from the obtained set of solutions in a special way. First, the minimum $f_j^{\min}$ and maximum $f_j^{\max}$ values of each objective function f_j are noted. Thereafter, for any solution x in the obtained set, the weight for each objective function is calculated as follows:

$$\bar{w}_j^{\mathbf{x}} = \frac{(f_j^{\max} - f_j(\mathbf{x}))/(f_j^{\max} - f_j^{\min})}{\sum_{k=1}^{M}(f_k^{\max} - f_k(\mathbf{x}))/(f_k^{\max} - f_k^{\min})}. \tag{2}$$

In the above calculation, minimization of objective functions is assumed. When a solution x is close to the individual minimum of the function f_j, the numerator becomes one, causing a large value of the weight for this function. For an objective which has to be maximized, the term $(f_j^{\max} - f_j(\mathbf{x}))$ needs to be replaced with $(f_j^{\mathbf{x}} - f_j^{\min})$. The division of the numerator with the denominator ensures that the calculated weights are normalized or $\sum_{j=1}^{M} \bar{w}_i^{\mathbf{x}} = 1$. Once the pseudo-weights are calculated, the local search procedure is simple. Begin the search from each solution x independently with the purpose of optimizing $F(\mathbf{x})$. Figure 1 illustrates this procedure. Since, the pseudo-weight

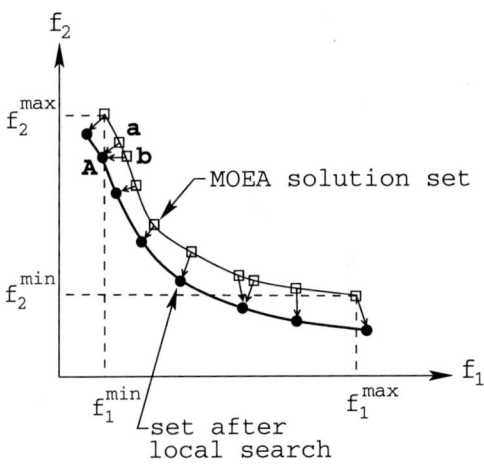

Fig. 1. The local search technique may find better solutions.

vector $\bar{\mathbf{w}}$ dictates roughly the priority of different objective functions at that solution, optimizing $F(\mathbf{x})$ will produce a Pareto-optimal or a near Pareto-optimal solution. This is true for convex Pareto-optimal regions. However, for non-convex Pareto-optimal regions, there exists no weight vector corresponding to Pareto-optimal solutions in certain regions. Thus, a different metric, such as Tchebysheff metric can be used in those cases. Nevertheless, the overall idea is that once NSGA-II finds a set of solutions close to the true Pareto-optimal region, we use a local search technique from each of these solutions with a differing emphasis of objective functions in the hope of better converging to the true Pareto-optimal front. Since independent local search methods are tried from each solution obtained using an MOEA, all optimized solutions obtained by the local search method need not be non-dominated to each other. Thus, we find the non-dominated set of solutions from the obtained set of solutions before proceeding further. Other studies as [11] use the local search method during a GA run. Each solution is modified with a local search method before including it in the population. The proposed approach is likely to have a lesser computational cost, however this will be a matter of future research to find a comparison between the two studies.

The complete procedure of the proposed hybrid strategy is shown in Figure 2. Starting from the MOEA results, we first apply a local search technique, followed by a non-domination check. After non-dominated solutions are found, a clustering technique is used to reduce the size of the optimal set, as discussed in the next subsection.

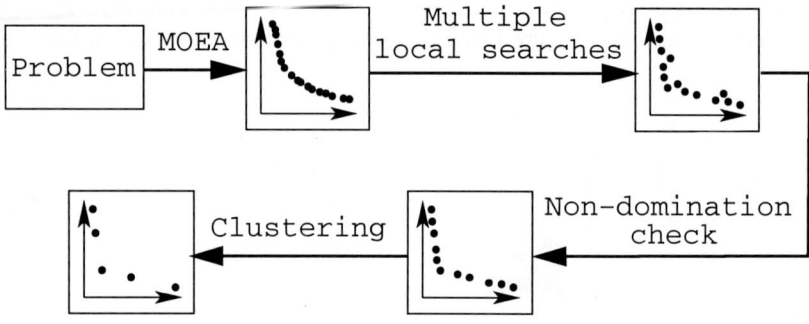

Fig. 2. The proposed hybrid procedure of using a local search technique, a non-domination check, and a clustering technique is illustrated.

2.2 Reducing the Size of Non-dominated Set

In an ideal scenario, an user is interested in finding a good spread of non-dominated solutions closer to the true Pareto-optimal front. From a practical standpoint, the user would be interested in a handful of solutions (in most cases, 5 to 10 solutions are probably enough). Interestingly, most MOEA studies use a population of size 100 or more, thereby finding about 100 different non-dominated solutions. The interesting question to ask is 'Why are MOEAs set to find many more solutions than desired?'

The answer is fundamental to the working of an EA. The population size required in an EA depends on a number of factors related to the number of decision variables, the complexity of the problem, and others [7,9]. The population cannot be sized according to the desired number of non-dominated solutions in a problem. Since in most interesting problems, the number of decision variables are large and are complex, the population sizes used in solving those problems can be in hundreds. Such a population size is mandatory for the successful use of an EA. The irony is that when an MOEA works well with such a population size N, eventually it finds N different non-dominated solutions, particularly if the niching mechanism used in the MOEA is good. Thus, we need to devise a separate procedure of identifying a handful of solutions from the large obtained set of non-dominated solutions.

One approach would be to use a clustering technique similar to that used in [17] for reducing the size of the obtained non-dominated set of solutions. In this technique, each of N solutions is assumed to belong to a separate cluster. Thereafter, the distance d_c between all pairs of clusters is calculated by first finding the centroid of each cluster and then calculating the Euclidean distance between the centroids. Two clusters having the minimum distance are merged together into a bigger cluster. This procedure is continued till the desired number of clusters are identified. Finally, with the remaining clusters, the solution closest to the centroid of the cluster is retained and all other solutions from each cluster are deleted. This is how the clusters can be merged and the cardinality of the solution set can be reduced. Figure 3 shows the MOEA solution set in open boxes and the reduced set in solid boxes. Care may be taken to choose the extreme solutions in the extreme clusters.

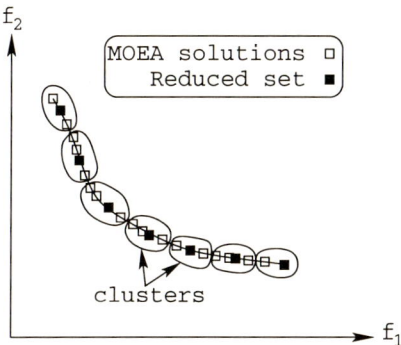

Fig. 3. The clustering method of reducing the set of non-dominated solutions is illustrated.

However, in many problems the local search strategy itself can reduce the cardinality of the obtained set of non-dominated solutions. This will particularly happen in problems with a discrete search space. For two closely located solutions, the pseudo-weight vectors may not be very different. Thus, when a local search procedure is started from each of these solutions (which are close to each other) with a $F(\mathbf{x})$ which is also similar, the resulting optimum solutions may be identical in a discrete search space problem. The solutions a and b in Figure 1 are close and after the local search procedure they may converge to the same solution A. Thus, for many solutions obtained using NSGA-II, the resulting optimum obtained using the local search method may be the same. Thus, the local search procedure itself may reduce the size of the obtained non-dominated solutions in problems with a finite search space. Figure 2 shows that clustering is the final operation of the proposed hybrid strategy.

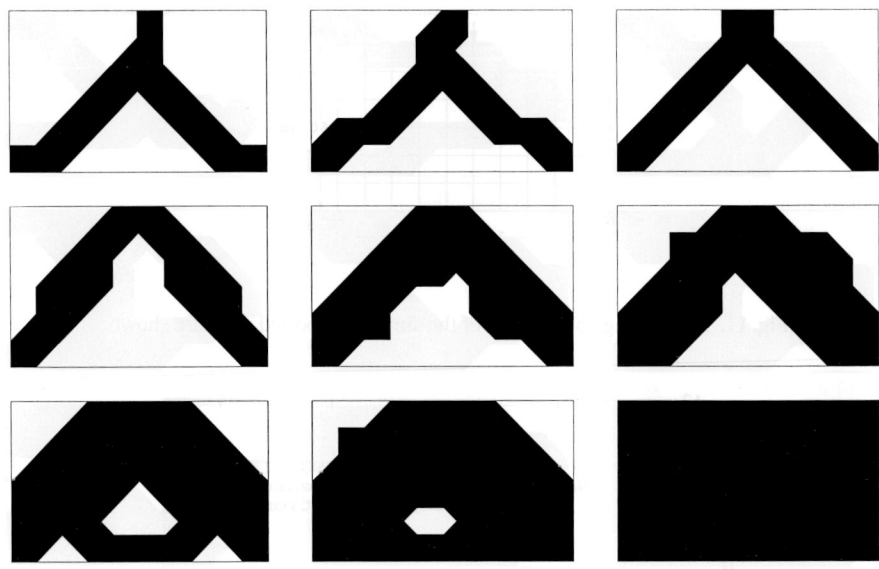

Fig. 13. Nine trade-off shapes for the simply-supported plate design.

and finally finding the complete rectangular plate having minimum deflection are all intuitive trade-off solutions. In the absence of any such knowledge, it is interesting how the hybrid procedure with NSGA-II is able to find the whole family of different trade-off solutions.

4.3 Bicycle Frame Design

Finally, we attempt to design a bicycle frame for a vertical load of 10 kN applied at A in Figure 14. The specifications are similar to that used elsewhere [13]. The plate is 20 mm thick and is restricted to be designed within in the area shown in Figure 14. The frame is supported at two places B and C. The point B marks the position of the axle of the rear wheel and the point C is the location of the handle support. The filled element is the location of the pedal assembly and is always present. The material yield stress is 140 MPa, Young's modulus is 80 GPa and Poisson's ratio is 0.25. The maximum allowed displacement is 5 mm.

Figure 15 shows the NSGA-II solutions and corresponding solutions obtained by the hybrid approach. Here, we are interested in finding four different trade-off solutions.

These four solutions obtained by NSGA-II are shown mounted on a sketch of a bicycle in Figure 16. The top-left solution is the minimum weight design. The second solution joins the two vertical legs to make the structure more stiff. The other two solutions make the legs more thick in order to increase the stiffness of the frame. The interior hole and absence of top-left elements are all intuitive. The proposed hybrid approach can evolve such solutions without these knowledge and mainly by finding and

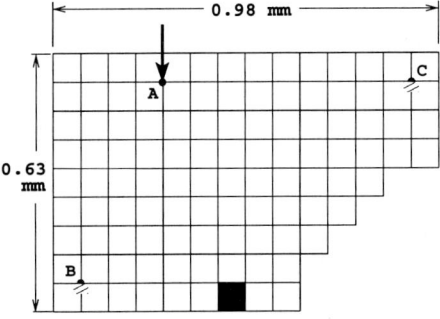

Fig. 14. The hybrid procedure is illustrated for the bicycle frame design.

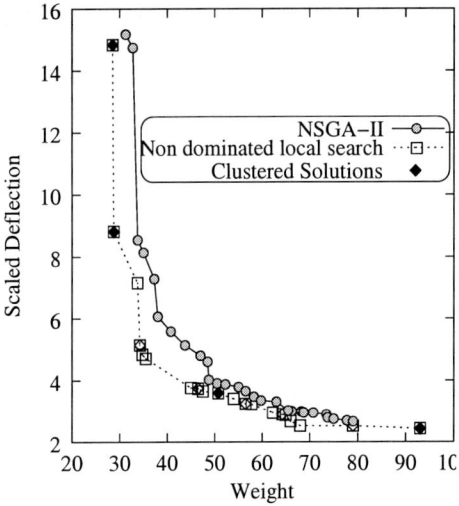

Fig. 15. The hybrid procedure is illustrated for the bicycle frame design.

maintaining trade-off solutions among weight and deflection. The presence of many such solutions with different trade-offs between weight and stiffness provides a plethora of information about various types of design.

5 Conclusion

The hybrid multi-objective optimization technique proposed in this paper uses a combination of an multi-objective evolutionary algorithm (MOEA) and a local search operator. The proposed technique ensures a better convergence of MOEAs to the true

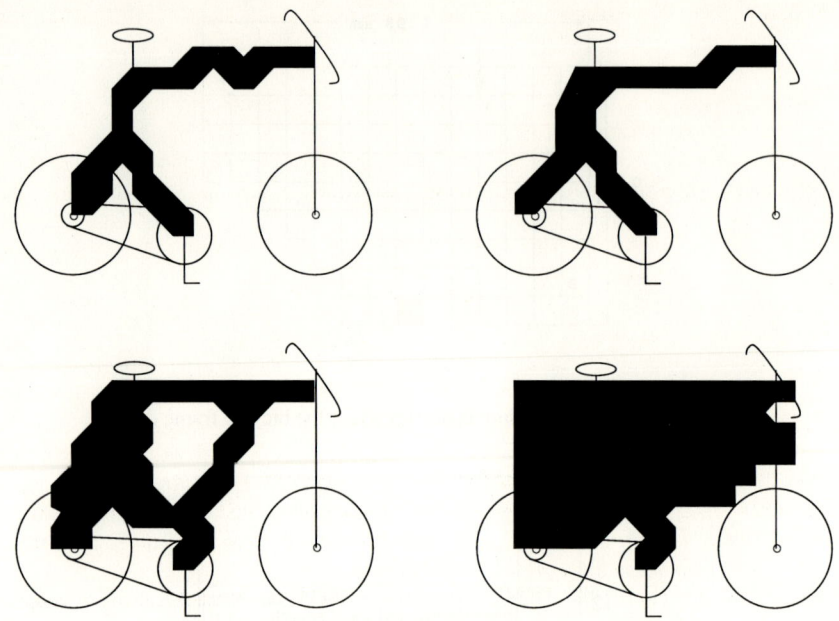

Fig. 16. Four trade-off shapes for the bicycle frame design.

Pareto-optimal region and helps in finding a small set of diverse solutions for practical reasons.

The efficacy of the proposed technique is demonstrated by solving a number of engineering shape design problems for two conflicting objectives—weight of the structure and maximum deflection of the structure. In all cases, the proposed technique has been shown to find a set of four to nine diverse solutions better converged than an MOEA alone. The results are encouraging and takes the evolutionary multi-objective optimization approach much closer to practice.

Acknowledgements. Authors acknowledge the support provided by Ministry of Human Resource Development (MHRD) for this study.

References

1. Chapman, C. D., Jakiela, M. J. (1996). Genetic algorithms based structural topology design with compliance and topology simplification considerations. *ASME Journal of Mechanical Design, 118*, 89–98.
2. Chapman, C. D., Saitou, K., Jakiela, M. J. (1994). Genetic algorithms as an approach to configuration and topology design. *ASME Journal of Mechanical Design, 116*, 1005–1012.
3. Deb, K., Pratap, A., Agrawal, S. and Meyarivan, T. (2000). A fast and elitist multi-objective genetic algorithm: NSGA-II. Technical Report No. 2000001. Kanpur: Indian Institute of Technology Kanpur, India.

4. Deb, K., Agrawal, S., Pratap, A., Meyarivan, T. (2000). A Fast Elitist Non-dominated sorting genetic algorithm for multi-objective optimization: NSGA-II. *Proceedings of the Parallel Problem Solving from Nature VI Conference*, pp. 849–858.
5. Duda, J. W. and Jakiela, M. J. (1997). Generation and classification of structural topologies with genetic algorithm speciation. *ASME Journal of Mechanical Design, 119*, 127–131.
6. Fonseca, C. M. and Fleming, P. J. (1993). Genetic algorithms for multi-objective optimization: Formulation, discussion, and generalization. *Proceedings of the Fifth International Conference on Genetic Algorithms*. 416–423.
7. Goldberg, D. E., Deb, K., and Clark, J. H. (1992). Genetic algorithms, noise, and the sizing of populations. *Complex Systems, 6*, 333–362.
8. Hamada, H. and Schoenauer, M. (2000). Adaptive techniques for evolutionary optimum design. *Proceedings of the Evolutionary Design and Manufacture.*, pp. 123–136.
9. Harik, G., Cantu-paz, E. Goldberg, D. E., and Miller, B. L. (1999). The gambler's ruin problem, genetic algorithms, and the sizing of populations. *Evolutionary Computation, 7*(3), 231–254.
10. Horn, J. and Nafploitis, N., and Goldberg, D. E. (1994). A niched Pareto genetic algorithm for multi-objective optimization. *Proceedings of the First IEEE Conference on Evolutionary Computation.* 82–87.
11. Isibuchi, M. and Murata, T. (1998). A multi-objective genetic local search algorithm and its application to flowshop scheduling. *IEEE Transactions on Systems, Man and Cybernetics—Part C: Applications and reviews, 28*(3), 392-403.
12. Jakiela, M. J., Chapman, C., Duda, J., Adewuya, A., abd Saitou, K. (2000). Continuum structural topology design with genetic algorithms. *Computer Methods in Applied Mechanics and Engineering, 186*, 339–356.
13. Kim, H., Querin, O. M., and Steven, G. P. (2000). Post-processing of the two-dimensional evolutionary structural optimization topologies. In I. Parmee (Ed.) *Evolutionary Design and Manufacture*, London: Springer. pp. 33–44.
14. Knowles, J. and Corne, D. (1999) The Pareto archived evolution strategy: A new baseline algorithm for multi-objective optimization. *Proceedings of the 1999 Congress on Evolutionary Computation*, Piscataway: New Jersey: IEEE Service Center, 98–105.
15. Sandgren, E., Jensen, E, and Welton, J. (1990). Topological design of structural components using genetic optimization methods. *Proceedings of the Winter Annual Meeting of the American Society of Mechanical Engineers*, pp. 31–43.
16. Srinivas, N. and Deb, K. (1995). Multi-Objective function optimization using non-dominated sorting genetic algorithms. *Evolutionary Computation*(2), 221–248.
17. Zitzler, E. and Thiele, L. (1998). An evolutionary algorithm for multi-objective optimization: The strength Pareto approach. *Technical Report No. 43 (May 1998)*. Zürich: Computer Engineering and Networks Laboratory, Switzerland.

Fuzzy Evolutionary Hybrid Metaheuristic for Network Topology Design

Habib Youssef, Sadiq M. Sait, and Salman A. Khan

Department of Computer Engineering, King Fahd University of Petroleum and Minerals
Dhahran 31261, Saudi Arabia
{youssef,sadiq,salmana}@ccse.kfupm.edu.sa
http://www.ccse.kfupm.edu.sa

Abstract. Topology design of enterprise networks is a hard combinatorial optimization problem. It has numerous constraints, several objectives, and a very noisy solution space. Besides the NP-hard nature of this problem, many of the performance metrics of the network can only be estimated, given their dependence on many of the dynamic aspects of the network, e.g., routing and number and type of traffic sources. Further, many of the desirable features of a network topology can best be expressed in linguistic terms, which is the basis of fuzzy logic. In this paper, we present a fuzzy evolutionary hybrid metaheuristic for network topology design. This approach is **dominance preserving** and scales well with larger problem instances and a larger number of objective criteria. Experimental results are provided.

1 Introduction

A typical enterprise network provides communication services to a large number of hosts, such as mainframe computers, mini systems, workstations, PCs, printers, etc., [1]. Network active elements such as routers, switches, and hubs are used to interconnect these computers and peripherals. The network topology is governed by several constraints. Geographical constraints dictate the breakdown of such internetworks into smaller parts or groups of nodes, where each group makes up what is called a LAN. A LAN consists of all network elements which do not include routers or layer-3 switches. Routers delineate the boundaries of LANs. Communication services of a modern organization are centered around a structured campus network, which consists of a backbone interconnecting a number of LANs via routers or layer-3 switches. Further, the nodes of a LAN may be subdivided into smaller parts, called *LAN segments* (see Fig. 1). Overdimensioning a network is easy; however, designing a cost-optimized network is always very hard. Hardness is a function of the size, the constraints, and obviously the cost parameters to tradeoff. Furthermore, with many cost parameters and constraints, the notion of optimality is not clear. A more reasonable approach is to seek a solution that possesses a set of desirable properties and do not violate some well established design principles. Examples of these principles are:

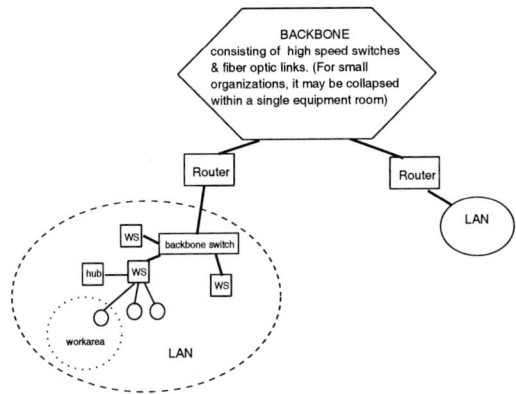

Fig. 1. A typical Campus Network (WS represents workgroup switch).

- There is a physical path between any two nodes.
- The number of hops between any two stations does not exceed a given threshold.
- Only a given small fraction of links have utilization levels below some threshold.

A category of algorithms that were found to be effective for such problems are iterative metaheuristics. These allow you to walk the state space of solutions while evaluating each solution against any desirable set of properties. These meta-heuristics are characterized by hill climbing property that allows occasional acceptance of inferior solutions [2]. Heuristics like genetic algorithm [3], simulated annealing [4], tabu search [5], simulated evolution [6], and stochastic evolution [7] are examples of stochastic iterative heuristics. Detailed description of these heuristics can be found in [2], and an interesting classification of some of them is given in [5].

In this work we propose a hybrid meta-heuristic for the topology design problem which follows the search strategy of Simulated Evolution (SE) algorithm. SE is a memoryless meta-heuristic, where the walk through the state space is heavily influenced by the allocation operator. The memoryless nature of the search usually results in partial revisiting of areas of the state space. To minimize the effect of such undesirable behavior, the allocation step of SE is implemented while following tabu search approach.

2 Background

Many combinatorial optimization problems can be formulated as follows [7]: *Given a finite set M of distinct movable elements and a finite set L of locations, a state is defined as an assignment function $S : M \rightarrow L$ satisfying certain constraints.* The topology design problem fits this generic model. For this problem, given a set of links $E = \{e_1, e_2, ..., e_n\}$ and a set of locations $L = \{0, 1\}$, where $L(e_i) = 1$ iff link e_i belongs to the topology and $L(e_i) = 0$ otherwise. We seek to find an assignment $S : E \to L$ which corresponds to feasible topology of desirable properties.

Unlike constructive algorithms, which produce a solution only at the end of the design process, iterative algorithms produce numerous solutions during the course of their search. In order to compare alternative topologies, the cost of each topology is estimated for the objectives under consideration. Important objectives are the minimization of monetary cost, network latency, and maximum number of hops between any source-destination pair. Most of the objectives and constraints depend on several aspects such as network flow dynamics, technology trends, strategic commercial goals, etc., that can best be expressed in linguistic terms, which is the basis of fuzzy logic. In this work, the cost function, constraints, as well as some of the SE algorithm operators are implemented using fuzzy algebra [8].

2.1 SE Algorithm

Simulated Evolution (SE) is a stochastic evolutionary search strategy that falls in the general category of meta-heuristics. It was first proposed by Kling and Banerjee in [6]. SE adopts the generic state model described above, where a solution is seen as a population of movable elements.

Starting from a given initial solution, SE repetitively executes the following three steps in sequence: **evaluation**, **selection**, and **allocation**, until certain stopping conditions are met. The pseudo-code of the SE algorithm is given in Fig. 2. The **evaluation** step estimates the **goodness** of each element in its current location. The goodness of an element is a ratio of its optimum cost to its actual cost estimate, and therefore belongs to the interval [0,1]. It is a measure of how near each element is to its optimum position. The higher the goodness of an element, the closer is that element to its optimum location with respect to the current configuration. In **selection** step, the algorithm probabilistically selects elements for relocation. Elements with low goodness values have higher probabilities of getting selected. A selection *bias* (B) is used to compensate for errors made in the estimation of goodness. Its objective is to inflate or deflate the goodness of elements. A high positive value of *bias* decreases the probability of selection and vice versa. Large selection sets also degrade the solution quality due to uncertainties created by large perturbations. Similarly, for high bias values the size of the selection set is small, which degrades the quality of solution due to limitations of the algorithm to escape local minima. A carefully tuned bias value results in good solution quality and reduced execution time [6].

Elements selected during the **selection** step are assigned to new locations in the **allocation** step with the hope of improving their goodness values, and thereby reducing the overall cost of the solution. Allocation is the step that has most impact on the quality of the search performed by the SE algorithm. A completely random allocation makes the SE algorithm behave like a random walk. Therefore, this operator should be carefully engineered to the problem instance

$Simulated\_Evolution(B, \Phi_{initial}, StoppingCondition)$
NOTATION
$B=$ Bias Value.
$\Phi=$ Complete Solution.
$e_i=$ Individual link in Φ.
$O_i=$ Lower bound on cost of ith link.
$C_i=$ Current cost of ith link in Φ.
$g_i=$ Goodness of ith link in Φ.
$S=$ Queue to store the selected links.
$ALLOCATE(e_i, \Phi_i)=$Function to allocate e_i in partial solution Φ_i
Begin
Repeat
    $EVALUATION$: **ForEach** $e_i \in \Phi$ **DO**
                      begin
                            $g_i = \frac{O_i}{C_i}$
                        end
    $SELECTION$: **ForEach** $e_i \in \Phi$ **DO**
                        begin
                            IF $Random > Min(g_i + B, 1)$
                            THEN
                                begin
                                  $S = S \cup e_i$; Remove e_i from Φ.
                            end
                        end
        Sort the elements of S
    $ALLOCATION$: **ForEach** $e_i \in S$ **DO**
                        begin
                            $ALLOCATE(e_i, \Phi_i)$
                        end
Until Stopping Condition is satisfied
Return Best solution.
End ($Simulated\_Evolution$)

Fig. 2. Structure of the simulated evolution algorithm.

and must include domain-specific knowledge. Different constructive allocation schemes are proposed in [6].

Though SE falls in the category of meta-heuristics such as simulated annealing (SA) and genetic algorithm (GA), there are significant differences between these heuristics (see [2]). A classification of meta-heuristics proposed by Glover and Laguna [5] is based on three basic features: (1) the use of adaptive memory where the letter A is used if the meta-heuristic employs adaptive memory and the letter M is used if it is memoryless; (2) the kind of neighborhood exploration, where the letter N is used if the meta-heuristic performs a systematic neighborhood search and the letter S is used if stochastic sampling is followed; and (3) the number of current solutions carried from one iteration to the next, where the digit 1 is used if the meta-heuristic maintains a single solution, and the letter P is used if a parallel search is performed with a population of solutions of cardinality P. For example, according to this classification, Genetic algorithm is M/S/P, tabu search is A/N/1, and both simulated annealing and simulated evolution are M/S/1. The heuristic proposed in this work is A/S/1.

Average Network Delay: The second objective is to minimize the average network delay, while considering the constraints and requirements. To devise a suitable function for average network delay, we approximate the behavior of a link and network device by an M/M/1 queue [9]. The delay per bit due to the network device between local sites i and j is $B_{i,j} = \mu b_{i,j}$, where $\frac{1}{\mu}$ is the average packet size in bits and $b_{i,j}$ is the delay per packet. If γ_{ij} is the total traffic through the network device between local sites i and j, then the average delay due to all network devices is:

$$D_{nd} = \frac{1}{\gamma} \sum_{i=1}^{d} \sum_{j=1}^{d} \gamma_{ij} B_{ij} \qquad (4)$$

where d is the total number of network devices in the network. Thus, the total average network delay is composed of delays of links and network devices and is given by [9]

$$D = \frac{1}{\gamma} \sum_{i=1}^{L} \frac{\lambda_i}{\lambda_{max,i} - \lambda_i} + \frac{1}{\gamma} \sum_{i=1}^{d} \sum_{j=1}^{d} \gamma_{ij} B_{ij} \qquad (5)$$

Maximum number of hops between any source-destination pair: The maximum number of hops between any source-destination pair is also another objective to be optimized. A hop is counted as the packet crosses a network device.

5 Proposed Algorithm and Implementation Details

This section describes our proposals of fuzzification of different stages of the SE algorithm. We confine ourselves to tree design. Trees are minimal and provide unique path between every pair of local sites. Further, the design of a general mesh topology usually starts from a near optimal constrained spanning tree.

5.1 Initialization

The initial spanning tree topology is generated randomly, while keeping into account the feasibility constraints mentioned earlier.

5.2 Proposed Fuzzy Evaluation Scheme

The **goodness** of each individual is computed as follows. In our case, an individual is a **link** which interconnects the devices of two local sites (at the backbone level) or two network devices (at the local site level). In the *fuzzy evaluation scheme*, monetary cost and optimum depth of a link (with respect to the root) are considered fuzzy variables. Then the goodness of a link is characterized by the following rule.

Rule 1: IF a link is *near optimum cost* AND *near optimum depth* **THEN** it has *high goodness*.

Here, *near optimum cost*, *near optimum depth*, and *high goodness* are linguistic values for the fuzzy variables cost, depth, and goodness. Using and-like compensatory operator [10], Rule 1 translates to the following equation for the fuzzy goodness measure of a link l_i.

$$g_{l_i} = \mu^e(l_i) = \alpha^e \times min(\mu_1^e(l_i), \mu_2^e(l_i)) + (1-\alpha^e) \times \frac{1}{2}\sum_{i=1}^{2}\mu_i^e(l_i) \qquad (6)$$

The superscript *e* stands for **evaluation** and is used to distinguish similar notation in other fuzzy rules. In (6), $\mu^e(l_i)$ is the membership in the fuzzy set of *high goodness links* and α^e is a constant. The $\mu_1^e(l_i)$ and $\mu_2^e(l_i)$ represent memberships in the fuzzy sets *near optimum monetary cost* and *near optimum depth*.

In order to find the membership of a link with respect to *near optimum monetary cost*, we proceed in following manner. From the cost matrix, which gives the costs of each possible link, we find the minimum and maximum costs among all the link costs. We take these minimum and maximum costs as the lower and upper bounds and call them "LCostMin" and "LCostMax" respectively and then find the membership of a link with respect to these bounds. Furthermore, in this work, we have normalized the monetary cost with respect to "LCostMax". The required membership function is represented as depicted in Fig. 3, where $x-axis$ represents $\frac{LCost}{LCostMax}$, $y-axis$ represents the membership value, $A = \frac{LCostMin}{LCostMax}$, and $B = \frac{LCostMax}{LCostMax} = 1$. This normalization enables us to use the same membership function for all topology design instances.

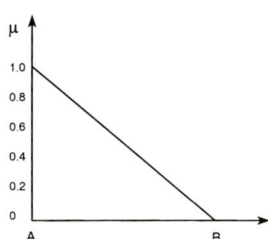

Fig. 3. Membership function for the objective to be optimized.

In the same manner, we can find the membership of a link with respect to *near optimum depth*. The lower limit, which we call "LDepthMin" is taken to be a depth of 1 with respect to the root. The upper bound, which we call "LDepthMax" is taken to be 1.5 times of the maximum depth generated in the initial solution or a maximum of a user specified limit.[1] For example, if in the initial solution, the maximum depth turns out to be 4, then "LDepthMax" for the depth membership function would be 6. This is done to give chance to links which may have more depth than the one in the initial solution. If we take the initial solution maximum depth as "LDepthMax", then in the following iterations some links with higher depths will have a membership value of zero (with respect to depth membership function) and thus they will not be able to play any role as far as depth is concerned. However, due to technological limitations, we have limited the maximum possible depth to 7, in the case when "LDepthMax" turns out to be more than 4. The reason for having the maximum depth of 7 is that the hop limit for RIP is 15. This means that if a maximum depth of 7 is taken, then in the worst case we would have a total of 14 hops from a source to a destination. The membership function with respect to *near optimum depth* can be represented as illustrated in Fig. 3, where $x-axis$ represents $LDepth$, $y-axis$ represents the membership value, $A = LDepthMin$, and $B = LDepthMax$.

5.3 Selection

In this stage of the algorithm, for each link l_i in current tree topology, where $i = 1,2,..., n\text{-}1$, a random number $RANDOM \in [0, 1]$ is generated and compared with $g_i + B$, where B is the selection bias. If $RANDOM > g_i + B$, then link l_i is selected for allocation and considered removed from the topology. Bias B is used to control the size of the set of links selected for removal. A bias methodology called *variable bias* [11] has been used in this paper. The *variable bias* is a function of *quality of current solution*. When the overall solution quality is poor, a high value of bias is used, otherwise a low value is used. Average link goodness g_i is a measure of how many "good" links are present in the topology. The bias value changes from iteration to iteration depending on the quality of solution. The *variable bias* is calculated as follows:

$$B_k = 1 - G_k$$

where B_k is the bias for k^{th} iteration and G_k is average goodness of all the links at the beginning of iteration k.

5.4 Proposed Fuzzy Allocation Scheme

During the **allocation** stage of the algorithm, the selected links are removed from the topology one at a time. For each removed link, new links are tried in such a way that they result in overall better solution. Before the allocation step starts, the selected links are sorted according to their goodness values in ascending order.

---

[1] This user specified limit may be a design constraint, e.g., if each hop represents a router that uses Routing Information Protocol (RIP) then a limit would be 7, i.e., a branch of the tree should not have more than 7 routers.

In the *fuzzy allocation scheme*, the three criteria to be optimized are combined using fuzzy logic to characterize a good topology. The reason for using fuzzy logic is that the characterization of a good topology with respect to several criteria is usually based on heuristic knowledge which is acquired through experience. Such knowledge is most conveniently expressed in linguistic terms, which constitute the basis of fuzzy logic. For the problem addressed in this paper, a good topology is one that is characterized by a low monetary cost, low average network delay, and a small maximum number of hops. In fuzzy logic, this can easily be stated by the following fuzzy rule:

Rule 2: **IF** a solution X has *low monetary cost* AND *low average network delay* AND *low maximum number of hops between any source-destination pair* **THEN** it is a *good topology*.

The words "low monetary cost", "low average network delay", "low maximum number of hops", and "good topology" are linguistic values, each defining a fuzzy subset of solutions. For example, "low average network delay" is the fuzzy subset of topologies of low average network delays. Each fuzzy subset is defined by a membership function μ. The membership function returns a value in the interval [0,1] which describes the degree of satisfaction with the particular objective criterion. Using the and-like ordered weighted averaging operator [10], the above fuzzy rule reduces to the following equation.

$$\mu^a(x) = \beta^a \times min(\mu_1^a(x), \mu_2^a(x), \mu_3^a(x)) + (1 - \beta^a) \times \frac{1}{3} \sum_{i=1}^{3} \mu_i^a(x) \qquad (7)$$

where $\mu^a(x)$ is the membership value for solution x in the fuzzy set *good topology* and β^a is a constant in the range [0,1]. The superscript a stands for allocation. Here, μ_i^a for $i = \{1,2,3\}$ represents the membership values of solution x in the fuzzy sets *low monetary cost*, *low average network delay*, and *low maximum number of hops between any source-destination pair* respectively. The solution which results in the maximum value for (7) is reported as the best solution found by the SE algorithm.

Below we describe how to get the membership functions for the three criteria mentioned above.

Membership Function for Monetary Cost. First, we determine two extreme values for monetary cost, i.e., the minimum and maximum values. The minimum value, "TCostMin", is found by using the Esau-Williams algorithm [12], with all the constraints completely relaxed. This will surely give us the minimum possible monetary cost of the topology. The maximum value of monetary cost, "TCostMax", is taken to be the monetary cost generated in the initial solution. The monetary cost is normalized with respect to "TCostMax". The corresponding membership function is shown in Fig. 3, where $x - axis$ represents $\frac{TCost}{TCostMax}$, $y - axis$ represents the membership value, $A = \frac{TCostMin}{TCostMax}$, and $B = \frac{TCostMax}{TCostMax} = 1$.

Membership Function For Average Network Delay. We determine two extreme values for average network delay. The minimum value, "TDelayMin", is found by connecting all the nodes to the root directly, ignoring all the constraints and then calculating the average network delay using (5). The maximum value of average delay, "TDelayMax", is taken to be the average delay generated in the initial solution. The average delay is normalized with respect to "TDelayMax". The membership function is shown in Fig. 3, where $x-axis$ represents $\frac{TDelay}{TDelayMax}$, $y-axis$ represents the membership value, $A = \frac{TDelayMin}{TDelayMax}$, and $B = \frac{TDelayMax}{TDelayMax} = 1$.

Membership Function For Maximum Number of Hops. Again, two extreme values are determined. The minimum value, "THopsMin", is taken to be 1 hop, which will be the minimum possible in any tree. The maximum value, "THopsMax", is taken to be the maximum number of hops between any source-destination pair generated in the initial solution. The membership function is shown in Fig. 3, where $x-axis$ represents $THops$, $y-axis$ represents the membership value, $A = THopsMin$, and $B = THopsMax$.

In the proposed allocation scheme, all the selected links are removed one at a time and trial links are placed for each removed link. We start with the head-of-line link, i.e. the link with the worst goodness. We remove this link from the topology. This divides the topology into two disjoint trees. Now the placing of trial links begins. In this work, the approach to place trial links is as follows. At most ten trial moves (i.e., trial links) are evaluated for each removed link. One point to mention is that for the ten moves, some moves may be invalid. However, we search for only four "valid" moves. Whenever we find four valid moves, we stop, otherwise we continue until a total of ten moves are evaluated (whether valid or invalid). The removal of a link involves two nodes P and Q, of which node P belongs to the subtree which contains the root node and node Q belongs to the other subtree. For the ten moves we make, five of them are greedy and five are random. For the greedy moves, we start with node Q and five *nearest* nodes in the other subtree are tried. For the random moves, we select any two nodes in the two subtrees and connect them. If all the ten moves are invalid, in which case the original link is placed back in its position. The valid moves are evaluated based on (7) and the best move among the ten moves is made permanent. This procedure is repeated for all the links that are present in the set of selected links.

We have implemented two variations of allocation schemes. The first one is the same as has been described above, which we call SE. In the second variation, Tabu Search characteristics have been introduced, details of which follow.

5.5 Tabu Search Based Allocation

Tabu Search (TS) is a general iterative heuristic that is used for solving combinatorial optimization problems. The algorithm was first presented by F. Glover [5].

Table 1. Characteristics of test cases used in our experiments. LCostMin, LCostMax, and TCostMin are in dollars. TDelayMin is in milliseconds. Traffic is in Mbps.

Name	# of Local Sites	LCostMin	LCostMax	TCostMin	TDelayMin	Traffic
n15	15	1100	9400	325400	2.14296	24.63
n25	25	530	8655	469790	2.15059	74.12
n33	33	600	10925	624180	2.15444	117.81
n40	40	600	11560	754445	2.08757	144.76
n50	50	600	13840	928105	2.08965	164.12

A key feature of TS is that it imposes restrictions on the search process, preventing it from moving in certain directions to drive the process through regions desired for investigation [5]. It searches for the best move in the neighborhood of the current solution.

In this work, we have modified the SE algorithm by introducing Tabu Search characteristics in the allocation phase. Recall that in the allocation phase, certain number of moves are made for each link in the selection set and the best move is accepted, making the move (i.e., link) permanent. This newly accepted link is saved in a *tabu list*. Thus our *attribute* is the link itself. The *aspiration criterion* adopted is that if the link that had been made tabu produces a higher membership value than the current one in the membership function "good topology", then we will override the tabu status of the link and make it permanent. This strategy prevents the selection and allocation operators from repetitively removing the same link and replacing it with a link of equal or worse goodness.

5.6 Stopping Criterion

In our experiments, we have used a fixed number of iterations as a stopping criterion. We experimented with different values of iterations and found that for all the test cases, the SE algorithm converges within 4000 iterations or less.

6 Results and Discussion

The SE algorithm described in this paper has been tested on several randomly generated networks. For each test case, the traffic generated by a typical local site was collected from real sites. Other characteristics, such as the number of ports on a network device, its type, etc. were assumed. However, the costs of the network devices and links were collected from vendors. The characteristics of test cases are listed in Table 1. The smallest test network has 15 local sites and the largest has 50 local sites. The hierarchies in which the devices are connected are that backbone switch is at the top, followed by routers, then workgroup switches, and then hubs.

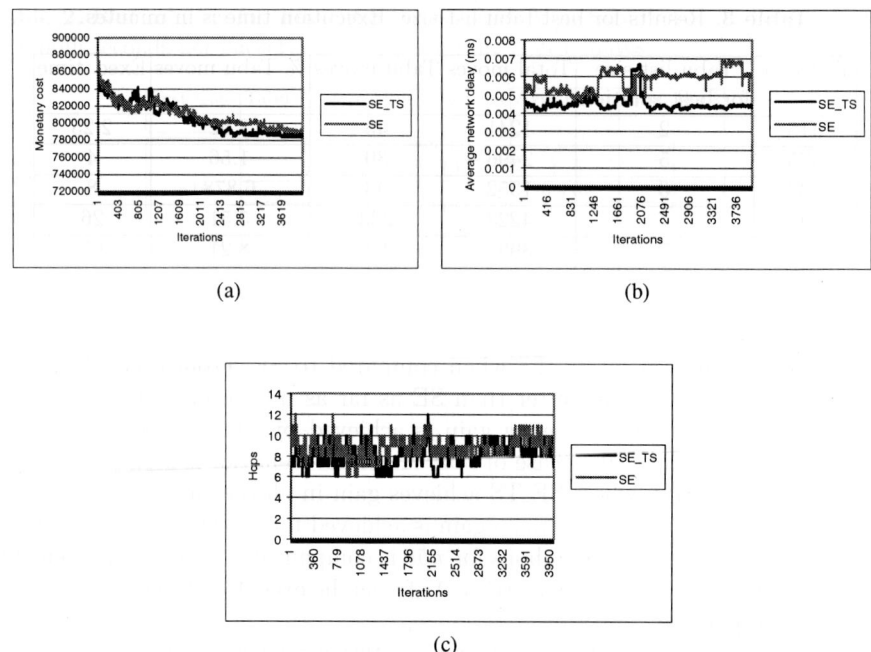

Fig. 4. Comparison of SE and SE_TS for n40.

Table 4. Comparison of SE and SE_TS. C = Cost in dollars, D = Delay in milli seconds per packet, H = hops, T = execution time in minutes, TL= Tabu list size. Percentage gain shows improvement achieved by SE_TS compared to SE.

Case	SE					SE_TS				% Gain		
	C	D	H	T	TL	C	D	H	T	C	D	H
n15	305500	4.135	7	1	2	297100	2.78	4	2.25	2.7	32.8	42.8
n25	512415	4.37	7	4.4	5	483210	3.537	6	4	5.7	19.1	14.28
n33	702815	5.319	7	17	6	682465	4.19	6	8	2.89	21.2	14.28
n40	789625	5.529	9	42	7	783970	4.441	9	26	0.72	19.7	0
n50	1042080	8.236	10	62	7	983020	5.245	11	65	5.67	36.3	-9.1

the allocation scheme. Results obtained for the test cases considered suggest that fuzzy simulated evolution algorithm with tabu search allocation is a robust approach to this problem, and was always able to find good quality feasible solutions.

Acknowledgments. Authors acknowledge King Fahd University of Petroleum and Minerals, Dhahran, Saudi Arabia, for all support.

References

1. Habib Youssef, Sadiq M. Sait, and Osama A. Issa. Computer-Aided Design of Structured Backbones. In *15th National Computer Conference and Exhibition*, pages 1–18, October 1997.
2. Sadiq M. Sait and Habib Youssef. *Iterative Computer Algorithms and their Application to Engineering*. IEEE Computer Society Press, 1999.
3. D. E. Goldberg. *Genetic Algorithms in Search, Optimization and Machine Learning*. Addison-Wesley Publishing Company, INC., 1989.
4. S. Kirkpatrick, C. Gelatt Jr., and M. Vecchi. Optimization by Simulated Annealing. *Science*, 220(4598):498–516, May 1983.
5. Fred Glover and Manuel Laguna. *Tabu Search*. Kluwer Academic Publishers, 1997.
6. Ralph M. Kling and Prithviraj Banerjee. ESP: Placement by Simulated Evolution. *IEEE Transactions on Computer-Aided Design*, 8(3):245–255, March 1989.
7. Y. Saab and V. Rao. Stochastic Evolution: A Fast Effective Heuristic for some Generic Layout Problems. In *27th ACM/IEEE Design Automation Conference*, pages 26–31, 1990.
8. L. A. Zadeh. Fuzzy Sets. *Information Contr.*, 8:338–353, 1965.
9. R. Elbaum and M. Sidi. Topological Design of Local-Area Networks Using Genetic Algorithm. *IEEE/ACM Transactions on Networking*, pages 766–778, October 1996.
10. Ronald Y. Yager. On Ordered Weighted Averaging Aggregation Operators in Multicriteria Decisionmaking. *IEEE Transactions on Systems, Man, and Cybernatics*, 18(1):183–190, Jan 1988.
11. Ali S. Hussain. *Fuzzy Simulated Evolution Algorithm for VLSI Cell Placement*. MS Thesis, King Fahd University of Petroleum and Minerals, 1998.
12. L. R. Esau and K. C Williams. On teleprocessing system design. A method for approximating the optimal network. *IBM System Journal*, 5:142–147, 1966.

A Hybrid Evolutionary Approach for Multicriteria Optimization Problems: Application to the Flow Shop

El-Ghazali Talbi[2], Malek Rahoual[1], Mohamed Hakim Mabed[1], and Clarisse Dhaenens[2]

[1] Institut d'Informatique / Université des Sciences et de Technologie Houari Boumèdienne, BP 32 El Alia, Bab Ezzouar / ALGER.
`rahoual@wissal.dz, mabed@lifl.fr`

[2] LIFL, Université de Lille1, Bât.M3 59655 Villeneuve d'Ascq Cedex FRANCE.
`{talbi, dhaenens}@lifl.fr`

Abstract. The resolution of workshop problems such as the Flow Shop or the Job Shop has a great importance in many industrial areas. The criteria to optimize are generally the minimization of the makespan or the tardiness. However, few are the resolution approaches that take into account those different criteria simultaneously. This paper presents an approach based on hybrid genetic algorithms adapted to the multicriteria case. Several strategies of selection and diversity maintaining are presented. Their performances are evaluated and compared using different benchmarks. A parallel model is also proposed and implemented for the hybrid metaheuristic. It allows to increase the population size and the number of generations, and then leads to better results.

Keywords: Genetic Algorithm, Multicriteria optimization, Flow Shop, Hybrid Metaheuristic, Local Search, Parallel Metaheuristic.

1 Introduction

The Flow Shop problem has received a great attention [6][16] since its importance in many industrial areas [13]. The proposed methods to its resolution vary between exact methods such as the branch & bound algorithm [3], specific heuristics [7][12][16] and metaheuristics [10][11]. However, the majority of these works study the problem in its single criterion form and aim mainly to minimize the makespan.

Population based algorithms such as genetic algorithms (GAs) have turned out to be of great efficiency to deal with multicriteria combinatorial optimization problems. The difficulty of the multicriteria case lies in the absence of a total order relation that links solutions of the problem. Considering the GAs, this insufficiency appears in the difficulty in designing a selection operator that assigns selection probabilities proportional to the desirability degree of the individuals in the population. Another difficulty is related to the balance between the exploration of the search space and the exploitation of the obtained Pareto frontier. Advanced mechanisms have been proposed to deal with this issue, such as combined sharing in the objective and the decision space, hybrid GAs with local search, and parallel model and implementation of the algorithm.

The next section of this paper presents the multicriteria flow shop problem we are interested in. We formulate the different objectives to optimize as well as the constraints to satisfy. In the third section, we will describe the application of genetic algorithms to the problem [16][17][18]. Different selection strategies are presented and their performances compared. We present the implemented diversity maintaining methods and their contribution in the quality of solutions. The fourth section is devoted to the presentation of the hybridization of multicriteria GAs with local search, and its contribution is underlined. In the fifth section, we describe and evaluate a parallel model for the proposed metaheuristic.

2 A Multicriteria Flow Shop Problem

The flow shop problem can be presented as a set of N jobs $\{J_1, J_2, ...J_N\}$ to schedule on M machines. The machines are critical resources: one machine cannot be assigned to two jobs simultaneously. Each job is composed of M consecutive tasks $J_i = \{t_{i1}, t_{i2},..., t_{iM}\}$, where t_{ij} represents the jth task of the job J_i requiring the machine m_j. Following this description, jobs have the same processing sequence on the machines. To each task t_{ij} is associated a processing time p_{ij} and each job J_i must be achieved before the due date d_i.

Scheduling of tasks on different machines must optimize certain regular criteria [16]. These criteria vary according the specificity of the treated problem, and generally consist in the minimization of the following objectives [16]:

C_{max} : makespan (total completion time);
$\overline{C}$: mean value of completion times of the jobs;
T_{max} : maximum tardiness; T: total tardiness;
U : number of jobs delayed with regard to their due date d_i;
F_{max} : maximum flow time; $\overline{F}$: mean flow time.

The optimization criteria taken into account are resumed into two objectives: minimizing the makespan and the total tardiness.

$$f1 = Cmax = \text{Max}\ (s_{iM} + p_{iM})$$

$$f2 = T = \sum [\text{max}\ (0, s_{iM} + p_{iM} - d_i)]$$

We are interested in the study of the permutation flow shop problem F/permu,d_i/(C_{max},T), where jobs must be scheduled in the same order on all the machines.

3 GA and Multicriteria Flow Shop Problem

The application of GAs to a given problem needs, first, a chromosomal representation of a solution (in our case the schedule of jobs). The processing sequence of jobs on the machines being identical, a schedule is then considered as a permutation defining the processing order of the jobs in the machines. The used coding is a jobs array. A position of a job defines its sequencement order.

Once a sequence of jobs is defined, all tasks are scheduled as early as possible (respecting precedence constraints between tasks of a same job and preventing any machine to be allocated to two tasks simultaneously). Then, starting time (s_{ij}) of each task of each job may be computed in a recursive manner, as follows, starting with the first job.

$$s_{ij} = \begin{cases} 0 & \text{if } J_i \text{ is the first job of the sequence and } j = 1. \\ s_{i(j-1)} + p_{i(j-1)} & \text{if } J_i \text{ is the first job of the sequence and } j \neq 1. \\ s_{i'j} + p_{i'j} & \text{if } J_i \text{ is not the first job of the sequence and } j = 1 \\ \max(s_{i(j-1)} + p_{i(j-1)}, s_{i'j} + p_{i'j}) & \text{otherwise}. \end{cases}$$

Where i' represents the job that immediately precedes job i in the sequence. This formula expresses the fact that a task t_{ij} cannot be planned unless: the machine m_j has finished to process the previous task $t_{i'j}$ and the previous task $t_{i(j-1)}$ of the same job is over.

Applying a GA method to a given problem requires also to define the genetic operators. The mutation operator consists in choosing randomly two points of the chromosome, inserting the last job before point 2, just after point 1 and shifting to the right jobs scheduled between the two points. The crossover operator, also called two points crossover, consists in generating one offspring from 2 parents [10]. Two points on Parent 1 are randomly chosen, defining two extremities that will constitute extremities of the Offspring. Then jobs that are not already selected in these two extremities, are selected in the order they appear in Parent 2, to fill the rest of the offspring.

3.1 Selection Operators

In this study, we have implemented 6 multicriteria selection strategies. The main differences between those methods consist in the way individuals of the population are ranked and the selection probabilities are calculated.

3.1.1 Selection by Weighted Sum of Objectives

It was one of the first methods used for the multicriteria optimization (used in [10] for example). Based on the transformation of the problem to a single criterion problem, this method consists in combining the different objective functions in one single function, generally in a linear manner.

$$f(S_i) = \sum_{k \in [1..2]} \lambda_k f_k(S_i)$$

The weights λ_k are taken in the interval [0..1] such as $\sum \lambda_k = 1$ (k=1..n). An individual S_i has then a probability to be selected equal to :

$$\pi(S_i) = f(S_i) / \sum_{j \in [1..tp]} f(S_j) \quad \text{where } tp : \text{population size}$$

3.1.2 Parallel Selection

This selection approach has been used in the VEGA algorithm. Half of the selected individuals are selected with regard to their makespan. The remaining $tp/2$ individuals are selected with regard to their tardiness.

3.1.3 NSGA Selection

In the NSGA selection [14] (Non-dominated Sorted Genetic Algorithm), the ranks of individuals are calculated in a recursive manner, beginning with the non-dominated individuals of the population. A rank equal to 1 is associated to the non-dominated set of individuals E_1 of the current population. Rank k is associated to the set of individuals E_k dominated only by individuals belonging to $E_1 \cup E_2 \cup ... \cup E_{k-1}$.

The selection probability of an individual S_i of rank n in the population follows Baker expression [1][16]:

$$\pi(S_i) = \frac{S(tp + 1 - R_i) + R_i - 2}{tp(tp - 1)}$$

Where S represents the selection pressure and

$$R_i = 1 + |E_i| + 2 * \sum_{j \in [1..i-1]} |E_j|$$

3.1.4 NDS Selection

In the NDS selection (Non-Dominated Sorting), the rank of an individual is equal to the number of solutions dominating this individual plus one [4].

$$Rank\ (S_i) = |\ S_j \in Population\ /\ S_j \text{ dominates } S_i | + 1$$

The selection probability is calculated by the same formula as for NSGA selection.

5 A Parallel Model and Its Implementation

We have adopted a parallel model which is based on the well known distributed island model of GAs. This approach is based on the subdivision of the population into sub-populations of equal size. Each processor executes the GA on the sub-population assigned to it. With a certain period in terms of the number of generations, the different GAs exchange some local Pareto individuals (migration model). We have implemented a ring communication topology in order to minimize the communication cost and also maintain the connected aspect of the graph, what guarantees that a good individual may spread to all sub-populations after a certain number of migrations. The algorithm has been implemented under the parallel programming environment C/PVM (Parallel Virtual Machine).

The evaluations are carried out on a cluster of workstations (Sun Ultra1). The speedup obtained with 2, 3, 4, 6, 9 and 12 machines are almost linear. This is due to the fact that the communication costs between machines are negligible compared to the computing costs. The parallel algorithm favor the increase of the population size as well as the number of generations of the hybrid metaheuristic in order to generate better Pareto fronts.

Table 3. Improvement of the quality of the solutions for (1): ma_ta11_bi and (2): ma_ta21_bi

Pb.	UB	Sequential GA with tp=200					Parallel GA with tp=300				
		MM	Dev %	MR	|PO|	Nb gen	MM	Dev %	MR	|PO|	Nb gen
(1)	1582	1586	0.25	1508	28	80000	1583	0.06	1431	32	300000
(2)	2297	2330	1,43	1062	32	200000	2305	0.34	1057	29	300000

UB : best known mekespan, MM: best obtained makespan, MR: minimal tardiness, PO: Pareto optimal set.

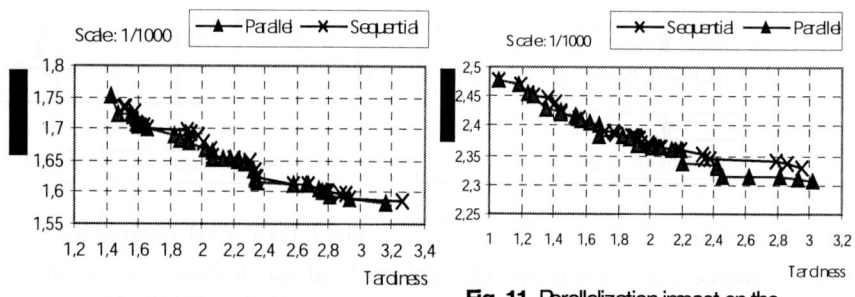

Fig. 10. Parallelization impact on the Pareto frontier for ma_ta11_bi

Fig. 11. Parallelization impact on the Pareto frontier for ma_ta21_bi

As shown in figures 10 and 11, the use of large populations distributed on different GAs and the increase of the number of generations improve the quality of the Pareto frontiers.

6 Conclusion and Perspectives

In this work, we have tried to construct our approach by a progressive introduction of concepts such as selection, diversity maintaining, hybridization and parallelization. At each stage we have shown the contribution of the introduced mechanisms, that's what allows us to formulate the following conclusions.
Pareto selection strategies (NSGA, NDS, WAR) seem to be well adapted to the multicriteria flow shop problem. The efficiency of such methods is improved with the introduction of elitism during the selection phase. The elitism may lead to premature search convergence though and then, parameters A (elitist pressure) and Q (selection pressure) have to be adequately chosen. However, the risk of a genetic drift is always present. The diversification strategies seem to be the privileged mean to prevent such problems. Three variants of the method, based on the sharing, were developed. The phenotypic sharing (diversification in the objective space) appears to be the most interesting. This interest is related to the fact that a larger and better dispersed Pareto frontier is desirable. However, the genotypic diversification may yield good results also. The combination of both concepts improves considerably the search.

Combining GAs with local search was used in order to refine the search. The idea is to run the GA first in order to get a first approximation of the Pareto frontier. The local search has the merit to improve the solutions (find the local optima of the search regions). The contribution of hybridization appears for problem of important size only.

The work presented here deals with a bi-criteria problem. However, the method proposed is able to be extended very easily to multi-criteria problems. Therefore, we now need to test the strength of the method for flow-shop problems with more than two criteria. A graphic comparison is impossible in this case, and more elaborated tests of performances methods must be used, such as the contribution notion [16] and entropy [16]. The extension of the method to the general flow-shop problem (not only permutation flow-shop) and to the job shop problems is also to be studied. Other hybridization schemes must be evaluated. Indeed, combining GAs with more advanced local search techniques such as tabu search may give better results.

The proposed hybrid metaheuristic is still very slow due to the advanced mechanisms introduced. A parallel model has been proposed and implemented for the algorithm to overcome this drawback. The obtained speedup favor the use of large populations and the increase of the number of generations, which improves the quality of the obtained Pareto frontier.

References

1. Baker, J.E. Adaptive selection methods for Genetic Algorithms. Proceeding of international conference on Genetic Algorithms and their application, Page 101, 1985.
2. Bentley, P.J., Wakefield, J.P. Find an acceptable Pareto-optimal solutions using multiobjective Genetic Algorithms. Springer Verlag, London, page 231, June 1997.
3. Brah, S.B., Hunsucker, J.L. Branch & Bound algorithm for the flow-shop with multiple processors. European Journal of Operational Research, Vol 51, page 88, 1991.
4. Fonseca, C. M., Fleming, P.J. Multiobjective genetic algorithms made easy: selection, sharing and mating restrictions. In IEEE Int. Conf On Genetic Algorithms in Engineering System: Innovations and Applications, Page 45, Sheffield, UK, 1995.
5. Fujita, K., Hirokawa, N., Akagi, S., Kimatura, S., Yokohata, H. Multi-objective optimal design of automotive engine using genetic algorithm. In Proceedings of DETC'98 - ASME Design Engineering Technical Conferences, page 11, 1998.
6. Gonzalez, T., Sahni, S. Flowshop and Job-shop Schedules: Complexity and Approximation. Operational Research, Vol 26, N°1, page 36, 1978.
7. Gupta, N.D. An improved Combinatorial Algorithm For The Flowshop-Scheduling Problem. Operational Research, Vol 19, page 1753, 1969.
8. Hajela, P., Lin, C.Y. Genetic search strategies in multicriterion optimal design. Structural Optimisation, (4) Page 99, 1992.
9. Heller, J. Some Numerical Experiments For a MxJ Flow Shop And Its Decision Theoretical Aspects. Operational Research, Vol 8, page 178, 1960.
10. Murata, T., Ishibuchi, H. A Multi-objectives Genetic Local Search Algorithm and Its Application Flow-shop Scheduling. IEEE Transaction System. Vol 28, N°3, pp 392, 1998.
11. Nowicki, E. The permutation Flow shop with buffers : A tabu search approach. European Journal of Operational Research, Vol 116, page 205, 1999.
12. Rajendran, C., Chaudhuri, D. An efficient heuristic approach to the scheduling of jobs in flow-shop. European Journal Of Operational Research, Vol 61, page 318, 1991.
13. Simon French, F., Phil, D. Sequencing and scheduling: An introduction to the mathematic of the Job-Shop. Department of Decision Theory, University of Manchester. John Wiley & Sons Edition, 1982.
14. Srinivas, N., Deb, K. Multiobjective optimisation using non-dominated sorting in genetic algorithms. Evolutionary Computation 2(8), page 221, 1995.
15. Taillard, E. Benchmarks for basic scheduling problems. European Journal of Operational Research, Vol 64, Page 278, 1993.
16. Talbi, E-G. Métaheuristiques pour l'optimisation combinatoire multi-objectifs: Etat de l'art. Rapport interne, Université de sciences et Technologies de Lille, France, Jun 1999.
17. Van Veldhuizen, D.A., Lamount, G.B. Multiobjective Evolutionary Algorithm Research : A History and Analysis.Technical Report 98-03, Department of Electrical and Computer Engineering, Air Force Institute of Technology, USA, Dec 1998.
18. Zitzler, E. Evolutionary Algorithms for Multiobjective Optimization : Methods and Application. Dissertation submitted to the Swiss Federal Institute of Technology Zurich for a degree of Doctor of technical science, Nov 1999.

The Supported Solutions Used as a Genetic Information in a Population Heuristic

Xavier Gandibleux[1], Hiroyuki Morita[2], and Naoki Katoh[3]

[1] LAMIH-ROI/ROAD – UMR CNRS 8530
Université de Valenciennes, Campus "Le Mont Houy"
F-59313 Valenciennes cedex 9, France
xavier.gandibleux@univ-valenciennes.fr
[2] Osaka Prefecture University
1-1 Gakuen-cho, Sakai
Osaka 599-8231, Japan
morita@eco.osakafu-u.ac.jp
[3] Kyoto University
Yoshidahonmachi,Sakyo ward
Kyoto 606-8501, Japan
naoki@archi.kyoto-u.ac.jp

Abstract. Population heuristics present native abilities for solving optimization problems with multiple objectives. The convergence to the efficient frontier is improved when the population contains 'a good genetic information'. In the context of combinatorial optimization problems with two objectives, the supported solutions are used to elaborate such information, defining a resolution principle in two phases. First the supported efficient solution set, or an approximation, is computed. Second this information is used to improve the performance of a population heuristic during the generation of the efficient frontier. This principle has been experimented on two classes of problems : the $1 \mid \mid (\Sigma C_i, T_{max})$ permutation scheduling problems, and the biobjective 0-1 knapsack problems. The motivations of this principle are developed. The numerical experiments are reported and discussed.

1 Introduction

For some combinatorial optimization problems with two objectives, a (sub)set of exact efficient solutions or bound sets on the efficient frontier can be computed [5]. The supported efficient solutions deserve a lot of attentions for two reasons. First, these solutions are characterized and can be generated using a convex combination of the objectives [7]. Second, the available experimental results [12, 16] show the number of supported efficient solutions which grows with the size of the problem, but still smaller than the number of non-supported efficient solutions. These observations conduct us to experiment the following solving principle; first, generation of supported solutions using an exact method or a heuristic method and second, usage of these solutions to improve the generation of the whole efficient frontier using a population heuristic -PH-. In this context, supported solutions are considered as a kind of 'good genetic information' available and used by the population heuristic. They allow a fast convergence to the efficient frontier and maintain the diversity among the efficient frontier. We experimented this principle in different ways on two biobjective combinatorial problems. This paper focuses on the experimental results obtained (a complete description of the population heuristic used is available in [11,12]). Section 2 introduces usual definitions, the two classes of problem and gives a short presentation of our population heuristic. Section 3 reports and analyses the numerical experiments. A conclusion with a discussion is given in section 4.

2 Background

2.1 The Multiple Objective Combinatorial Optimization Problems

Given a finite set X and several objective functions $z^j : X \to \mathbb{R}$, $j = 1\ldots P$, a multiobjective combinatorial optimization (MOCO) problem is defined as [3] :

$$" \min_{S \in X} " (z^1(S), \ldots, z^P(S)) \qquad \text{(MOCO)}$$

An element $S \in X$ is a feasible *decision* and X is called the *decision space*. An vector $z(S) = (z^1(S), \ldots, z^P(S)), z(S) \in Z$ is a *performance* and Z is called the *objective space*. Typically two types of objective functions are considered, namely the sum and the bottleneck objective. The problem is then to solve (MOCO) where the meaning of "min" has still to be defined. Since objectives cannot be minimized simultaneously in general, most often the minimization in (MOCO) is understood in the sense of efficiency (or Pareto optimality).

Definition 1: A solution $S \in X$ is called efficient if there does not exist another feasible solution $S' \in X$ such that $z^j(S') \leq z^j(S)$ for all $j = 1\ldots P$ with strict inequality for at least one of the objectives.

Definition 2: The corresponding vector $z(S)$ is called non-dominated. The set of Pareto optimal (efficient) solutions of (MOCO) will be denoted by E, the set of *non-dominated vectors* by ND in the sequel.

Definition 3: The *efficient frontier* is the lower left part of the shortest curve that connects all ND vectors.

2.2 Resolution Difficulties

In the worst case, three factors are united and contribute to the (MOCO) resolution difficulties.

(A) Computational Complexity

A (MOCO) is $\mathcal{NP}$-hard if the single objective problem is $\mathcal{NP}$-hard [3]. For instance, the knapsack problem is one of the fundamental $\mathcal{NP}$-complete combinatorial optimization problems. Thus the biobjective 0-1 knapsack problem is $\mathcal{NP}$-hard as is the single objective case. A similar result exists for scheduling problems. As soon as one objective is recognized as $\mathcal{NP}$-hard to be optimized, then the multicriteria problem is also $\mathcal{NP}$-hard [2]. Thus a single machine scheduling problem with 2 objectives is difficult to solve with an exact method in reasonable computational time in practice when one objective is $\mathcal{NP}$-hard to be optimized.

(B) Solutions sets

In the convex case (the P functions z^j are convex and X is a convex set), E coincides with the set SE of *supported efficient solutions*. Given the weights λ^j, $j = 1\ldots P$, a supported efficient solution is the optimal solution of a single objective problem (P_λ) corresponding to the maximization of the weighted sum of the objective functions z^j:

$$\left[\max\left\{\sum_{j=1}^{P}\lambda^j z^j(x) \mid x \in X, \sum_{j=1}^{P}\lambda^j = 1, 0 \leq \lambda^j \leq 1, j = 1\ldots P\right\}\right] \quad (P_\lambda)$$

Due to the discrete structure of the (MOCO) problem, the feasible domain X is generally non-convex (see Figure 1), so that $E = SE \bigcup NE$ where NE denotes the set of *non-supported efficient solutions*.

(C) Number of solutions

The numerical results available on the biobjective knapsack problem [16] show that the number of supported solutions grows linearly with the problem size, but the number of non-supported solutions grows following an exponential function. We noticed the same observation (see Figure 2) for a class of biobjective permutation scheduling problems [12]. Other observations mention that the number of efficient solutions grows also with the number of objectives.

To summarize, (MOCO) is in the worst case an $\mathcal{NP}$-hard problem, with a non convex feasible domain X and a huge number of solutions for large multiple objective instances! In this context, an approximation approach is a reasonable alternative to exact methods. A solving procedure based on a (meta)heuristic is able to cope with the factors (A) and (B) discussed before. It yields a good tradeoff between the quality of an approximation $\widehat{E} = \widehat{SE} \cup \widehat{NE}$ of the exact efficient solution set E, and the time and memory requirements.

2.3 The $1 \mid\mid (\Sigma C_i, T_{max})$ Permutation Scheduling Problems

We consider a set of n independent jobs to be scheduled on a single machine, which can handle no more than one job at a time. The machine is assumed to

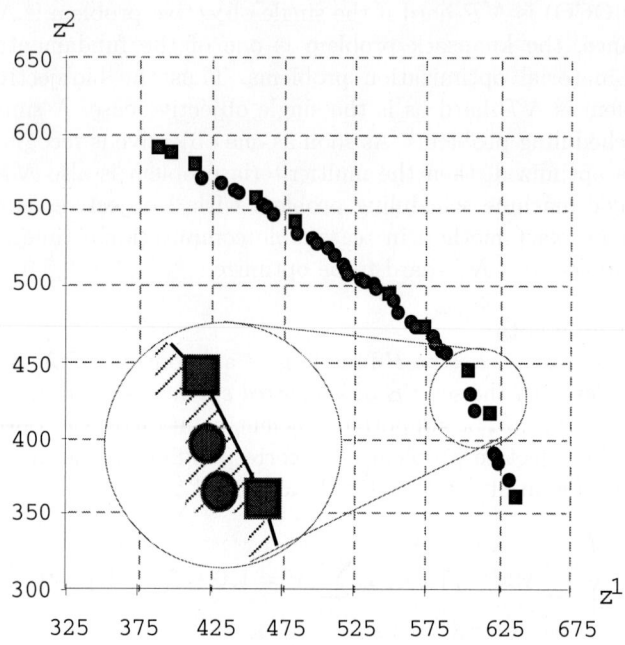

Fig. 1. Supported (square) and non-supported (circle) efficient solutions

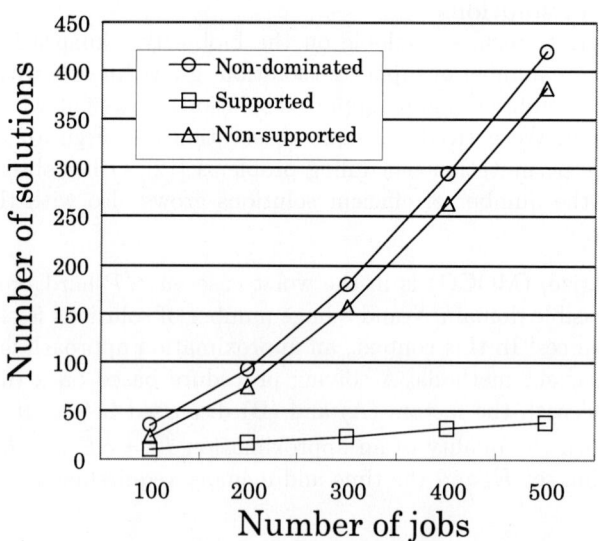

Fig. 2. Input size and numbers of efficient solutions

be available at time 0. Job J_i ($i = 1, \ldots, n$) requires processing during a given non-preemptive time p_i; to each job is assigned a due date d_i, at which J_i should be finished. Moreover if a relative importance between jobs has to be considered, a weight w_i is assigned to each job. It is assumed that all values p_i, d_i and w_i are positive integers. A schedule σ defines for each job J_i a completion time $C_i(\sigma)$ such that technical constraints (capacity and availability) of the machine are respected. We consider that the quality of a schedule is measured in term of two objectives $z^j : \sigma \to I\!N$, $j = 1, 2$, both are assumed to be minimized. z^1 and z^2 are two *regular performance criteria* defined on σ :

- the total flow time : ΣC_i defined as $\Sigma_{i=1,n} C_i(\sigma)$
- the maximum tardiness : T_{max} defined as $\max_{1 \leq i \leq n} T_i$

where tardiness T_i of job i is defined to be $\max\{C_i(\sigma) - d_i, 0\}$.

According to the classical notation of [13], this class of biobjective single-machine scheduling problems is denoted by $1 \mid\mid \Sigma C_i, T_{max}$ (biobjective single machine permutation scheduling problems minimizing both total flow time and maximum tardiness simultaneously). Since we are considering regular performance measures, a schedule is determined by a permutation σ over $\{1, 2, \ldots, n\}$. Given a schedule σ, completion time $C_i(\sigma)$ of job i is uniquely determined and is denoted by C_i for simplicity in the sequel.

This problem has received a lot of attention. Van Wassenhove and Gelders [15] have proposed a pseudo-polynomial algorithm for generating the exact set of efficient solutions. This class of problems has been chosen in this paper, not to put in competition a heuristic method with an exact method but, to give an experimental feedback easily.

2.4 The Biobjective Knapsack Problems

We consider the 0-1 MultiObjective Knapsack (0-1MOKP) where coefficients c_i^j, w_i and ω are nonnegative constants. Since all coefficients are nonnegative, if $w_i > \omega$ or $c_i^j = 0$, $j = 1 \ldots P$ then the variables x_i can be fixed at 0 in any efficient solution $x \in E$, while if $\sum_{i=1}^{N} w_i \leq \omega$, $E = \{e\}$ where $e = (1, 1, \ldots, 1)^t$ is the vector of all 1's. The biobjective case ($P = 2$) is denoted by (BiKP).

$$\left[\begin{array}{ll} \text{"max"} & z^j(x) = \sum_{i=1}^{N} c_i^j x_i \quad j = 1 \ldots P \\ \text{s.t.} & \sum_{i=1}^{N} w_i x_i \leq \omega \\ & x_i \in \{0, 1\} \quad\quad i = 1 \ldots N \end{array}\right] \quad \text{(0-1MOKP)}$$

With one or multiple objectives, the 0-1 knapsack problem has received quite some consideration by researchers in the last two decades. All the papers concerning the resolution of the (0-1MOKP) deal with the problem of identifying or approximating E. The proposed algorithms are either based on implicit enumeration methods such as dynamic programming, branch and bound, or apply heuristic procedures, especially metaheuristics like Simulated Annealing, Tabu Search, Genetic Algorithms, and others to approximate E (see [4]).

2.5 Overview of Our Population Heuristic

A genetic algorithm is a population heuristic built on iterative solution techniques that handle a *population* of *individuals* and make them evolve according to some specified rules. Self-adaptation and co-operation are the fundamental mechanisms applied on the individuals at each iteration [9]. Genetic algorithms have been recognized to be well suited to multiobjective optimization in nature since they are keeping multiple solutions in parallel.

Classical GA did not incorporate any problem specific knowledge. However incorporating knowledge improves the computational effectiveness of the heuristic. The term hybrid genetic algorithms was introduced to distinguish from a classical GA. An equivalent term is memetic algorithm. A hybrid GA is also sometimes called a genetic local search algorithm [1].

GA proceeds in general by keeping a set of solutions (the population), and by performing crossover, mutation, and selection operations [8]. The major difference between single-objective and multiobjective GA's lies in the way of selecting individuals for the next generation. In a single-objective GA an individual can be evaluated according to the single objective function. In a multiobjective GA it is desired to obtain a set of individuals uniformly distributed over the efficient frontier and well approximates the set of non-dominated vectors.

Considering the results presented in [11], the main strategies of our population heuristic (called MGK for Morita/Gandibleux/Katoh) are described briefly (a complete description of the population heuristic used is given in [11,12]):

a) All solutions ranked one by Goldberg's method [8] are kept for the next generation, because computing them again is expensive in CPU time.
b) (Schaffer's strategy) When selection is performed, a few good solutions in the current population with respect to each objective are kept for the next generation even if their ranks are more than one.
c) Among the solutions not selected in (a) and (b), solutions kept for the next generation are determined by tournament selection based on domination relationship with *sharing*. We use this strategy hoping to keep the diversity of individuals.
d) (Seeding strategy) In addition to the set of randomly generated individuals, we initially add a few solutions to the current population that are good with respect to each objective. An exact or heuristic algorithm depending on the computational difficulty of the corresponding single-objective scheduling problem computes such good solutions. Using this strategy, we can obtain good solutions more quickly.
e) To improve the convergence to non-dominated vectors, a local search method is incorporated into our algorithm. At any generation, the local search

method is applied to all elite individuals except for which local search has
been already applied in the previous generations. In this local search method,
all of the neighborhood solutions which is not dominated by a tentative approximate non-dominated solution are kept. So it is possible to keep the
numbers of approximate non-dominated solutions at a time.

3 Numerical Experiments

In these experiments, a standard PC (Celeron 300A) is used and compilation is
achieved with gcc (ver.2.5.8). All the numerical instances are available on our
web site [17]. We report several information, especially the *detection ratio* e.g.
the number of efficient solutions that MGK can find in E. The seeding solutions
used in the sequel come from the supported efficient solution set.

3.1 The 1 | | (ΣC_i , T_{max}) Permutation Scheduling Problems

We have randomly generated 10 problem instances for each of five different
numbers of jobs, i.e., $n = 100, 200, 300, 400, 500$. Exact efficient solutions are
computed using the algorithm of [15]. To achieve our analysis, the MGK has
been 'forced' to detect all efficient solutions. For all instances, our population
heuristic is able to find all efficient solutions, often in reasonable computational
time up to 500 jobs. Nevertheless, the generation of E seems more difficult for
the sets Dat3 and Dat4. An explanation could be due to the large number of
efficient solutions, especially the non-supported solutions, to generate.

Resolution with Two Seeding Solutions. For 1 | | $\Sigma C_i, T_{max}$, we can compute an exact solution for each objective [15]. 1 | | ΣC_i and 1 | | T_{max} can be
optimally solved by the Smith's rule and by the EDD rule, respectively. So these
solutions are incorporated in the initial solution set. Tab. 1 illustrates the results
for all instances using two seeding solutions. Fig. 3 and Fig. 4 illustrate individuals after 50 generations and 500 generations respectively. As figures show, the
algorithm can spread individuals to entire efficient frontier, because seeding solutions can propagate their superior genetic information to other individuals with
a weak number of generations. Fig. 5 makes a zoom in dotted square plotted in
Fig. 4. The individuals come near the efficient frontier, and most of individuals
are efficient solutions.

Resolution using whole SE as Seeding Solutions. In computing the convex
hull for E, the subsets SE and NE are identified. We experiment now the case
when whole SE is used as seeding solutions. Tab. 2 illustrates the results for all
instances when all SE solutions are used in the initial population. It shows that
the good genetic information of SE solutions have much impacts for finding all
efficient solutions. We notice that number of generations and computational time
become about half of them using two seeding solutions. From these results, we
can see that it is very effective to use these genetic information of SE solutions
as much as possible.

2 Problem Statement

The permutation flow-shop problem consists of a set **J** of n jobs that must be processed in a set of machines **M**. Each job $j \in \mathbf{J}$ has $m = |\mathbf{M}|$ operations. Each operation O_{kj}, representing the k-th operation of job j, has an associated processing time t_{kj}. Each machine must finish the operation once it is started to be processed (no preemption allowed). No machine can process more than one operation at the same time. No operation can be processed by more than one machine at a time. Each job j is assigned a readiness time r_j, and due date d_j. All jobs must have the same routing through all machines. The goal is to find a permutation of jobs that minimizes a given objective function (since the order of machines is irrelevant).

In order to understand the objective functions we want to optimize we need to set up some notation first. Let us denote the starting time of operation O_{kj} by s_{kj}, its completion time by c_{kj}. Define $\mathbf{K}_m$ as the set $\{1, 2, \cdots, m\}$. With this notation a feasible solution holds the following conditions:

$$s_{kj} > r_j \quad \forall k \in \mathbf{K}_m, j \in \mathbf{J} , \tag{1}$$

$$s_{kj} + t_{kj} \leq s_{(k+1)j} \forall k \in \mathbf{K}_{m-1}, j \in \mathbf{J} . \tag{2}$$

All pairs of operations O_{kj} and O_{ri} processed on the same machine must satisfy:

$$\begin{aligned} s_{kj} + t_{kj} \leq s_{ri} \quad &\text{or} \\ s_{ri} + t_{ri} \leq s_{kj} \quad &\text{for each machine in } \mathbf{M}, k \neq r \text{ or } j \neq i . \end{aligned} \tag{3}$$

Now we are in position of defining the objective functions. First we consider the makespan, which is the completion time of the latest job, i.e.

$$f_1 = \max_j \{s_{mj} + t_{mj}\} . \tag{4}$$

The mean flow time, representing the average value of the time during which the jobs remain in the shop, is the second objective.

$$f_2 = \bar{fl} = (1/n) \sum_{j=1}^{n} fl_j , \tag{5}$$

where $fl_j = \max_j \{s_{mj} + t_{mj}\} - r_j$, i.e. the time job j spends in the shop after it is released. The third objective is the mean tardiness, i.e.

$$f_3 = \bar{T} = (1/n) \sum_{j=1}^{n} T_j , \tag{6}$$

where $T_j = \max\{0, L_j\}$, and $L_j = s_{mj} + t_{mj} - d_j$.

Thus, we have the following MO problem:

$$\min(f_1, f_2, f_3)$$
$$\text{subject to } (1) - (3) \ . \tag{7}$$

3 GA's Approach to MO Scheduling Problems

There are many approaches for solving the general MO problem by using GA's. Surveys on the exiting GA's methodologies can be found in [10], [1], [4], and references therein. Almost any application uses the methodologies described in these surveys.

Since this is a new research area, there are still many fundamental questions to be answered. Specially, in the field of MO-COP's, everything is to be done. To date, one of the most pragmatic question to answer is how to fairly compare two given methodologies, or in the best case, how to judge any given methodology.

The application of GA's to MO scheduling problems has been rather scarce. Two interesting ideas are those presented in [8], and [9].

In [8] the scheduling of identical parallel machines, considering as objective functions the maximum flow time among machines and a non-linear function of the tardiness and earliness of jobs, is presented. In [9] a natural extension of NSGA [10] is presented and applied to flow-shop and job-shop scheduling problems. Another, totally different approach is that presented by Isibuchi and Murata [6]. They use a local search strategy after the genetic operations without considering non-dominance properties of solutions. Their method is applied to the MO flow-shop problem.

The main idea when solving MO scheduling problems is to apply the existing GA's methodologies to the problem to solve. However, there are no traces of studies on how adequate these methodologies may be. Again, the lack of a fair methodology for comparing the results does not help to improve this situation.

In order to design adequate genetic operators we need to know the properties of solutions and to understand the problem-algorithm landscape. The following questions are of much interest:

1. Are close solutions in the objective function space close in the domain space?
2. Are close Pareto optimal solutions (in the objective function space), close in the domain space?
3. Does crossover of non-dominated solutions generate mostly non-dominated solutions?
4. What type of crossover or mutations favours the creation of non-dominated solutions from non-dominated solutions?

These questions, related to the problem-algorithm landscape have received very little attention, although they are of primary importance.

When we design move-operators to deal with neighbourhood construction for multi-objective optimization problems, there are also fundamental questions we need to answer in order to choose the right operator. In the generated neighborhood:

5. Is there always at least one non-dominated neighbour?
6. Is there a high percentage of non-dominated solutions among the neighbours?
7. Is there any type of neighborhood that favours the generation of non-dominated solutions? at least one, (almost always) one, or many?

There is no trace of research addressing these questions for MO problems. In the case of single objective scheduling problems such questions are answered in many works related to landscape study as well as neighbourhoods study (see for example [7] and references therein).

4 The Proposed Algorithm

The algorithm we propose here is just the standard GA for MO problems as suggested in [10], with a minor modification. The contribution we try to make is in the analysis of genetic operators in order to choose the adequate set for a given problem. The proposed procedure is in its preliminary stage. Therefore, more questions than answers will be highlighted.

The specific MOGA we use here as a framework is stated as follows.
Algorithm 1. Multi-objective GA.

Step 1. Set $r = 0$. Generate an initial population $POP[r]$ of g individuals.
Step 2. Classify the individuals according to a non-dominance relation. Assign a dummy fitness to each individual.
Step 3. Modify the dummy fitness by fitness sharing.
Step 4. Set $i=1$.
Step 5. Use RWS to select two individuals for crossover according to their dummy fitness. Perform crossover with probability p_c.
Step 6. Perform mutation of individual i with probability p_m.
Step 7. Set $i = i+1$. If $i = g$ then go to Step 8 otherwise go to Step 5.
Step 8. Set $r = r + 1$. Construct the new generation $POP[r]$ of g individuals. If $r = r_{max}$ then STOP; otherwise go to STEP 2.

The procedures involved at each step of this algorithm are explained in the following subsections.

4.1 Individual Representation and Decoding

Each individual is represented by a string of integers representing job numbers to be scheduled. In this representation individual r looks like:

$$\mathbf{i}_r = (i_1^{(r)} i_2^{(r)} \cdots i_n^{(r)}), \quad r = 1, 2, \cdots, g ,$$

where $i_k^{(r)} \in \mathbf{J}$.

The schedule construction method for this individual is as follows:

1) Enumerate all machines in $\mathbf{M}$ from 1 to m.
2) Select the first job $(i_1^{(r)})$ of $\mathbf{i}_r$ and route it from the first machine (machine 1) to the last (machine m).
3) Select iteratively the second, third, $\cdots$, n-th job and route them through the machines in the same machine sequence adopted for the first job $i_1^{(r)}$ (machines 1 to m). This must be done without violating the restrictions imposed in (1) to (3).

4.2 Genetic Operators

The selection operator we use here is standard to GA's, like those proposed elsewhere [5]. Two selection processes are distinguished here.

Selection for mating (Step 5). This is the way we choose two individuals to undergo reproduction (crossover and mutation). In our algorithm the so called roulette wheel selection (RWS) is used. This selection procedure works based on the dummy fitness function assigned to each individual. The way to compute the dummy fitness (Step 2) and the way to do the fitness sharing (Step 3) are standard (see [9]).

Selection after reproduction (Step 8). This is the way to choose individuals to form the new generation from a set given by all parents and all offsprings. In this paper, the best elements are selected from the pool of parents and offsprings.

To define "the best", g individuals are sorted according to those belonging to the non-dominated front, among these, individuals with better makespans have higher priority followed by tardiness, and finally by the mean flow time. After sorting all individuals in this front they are erased from the population. The same procedure is applied to the remaining individuals, until we complete g sorted individuals.

If there are repeated individuals (considering the objective functions), then these are erased (only one copy of each type, at each step of the sorting process, is left) and replaced by randomly selected individuals from the pool of parents and children that where not sorted.

We need to explain now the crossover and mutation operators to be used. Three different types of crossover and mutation operators are considered.

We start explaining the crossover operators (Step 5).

OBX. This is the well known order-based crossover (see [3]) proposed by Syswerda. The position of some genes corresponding to one of the parents are preserved in the offspring.

PPX. Precedence-based crossover. A subset of precedence relations of the parents genes are preserved in the offspring.

TPX. Two point crossover. This is a special case of OBX with the difference that two segments of one of the parents are always copied into the offspring.

The mutation operators for the flow-shop problem can be considered as move operators in a neighborhood since, in average, the mutated solution is not far away from the original solution. The following mutation operators are used (Step 6).

SWAP1. A single swap of two adjacent genes is performed. The locus to swap is randomly selected.

SWAP2. Two loci are randomly selected and their alleles interchanged.

SWAP3. Two loci (l_1, l_2) are randomly selected if $l_1 < l_2$ then the allele corresponding to l_1 is placed on l_2 and all genes from $l_1 + 1$ to l_2 are shifted one position towards l_1. If $l_1 > l_2$ then the opposite operation is performed.

Before actually using any of these operators in Algorithm 1 we would like to know about their effects on the non-dominance and distance relations between the parents and the offsprings. It is also important to know the length of the jumps of each move (mutation) operator in order to understand which is the most appropriate for the problem to solve.

To do this we start by defining the distance measure we are going to deal with.

5 Distance Measures

In the classical permutation flow-shop problem, the solution is totally defined by the sequence of jobs numbers. Therefore, the distance measure gives an idea of how different two such sequences are. To compute this difference, each sequence $\mathbf{s}=(j_1, j_2, \cdots, j_n)$ is associated with an $n \times n$ matrix for whose elements we define $a_{ij}(\mathbf{s})=1$ if job j is scheduled before job i, and zero otherwise. Thus, the difference between schedules $\mathbf{s}r$ and $\mathbf{s}k$ is given by

$$d(\mathbf{s}k, \mathbf{s}r) = \sum_{j=1}^{n} \sum_{i=1}^{n} a_{ij}(\mathbf{s}r) \oplus a_{ij}(\mathbf{s}k) \qquad (8)$$

where $\oplus$ represents the exclusive-or logical operation. To normalize the distance (8) we divide it by the maximum number of different elements between two given associated matrices, i.e.

$$dn(\mathbf{s}k, \mathbf{s}r) = d(\mathbf{s}k, \mathbf{s}r)/n(n-1) \ . \qquad (9)$$

We call this the *domain distance* since it uniquely represents the solution which is mapped into the objective function space. This type of distance measure definition can be found elsewhere [7].

5.1 Objective Function Distance

We define the *objective function distance (ofd)* between solutions $\mathbf{s}r$ and $\mathbf{s}k$ as the Euclidean distance of their mappings, i.e.

$$ofd(\mathbf{sk}, \mathbf{sr}) = (\sum_{j=1}^{q}(f_j(\mathbf{sk}) - f_j(\mathbf{sr}))^2)^{1/2} ,\qquad(10)$$

for a problem with q objective functions f_j ($j = 1, 2, \cdots, q$).

In the case of continuous function optimization the Pareto optimal solutions are close to each other. Then, if we want to reach any neighbour Pareto solution from a given Pareto solution, we need to move as little as possible. However, for discrete domain problems, this continuity property does not hold. Thus, we need to know how far the Pareto solutions are from one another. We need also to know what type of move operator is needed to go from one Pareto solution to another. This is important from the application point of view, since it will allow to increase the number of solutions available to the decision maker. Studies aimed to address this issue are exposed in the next section.

6 Experimental Setup and Results

This section is devoted to the study of the genetic operators: mutation (move), and crossover. We specially emphasize on the distance-dominance relations of these operators.

The specific problem we are dealing with is a 49-jobs 15-machines flow-shop problem with three objective functions. This problem was proposed in [9], and its solution space size is 6.08×10^{62}. Experiments and results related to the move operators are presented.

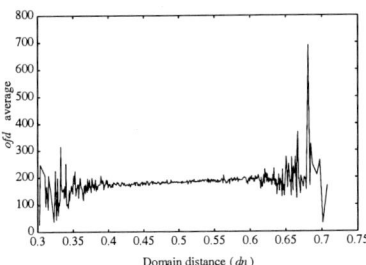
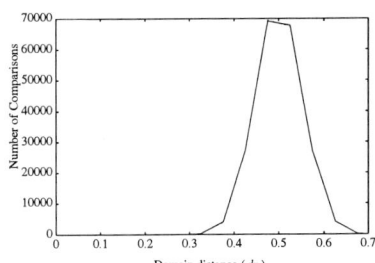

Fig. 1. Average of *ofd* for random solutions.

Fig. 2. Domain distance distribution. Random solutions.

6.1 Move Operators

The first experiment is aimed to study relations between the domain distance and the objective function distance. To do this, a set of 500 random solutions is generated, and for each domain-distance value (generated by comparing all against all solutions), the average on the objective function distance pis computed. The experiment is repeated 100 times.

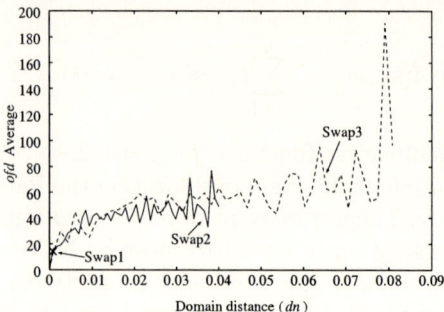

Fig. 3. Average of *ofd* for SWAP1, SWAP2, and SWAP3. The operators are applied to random solutions.

Table 1. Dominance relations. The operators are applied to random solutions.

Operator	$(o \succ n)$	$(o = n)$	$(o \prec n)$	$(o \succ\prec n)$
Swap1	42.46	6.04	42.56	8.94
Swap2	34.10	0.43	34.08	31.39
Swap3	33.55	0.41	33.65	32.39

Figure 1 shows the results of this experiment. All averages for all domain-distance values tend to the same constant objective function value. This tells us that close/far random solutions, in average, produce similar distances in the objective function space. The noisy behaviour in both extremes of the curve is due to the small number of individuals that are present for these values of domain distance, as it is shown in Figure 2.

Table 2. Dominance relations. The operators are applied to non-dominated solutions only.

Operator	$(o \succ n)$	$(o = n)$	$(o \prec n)$	$(o \succ\prec n)$
Swap1	52.23	4.59	36.11	10.83
Swap2	50.18	0.33	17.24	32.35
Swap3	44.70	0.36	19.29	35.65

This result does not give much information on how close/far solutions in the domain space are mapped in the objective function space. One would, at first glance, expect that close/far solutions in the domain space produce close/far mappings in the objective function space. But, this is not the case for randomly generated solutions.

Table 3. Dominance relations. The operator are applied to the non-dominated solutions obtained after one GA run.

Operator	$(o \succ n)$	$(o = n)$	$(o \prec n)$	$(o \succ\prec n)$
Swap1	95.74	0.28	0.40	3.58
Swap2	98.57	0.01	0.02	1.40
Swap3	97.26	0.01	0.05	2.68

The objective of the second experiment is to study how move operators in the domain space move in the objective function space. A set of 2000 random solutions is generated, and the move operators are applied to each solution. The distance between the original solution o (origin) and the new solution n (neighbour) is measured along with their objective function distances.

Figure 3 shows the results for each move operator defined in section 4. We see that, as expected, the move operator SWAP1 (a single step in the domain space) produces neighbours which are close to each other in the objective function space. The point to learn is that if we need to go few steps in the objective function space we can use SWAP1, or to choose those solutions generated by SWAP2 or SWAP3 which are close to their origins in the domain space.

Now, we just need to know about the non-dominance relations generated by these move operators. To study these relations we propose the following experiment. Again, a set of 2000 random solutions is generated. Each solution is modified with each of the three move operators, then the dominance relation between the original and the modified solution is counted. The experiment is repeated 100 times and the mean is computed.

If the origin o dominates the neighbour n, then $(o \succ n)$ is used. If o and n produce the same objective function values, then $(o = n)$ is used. Neighbour dominance of the origin and non-dominance of neither the origin nor the neighbour are expressed by $(o \prec n)$ and $(o \succ\prec n)$, respectively.

Table 1 shows the results for random solutions. We see that the three operators behave similarly except for the number of solutions where no dominance relation can be determined. Swap2 and Swap3 produce higher values than Swap1.

Table 2 shows the results when only non-dominated solutions are selected (from the set of random solutions) as origin points. Here we observe that Swap3 produces more promising results than the other operators. This is because Swap3 accounts for 54.95% for cases where $(o \prec n)$ and $(o \succ\prec n)$, while the others do not reach 50%.

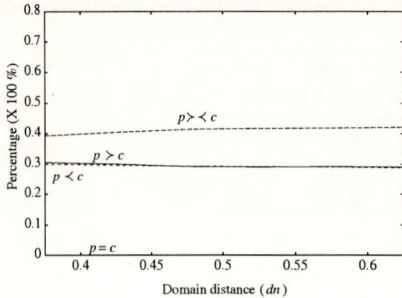 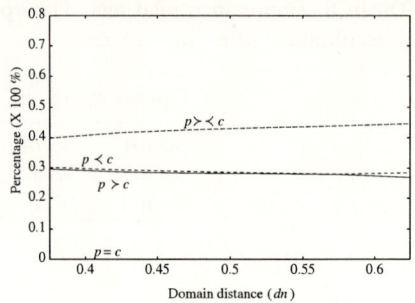

Fig. 4. Distance and dominance relations. The OBX operator is applied to random solutions.

Fig. 5. Distance and dominance relations. The PPX operator is applied to random solutions.

Table 3 presents the results when the move operators are applied to non-dominated individuals obtained after a 1000-generations-run of a GA. The results are the average over 100 runs. These results show how good or bad the used GA is. If we can easily find any dominating neighbour after a GA run, then it means that the algorithm performs poorly. However, if it is difficult to find new dominating solutions then it means that our algorithm performs well (i.e. converges to the Pareto-optimal set). Table 3 shows that it is difficult to find a dominating solution by using Swap2 or Swap3.

The analysis of move operators presented here gives us the idea of exploiting what can be called the "non-dominated local search" procedure. Here, move operators as well as move decisions can be studied to see their influence in the quality of the final set of non-dominated solutions.

6.2 Crossover Operators

Crossover operators are in charge of information interchange among individuals. Therefore, it is important to know which individuals are to be chosen, and how the information should be interchanged among these chosen individuals.

As a first step in the study of crossover operators we analyze the relations between the domain distance of the parents and the dominance relations between the parents and the offsprings. For doing this we use a set of randomly generated solutions, non-dominated solutions from the set of random solutions, and a set of non-dominated solutions of the last generation after a 1000-generations-run of a GA. In all cases the experiment is repeated 100 times and the average is computed.

Figures 4 to 6 show the relations of dominance against the parental distance when the parents come from the set of random solutions. The three operators, i.e. OBX, PPX, and 2PX have similar characteristics. For all distance values, cases where non-dominance relation can be establish between the offspring and at least one of the parents, are always greater than the other cases.

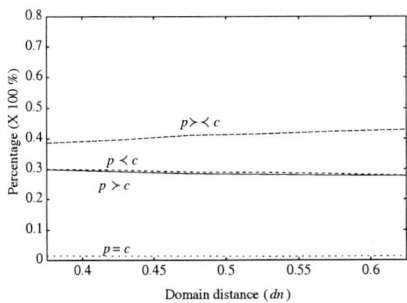

Fig. 6. Distance and dominance relations. The 2PX operator is applied to random solutions.

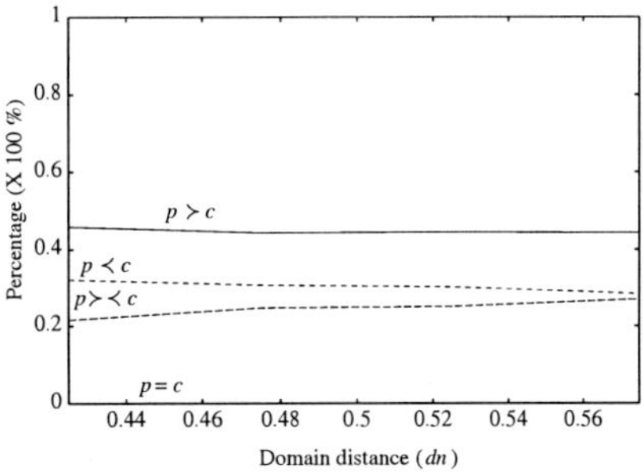

Fig. 7. Distance and dominance relations. The OBX operator is applied to non-dominated solutions only.

Table 4 shows the overall average results for each operator. We see that all operators have very similar averages.

Figures 7 to 9 show the results when the parents come from non-dominated solutions in the set of random solutions.

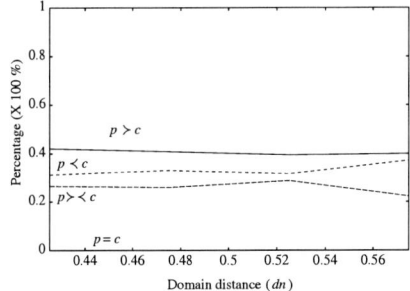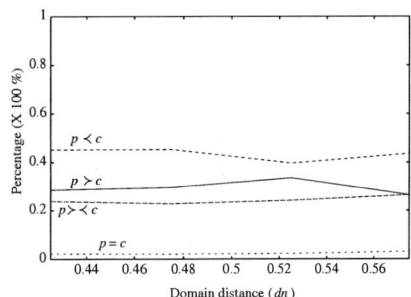

Fig. 8. Distance and dominance relations. The PPX operator is applied to non-dominated solutions only.

Fig. 9. Distance and dominance relations. The 2PX operator is applied to non-dominated solutions only.

Table 5 corresponds to the average results regardless the distance between the parents. It is observed from this table and Figure 9 that the superior characteristic of 2PX crossover over the other two types is clearly appreciated. The number of cases where the child dominates at least one parent is larger over all domain distance values. This could be the reason to explain why Isibuchi and Murata [6] found that this operator was adequate when dealing with the MO flow-shop problem.

Figures 10 to 12 and Table 6 present the results when the parents come from non-dominated solutions after one GA run.

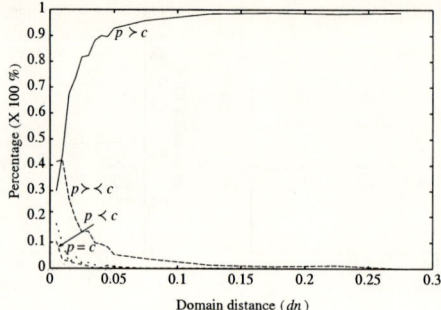

Fig. 10. Distance and dominance relations. The OBX operator is applied to non-dominated solutions obtained after one GA run.

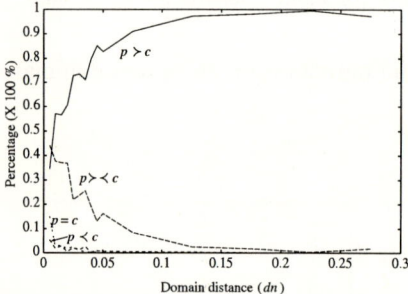 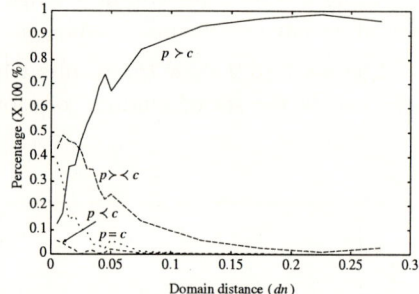

Fig. 11. Distance and dominance relations. The PPX operator is applied to non-dominated solutions obtained after one GA run.

Fig. 12. Distance and dominance relations. The 2PX operator is applied to non-dominated solutions obtained after one GA run.

We can see that for these experiments the number of cases where at least one parent dominates the offspring increases with increasing values of the domain distance. Again, the 2PX operator seems to outperform the others as it is also shown in Figure 12 and Table 6.

6.3 Comparative Results

Based on the experiments outcome in the previous subsections, we select the appropriate operators to use in Algorithm 1, and compare our results with those presented in [9].

It is worth mentioning at this point that the operators used in [9] are different from those used here. In this reference one-point crossover, SWAP1, and stochastic remainder selection with respect to the dummy fitness are the used crossover, mutation and selection operators, respectively.

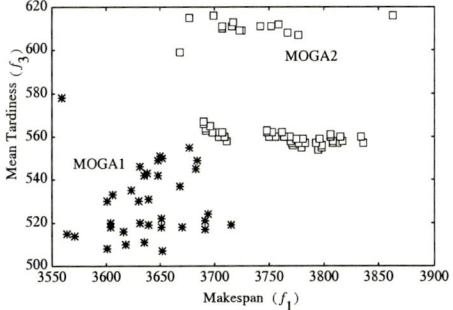

Fig. 13. Tardiness-Makespan relations. Non-dominated solutions in the last generation.

Table 4. Dominance relations. The operators are applied to random solutions.

Crossover	$(p \succ c)$	$(p = c)$	$(p \prec c)$	$(p \succ\prec c)$
OBX	29.23	0	29.25	41.53
PPX	28.31	0	28.81	42.88
2PX	28.65	1.31	28.84	41.19

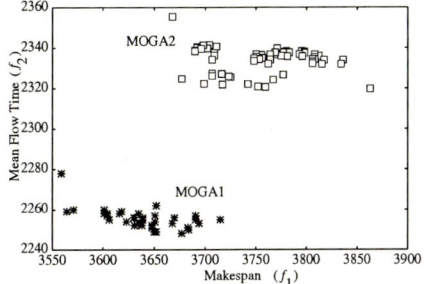
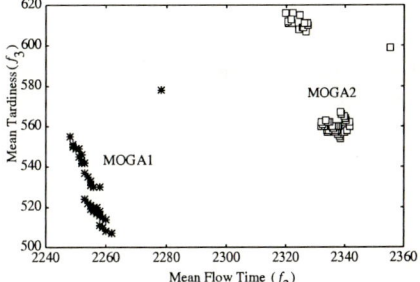

Fig. 14. Mean Flow Time-Makespan relations. Non-dominated solutions in the last generation.

Fig. 15. Tardiness-Mean Flow Time relations. Non-dominated solutions in the last generation.

Swap3 and 2PX are selected as the genetic operators. The population size is set as $g = 100$ individuals. The maximum number of generations is $r_{max} = 1000$. The crossover and mutation rates are $p_c = 1.0$ and $p_m = 0.01$, respectively.

Table 7 shows comparative results for the best (Swap3-2PX) and the worst (Swap1-PPX) combinations of operators. The number of non-dominated solutions, the average domain distance and objective value distance are shown. As expected the combination Swap3-2PX gives better results than those given by Swap1-PPX.

Table 5. Dominance relations. The operators are applied to non-dominated solutions only.

Crossover	$(p \succ c)$	$(p = c)$	$(p \prec c)$	$(p \succ\!\prec c)$
OBX	42.52	0	29.82	27.66
PPX	42.42	0	33.26	24.32
2PX	29.00	2.04	53.05	15.91

Table 6. Dominance relations. The operators are applied to non-dominated solutions obtained after one GA run.

Crossover	$(p \succ c)$	$(p = c)$	$(p \prec c)$	$(p \succ\!\prec c)$
OBX	93.59	0.86	0.46	5.09
PPX	89.77	0.67	0.60	8.96
2PX	79.45	4.11	0.83	15.61

Table 7. Comparison of non-dominated solutions (NDS) for the best and the worst combination of operators.

Crossover	% of NDS	$\overline{dn}$	$\overline{ofd}$
Swap3-2PX (best)	58.42	0.048	47.75
Swap1-PPX (worst)	26.16	0.016	27.36

Finally we compare our results with those reported in [9]. The projection of the solutions are shown in Figures 13 to 15. This are solutions that belong to the non-dominated front of the last generation of a single GA run. We can see that our results (MOGA1) clearly outperforms those of Bagchi (MOGA2) [9].

7 Conclusions

A detailed analysis of mutation and crossover operators is presented for a multi-objective flow-shop problem. The analysis is focused on how the operators influence the generation of non-dominated solutions according to the parental distance.

Based on this analysis we are able to design a high performance GA and applied it to a problem presented by Bagchi in [9]. The simulation results show that our results clearly outperform (in solution quality) the ones presented in [9]. The relevance of this work is in the procedure proposed for choosing the right operators to use and not much in the superiority of our results over those presented in [9]. The analysis of mutation operators also gives some insight on how to perform effective moves when a "non-dominated local search" is to be designed.

There are still many open questions related to the landscape of multi-objective combinatorial optimization problems, specifically for multi-objective scheduling problems. Our results present just a little but motivating step in answering the open questions.

References

1. Coello Coello, C. A. A comparative survey of Evolutionary-Based Multiobjective Optimization Techniques. Unpublished Document.
2. Deb, K. Multi-objective Genetic Algorithms: Problem Difficulties and Construction of Test Problems. *Evolutionary Computation*, 7(3), pp:205-230 (1999).
3. Gen, M. and Cheng, R. *Genetic Algorithms & Engineering Design*. John Wiley & Sons (1997).
4. Gen, M. and Cheng, R. *Genetic Algorithms & Engineering Optimization*. John Wiley & Sons (1997).
5. Golberg, D. *Genetic Algorithms in Search, Optimization, and Machine Learning*, Addison-Wesley (1989).
6. Isibuchi, H. and Murata, T. Multi-objective Genetic Local Search Algorithm. Proceedings of the 1996 International Conference on Evolutionary Computation, pp:119-124, (1996).
7. D. C. Mattfeld. *Evolutionary Search and the Job Shop*. Physica-Verlag, Heidelberg (1996).
8. Tamaki, H., and Nishino, E. A Genetic Algorithm approach to multi-objective scheduling problems with regular and non-regular objective functions. Proceedings of IFAC LSS'98, pp:289-294 (1998).
9. Bagchi, T. P. *Multiobjective Scheduling by Genetic Algorithms*. Kluwer Academic Publishers (1999).
10. Srinivas, N. and Deb, K. Multi-Objective function optimization using non-dominated sorting genetic algorithms. *Evolutionary Computation*, 2(3), pp:221-248 (1995).

Pareto-Optimal Solutions for Multi-objective Production Scheduling Problems

Tapan P. Bagchi

Indian Institute of Technology Kanpur, India 208016
Bagchi@iitk.ac.in

Abstract. This paper adapts metaheuristic methods to develop Pareto optimal solutions to multi-criteria production scheduling problems. Approach is inspired by enhanced versions of genetic algorithms. Method first extends the Nondominated Sorting Genetic Algorithm (*NSGA*), a method recently proposed to produce Pareto-optimal solutions to numerical multi-objective problems. Multi-criteria flowshop scheduling is addressed next. Multi-criteria job shop scheduling is subsequently examined. Lastly the multi-criteria open shop problem is solved. Final solutions to each are Pareto optimal. The paper concludes with a statistical comparison of the performance of the basic NSGA to NSGA augmented by elitist selection.

1 Multi-criteria Shop Scheduling

Managing a production rarely implies "getting the orders out the fastest way possible." Scheduling is an optimization process by which limited resources are allocated over time among parallel and sequential activities, has been now formally studied over five decades. The task can quickly become complex, limiting the practical utility of combinatorial, mathematical programming and other analytical methods (Baker, 1974; French, 1982; Sannomiya and Iima, 1996). Nevertheless, typically, a production manager is looking for ways to *simultaneously* minimize tardiness of the jobs from committed shipping dates, maximize the use of expensive presses, furnaces, reactors and rolling mills, and human resources, minimize the mean flow time of jobs, etc. etc. Such scheduling situations are multi-criteria or *multi-objective* (Table 1). Ironically, most formal scheduling techniques confine to single objective optimization (Pinedo, 1995). In multi-objective scheduling, the objectives are often conflicting. Such problems are called multiple criteria decision making (MCDM) problems.

MCDM assumes that all the objectives are dispensable and that all can be traded off although some may be more important than the rest. Adulbhan and Tabucanon (1980) classified these techniques based on the way the initial multi-objective problem is transformed into a mathematically manageable format using conversion of secondary objectives into constraints, development of a single combined objective function, or treatment of all objectives as constraints. Others have emphasized the *stage* at which the analyst needs information from the decision-maker. Recently, heuristic methods, which engage the Pareto-optimality concept to solve multiobjective problems have been proposed. In multi-objective optimization there usually exist many solutions that are optimal in the *Pareto* sense, a concept existing in economics. From this viewpoint, due to the plurality of optimal decisions, the most desirable decision may be selected *after* one has generated the nondominated solutions. The final solution thus selected is called the *preferred solution*.

One approach proposed to search for Pareto optimal solutions is the Nondominated Sorting Genetic Algorithm (NSGA), created by Srinivas and Deb (1995). NSGA is based on concept of niche formation and speciation seen in natural biological evolution and is actually a clever extension of Simple Genetic Algorithm (SGA), the original method created by Holland (1975) to optimize a single objective. This paper accepts *Pareto optimality* as the basis for rational choice in multi-objective decision making investigates the efficacy of variations of NSGA to solve multi-objective shop scheduling problems. This needs no prior specification of the decision maker's preferences, which may be easier to express once the nondominant solutions are at hand. Indeed, the use of weighted-sum, if preferred, is a much easier task: it requires only a direct application of SGA.

In multi-objective optimization with conflicting objectives, there is *no* unique optimal solution. A simple optimal solution may exist here only when the objectives are non-conflicting. For conflicting objectives one may at best obtain what economists call "efficient" or nondominant solutions (Figure 1). An *efficient* solution x^* (also called a *Pareto optimal* solution) is one in which no increase can be obtained in any of the objectives without causing a simultaneous *decrease* in at least one of the remaining objectives (Keeney, 1983). However, although the concept of Pareto optimality is now over 90 years old (Pareto, 1906), methods for finding nondominated solutions are relatively few. When the factors and the constraints are well-behaved, a procedure known as the ε-method (Seo and Sakawa, 1988) may find the nondominated solutions. For other problems, search methods are used. Shop scheduling decisions involve sequences (of jobs, machines, etc.), hence the constraints and objectives of relevance here are typically not well-behaved.

Table 1. Multiple management objectives in an enterprise

Department	Objective(s)
Budget	Cost minimization
Production	• Production output maximization • Production time minimization • Resource utilization
Quality Control	• Product quality maximization • Rework minimization
Personnel	Minimization of hiring and firing
Marketing	Uninterrupted supply of products to customers

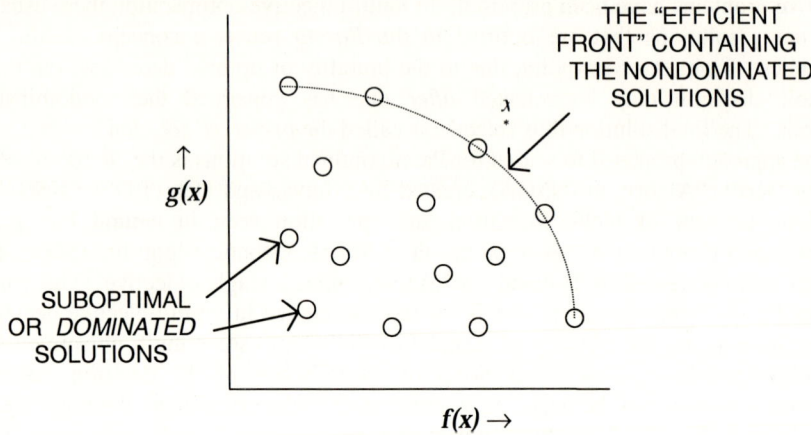

Fig. 1. Nondominated multi-objective maximizing solutions on the efficient front

2 Machine Scheduling Preliminaries

A flowshop requires unidirectional flow of work with a variety of jobs all being processed sequentially in the same order, in a one-pass manner. Each job follows identical routing through the processing stages or "machines." Unlimited storage exists between the machines. The challenge here is to determine the *optimum sequence* in which the jobs should be processed in order that one or more performance measure, such as the total time to complete all the jobs, the average mean flow time, or the mean tardiness of jobs from their committed dates is minimized. A job shop, on the other hand, involves processing of the jobs on several machines without any "series" routing structure (Uckun et al, 1993). Open shops are similar to job shops with the exception that there is no *a priori* order of the operations to be done *within* a single job (Pinedo, 1995). Many shop scheduling problems belong to the *NP-hard* class (Lenstra et al., 1977, French, 1982).

Recent advances in *metaheuristic search* methods that help conduct directed "intelligent" search of the solution space have brought new possibilities to rapidly find efficient and economic schedules, even if not optimal. These methods are context-independent and can be applied even when very little in known about the (mathematical) structure of the response functions. These methods are stochastic and now include genetic algorithms (GAs), tabu search, threshold acceptance and simulated annealing.

The genesis of GAs was an insightful observation by John Holland (1975) that some aspects of natural evolution, in particular *adaptation*, could be cast into useful algorithms to seek out solutions to the more difficult global optimization problems. GAs discover solutions to global optimization problems *adaptively*, looking for small, local improvements rather than big jumps in solution quality (Goldberg, 1989). While most stochastic search methods operate on a single solution to the problem at hand,

GAs operate on a *population* of solutions. To use GA, however, one must first *encode* the solutions to the problem in a chromosome-like *structure* (Goldberg, 1989). The procedure then applies *crossover* and *mutation* and other processes inspired by natural evolution to the individuals in the population to generate new individuals (solutions). The GA uses various selection criteria to pick the best individuals for *mating* so as to produce superior solutions by combining parts of parent solutions akin to genetically breeding race horses or superior strains of food crop. The objective function of the problem being solved determines how "good" each individual is.

Two phenomena closely linked to evolution are known as *niche formation* and *speciation* (Smith, 1995). Niches are behavior patterns that develop when organisms compete with each other for limited resources, or they attempt to survive in unfavorable environmental conditions. Speciation is the process by which new and stable species evolve in natures. NSGA uses the notions of nondomination (Figure 1) and niche formation. However, the performance of NSGA is significantly enhanced if one incorporates elitism in it. This speeds up the rate at which Pareto-optimal solutions are discovered. The elitist enhancement of NSGA, called *ENGA* in this paper, incorporates an additional nondominated sorting step to decide which solutions would form the parents.

While the methods presented in this paper solve only "static" flowshop, job shop and open shop problems, solutions to multi-criteria scheduling problems even for the static case are not easy to develop. However, the method can be extended to a dynamic flow, job or open shop. Solving the multi-criteria static open shop may be extended to multi-criteria classroom scheduling, a problem as common as universities and colleges that abound.

3 Genetic Algorithms for Sequencing Jobs in a Flowshop

Genetic algorithms (GAs) belong to the class of heuristic optimization techniques that utilize randomization as well as directed *smart* search to seek the global optima. Increasingly, GAs are being found to be more general and abstract (context independent) than other popular heuristic techniques presently available. As a result, many researchers have already turned to GAs to solve the more difficult and large sequencing, lot sizing and even classroom scheduling problems (Carter, 1997). Pinedo (1995) provides a clear rendering of GAs in the context of scheduling. The description goes briefly as follows and this is the broad approach followed by many researchers.

When applied to flowshop scheduling, GAs can view job sequences directly as "chromosomes" (the candidate schedules or solutions), which then constitute the members of a GA population. Subsequently, each individual (a schedule) is characterized (merited) by its fitness (e.g. by its makespan value). For a flowshop a chromosome would represent a job sequence on a machine, such as [1 3 2 4 5]. Fitness evaluation for a sequence would go, for instance, as the *smaller* its makespan, the "fitter" it is. As the GA executes, in each generation the "fittest" chromosomes are encouraged to reproduce while the least fit "die." A *mutation* in a parent chromosome may be an adjacent pairwise interchange of jobs or some variation of it in the corresponding sequence. Mutation here is designed to perform random exchange of jobs ("genes") between two randomly selected positions on a target chromosome. An example is shown below.

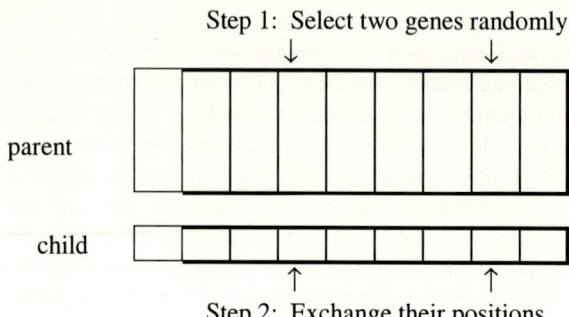

A *crossover* may combine some features of two parent chromosomes to create progenies inheriting some characteristics from each parent. A "repair" scheme may be set up to ensure that only feasible progeny sequences are produced. Thus the child [1 3 2 2 5] would require some "repair" (replacement of the repeated job (2) by the missing job (4)). An example of position-based crossover that shows the transfer of jobs from two "parents" to form a feasible "child" would be

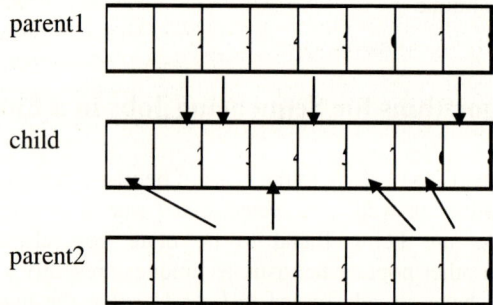

4 NSGA Solutions to the Multi-objective Flowshop Problem

Work on the multi-objective flowshop problem has begun relatively recently. Murata, Ishibuchi and Tanaka (1996) provide a GA-based approach to produce a tentative set of *Pareto optimal* solutions to the problem combining the minimization of makespan, the minimization of total tardiness in a flowshop, and the minimization of total flow time. The method minimizes a *weighted sum* of multiple objectives $\{f_i(x)\}$ given by

$$f(x) = w_1 f_1(x) + w_2 f_2(x) + \ldots + w_n f_n(x)$$

Fitness is calculated from the weighted sum objective function. The weights $\{w_i\}$ in this weighted sum are not the same for each solution; rather they are randomly varied to realize probing of various search directions. The investigators qualitatively compare their solutions to those obtained using VEGA (see Srinivas and Deb, 1995). Tamaki et al (1995) perform multi-criteria optimization of hot rolling by decomposing the problem first into mathematical programming subproblems. GA combined with local search is used to develop solutions. Thus, GA seems to have been exercised well in the past five years to tackle the single-objective flowshop. These results appear to be comparable to the best-performing heuristics available. Multi-criteria flowshops, however, are sequenced till now as weighted-average objective problems.

The kernel of NSGA (Srinivas and Deb, 1995), the subject of the present work, is the *ranking selection* method it uses to emphasize the Pareto optimal regions where the desirable solutions reside. NSGA also uses a niche forming procedure to maintain a stable population of good solutions (thus avoiding genetic drift). Thus NSGA differs from SGA in the manner the *selection* operator works. The crossover and mutation operators in NSGA work as they do in SGA, so the representation (solution coding) schemes can be identical in SGA and NSGA. However, before selection is performed in NSGA, the population is ranked on the basis of the nondominated sorting concept (Figure 2), to emphasize Pareto-optimality. An illustration follows.

Suppose we wish to minimize makespan, mean flow time as well as the mean tardiness of jobs in a flowshop *simultaneously*. In this illustration the solutions (permutation representation of job sequences) are coded by the procedure described in Section 2 above. Crossover is one-point while mutation is adjacent job-interchange.

Reproduction would give preference to nondominated members identified by nondomination ranking of all solutions in the population. Sharing to encourage niche formation would be phenotypic, derating the dummy fitness values of the solutions by dividing it by *niche count* (see Srinivas and Deb, 1995). NSGA is initiated with a randomly picked collection (population) of permutations involving *n* jobs. Table 2 displays partial list of the processing times and the due dates for a typical flowshop problem in which 49 jobs are to be optimally sequenced in a 15-machine flowshop. 49! different solutions are possible here from which the nondominating solutions must be separated. The three objectives to be simultaneously minimized are makespan, mean flow time and mean tardiness of jobs.

The NSGA may be optimally pameterized using a *design-of-experiments* (DOE) approach (Bagchi and Deb, 1996) employing pilot GA runs. This methodology is new in the domain of GA parameterization and is therefore briefly recalled here. A critical difficulty in applying GAs is that the various parameters must be correctly chosen to ensure the GA's satisfactory on-line and off-line performance. Using NSGA satisfactorily would be no exception. Further, the answers found here (the optimum values for the GA parameters p_s, p_c, p_m, etc.) are often problem-dependent. Also, one notes that crossover and mutation effects can interact and "support each other in important ways" and observes that a judicious blend of mutation and crossover does better than either one alone to strike a good balance between exploration of the total solution space and exploitation of good solutions currently at hand. It is easy to see, therefore, that the optimization of different GA parameters itself is a global search problem and one that must be tackled in the problem domain of interest, before we apply the GA in a "production run".

Table 2. Partial list of processing times and due dates for a 15 m/c-49 job flowshop

	m/c # 1	2	3	4	5	6	7	8	9	10	11	12	13	14	15	Due Date
Job1	74	72	54	57	52	60	4	8	40	8	85	45	74	67	48	80
2	99	77	58	50	31	67	19	96	93	29	27	6	85	22	48	160
3	15	10	85	2	92	53	60	63	11	94	44	71	19	99	94	240
...	...															
49	0	77	72	40	0	64	38	34	76	79	39	1	64	23	17	3920

In multi-objective optimization, which here is the domain of NSGA, we would be interested in a good *set* of Pareto-optimal solutions. Good Pareto optimal solutions have two properties: (1) They lie on the Pareto-optimal front, and (2) They are well-dispersed on the front (the solutions should not form "clumps").

To parameterize NSGA, the concept of nondominance is used to select the response factor, as follows. All solutions obtained in a fixed number of GA iterations by conducting a set of DOE experiments are pooled together and then subjected to non-dominated sorting. The *count* of *distinct Front 1 members contributed* by each experiment in the DOE matrix was used as the response. Dispersion was evaluated subjectively.

The effectiveness of NSGA is observed to be affected by the mutation and crossover probabilities p_m and p_c, the population size p_s, and the NSGA parameter σ_{share}, which controls the precision with which the Pareto optimal front is combed for the existence of a possible optimal point or peak. For this illustration we use again the 15 machine-49 jobs flowshop problem, the objectives being minimization of *makespan*, minimization of *mean flow time*, and minimization of *mean tardiness*.

Figure 3 displays the factor effects and the interaction between the parameters, indicating the relatively strong impact on convergence of solutions to the Pareto front of population size (p_s), probability of mutation (p_m) and interaction between p_s and p_m. The results also indicated that a *high* population size (p_s), *low* σ_{share} (= ds), *low* probability of mutation and a *high* probability of crossover would be the best parameter combination for NSGA.

Subsequently, in implementation of NSGA these parameters would be set at values p_s = 200, ds (= σ_{share}) = 1, Pc = 0.9 and p_m = 0.0. We note that DOE-based parameterization appears to work considerably better than the use of ad hoc values arbitrary "rules-of-thumb" parameterization guidelines culled from GA literature. We also note that the DOE method is exploratory, giving the analyst freedom to make parameterization *problem-specific*.

5 How NSGA Produced Pareto Optimal Job Sequences

Table 3 presents a partial display of the results of running NSGA for 250 generations with 37 different initial random populations. As Table 3 shows, in this instance about 22% of the population resulting at the end of 250 generations were nondominated or Pareto optimal. The actual number of Pareto optimal solutions discovered by NSGA is a function of the *Pareto optimal landscape* (the number of Pareto optimal solutions that exist for a problem), and population size (p_s).

Experimental evidence indicates that the larger is the number of jobs being sequenced, the higher should be the value of p_s to facilitate the finding of as many Pareto optimal solutions as possible for a given number of NSGA generations executed. We record here that the average time to execute NSGA for the 49-job 15-machine problem on a HP 9000/850 system running C++ was 34 seconds. We recount briefly how NSGA achieved this and then mention how the quality of the solutions may be further improved. The two key strategies employed by NSGA are (1) the use of nondominated ranking to seek individuals with the best Pareto ranks, and (2) the use of fitness sharing to obtain as many nondominated solutions as possible.

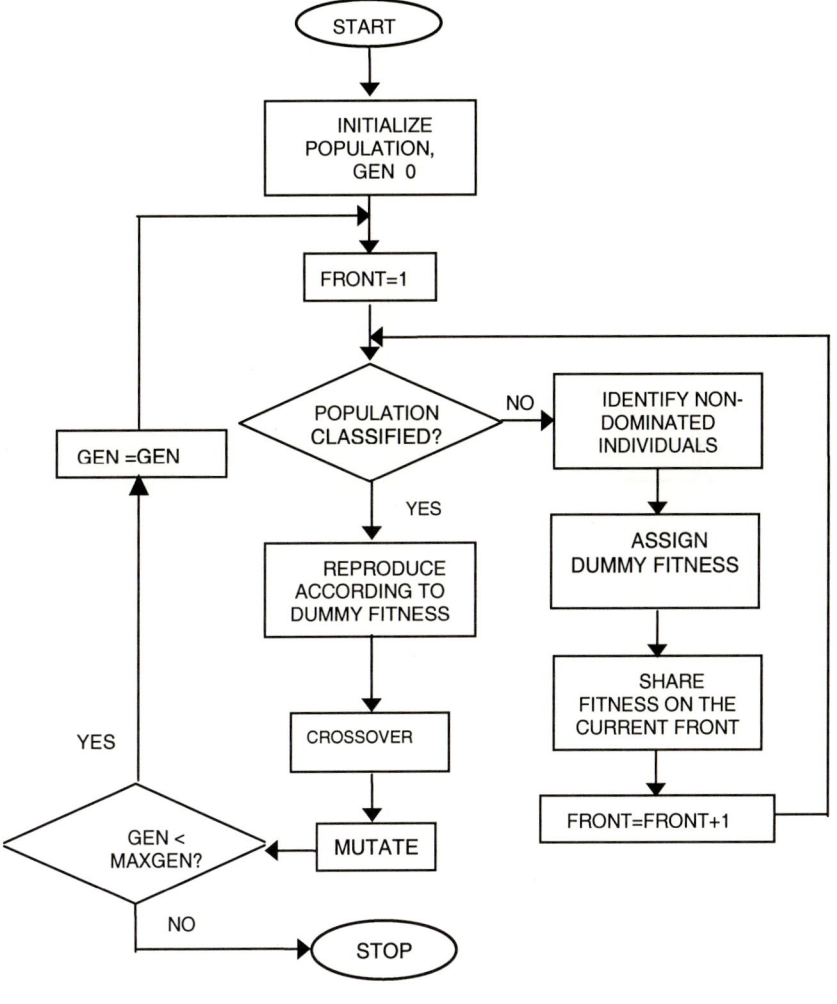

Fig. 2. The Nondominated Sorting GA

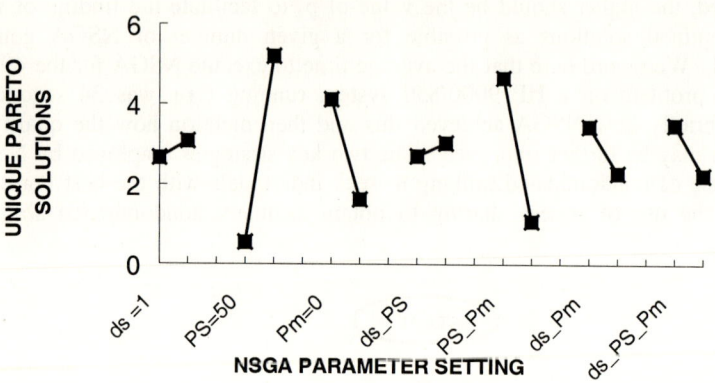

Fig. 3 Factor effects on the discovery of Pareto optimal solutions by NSGA

Table 3 Partial list of the count of Pareto optimal solutions found by NSGA for the 15 m/c-49 Job flowshop

Generation→	0	25	50	75	100	125	150	175	200	225	250
Seed # 1	7	7	16	15	25	36	29	26	27	27	27
Seed # 2	2	8	16	10	15	26	25	23	26	23	23
Seed # 3	3	11	3	18	20	21	11	16	12	11	12
...	...										
Seed # 37	4	12	20	17	32	32	28	26	27	30	33
Average Count	6.2	10.9	11.7	15.4	17.7	22.4	23.7	22.8	22.3	22.4	22.2

For each NSGA iteration, an initial set of solutions (randomly generated by a seed) started the process. The nondominated individuals among these solutions were first identified from the initial population of solution. These initial nondominated individuals constitute the *first nondominated front* in the population and assigned a dummy fitness value proportional to the population size (dummy fitness = p_s). This dummy fitness is intended to give an equal reproductive chance to all these initial *nondominated* individuals. Diversity among the solutions was maintained by inducing niche formation by *sharing* fitness among neighboring individuals. Next, the individuals in the first front were ignored temporarily and the rest of the population was attended to, as follows. Individuals on the *second* nondominated front were identified. These second front solutions were then assigned a new and equal fitness value kept *smaller* than the shared dummy fitness value of the solutions on the first front. This was done to differentiate between the members of the first front and the members of the second front. Then, sharing was again done within the second front, and the process went on till whole population had been evaluated, and classified into successive fronts. This process led to the creation of several successive fronts of "nondominated" individuals.

Next, individuals in the whole population were reproduced according to their relative (shared dummy) fitness value. This approach facilitates the search for the nondominated regions of the Pareto-optimal fronts. This results in the quick convergence of the population towards the nondominated region while sharing helps to distribute the individuals over the entire nondominated region.

The efficiency of NSGA in seeking Pareto optimal solutions lies in the manner it reduces multiple objectives to be optimized to *a single dummy fitness function* using the nondominated sorting procedure. If a solution is locally dominated, it is globally dominated. However the converse is not true (Srinivas and Deb, 1995). In each generation NSGA finds locally nondominated points only. However, if there exists any globally nondominated member, NSGA increases the likelihood its survival. This is because in the NSGA scheme, any such individual will have the highest possible dummy fitness. Thus the power of NSGA lies in the successful transformation of a multi-objective problem, no matter how many objectives are there, to a single function problem, without losing the perspective of vector optimization. By selecting nondominated points, NSGA actually processes the schemata that represent Pareto-optimal regions. Therefore, the building blocks for NSGA will be those schemata that represent characteristics of globally nondominated individuals.

NSGA implemented here did not incorporate *mating restriction*, a phenomenon that encourages *speciation* (formation of species among solutions) in nature. Deb and Goldberg (1989) have shown that speciation improves the discovery of multiple peaks. Yet another method based on *elitism* may improve the rate at which the Pareto optimal solutions are found. This method is outlined in the next section. Also in it this new method is statistically compared with NSGA using an assortment of tri-objective flowshop problems as test beds.

6 An Elitist Multi-objective GA for Sequencing the Flowshop

This section describes a different genetic algorithm that obtains Pareto optimal solutions faster than NSGA. When NSGA is used, it is often seen that it lacks somewhat in both on-line performance (rapid convergence to good solutions) and off-line performance (superior quality of the final solutions). One key reason for this is that NSGA *does not* preserve the good solutions found from one generation to the next generation. Thus, good (near-optimal) solutions lost in one generation have only a probabilistic chance in NSGA to reappear in the future. Also, the *number* of final solutions on the Pareto front in NSGA often remains relatively low even with good choice of parameters and even after many generations, unless large population sizes (> 250) are used. To deal with this deficiency, an enhancement may be devised for NSGA that effects a significant performance improvement. Such a GA would be *elitist* in that it would consciously preserve a controlled fraction of the best structures or solutions present in a generation to the next generation. The value of preserving the elite (high fitness solutions) in GA is well-recognized.

Citing studies done in 1975 Goldberg (1989) notes that an elitist plan significantly improves both on-line and off-line performance of GAs on unimodal surfaces. Goldberg also notes that elitism improves local search albeit at some expense of global perspective. The enhanced version of NSGA that we describe here is called *ENGA* (the *Elitist* Nondominated Sorting GA, Figure 4). Unlike NSGA, ENGA is designed so as not to mercilessly discard the old population and replace it completely by the progeny. Like NSGA, ENGA first produces the progeny through crossover and mutation (equal in number to parents). But it uses a different selection procedure. It first ranks the candidate constituents of the next generation by performing an *additional* nondominated sorting of the *combined parents + progeny* pool. A controlled fraction (the top 50%) of the individuals in this combined pool is then selected to form the next generation, to mate and propagate their nondominating schema characteristics. Thus each generation may contain several members of the *parent* chromosomes *without modification* if these parents are good enough to be able to outrank (in the nondomination sense) some of the newly-created progeny. Beyond this, in subsequent iterations it improves the solutions by the combined effects of recombination, mutation and fitness sharing.

In order to evaluate their relative performance, both NSGA and ENGA were coded in the C++ language running on HP 9000/850 and then tested on a number of different flowshop problems. The overall quality of solutions produced by ENGA was indistinguishable from those produced by NSGA, i.e. both ENGA and NSGA produced solutions with comparable degree of convergence to the Pareto front—with identical computational effort employed. But, for majority problems tested ENGA produced *many more* Pareto solutions for the same effort.

The runs hinted that ENGA might be populating the Pareto front faster. To verify this the Wilcoxon signed-rank statistical test was applied to the results obtained. Of 21 randomly selected flowshop problems tested using identical GA population size, in 19 problems ENGA produced numerically larger number of Pareto optimal schedules; in two NSGA produced more Pareto optimal solutions. Statistically speaking, therefore, ENGA outpaced NSGA in the discovery of Pareto optimal solutions. Figures 5, 6 and 7 display typical Pareto optimal and the dominated solutions found by 100 iterations of ENGA for a 49-job 15 m/c flowshop, a 10-job 5 m/c job shop and a 10-job 10 m/c open shop (Jayaram, 1997). No other method known produces such solutions.

7 Conclusions

This paper has demonstrated how Pareto optimal shop schedules may be developed. In particular it has shown that nondominated sorting augmented by elitism (modeled here by ENGA) statistically improves the speed of search to seek out multiple Pareto optimal solutions. The method applies to the optimization of arbitrary number of conflicting objectives. In this work the GAs were all optimally parameterized using the design of experiments procedure, also done for the first time for GAs. This approach proved to be highly productive. This paper used three-objective flowshops (Figure 5), three-objective job shops (Figure 6) and two-objective open shop problems (Figure 7) to demonstrate the techniques. We note here that open shop multi-objective problems thus solved can be easily extended to multi-criteria school timetabling. Computationally, problems involving up to 15 machines and 50 jobs were solved in the present work in one minute or less on systems comparable to an HP 9000/850 machine using C++ coding. More extensive tests of the procedure using larger problems are suggested.

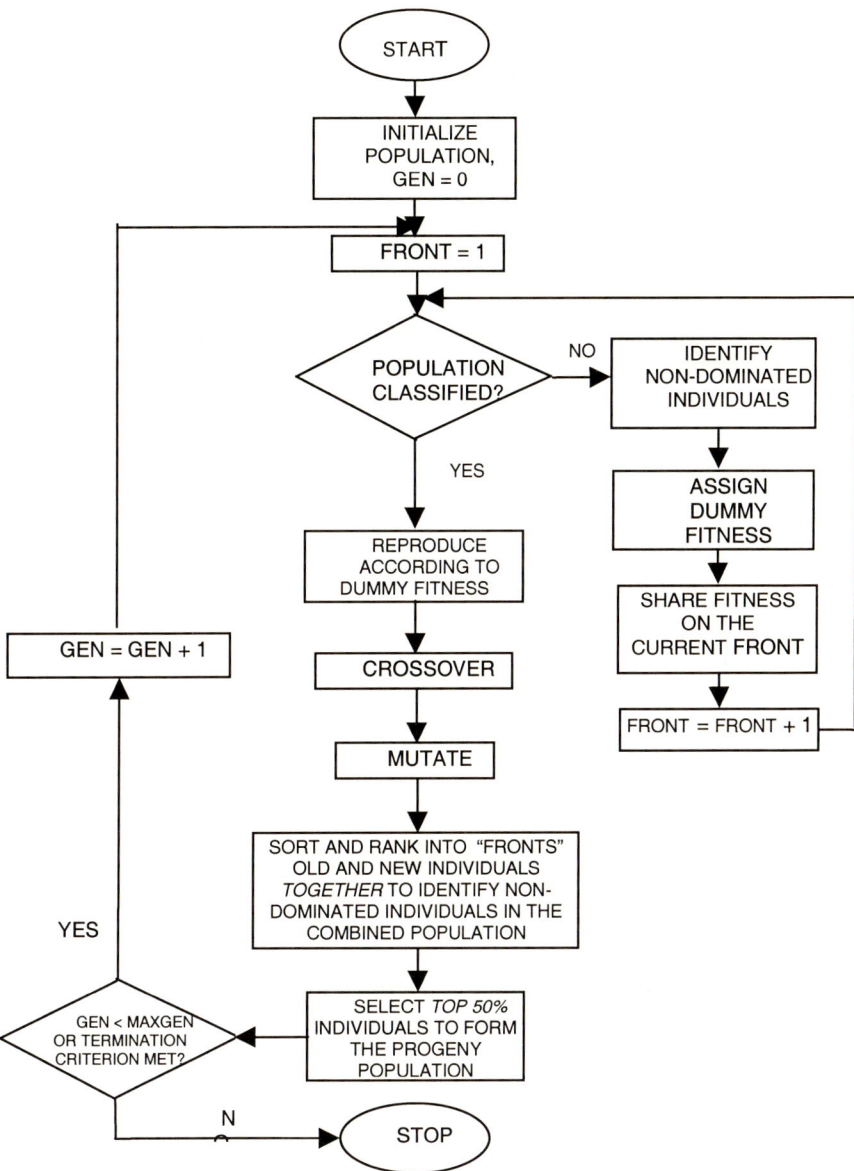

Fig. 4 The elitist nondominated sorting GA

4 Hybrid Genetic Algorithms (MPGA, MOGA) for Parallel Machine Scheduling Problems

For multiple objective parallel machine scheduling problems such as the one we are considering in this paper, several approximation algorithms have been proposed. In this paper, we use two genetic algorithms: Multi-Population Genetic Algorithm (MPGA) developed by Cochran et al. [2] and Multi-Objective Genetic Algorithm (MOGA) suggested by Murata et al. [9].

Schaffer [12] proposed the Vector Evaluated Genetic Algorithm (VEGA) method to find the efficient frontier for general multiple objective problems. Using this method, a population is divided into disjointed sub-populations and a sub-population is optimized with respect to each objective. VEGA, by the nature of its disjointing approach (vector optimization), tends to form the extreme solutions of the efficient frontier since its search direction is unidirectional.

Murata et al. [9] proposed the Multi-Objective Genetic Algorithm (MOGA), which selects individuals for a crossover operation, based on a weighted sum of linear objective functions with variable weights that are not constant but randomly specified for each generation. That is, the fitness value of solution x in each generation, $f(x)$ is equal to

$$f(x) = \sum_{i=1}^{K} \alpha_i f_i(x), \qquad (3)$$

where K is the number of objectives, α_i is the weight for each objective i, and $f_i(x)$ is the ith objective value of solution x, and x is in the set X of all solutions considered. The selection probability function is equal to

$$\frac{(f(x) - f^*)}{\sum_{x \in X}(f(x) - f^*)}, \qquad (4)$$

where f^* is the minimum fitness value in each generation. With these variable weights, MOGA is essentially searching in random directions. The authors argued that the method generally produces more diverse non-dominated solutions than VEGA, providing a broader choice of solutions to the decision maker. They applied MOGA to solve a multiple criteria flow shop scheduling problem with two and three objectives (makespan, total weighted tardiness, and total weighted completion time). The solution quality of a set of near Pareto-optimal solutions generated by MOGA is compared with that of VEGA. The authors showed that MOGA generates better Pareto fronts than VEGA by visual comparison. However, as pointed out by Coello [3], MOGA has the disadvantage of missing non-supported solutions of the efficient frontier since the weighted sum of objectives is used as a fitness function. Also, Cochran et al. [2] pointed out that it is focusing on the generation of diverse solutions but does not consider the convergence of the solutions; hence it may be time consuming to obtain good approximate solutions.

MPGA is a hybrid genetic algorithm to solve multiple criteria parallel machine scheduling problems with sequence dependent setups. The genetic algorithm is used to assign jobs to machines, after which dispatching rules such as setup avoidance and apparent tardiness cost with setups are used to sequence the jobs assigned on the individual machines [11]. The method combined an aggregating function method and the VEGA method. The motivation of MPGA is to achieve diversity and convergence of solutions by using two phases. In [2], six aggregating functions were tested that was focused on generating good compromised solutions. In the first phase, a multiplying function $f(x)$ was selected through experiments. The aggregating function $f(x)$ is defined as

$$f(x) = \prod_{i=1}^{K} \frac{f_i(x)}{f_i^*}, \qquad (5)$$

where $f(x)$ is the fitness value of the solution x, $f_i(x)$ is i^{th} objective value of solution x, and f_i^* is the minimum value of the ith objective in each generation. Then, the selection probability function, which is equal to

$$\frac{(f^w - f(x))^2}{\sum_{x \in X} (f^w - f(x))^2}, \qquad (6)$$

where f^w is the worst fitness value in each generation, is used to select chromosomes for a crossover operation in that generation. In the second stage, MPGA adopts the approach of VEGA to increase diversity of solutions from good compromised solutions. To do this, the solutions of the first stage are rearranged and divided into $p + 1$ sub-populations (p for each objective and one for the combined objective function), which are the initial populations of the second stage. Each sub-population evolves separately (similar to the VEGA approach). Cochran et al. [2] also sought to find the best time to change between the two stages (called the turning point). Through experimentation, they determined that a fixed turning point of 2000 generations work well [2]. MPGA outperformed MOGA when compared using the measures of 1) number of Pareto-optimal solutions and 2) number of combined Pareto-optimal solutions [2].

In both MPGA and MOGA, a tentative set of non-dominated solutions is stored and updated at every generation. A certain number of individuals randomly selected from the tentative set of non-dominated solutions are preserved as elite individuals. After the stopping criterion is met, a final set of non-dominated solutions remains.

5 Experiments and Results

In this section we provide a description of the parameter settings used for our experiments, the problem instances generated for testing, and the performance measures evaluated. We then present the results of our analysis.

5.1 Parameter Settings for Each Algorithm

In [2], preliminary experiments were performed to find the best parameter settings for both algorithms, since the performance of each algorithm is dependent on the parameter settings used. We use the same parameter settings as in [2], listed as follows:
- Crossover probability: 0.6
- Mutation probability: 0.01
- Population size: 20
- Elitism: three elite solutions are selected from the tentative set of non-dominated solutions
- Stopping criteria: 5000 generations

For MPGA, the turning criterion is set at the 2000th generation. After the turning criterion has been reached, the population is divided into three sub-populations, one for each of the two objectives and one for the combined objective function.

5.2 Problem Instances

As shown in Table 1, four factors are used to generate problem instances with 100 jobs each. A total of 36 sets of problems are generated using these four factors. Ten problem instances are generated randomly in each set, resulting in 360 test problems. All problems are replicated 10 times due to the inherent randomness of the genetic algorithms, yielding 3600 total instances.

5.3 Performance Measures

Our purpose is not only to compare the solution quality of two competing algorithms to solve a strongly NP-hard problem, but also to compare the results by different measures and provide evidence of the effectiveness of the ICP functional. Since no efficient algorithms to generate the true efficient frontier for the considered scheduling problem exist, measures based on this [4, 5, 6, 15] are not applicable, whereas the ICP measure has been developed based on this exact premise. The following two types of ICP measures are used in this study: (1) ICP with non-scaled objective values (ICP_U) and (2) ICP with scaled objective values (ICP_U_S). For analysis purposes, the comparison results given by [2] are provided in the same table. When using ICP measures, we assume that the decision maker's value function can be represented as a convex combination of objective values (even if the exact weight for each objective is not known) and the decision maker's weight for each objective is uniformly distributed. We note, however, that the use of the ICP measure is not restricted by the assumption of weighted sum scalarization; the measure can be used with any type of fitness value structure.

Table 1. Four factors and levels to generate 36 ($2^2 \times 3^3$) problem sets (Adapted from [2])

Factors	Levels	Description
Range of weights	1 (Narrow) 2 (Wide)	$U(1,10)$ $U(1,20)$
Range of due dates	1 (Narrow) 2 (Wide)	Release time + $U(-1,2)$ × total process time. Release time + $U(-2,4)$ × total process time.
Ratio ($\overline{p}/\overline{s}$)	1 (High) 2 (Moderate) 3 (Low)	50/10, $p = 50 + U(-9,9)$, $s = 2^* U(3,7)$. 30/30, $p = 30 + U(-9,9)$, $s = 6^* U(3,7)$. 10/50, $p = 10 + U(-9,9)$, $s = 10^* U(3,7)$.
WIP status	1 (High) 2 (Moderate) 3 (Low)	All jobs are ready at time 0 50% of jobs are ready at time 0 and the others are ready at time $U(0,720)$ All jobs are ready at time $U(0,720)$

Notes)
$U(a, b)$: random number generated from uniform distribution between a and b
$\overline{p}$: average process time.
$\overline{s}$: average setup times.
Total process time: process time plus average setup times.
Average setup times of job j: $(1/n) \times (\Sigma s_{ik})$, where s_{ik} is setup times from job with family i to job with family k, i is the family of job j, and $k = 1,2,...,n$.

5.4 Experimental Results and Analysis

In this section, we provide comparison results of the ICP measure and other measures in the literature first. We then present the results and analysis of the comparison of the two genetic algorithms.

5.4.1 Comparison of ICP and Other Measures

Here we provide a summary of the results presented in [7] on the comparison of the ICP measure to other measures in the literature. We believe that the results of any comparison method should be consistent with the results of visual comparison on clear instances.

Consider the following 10 replicates (i.e., solutions using different seed values) using each of the two heuristics, shown in Figure 2. Set 1 and set 2 represent approximations of the efficient frontier generated by MOGA and MPGA, respectively. Based on a visual comparison, we can conclude that set 1 is better than set 2 in replicates 5, 6, and 9; set 2 is better than set 1 in replicates 1 and 4; these replicates will be referred to as "clear cases". In replicates 2, 3, and 7, set 1 has better solution points in the tail areas (for each objective) while set 2 has better solution points in the elbow area (for compromised objectives). In replicate 8, set 1 has better solutions for objective 2, and in the elbow area, set 1 and set 2 have similar solutions for objective 1, so it is hard to say which solution set would be preferable. In replicate 10, the solution sets are very close, resulting in a (visual) tie. Hence, replicates 2, 3, 7, 8, and 10 will be referred to as "unclear cases".

Table 2 summarizes the results of comparing a visual method, the ICP measure, and measures that count the number of Pareto-optimal solutions (# of POS) and the number of Pareto-optimal solutions from that algorithm when the two solution sets are combined (also called number of combined Pareto-optimal solutions, or # of CPOS). The symbol '?' is used to indicate that the result of the visual comparison is not clear. Note that sets of solutions with high values of the two cardinality measures (# of POS and # of CPOS) and low values of the ICP measure are preferred. The preferred solution set found by each comparison method is marked in bold.

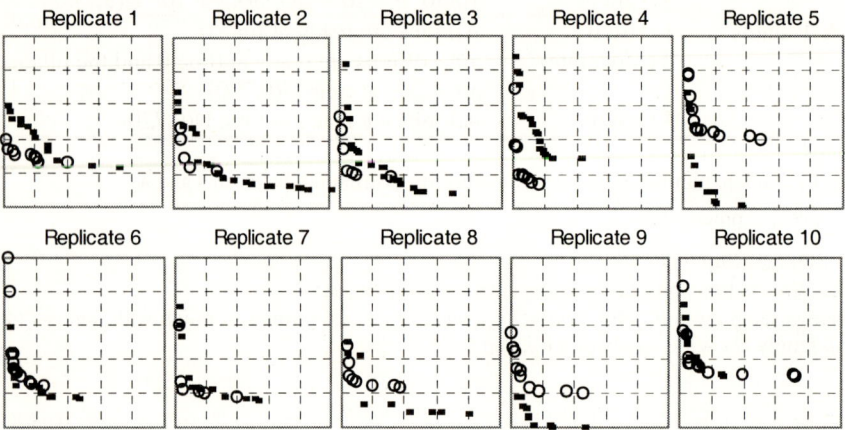

Fig. 2. Sets of near Pareto-optimal solutions generated by the two algorithms for 10 replications of one specific problem instance. '-' represents set 1, 'o' represents set 2 and each grid is 0.2. (Adapted from [7])

An analysis of Table 2, as in [7], shows that the two cardinality-based measures have pretty clear shortcomings. They can miss the general structure of the solution set, and, especially in the case of # of POS, can favor a solution set that isn't necessarily closer to the true efficient frontier.

In all of the clear cases, ICP gives results that are identical to those with visual comparison. For replications that yield a tie with the visual comparison (e.g., Replication #10), the ICP measures of set 1 and set 2 are very close. Also, from the fact that the minimum value of ICP is 0.05, the best set of near Pareto-optimal solutions among the 20 can be easily found in Table 2 (see the graph of set 1 in replicate 9 in Figure 2). In the not-so-clear cases, set 1 wins in replicates 8, 10 and set 2 wins in replicates 2, 3, and 7 according to the ICP measure.

From above results, we can see that the ICP measure provides the closest comparison result to the visual comparison, with a scalar value. One shortcoming of ICP measure presented here is that it uses only efficient extreme points in evaluating a set of Pareto-optimal solutions, since it assumes the decision maker's value function is a convex combination of objective functions. In this case, ICP cannot differentiate sets of Pareto-optimal solutions, which have the same efficient extreme solutions and different non-supported solutions. This limitation can be overcome by using different types of value functions. For example, if a weighted Tchebycheff metric is assumed, then all Pareto-optimal solutions in a set will be considered in evaluating the solution quality of sets of near Pareto-optimal solutions.

Table 2. Numerical comparison of solution set 1 and set 2 (Adapted from [7])

Rep. #	Visual Comparison	# of POS		# of CPOS		ICP		ICP-difference
		set 1	set 2	set 1	set 2	set 1	set 2	
1	Set 2	14	9	2	9	0.21	**0.17**	0.04
2	?	21	5	15	4	0.17	**0.16**	0.01
3	?	22	8	10	7	0.15	**0.12**	0.03
4	Set 2	18	9	1	9	0.21	**0.10**	0.11
5	Set 1	11	12	11	0	**0.08**	0.26	-0.18
6	Set 1	14	13	14	0	**0.14**	0.18	-0.04
7	?	12	7	7	5	0.15	**0.13**	0.02
8	?	9	8	8	4	**0.12**	0.16	-0.04
9	Set 1	10	10	10	2	**0.05**	0.15	-0.10
10	?	11	12	10	5	**0.20**	0.21	-0.01
Wins	-	7	2	8	2	5	5	5

5.4.2 The Use of ICP to Compare MPGA and MOGA

Our experimental results are summarized in Table 3, where the column 'Problem instance set' represents the combination of the levels of the four factors. Values in both 'ICP_*' columns are the number of wins of MPGA (over MOGA) out of 100 comparisons (10 randomly generated problems * 10 replicates) in the corresponding problem sets using the corresponding ICP measures. In the '# Pareto-optimal' column, the average number of non-dominated solutions generated by MPGA and MOGA are shown respectively. In the '# combined Pareto' column, the average combined number of non-dominated solutions generated by MPGA and MOGA are shown. In the 'win' column, '1' means MPGA wins over 50 times, '0' means MOGA wins over 50 times, and '0.5' means the algorithms are tied. In the 'Total' row, values in the two 'ICP' columns indicate the number of wins of MPGA out of 3,600 comparisons, and values in the four 'win' columns indicate the number of wins of MPGA out of 36 problem instance sets. Finally in the 'Ratio' row, the values given are the ratio of the number of wins of MPGA out of the total number of comparisons.

We can first see that MPGA outperforms MOGA in overall performance, since all six ratios in Table 3 are greater than or equal to 0.5. However, the number of wins for MPGA is much different for the various measures. When the number of Pareto-optimal solutions measure is used, MPGA wins in all of the problem sets, but when the combined number of Pareto-optimal solution measure is used MPGA wins in 32 problem sets. These indicate that the VEGA method embedded in the MPGA plays a role in increasing the number of non-dominated solutions. When ICP_U is used, MPGA wins in 32 of the problem sets, and when ICP_U_S is used MPGA wins 19 times out of 36.

These results indicate that the problem instances employed in this study have scaling differences. For example, the makespan objective values range from 1,000 to 1,300, the total weighted tardiness objective values range from 200,000 to 350,000 for problem instance set 2212. Hence, when ICP_U is used, the narrow ranged objective (makespan) is nullified by the wide ranged objective (total weighted tardiness). We can see that MPGA tends to generate solutions that are better in the total weighted tardiness objective. Also, as indicated in [7], ICP_U_S gives results that are closer to results from visual comparisons than the other measures. Hence, further analysis is performed using the results by ICP_U and ICP_U_S.

Table 3. Number of wins of MPGA out of 100 comparisons in each 36 problem instance set

Problem instance set	ICP comparison results				Comparison results by *Cochran et al.*[2]					
					# Pareto-optimal			# combined Pareto		
	ICP_U	win	ICP_U_S	win	MP-GA	MO-GA	win	MP-GA	MO-GA	win
1111	62	1	50	0.5	11.9	9.4	1	6.6	5.2	1
1112	67	1	61	1	12.0	9.5	1	8.2	4.2	1
1113	81	1	77	1	10.7	8.4	1	7.9	2.5	1
1121	55	1	47	0	11.7	8.8	1	6.0	5.4	1
1122	65	1	56.5	1	5.7	4.6	1	3.5	1.9	1
1123	66	1	62	1	7.0	5.4	1	4.4	2.3	1
1131	52	1	46	0	12.6	8.6	1	5.8	5.6	1
1132	44	0	42	0	1.8	1.2	1	0.6	0.7	0
1133	58	1	44	0	2.9	1.9	1	1.2	1.1	1
1211	59	1	53	1	12.0	9.7	1	7.3	4.7	1
1212	70	1	62	1	14.1	9.5	1	9.5	4.0	1
1213	84	1	69	1	10.8	8.9	1	7.9	3.0	1
1221	57	1	45	0	11.4	9.2	1	6.0	5.4	1
1222	73	1	60	1	6.7	4.8	1	3.9	2.0	1
1223	71	1	55	1	7.9	5.3	1	5.0	2.5	1
1231	52	1	49	0	12.5	9.5	1	5.7	6.1	0
1232	54	1	49	0	2.1	1.3	1	0.8	0.7	1
1233	48	0	38	0	4.3	1.9	1	1.5	1.3	1
2111	58	1	53	1	12.2	8.9	1	6.6	4.8	1
2112	76	1	71	1	13.3	9.4	1	9.7	3.7	1
2113	72	1	62	1	11.5	8.8	1	7.1	3.7	1
2121	64	1	54	1	11.9	8.7	1	6.7	4.5	1
2122	74	1	62.5	1	5.7	4.8	1	3.7	1.4	1
2123	63	1	55	1	8.1	5.4	1	4.5	2.7	1
2131	55	1	41	0	12.6	9.3	1	6.0	6.0	0
2132	51	1	47	0	1.6	1.3	1	0.7	0.7	1
2133	62	1	44	0	3.3	2.0	1	1.3	1.1	1
2211	53	1	51	1	12.4	9.3	1	6.9	4.9	1
2212	82	1	73	1	13.8	9.4	1	10.7	3.2	1
2213	80	1	74	1	11.9	9.2	1	8.9	3.0	1
2221	59	1	49	0	11.7	8.6	1	6.5	4.7	1
2222	73	1	65	1	7.5	5.2	1	5.0	1.8	1
2223	60	1	49	0	8.5	5.6	1	4.7	2.9	1
2231	45	0	45	0	12.9	9.6	1	6.1	6.0	1
2232	45	0	41	0	1.9	1.3	1	0.7	0.8	0
2233	59.5	1	39.5	0	4.2	2.0	1	1.8	1.0	1
Total	2249.5	32	1941.5	19.5	-	-	36	-	-	32
Ratio	0.62	0.89	0.54	0.54	-	-	1.00	-	-	0.89

In Table 3, we see a clear trend. When the process time/setup time ratio (3rd factor) is '3' (low), MPGA always loses. This indicates that the solution qualities of MPGA and MOGA depend on the factor levels. To perform the factor analysis, Table 4 is derived from Table 3. As shown in Table 4, MPGA wins across all weight range and due date range factor levels. However, when the level of the process-setup ratio is '3' and the WIP ratio factor is '1', we cannot say that MPGA outperforms MOGA.

Table 4. Number of wins of MPGA by levels of the four factors

Factor	Level	ICP_U	Win	ICP_U_S	win
Weight	1	1118	1	965.5	1
Range	2	1131.5	1	976	1
Due date	1	1125	1	975	1
Range	2	1124.5	1	966.5	1
Process-	1	844	1	756	1
Setup Ratio	2	780	1	660	1
	3	625.5	1	525.5	0
WIP Ratio	1	671	1	583	0
	2	774	1	690	1
	3	804.5	1	668.5	1

Note: In 'win' columns, '1' means MPGA wins over 900 times out of 1800 comparisons in weight range factor and due date range factor rows and MPGA wins over 600 times out of 1200 comparisons in the process-setup time ratio factor and the WIP status factor rows.

For further analysis of the relation between the performance of MPGA and process-setup and WIP ratio factors, we develop Table 5 from the information in Table 3. As shown in Table 5, when the level of the process-setup ratio is '3'(low), or when the level of WIP ratio is '1', it is difficult to say which one generates better Pareto fronts. MOGA outperformed MPGA when the level of process-setup ratio is '3'(low) and the level of WIP ratio is '2'(moderate) in ICP_U_S measure. In all other cases, MPGA outperforms MOGA in both ICP_U and ICP_U_S measures.

One of the reasons for this problem-dependent performance can be inferred from the algorithmic characteristics of MPGA. In MPGA, the population is divided after the specified turning criteria (2000th generation). After this point, the subpopulation evolves for the improvement of the objective function assigned to it. This is one of the reasons that MPGA outperformed MOGA in most cases under the cardinality measure. However, once one objective reaches the optimal or near optimal solution before the stopping criterion of the algorithm (the $5,000^{th}$ generation), then the subpopulation assigned to that objective has less chance to improve the solutions. This happens, for instance, when one objective is significantly easier to optimize than the others.

As shown in Table 1, when the level of process time and setup time ratio is 3(low), the process times of 100 jobs are randomly generated from $U(1,19)$ and the setup times are generated from $U(30,70)$. In this case, the makespan objective is much more dependent on the setup times than process times. As stated before, there are 4 families of jobs and 5 identical machines. Thus, near optimal schedules for the makespan objective can be obtained easily (i.e., assigning the jobs with the same family to the same machine). Once a job sequence that satisfies this condition approximately is determined by the genetic algorithm (crossover or mutation operation), then there is not much room for improving the solutions. There is then more of a chance that the makespan objective reaches a (near) optimal solution within fewer generations for level 3 than for level 1 or level 2 of the ratio of processing time to setup time.

Similarly, when the level of WIP is 1(high), the release time (r_j) of all 100 jobs is 0. In this case, the makespan objective without release times seems to be easier than with release times.

Table 5. Number of wins of MPGA by levels for sensitive factors

Process-Setup time Ratio	WIP Ratio	ICP_U	win	ICP_U_S	win
1	1	232	1	207	1
1	2	295	1	267	1
1	3	317	1	282	1
2	1	235	1	195	0
2	2	285	1	244	1
2	3	260	1	221	1
3	1	204	1	181	0
3	2	194	0	179	0
3	3	227.5	1	165.5	0

Note: In 'win' columns, '1' means MPGA wins over 200 times out of 400 comparisons.

6 Conclusions and Future Research

We have provided a detailed, real-world example that illustrates the use of a new quantitative measure for comparing heuristic algorithms for multiple objective combinatorial optimization problems. Two competing genetic algorithms (MPGA and MOGA) are compared by two ICP measures for 360 problem instances of bi-criteria scheduling problems. Overall, both ICP and cardinality measures indicate that MPGA generates better sets of approximate solutions than MOGA. Also, we can see that the solution qualities of both algorithms are dependent on the problem instances. Even though MPGA generates a larger number of non-dominated solutions in most cases, MOGA can work better than MPGA in terms of the scaled ICP measure when one objective can be optimized more easily than the other objective. Because MPGA may waste sub-populations assigned to improve an objective function for which a near optimal solution has already been found.

In developing multiple objective genetic algorithms, parameter tuning and choosing a selection scheme are often performed for a limited number of problem instances. From our experimental results, we conclude that preliminary experiments on a wide range of problem instances are needed since different parameters can perform better for different problem instances. Hence, automatic (dynamic) parameter setting and automatic choosing of a selection scheme depending on the problem and problem instances in a computerized decision support system seems to be a promising avenue of research. However, reliable and easily obtainable performance measures are required to do this.

Further research in this area is expected to be fruitful. One example is that various selection schemes reviewed in [3] need to be compared for different practical multiple objective optimization problems under measures that have appeared in literature. To do these experiments and to tune parameters in each setting, experimental design and response surface optimization methodology [10], which are frequently used in stochastic process design and optimization, will likely be extremely useful. This study will build intuition about the strengths and weaknesses of various selection scheme and solution quality measures.

References

1. Aksoy, Y., T. W. Butler and E. D. Minor, (1996) Comparative studies in interactive multiple objective mathematical programming, European Journal of Operational Research, 89, 408-422.
2. Cochran, J. K., S.-M. Horng, and Fowler, J. W., (2000), A Multi Population Genetic Algorithm to Solve Parallel Machine Scheduling Problem with Sequence Dependent Setups. Working Paper, Arizona State University, Department of Industrial Engineering.
3. Coello, C. A. C., (Aug., 1999), A Comprehensive Survey of Evolutionary-Based Multiobjective Optimization Techniques, An International Journal of Knowledge and Information Systems, 1(3):269-308.
4. Czyzak, P. and A. Jaszkiewicz, (1998). Pareto Simulated annealing – A Metaheuristic Technique for Multiple-Objective Combinatorial Optimization. Journal of Multi-Criteria Decision Analyses, 7, 34-47.
5. Daniels, R. L., (1992). Analytical evaluation of multi-criteria heuristics. Management Science, 38(4), 501-513.
6. De, P., J. B. Ghosh, and C. E. Wells, (1992), Heuristic Estimation of the Efficient Frontier for a Bi-Criteria Scheduling Problem. Decision Sciences, 23, 596-609.
7. Kim B., E. S. Gel, M. W. Carlyle, J. W. Fowler, (2000), A new technique to compare algorithms for bi-criteria combinatorial optimization problems, submitted to Proceedings of the 15th International Conference on Multiple Criteria Decision Making, to appear.
8. Louis, S. J. and G. J. E. Rawlins, (1993), Pareto-optimality, GA-easiness and Deception. Proceedings of the International Conference on Genetic Algorithms, 118-123.
9. Murata, T., H. Ishibuchi, and H. Tanaka, (1996), Multi-objective genetic algorithm and its application to flowshop scheduling. Computers and Industrial Engineering, 30(4), 957-968.
10. Myers, R. H. and D.C. Montgomery, (1995), Response Surface Methodology – Process and Product Optimization Using Designed Experiments, John Wiley & Sons, New York.
11. Pinedo, M., (1995), Scheduling: Theory, Algorithms, and Systems. Prentice Hall, New Jersey.
12. Schaffer, J.D., (1985), Multiple objective optimization with vector evaluated genetic algorithms. Proceedings of the First International Conference on Genetic Algorithms, 93-100.
13. Serifoglu, F. S. and G. Ulusoy, (1999), Parallel Machine Scheduling with Earliness and Tardiness Penalties. Computers & Operations Research, 26, 773-787.
14. Valenzuela-Rendon, M. and E. Uresti-Charre, (1997), A Non-Generational Genetic Algorithm for Multiobjective Optimization, Proceedings of the International Conference on Genetic Algorithms, pp 658-665
15. Viana, A. and J. P. Sousa, (2000). Using metaheuristics in multi-objective resource constrained project scheduling. European Journal of Operations Research, 120, 359-374.
16. Zitzler, E. and L. Thiele, (Sep. 1998), Multiobjective Optimization Using Evolutionary Algorithms – A Comparative Case Study. In A. E. Eiben, Editor, Parallel Problem Solving from Nature V, 292-301, Amsterdam, Springer-Verlag.

A Bi-Criterion Approach for the Airlines Crew Rostering Problem

Walid El Moudani[1,2], Carlos Alberto Nunes Cosenza[3], Marc de Coligny[4], and Félix Mora-Camino[1,2]

[1] LAAS du CNRS, 7 Avenue du Colonel Roche, 31077 Toulouse, France.
{wmoudani, mora}@laas.fr
http://www.laas.fr
[2] Air Transportation Department, ENAC/DGAC, 7 Avenue Edouard Belin, 31055 Toulouse, France.
http://www.enac.fr
[3] APIT, COPPE/UFRJ Centro de tecnologia, Rio de Janeiro, Brazil.
Cosenza@pep.ufrj.br
[4] MIRA, Université de Toulouse II, France.
coligny@univ-tlse2.fr

Abstract. In this communication a bi-criterion approach for the nominal Airlines Crew Rostering Problem is developed. The nominal Crew Rostering Problem considers the assignment of the crew staff to a set of pairings covering all the scheduled flights so that operations costs are minimized while its solution must meet hard constraints resulting from the safety regulations of Civil Aviation as well as from the airlines internal agreements. Another goal is of the highest interest for airlines: since the overall satisfaction of the crew staff may have important consequences on the quality and on the economic return of the operations.
In this communication, a new mathematical formulation of the crew scheduling problem which takes into account the satisfaction of the crew members is proposed. A heuristic approach, combined with a genetic algorithms technique, is adopted to produce reduced cost solutions associated to acceptable satisfaction levels for the crew staff. The application of the proposed approach to a medium size Airline Crew Rostering Problem is evaluated.

Keywords. Multi-Criterion Optimization, Heuristics, Genetic Algorithms, Airlines Operations, Crew Scheduling, Crew Rostering.

1 Introduction

For more than three decades now the Airlines Crew Scheduling Problem (ACSP) has retained the attention of the Management and Operations Research community since crew costs in air transportation are extremely high, amounting 15-20% of total airlines operations costs. Therefore, airlines consider that the efficient management of their crew staff is a question of the highest economic relevance. Unfortunately, the exact numerical solution of the associated large scale combinatorial optimization problem is very difficult to obtain. Early rules of thumb [13] have been quickly overrun by the size of the practical problems encountered (hundreds or thousands of crew members to be assigned to at least as many pairings) and by the complexity of the set of constraints to be satisfied, leading very often to poor performance solutions. More recently, with the enhancement of computer performances, optimization approaches have been proposed to solve this problem : mathematical programming methods (large scale linear programming and integer programming techniques) (see [3], [4], [9], [10]), artificial intelligence methods (logical programming, simulated annealing, neural networks, fuzzy logic and genetic algorithms) as well as heuristic approaches and their respective combinations [11], [12], [15], [20]. Many studies refer to the nominal ACSP which is a static decision problem, based on a monthly table of flights, and devoted exclusively to the minimization of airlines operations costs. This problem is in general split in two sub problems: a crew pairing problem where the set of pairings covering the programmed flights is defined and a crew rostering problem where the retained pairings are nominally assigned to the crew staff.

In this paper, after discussing the airlines crew scheduling problem, the nominal Airline Crew Rostering Problem (ACRP) is introduced as a bi-criterion decision problem where the main decision criterion is the crew operations cost of the airline and the secondary decision criterion is relative to the crew staff overall degree of satisfaction. The solution approach proposed here is composed of two steps: in the first one a heuristic approach is designed to get a first set of high satisfaction assignment solutions and in the second one, an optimization process, based on genetic algorithms (GA) is developed. The application of this solution approach to a medium size ACRP is displayed.

2 The Airline Crew Scheduling Problem

The Airline Crew Scheduling Problem (ACSP) is treated in general once the schedule of the flights has been established for the next month and once the available fleet has been assigned to the scheduled flights. Two classes of constraints are considered in order to produce the *"line of work"* for the crew staff over the planning period : *hard constraints* whose violation impair the security of the flight (crew qualifications, national regulations concerning duration of work and rest times, medical clearances, training and license renewal requirements) and *soft constraints* (internal company rules, agreements with unions regarding the crew's working and remuneration conditions, office duties, holidays and declared assignment preferences by the crew staff) which are relevant to build the crew schedule but whose relaxation may lead to lower cost solutions. While some of these soft constraints are common to most airlines, others are only relevant for some classes of airlines and some few are specific to a given airline. The primary objective sought by airlines at this level of decision making is to minimize the crew related operations costs, so in most research studies, the ACSP has been formulated as a mono-criterion minimization problem.

A sub optimal but widely accepted approach to tackle more efficiently the ACSP, which is of the NP-hard computational complexity class [8], consists in decomposing it in two sub-problems of lower difficulty. The first sub-problem, the Airline Crew

Pairing Problem (ACPP), involves the construction of an efficient set of pairings (a pairing is a sequence of flights which starts and ends at the same airline base while meeting all relevant legal regulations) which covers the whole programmed flights. The second sub-problem, the Airlines Crew Rostering Problem (ACRP), considers the nominal assignment of the airline crew to the generated set of pairings over the planned period so that an effective "line of work" is obtained for each staff member (Fig. 1).

To get in a simple way a solution to the ACRP, most North American airlines have adopted in a first place some heuristics such as bidding processes where the crew staff is arranged by decreasing seniority and each crew member builds at his turn his own monthly "*line of work*" from the remaining pairings. This greedy heuristic approach generates too often uneven workloads and so induce repeated dissatisfaction among the crew staff. More recently, global approaches, based on Mathematical Programming techniques, have been proposed to tackle the ACRP [9].

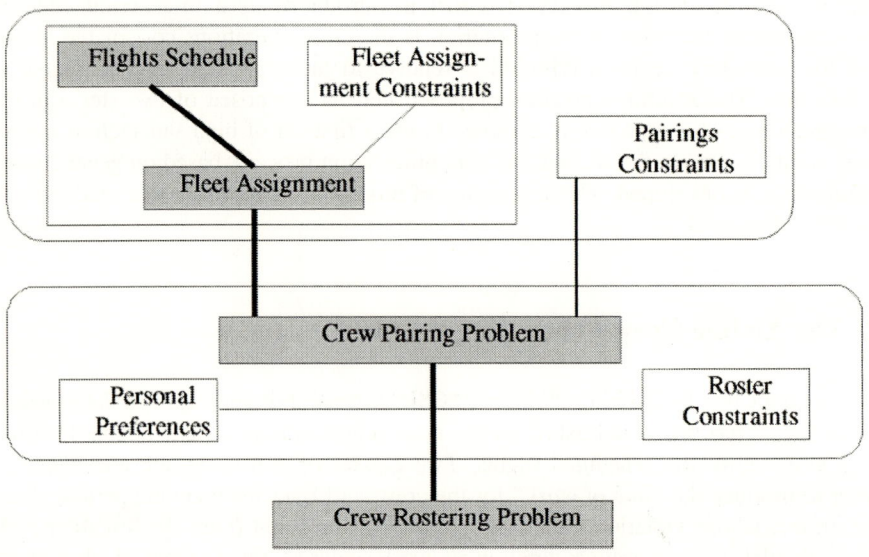

Fig. 1. The Airline Crew Scheduling Problem

3 A Mathematical Programming Approach of the Nominal ACRP

In this section a standard formulation of the ACRP is analyzed and the direct introduction of a crew satisfaction index is considered.

3.1 Analysis and Formulation of the Nominal ACRP

The nominal Airline Crew Rostering Problem has been formulated as a zero-one integer mathematical programming problem [1], [13] where the crew operations cost is the criterion to be minimized under a finite set of hard constraints:

$$\text{Minimize} \sum_{j \in J} \sum_{i \in A_j} c_{ij} x_{ij}$$

Subject to

$$\sum_{i \in A_j} x_{ij} \geq 1, \quad \forall j \in J \tag{1}$$

$$\left(x_{ij_1} + x_{ij_2}\right) \leq 1 \quad \forall j_1 \in J, j_2 \in O_{j_1}, i \in I \tag{2}$$

$$\sum_{j=1}^{m} d_j x_{ij} \leq LH \quad \forall i \in I \tag{3}$$

$$\sum_{j=1}^{m} x_{ij} \leq R_{\max}^i \quad \forall i \in I \tag{4}$$

$$x_{ij} \in \{0,1\} \quad \forall i \in I, \forall j \in J \tag{5}$$

where c_{ij} is the cost resulting from the assignment of pairing "j" to pilot "i", I is the set of the n pilots, J is the set of the "m" pairings to be covered during the planning period, A_j is the set of the pilots able to fly pairing "j", O_{j_1} is the set of the pairings overlapping with pairing "j_1", d_j is the amount of flying hours associated to pairing "j", x_{ij} are binary variables such that $x_{ij} = 1$ if pairing "j" is assigned to pilot "i", $x_{ij} = 0$ otherwise.

The first set of constraints ensures that to each pairing is assigned a unique crew, the inequality sign allowing crew deadheading (a transfer of crew members out of duty to another base in order to carry out a planned flight). The second set of constraints ensures that the same crew is not assigned to two overlapping pairings. The third set of constraints ensures that the number of hours flown by a pilot during the rostering period (a month in general) does not exceed an upper limit LH and in the fourth set of constraints, $R_{\max}^i$ is the maximum number of pairings that can be assigned to crew member "i" over a rostering period.

4 A New Solution Approach for the Nominal Airline Crew Rostering Problem

It has been observed in section 3 that the cost structure of the ACRP is not separable with respect to pairings since to estimate accurately the operations cost associated with a given crew, his entire "line of work" for the next planning period must be known. Quite the same can be said about the crew individual and global satisfaction levels over a planning period. At this point, Genetic Algorithms techniques [15], [16], [17], [18] which manipulate complete solutions, appear interesting. [19] have used this property of AG to propose a new solution approach for the ACPP. Recall that the solution search process of GA is based on the stochastic improvement of a set of initial solutions ("the initial population") through the use of operators recalling the selection and evolution processes of natural species.

Here the ACRP is considered to present a main criterion, the airline operations cost, and a secondary criterion, the overall satisfaction degree of the staff. So, in order to generate a representative set of near Pareto solutions, it is proposed to start from a population of solutions obtained from a heuristic whose aim is to maximize the overall degree of satisfaction regardless of the operations cost, then, genetic techniques are used to generate new solutions sets with reduced operations cost over a sequence of « generations ». During this process, the levels of the degrees of crew satisfaction of the solutions composing the current generation suffer some abatement. The three mains stages of the proposed approach are showed in (Fig. 2).

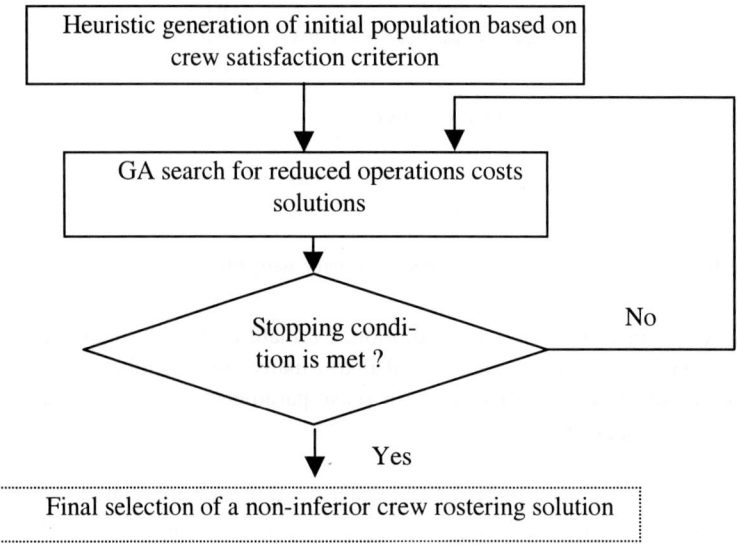

Fig. 2. The proposed GA-based solution approach

4.1 A Greedy Heuristic Approach to Build the Initial Population

Once a representative index of the satisfaction of each crew member has been made available [22] for an approach based on fuzzy set theory), the initial solutions set can be built from a unique greedy heuristic technique (the Crew Satisfaction Heuristic-CSH) applied to different arrangements of the set of crews. Note that the satisfaction of the crew builds up also from past assignments to different activities. It is supposed here that at the beginning of a new planning period each crew is characterized by a degree of satisfaction belonging to a qualitative scale such as $\{VeryLow, Low, Medium, Fair, High, VeryHigh\}$. With the proposed greedy heuristic, the preferences of crew members of lowest satisfaction degree are considered first. This heuristic technique is divided in three stages. In the first stage, the crew staff of lowest satisfaction are assigned, when possible, to their requested pairings. In the second stage, the remaining rejected pairings are assigned to the crew staff who do not bother with them. However, if all remaining crew members wish to avoid some pairings, then the crew members with the highest satisfaction level will have to cope with them. Finally, the rest of the remaining pairings is assigned to the crew members in order to complete in a balanced way their *"line of work"*.

To build the initial population this process must be repeated until the desired size of the population has been reached. Since in each satisfaction class 'i' of cardinal "n_i", there are $n_i!$ different arrangements and since "n_i" can be high, the number of different arrangements can be an extremely high number $\left(\prod_{i=1}^{5} n_i!\right)$. So, if sometimes, the heuristic assignment process produces identical assignment solutions, it is easy to find by random another arrangement in each satisfaction class, so that a new initial assignment solution is obtained.

Since the set of pairings can be structured as a directed graph when temporal precedence, reachability and overlapping constraints are considered (see table 1 where X^- and X^+ represent the starting and the completion times of pairing X), each element of the initial population can be represented by a set of "n" independent paths in the corresponding pairings precedence graph (PPG).

Table 1. Precedence and overlapping relations

Relation	Temporal conditions
X before Y	$X^- \prec X^+ \prec Y^- \prec Y^+$
X meets Y	$X^- \prec X^+ = Y^- \prec Y^+$
X overlaps Y	$X^- \prec Y^- \prec X^+ \prec Y^+$
Y finishes X	$X^- \prec Y^- \prec X^+ = Y^+$
Y during X	$X^- \prec Y^- \prec Y^+ \prec X^+$
X starts Y	$X^- = Y^- \prec X^+ \prec Y^+$
X equals Y	$X^- = Y^- \prec X^+ = Y^+$

4.2 A New GA Solution Strategy for the Nominal Airline Crew Rostering Problem Using GA

Genetics Algorithms process sets of feasible solutions, called "populations" where each element of these populations is similar to a chromosome composed of genes, each gene corresponding to a particular parameter of the problem. The chromosomes can be represented using either a binary or a non-binary codification.

For the problem tackled here, a non-binary codification has been adopted for the representation of the solutions composing a given population. The ith component of a « chromosome » indicates which crew member is assigned to the ith pairing (Fig. 3). The size of a population considered in the case studied here is equal to 30 chromosomes.

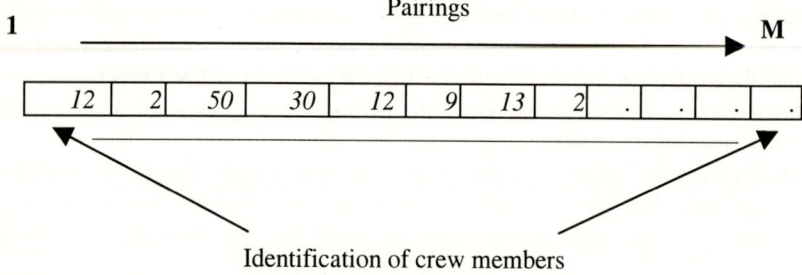

Fig. 3. The adopted coding for an ACRP solution

This codification has been chosen in order to minimize the memory requirements to codify a whole population, however, it is straightforward to associate by inspection, to each of these chromosomes, a set of "n" independent paths in the PPG (Fig. 4).

An operator select the chromosomes which become parents according to their evaluation values. A roulette wheel method picks two chromosomes to which classical GA operators like crossover, inversion and mutation are applied according to a chosen probability.

Classical genetics operators (crossover, mutation and inversion) have been adapted to the context of the present assignment problem to produce in a progressive way improved new generations. Each elementary genetic operation processes indirectly the pairing assignments of two different pilots. When a noticeable improvement of the operations cost is obtained, the local modification is retained, but when local solution costs are equivalent, the respective degrees of satisfaction of the two pilots are taken into consideration to make the choice.

The stopping rule adopted here is based not only, as usual with GA, on the allowed maximum number of non-improving successive generations, but also on the proportion in the current population of the cost effective solutions for which too many individual degrees of satisfaction are insufficient ("*Low*" or "*Very Low*"), since these cost effective solutions will represent a high risk for the airline.

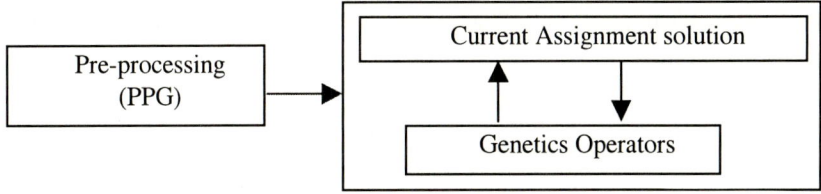

Fig. 4. Feasibility of the assignment solutions

4.3 Problem-Specific Mutation and Inversion Operators

In order to speed up the discovery of the most promising assignment solutions, a local heuristic technique has been introduced to restrict the search space of the mutation and the inversion processes. It consists in dividing the crew members in two sets, WL_L and WL_H, corresponding to workloads less and higher than the current average workload which is computed for the whole staff.

With respect to the mutation operator, a position "k" is chosen at random in the chromosome, the crew member "x" found in the position "k" will be replaced by another crew member "y" chosen randomly in set WL_L and for which this new task is feasible (the precedence constraints in the PPG must be checked for that). The idea is here to contribute actively to a more balanced workload over the whole staff. Then the solution under treatment by the mutation operator suffers the following modification:

$$\forall\, x \in WL_L \cup WL_H \quad \text{then} \quad \exists\, y \in WL_L\, /\, SOL[k] = y \quad \text{with}$$

$$WL_x = WL_x - dr^k, \quad WL_y = WL_y + dr^k \qquad \forall k \in \{1, \cdots N\},\, x \neq y \tag{10}$$

where $SOL[k] = y$ means that the assignment solution array SOL contains the crew member "y" in the position "k" and dr^k is the duration of the pairing "k", i.e. the pairing "k" of duration dr^k is non assigned to crew member "y".

In the case of the inversion operator, a position "i" corresponding to pairing "i" is chosen randomly, if it is assigned to a crew member "x" belonging to WL_H then a crew "y" belonging to WL_L is chosen at random, while a pairing "j" performed initially by crew member "y" is selected randomly. If its duration is less than the duration of pairing "i" and the inversion is feasible, then a new solution is produced where:

$$SOL[i] = y \quad \text{and} \quad SOL[j] = x \qquad \forall i, j \in \{1, \cdots N\},\, i \neq j,\, x \neq y \tag{11}$$

In the case where position "i" corresponds to a crew belonging to WL_L, the above process is inverted.

The final assignment solutions are arranged in accordance with the criterion cost and, when equality of costs, with respect to an overall satisfaction degree criterion.

5 Case Study

This solution approach has been already applied to a medium size problem where 75 crew members must be assigned to 275 pairings corresponding to a total amount of 4250 flight hours. Some learning can be obtained from this first application with respect to the computer effectiveness of the proposed approach and with respect to the quality of the solution set obtained.

5.1 Computer Effectiveness Results

When applying the first step of the proposed approach, the satisfaction-based computation of the initial population through the CSH greedy heuristic is immediate and requires a relatively short computing effort. In relation to the subsequent genetic algorithm, it appears that the different genetic operators do not present equivalent performances: the crossover operator, which is quite computer time consuming, does not contribute too much to produce new promising assignment solutions since the set of constraints to be checked is very large, while the mutation and inversion operators appear to be more efficient to generate new assignment solutions with relatively moderate computing times. The genetic operators are applied according to the probabilities $p_c = 0.20$, $p_m = 0.40$ and $p_i = 0.40$.

To compare the efficiency of the different genetic operators combinations over the final cost effective solutions, some parameters have been introduced: the workload deviation index σ_{WL} and the proportion of non satisfied crews τ_{NS}. Here σ_{WL} is given by:

$$\sigma_{WL} = \sqrt{\sum_{i=1}^{n}(WL_i - \overline{WL})^2 / n} \tag{12}$$

where WL_i is the workload assigned to crew member "i", $\overline{WL}$ is the average workload for the rostering period and "n" is again the size of the staff. The comparative results are displayed in Table 2.

Table 2. Effectiveness of CSH and genetic processes for the ACRP

	Minimum Cost Solution	σ_{WL} Cost Dispersion	Proportion of non satisfied crews (τ_{NS})
CSH Heuristic	4312:30	4:50	0.13
Crossover alone	4304:15	4:35	0.25
Mutation alone	4290:30	3:30	0.49
Inversion alone	4291:00	3:35	0.42
Crossover + Mutation + Inversion	4278:30	3:20	0.48

5.2 Analysis of the Assignment Solutions

In figure 5, the total cost and the proportion of non satisfied crew distributions of the final population are displayed to provide an idea of the non inferior set of solutions to the considered problem. In figures 6.a and 6.b, the individual workloads and satisfaction degrees corresponding to its extreme solutions are displayed: S_L and S_H. In figures 7.a and 7.b, the individual workload distribution generated after using CSH and GA are displayed.

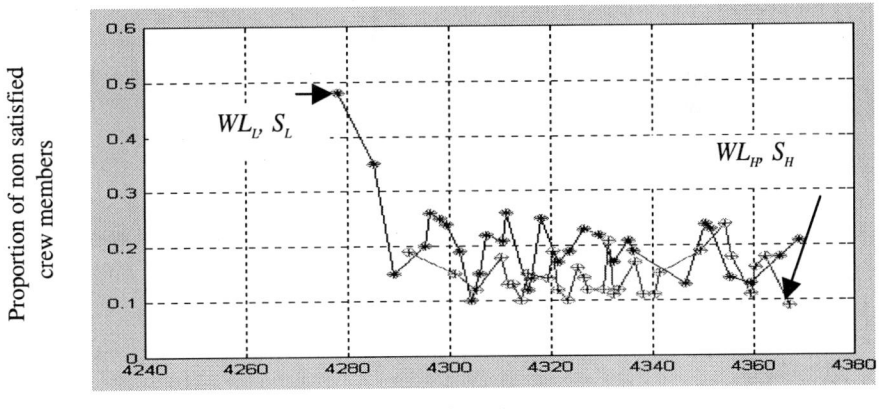

Fig. 5. Final solution set

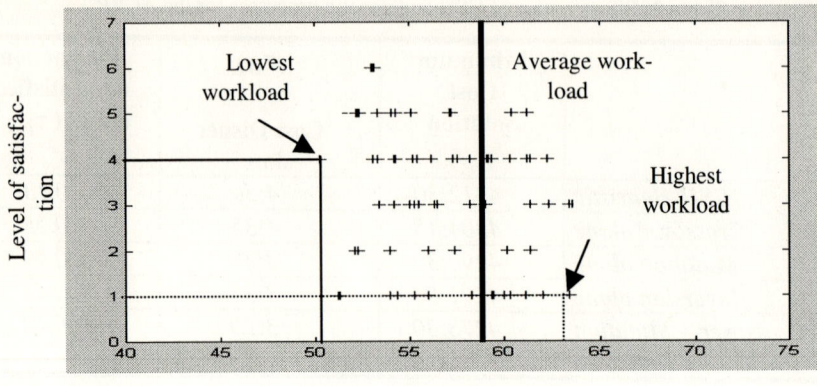

Fig. 6.a. Solution corresponding to S_H

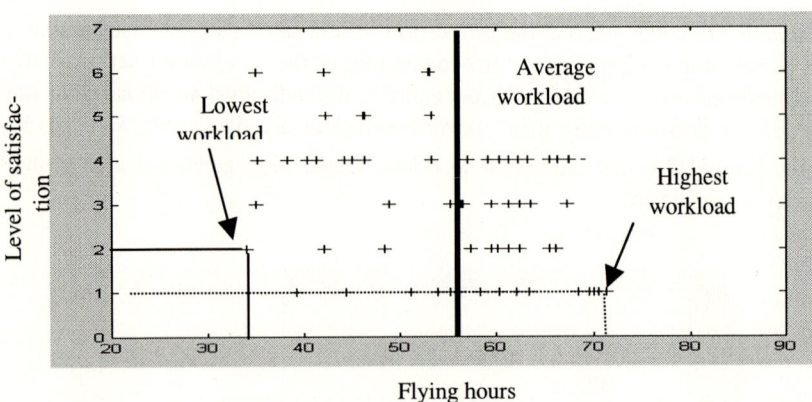

Fig. 6.b. Solution corresponding to S_L

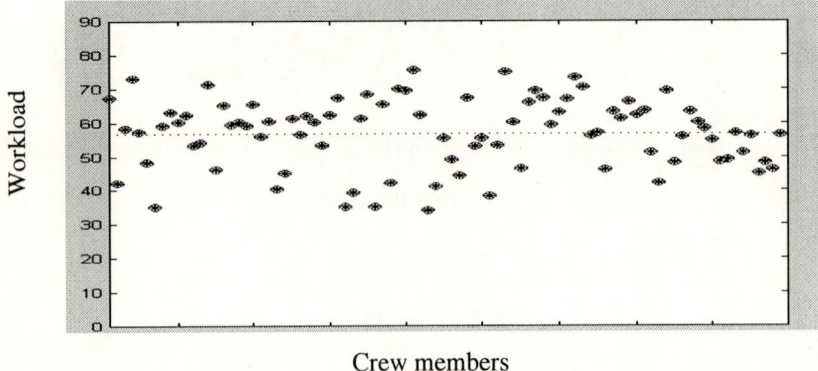

Fig. 7.a. Example of workload distribution corresponding to the greedy heuristic

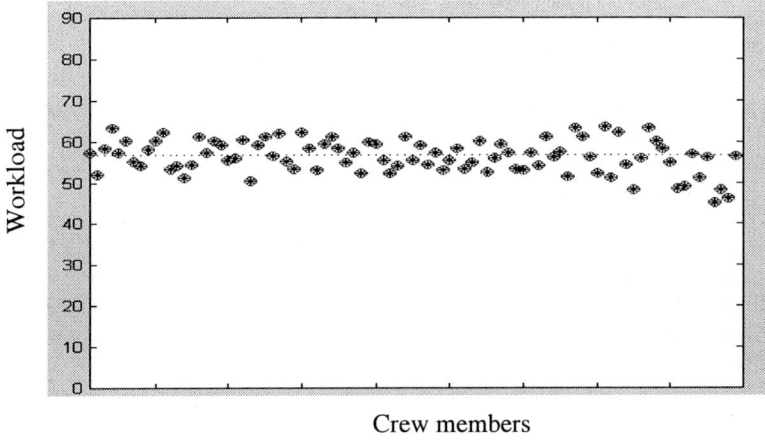

Fig. 7.b. Example of final workload distribution with GA (S_L solution)

6 Conclusion

In this communication, one of the main operations decision problem faced by airlines has been dealt with using Mathematical Programming and Computational Intelligence methods. The proposed approach does not produce an exact solution in pure mathematical terms but appears to be quite adapted to give a real support to decision making, by providing, through a comprehensive process, an improved approximation of the set of non inferior solutions attached to this bi-criterion decision problem.

References

1. Arabeyre, J.P., Fearnley, J., Steiger F.C.: The Airline Crew Scheduling Problem : A Survey. Transportation Science, 1 (1969) 140-163.
2. Baker, E., Bodin, L., Fisher, M.: The Development and Implementation of a Heuristic Set Covering Based System for Air Crew Scheduling. Management Science and Statistics Working Paper 80-015, College of Business and Management, University of Maryland (1980).
3. Ball, M. and Roberts, A.: A Graph Partitionning Approach to Airline Crew Scheduling. Transportation Science, 19/2 (1985) 107-126.
4. Desaulniers, G., Desrosiers, J., et al.: Crew pairing at Air France. European Journal of Operational Research, (1997) 245-259.
5. EL Moudani, W. and Mora-Camino, F.: Airlines Fleet Operations Management : An Integrated Solution Approach. XV IEE on CAD/CAM, Aguas de Lindoia, SP, Brazil (1999).

6. EL Moudani, W. and Mora-Camino, F.: A Dynamic Approach for Aircraft Assignment and Maintenance Scheduling by Airlines. Journal of Air Transport Management, 6/4, (2000) 233-237.
7. EL Moudani, W. and Mora-Camino, F.: A Fuzzy Solution Approach For The Roster Planning Problem. The 9th IEEE International Conference on Fuzzy Systems, San Antonio, Texas, USA (2000).
8. Mora-Camino, F.: Modélisation, Optimisation, complexité et Algorithmes. Lectures Notes, CNAM-Toulouse, France (1997).
9. Gamache, M., Soumis, F., Marquis, G., Desrosiers J.: A Column Generation Approach for Large Scale Aircrew Rostering Problems. Les Cahiers du GERAD, G94-20, Ecole des Hautes Etudes Commerciales, Montréal, Canada, H3T 1V6 (1994).
10. Crainic, T.G. and Rousseau, J.M.: The Column Generation Principle and the Airline Crew Scheduling Problem. INFOR 25 (1987) 136-151.
11. Byrne, J.: A Preferential Bidding System for technical Aircrew (QANTAS AUSTRALIA). The 28th AGIFORS Symposium, Massachusetts, US (1988).
12. Sarra, D.: The automatic assignment model (SATURN - ALITALIA). The 28th AGIFORS Symposium, Massachusetts, US (1988).
13. Rubin, J.: A Technique for the Solution of Massive Set Covering Problems, with Application to Airline Crew Scheduling. Transportation Science, 7 (1973) 34-48.
14. Campello, R.E. and Maculan, N.: Algoritmos e heuristicas. Ed. Universidade federal Fluminense, Niteroi, Brazil (1994).
15. Teodorovic, P. and Lucic, P.: A fuzzy set theory approach to the aircrew rostering problem. Fuzzy Sets and Systems, 95 (1998) 261-271.
16. Holland, J.H.,: Adaptation in Natural and Artificial Systems. University of Michigan Press, Ann Arbor (1975).
17. Goldberg, D.E.: Genetic Algorithms in search, optimization and machine learning. Addison-Wesley, Reading (1989).
18. Bridges, C.L. and Goldberg, D.E.: An Analysis of Reproduction and Crossover in a Binary-Coded Genetic Algorithm. Proceedings of the second International Conference on Genetic Algorithm. ICGA (1987).
19. Bridges, C.L. and Goldberg, D.E.: An Analysis of Multipoint Crossover. Proceedings of the Foundation of Genetic Algorithms. FOGA (1991).
20. Beasly, J.E. and Chu, P.C.: A Genetic Algorithm for the Set Covering Problem. European Journal of Operational Research, 94 (1996) 392-404.
21. Guerinik, N. and Van Caneghem, M.: Solving Crew Scheduling Problem by Constraint Programming. Proceedings of the 1st Conference of Principles and Practice of Constraint Programming, (1995) 481-498.
22. EL Moudani, W. and Mora-Camino, F.: Evaluation par la Logique Floue de la satisfaction du personnel navigant technique dans les compagnies aériennes. Rapport LAAS-CNRS, Toulouse, France, Déc. 2000.

Halftone Image Generation with Improved Multiobjective Genetic Algorithm

Hernán E. Aguirre, Kiyoshi Tanaka, Tatsuo Sugimura, and Shinjiro Oshita

Dept. of Electrical and Electronic Engineering, Faculty of Engineering,
Shinshu University, 4-17-1 Wakasato, Nagano, 380-8553 Japan
{{ahernan, ktanaka, tsugimu}@gipwc,oshita@oshan}.shinshu-u.ac.jp

Abstract. A halftoning technique that uses a simple GA has proven to be very effective to generate high quality halftone images. Recently, the two major drawbacks of this conventional halftoning technique with GAs, i.e. it uses a substantial amount of computer memory and processing time, have been overcome by using an improved GA (GA-SRM) that applies genetic operators in parallel putting them in a cooperative-competitive stand with each other. The halftoning problem is a true multiobjective optimization problem. However, so far, the GA based halftoning techniques have treated the problem as a single objective optimization problem. In this work, the improved GA-SRM is extended to a multiobjective optimization GA to generate simultaneously halftone images with various combinations of gray level and spatial resolution. Simulation results verify that the proposed scheme can effectively generate several high quality images simultaneously in a single run reducing even further the overall processing time.

Keywords: multiobjective genetic algorithm, multiobjective optimization, halftoning problem, cooperative-competitive genetic operators.

1 Introduction

The multiobjective nature of most real-world problems makes multiobjective optimization (MO) a very important research topic. Evolutionary algorithms (EAs) seem particularly desirable to solve MO problems because they evolve simultaneously a population of potential solutions to the problem in hand, which allows to search for a set of Pareto optimal solutions concurrently in a single run of the algorithm. Many authors have been increasingly investigating MO using EAs in recent years and the number of applications has been rapidly growing [1,2,3,4]. In the signal processing area, application methods using EAs, especially genetic algorithms (GAs), are also steadily being developed[5].

In this work, we especially focus on the image halftoning technique using GAs. Kobayashi et al.[6,7] use a GA to generate bi-level halftone images with quality higher than conventional techniques such as ordered dithering, error diffusion and so on[8]. However, it uses a substantial amount of computer memory and processing time[6,7]. Recently, Aguirre et al.[9,10] have proposed an improved GA (GA-SRM) to overcome these two drawbacks of the conventional halftoning technique with GAs. GA-SRM is based on an empirical model of GA that applies genetic operators in parallel putting them in a cooperative-competitive stand with each other[11,12,13,14]. The improved GA-SRM, extended to the halftoning problem, can generate high quality images achieving a 98% reduction in the population size and an 85%-70% reduction in processing time.

The halftoning problem is a true MO problem in which high gray level and high spatial resolution must be sought to achieve high quality images. The GA based halftoning techniques mentioned above, however, treat the problem as a single objective optimization problem and can generate only one image at a time.

In this work, the improved GA-SRM[9,10] is extended to a multiobjective optimization GA to generate simultaneously halftone images with various combinations of gray level and spatial resolution. The simulations results show that the proposed scheme can effectively generate several images in a single run reducing even further the overall processing time.

2 Halftoning Problem with GAs

Digital halftoning, a key component of an image display preprocessor, is the method that creates the illusion of continuous tone pictures on printing and displaying devices that are capable of producing only binary picture elements. The fast growing computer and information industry requires each time higher image quality and demands higher resolution devices. The halftoning algorithms capable of delivering the appropriate image quality for such devices are also needed.

Kobayashi et al.[6,7] use a GA to generates bi-level halftone images with quality higher than traditional techniques such as ordered dithering, error diffusion and so on[8]. An input gray tone image of R gray levels is divided into non-overlapping blocks of $n \times n$ pixels, and then the 2-dimensional optimum binary pattern for each image block is searched using a GA[6,7]. The GA uses a $n \times n$ 2-dimensional binary representation for the individuals. Crossover interchanges either sets of adjacent rows or columns between two individuals and mutation inverts bits with a very small probability per bit after crossover similar to canonical GA[15,16]. Individuals are evaluated for two factors required to obtain visually high quality halftone images. (i) One is high gray level resolution (local mean gray levels close to the original image), and (ii) the other is high spatial resolution (appropriate contrast near edges)[6,7]. The gray level resolutions error is calculated by

$$E_m(\boldsymbol{x}_i^{(t)}) = \frac{1}{n^2} \sum_{(j,k) \in block} | p(j,k) - \hat{p}_b(j,k) | \qquad (1)$$

where $\boldsymbol{x}_i^{(t)}$ is i-th individual at t-th generation, $p(j,k)$ is the gray level of the (j,k)-th pixel in the original image block, and $\hat{p}_b(j,k)$ is the estimated gray level associated to the (j,k)-th pixel from the generated binary block. To obtain $\hat{p}_b(j,k)$, a reference region around the (j,k)-th binary pixel (for example 5×5 pixels) is convoluted by a gaussian filter that models the correlation among pixels. On the other hand, the spatial resolution error is calculated by

$$E_c(\boldsymbol{x}_i^{(t)}) = \frac{1}{n^2} \sum_{(j,k) \in block} \mid (p(j,k) - \bar{p}_s(j,k)) - (q(j,k) - \frac{1}{2})R \mid \quad (2)$$

where $\bar{p}_s(j,k)$ is the local mean gray level around the (j,k)-th pixel (within a reference region) in the original image block, and $q(j,k)$ is the binary level of the (j,k)-th pixel in the generated image block. These two errors are combined into one single objective function as

$$e(\boldsymbol{x}_i^{(t)}) = \omega_m E_m(\boldsymbol{x}_i^{(t)}) + \omega_c E_c(\boldsymbol{x}_i^{(t)}) \quad (3)$$

where ω_m and ω_c are the weighting parameters for gray level and spatial resolution errors, respectively. The individuals' fitness is assigned by

$$f(\boldsymbol{x}_i^{(t)}) = e(\boldsymbol{x}_W^{(t)}) - e(\boldsymbol{x}_i^{(t)}) \quad (4)$$

where $e(\boldsymbol{x}_W^{(t)})$ is the combined error of the worst individual at t-th generation. The high image quality that can be achieved is the method's major strength. However, it uses a substantial amount of computer memory and processing time. High quality, visually satisfactory, halftone images are obtained with 200 individuals and 200 generations (totally 40,000 evaluations) per image block[6,7].

Recently, Aguirre et al.[9,10] have proposed an improved GA (GA-SRM) to overcome these two drawbacks of the conventional halftoning technique with GAs. GA-SRM is based on an empirical model of GA that applies genetic operators in parallel putting them in a cooperative-competitive stand with each other[11,12,13,14]. GA-SRM is applied to the halftoning image problem using genetic operators properly modified for this kind of problem(see **4.3**). GA-SRM with parallel adaptive dynamic block (ADB) mutation impressively reduces processing time and computer memory to generate high quality images. For example, GA-SRM with qualitative ADB using a 2 parent 4 offspring configuration needs about 6,000-12,000 evaluations per image block, depending on the input image, to obtain results similar to those achieved by the conventional image halftoning technique using GAs. These data represent a 98% reduction in the population size and an 85%-70% reduction in processing time.

3 Multiobjective Optimization (MO)

MO methods deal with finding optimal solutions to problems having multiple objectives. Let us consider, without loss of generality, a minimization multiobjective problem with M objectives:

$$minimize \; \boldsymbol{g}(\boldsymbol{x}) = (g_1(\boldsymbol{x}), \cdots, g_M(\boldsymbol{x})) \quad (5)$$

where $\boldsymbol{x} \in \boldsymbol{X}$ is a solution vector in the solution space $\boldsymbol{X}$, and $g_1(\cdot), \cdots, g_M(\cdot)$ the M objectives to be minimized. Key concepts used in determining a set of solutions for multiobjective problems are dominance, Pareto optimality, Pareto set, and Pareto front. These concepts can be defined as follows.

A solution vector $\boldsymbol{y} \in \boldsymbol{X}$ is said to *dominate* a solution vector $\boldsymbol{z} \in \boldsymbol{X}$, denoted by $\boldsymbol{g}(\boldsymbol{y}) \preceq \boldsymbol{g}(\boldsymbol{z})$, if and only if $\boldsymbol{y}$ is partially less than $\boldsymbol{z}$, i.e., $\forall j \in \{1, \cdots, M\}$, $g_j(\boldsymbol{y}) \leq g_j(\boldsymbol{z}) \wedge \exists j \in \{1, \cdots, M\} : g_j(\boldsymbol{y}) < g_j(\boldsymbol{z})$.

The objective values are calculated once for each individual in the offspring population. However, we keep as many fitness values as defined search directions. A combined objective value is calculated for each $\boldsymbol{\omega}^k$ ($k = 1, 2, \cdots, N$) by

$$g^k(\boldsymbol{x}_i^{(t)}) = \sum_{j=1}^{M} \omega_j^k g_j(\boldsymbol{x}_i^{(t)}) = \omega_1^k g_1(\boldsymbol{x}_i^{(t)}) + \omega_2^k g_2(\boldsymbol{x}_i^{(t)}) \tag{14}$$

and the individuals' fitness in the k-th search direction is assigned by

$$f^k(\boldsymbol{x}_i^{(t)}) = g^k(\boldsymbol{x}_W^{(t)}) - g^k(\boldsymbol{x}_i^{(t)}) \tag{15}$$

where $g^k(\boldsymbol{x}_W^{(t)})$ is the combined objective value of the worst individual in the k-th search direction at the t-th generation.

For each search direction $\boldsymbol{\omega}^k$, CM creates a corresponding λ_{CM}^k number of offspring. Similarly, SRM creates λ_{SRM}^k offspring (see detailed information about CM and SRM implementation for halftoning problem in **4.3**). Thus, the total offspring number for each search direction is

$$\lambda^k = \lambda_{CM}^k + \lambda_{SRM}^k. \tag{16}$$

The offspring created for all N search directions coexist within one single offspring population. Hence the overall offspring number is

$$\lambda = \sum_{k=1}^{N} \lambda^k. \tag{17}$$

SRM's mutation rates are adapted based on a normalized mutants survival ratio. The normalized mutant survival ratio used in [9,10] is extended to

$$\gamma = \frac{\sum_{k=1}^{N} \mu_{SRM}^k}{\sum_{k=1}^{N} \lambda_{SRM}^k} \cdot \frac{\lambda}{\sum_{k=1}^{N} \mu^k} \tag{18}$$

where μ^k is the number of individuals in the parent population of the k-th search direction $P^k(t)$, μ_{SRM}^k is the number of individuals created by SRM present in $P^k(t)$ after extinctive selection, λ_{SRM}^k is the offspring number created by SRM and λ is the overall offspring number as indicated in **Eq. (17)**.

We chose (μ, λ) Proportional Selection[17] to implement the extinctive selection mechanism. Since we want to search simultaneously in various directions, selection to choose the parent individuals that will reproduce either with CM or SRM is accordingly applied for each one of the predetermined search directions. Thus, selection probabilities for each search direction $\boldsymbol{\omega}^k$ are computed by

$$P_s^k(\boldsymbol{x}_i^{(t)}) = \begin{cases} f^k(\boldsymbol{x}_i^{(t)}) / \sum_{j=1}^{\mu^k} f^k(\boldsymbol{x}_j^{(t)}) & (1 \leq i \leq \mu^k \leq \lambda^k) \\ 0 & (\mu^k < i \leq \lambda) \end{cases} \tag{19}$$

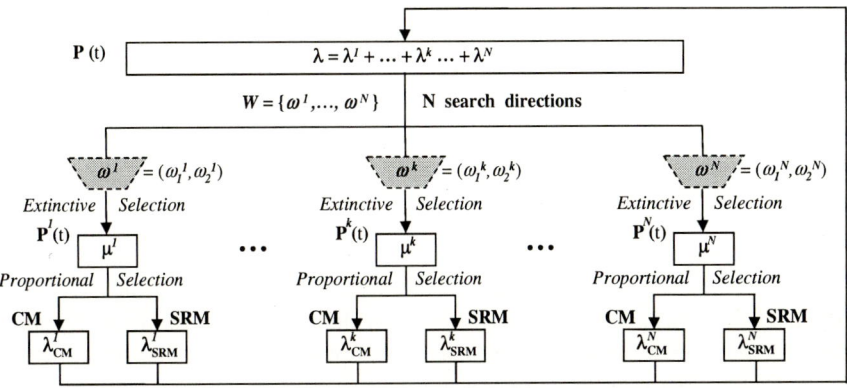

Fig. 1. Block diagram of the extended multiobjective GA-SRM

where $x_i^{(t)}$ is an individual at generation t which has the i-th highest fitness value in the k-th search direction $f^k(x_i^{(t)})$, μ^k is the number of parents and λ^k is the number of offspring in the k-th search direction, and λ is the overall number of offspring.

Note that for each search direction only $\lambda^k < \lambda$ individuals are created. However, the parent population μ^k is chosen among the overall λ offspring population. In this way information sharing is encourage among individuals created for neighboring search directions provided that the neighbors' fitness are competitive with the locals'. **Fig. 1** presents the block diagram of the extended multiobjective GA-SRM.

Once the offspring has been evaluated, a set of non-dominated solutions is sought for each search direction, i.e. for the k-th search direction non-domination is checked only among the offspring created for that search direction. Two secondary populations keep the non-dominated solutions. $P_{cur}(t)$ keeps the non-dominated solution obtained from the offspring population at generation t and P_{nds} keeps the set of the non-dominated solutions found through the generations. P_{nds} is updated at each generation with $P_{cur}(t)$. In the halftoning problem an image is divided into blocks and the GA is applied to each image block. Hence, the GA would generate a set of non-dominated solutions for each image block. Since we are interested in generating simultaneously various Pareto optimal "whole" images, a decision making process is integrated to chose only one solution for each search direction in each image block. Thus, among the various non-dominated solutions found for a given search direction, we chose the one that minimizes the combined error E_m and E_c in that particular direction.

4.3 CM and SRM for Halftoning Problem

In the halftoning problem an individual is represented as a $n \times n$ two-dimensional structure. In this work we use the same two-dimensional operators, CM (Crossover and Mutation) and SRM-ADB (Self Reproduction with Mutation - Adaptive Dynamic Block), presented in [9,10] to create offspring.

CM first crosses over two previously selected parents interchanging either their rows or columns, similar to [6,7], and then it applies standard mutation inverting bits with a small mutation probability per bit, $p_m^{(CM)}$, analogous to canonical GAs. Thus, mutation in CM is of a quantitative nature after which the number of 0s and 1s may change. It may be worth trying more specialized approaches to implementing crossover, however this point will not be discussed in this work.

SRM, on the other hand, first creates an exact copy of a previously selected individual from the parent population and then applies mutation only to the bits inside a mutation block. SRM is provided with an Adaptive Dynamic-Block (ADB) mutation schedule similar to Adaptive Dynamic-Segment mutation (ADS)[12,14]. With ADB mutation is directed only to a block (square region) of the chromosome and the mutation block area $\ell \times \ell$ is dynamically adjusted to $\ell/2 \times \ell/2$ every time the normalized mutants survival ratio γ by **Eq. (18)** falls under a threshold τ. The block's side length ℓ varies from n to 2, $[n,2]$. The offset position of the mutation block is chosen at random for each chromosome. The adaptive mechanism in SRM is designed to control the required exploration-exploitation balance during the search process.

The effect of ADB's mutation on the distribution of 0s and 1s within an individual could be of a qualitative or quantitative nature. It has been verified in [9,10] that for the halftoning problem ADB with qualitative mutation shows superior performance than ADB with quantitative mutation (i.e. bit flipping mutation). Since qualitative mutation do not change the number of 0s and 1s within an individual it has an impact only on the spatial resolution error E_c, while quantitative mutation has an impact on both E_m and E_c in **Eq. (3)** and **(14)**. Thus, qualitative mutation is less disruptive and can take better advantage of the high correlation among contiguous pixels in an image[25] contributing to a more effective search. Therefore, in this work we use ADB with qualitative mutation, which is implemented as a bit swapping process. Note that there is no need to set a mutation probability in qualitative mutation since all pairs of bits within the mutation block are simply swapped.

5 Experimental Results and Discussion

We observe and compare the performance of four kinds of GAs generating halftone images: (i) a simple GA that uses CM and proportional selection, similar to [6,7], (denoted as cGA) (ii) an extended cGA using the same multiobjective technique described in **4.2** (denoted as a moGA), (iii) a GA with SRM that uses CM, SRM and (μ, λ) proportional selection[9,10] (denoted as GA-SRM), and (iv) the extended multiobjective GA-SRM (denoted as moGA-SRM).

The GAs are applied to SIDBA's benchmark images in our simulation. The size of the original image is 256×256 pixels with $R = 256$ gray levels. An image is divided into 256 non-overlapping blocks, each one of size $n \times n = 16 \times 16$ pixels. For each block, the algorithms were set with different seeds for the random initial population.

We define 11 search directions, $N = 11$, setting $\boldsymbol{W} = \{\boldsymbol{\omega}^1, \boldsymbol{\omega}^2, \cdots, \boldsymbol{\omega}^{11}\} = \{(0.0, 1.0), (0.1, 0.9), \cdots, (1.0, 0.0)\}$. With $\boldsymbol{\omega}^1 = (0.0, 1.0)$ the search focuses exclusively in E_c's space and with $\boldsymbol{\omega}^{11} = (1.0, 0.0)$ in E_m's; whereas with $\boldsymbol{\omega}^k$, $2 \leq k \leq 10$, the search focuses in the combined space of E_c and E_m. moGA and moGA-SRM generate simultaneously one image for each direction in a single run. On the other hand, to generate the 11 images with either cGA or GA-SRM an equal number of separate runs are carried out, each one using a different $\boldsymbol{\omega}^k$ as weighting parameter. Unless stated otherwise, the GAs are set with the parameters detailed in **Table 1**[1]) and the experimental image used is "Lenna". The values set for crossover and mutation probabilities in cGA are the same used in [6,7]. The image quality attained by the cGA with a 200 parent population and the same $T = 4 \times 10^4$ evaluations used in [6,7] are taken as a reference for comparison in our study. The number of generations performed for each algorithm is calculated as T/λ.

Table 1. Genetic algorithms parameters

Parameter	cGA	moGA	GA-SRM	moGA-SRM
Selection	Proport.	(μ, λ) Proport.	(μ, λ) Proport.	(μ, λ) Proport.
Mating	$(\boldsymbol{x}_i, \boldsymbol{x}_j), i \neq j$	$(\boldsymbol{x}_i, \boldsymbol{x}_j), i \neq j$	$(\boldsymbol{x}_i, \boldsymbol{x}_j), i \neq j$	$(\boldsymbol{x}_i, \boldsymbol{x}_j), i \neq j$
p_c	0.6	0.6	1.0	1.0
$p_m^{(CM)}$	0.001	0.001	0.001	0.001
$\mu^k : \lambda^k$	-	1 : 1	1 : 2	1 : 2
$\lambda_{CM}^k : \lambda_{SRM}^k$	-	-	1 : 1	1 : 1
τ	-	-	0.40	0.40

Table 2 shows the average in all image blocks of the non-normalized combined errors $e^k(\boldsymbol{x}) = \omega_1^k E_m(\boldsymbol{x}) + \omega_2^k E_c(\boldsymbol{x})$ by cGA(200) after T evaluations for each search direction $\boldsymbol{\omega}^k$, $1 \leq k \leq 11$, under column $\boldsymbol{W}$. For the other algorithms under $\boldsymbol{W}$ we present the fraction of T at which the algorithm reach similar image quality (for cGA(200) these values are all 1.00 and are shown right below the combined error). Column $T^{\boldsymbol{W}}$ indicates the overall evaluations needed to generate the 11 images. Since the cGA generates one image at a time, it needs $11T$[2]) evaluations to generate all 11 images. The first moGA row show results by the multiobjective simple GA with a $\mu^k = 18$ parents and a $\lambda^k = 18$, $\lambda = 198$ offspring configuration. moGA simultaneously generates the 11 images and needs approximately $2.43T$[3]) to guarantee that all images would have at least the same quality as cGA(200). moGA's second row show results by moGA with a $\mu^k = 4$ parents and a $\lambda^k = 4$, $\lambda = 44$ offspring configuration. In this case population size reduction in moGA accelerates a little bit more the overall convergence and still produces better images than cGA(200). It should be noticed that population

---

[1]) GA-SRM search only in one direction at a time and the population related parameters μ^k, λ^k, λ_{CM}^k, and λ_{SRM}^k should be read without the index k

[2]) The entire number of evaluations required by the single objective GAs to generate all 11 images are given by the sum of the evaluations expended in each direction

[3]) In the case of multiple objective GAs, due to the concurrent search, the maximum number of the evaluations among all search directions determines the overall number of evaluations needed to generate all 11 images

reductions in cGA accelerates convergence but it is affected by a lost of diversity and the final image quality is inferior than cGA(200)'s[6,7]. moGA benefits from the information sharing induced by selection (see explanation below for **Fig. 2**) and can tolerate population reductions. Compared with cGA, the results by moGA represents an enormous reduction in processing time and illustrates the benefits that can be achieved by including multiobjective techniques within GAs.

Table 2. Evaluations to generate high quality images (Lenna)

Algorithm	$W = \{\omega^1, \omega^2, \cdots, \omega^{11}\}$											T^W
	ω^1	ω^2	ω^3	ω^4	ω^5	ω^6	ω^7	ω^8	ω^9	ω^{10}	ω^{11}	
combined error	121.0	111.4	100.6	89.5	78.2	66.9	55.5	44.2	32.8	21.5	10.1	-
$cGA(200)$	1.00	1.00	1.00	1.00	1.00	1.00	1.00	1.00	1.00	1.00	1.00	$11T^{2)}$
$moGA(18, 198)$	1.43	2.43	1.65	1.27	1.21	1.00	0.86	0.76	0.70	0.65	0.72	$2.43T^{3)}$
$moGA(4, 44)$	1.12	2.30	1.44	1.36	1.20	1.02	0.85	0.79	0.73	0.66	0.79	$2.30T^{3)}$
$GA\text{-}SRM(2, 4)$	0.40	0.23	0.15	0.13	0.12	0.11	0.10	0,09	0.09	0.08	0.08	$1.58T^{2)}$
$moGA\text{-}SRM(9, 198)$	1.12	1.07	0.58	0.44	0.30	0.27	0.24	0.23	0.22	0.21	0.21	$1.12T^{3)}$
$moGA\text{-}SRM(2, 44)$	1.56	1.03	0.50	0.30	0.20	0.16	0.15	0.13	0.12	0.12	0.12	$1.56T^{3)}$
$moGA\text{-}SRM^*(2, 44)$	0.96	0.92	0.40	0.31	0.22	0.17	0.15	0.14	0.13	0.13	0.13	$0.96T^{3)}$

Row GA-SRM(2,4) presents results by GA-SRM with a 2 parents and 4 offspring configuration. GA-SRM even with a very scaled down population configuration considerably reduces processing time to generate high quality images for all combinations of weighting parameters. GA-SRM, for this particular image, would need approximately $1.58T^{2)}$ to generate all 11 images. Note that GA-SRM sequentially generating the 11 images is faster than moGA.

The first moGA-SRM row show results by the multiobjective proposed GA-SRM with a $\mu^k = 9$ parents and a $\lambda^k = 18$, $\lambda = 198$ offspring configuration. Compared with moGA we can see that the inclusion of SRM notoriously increases the multiobjective algorithm's performance needing no more than $1.12T^{3)}$ to generate the 11 images, which is faster than GA-SRM. Results by a scaled down population configuration is shown in row moGA-SRM(2,44) that represents a $\mu^k = 2$ parents and a $\lambda^k = 4$, $\lambda = 44$ offspring configuration. The population size reduction in moGA-SRM accelerates convergence in all but one search direction (see under ω^1) and the overall evaluation time is similar to GA-SRM. From GA-SRM and moGA-SRM results we see that parallel mutation SRM can greatly improve the performance of single objective as well as multiobjective genetic algorithms in the halftoning problem.

We observe that moGA(2,44), which uses CM but not SRM, only for ω^1 produces faster convergence than moGA-SRM ($e^1 = 0.0E_m + 1.0E_c$). It seems that CM alone is particularly useful for searching in E_c's search space. However, when the search involves both E_m's and E_c's spaces the interaction of CM and SRM produce better results. We conduct an experiment in which we favor CM's offspring over SRM's only in the ω^1 direction. In row moGA-SRM*(2,44) we show results using a configuration that creates offspring in ω^1 direction only with CM, i.e. $\lambda^1_{CM} = 4$, $\lambda^1_{SRM} = 0$ and $\lambda^k_{CM} = 2$, $\lambda^k_{SRM} = 2$ for $2 \leq k \leq 11$. This has the effect of accelerating convergence in ω^1 search direction and therefore reducing the overall evaluation time to $0.96T$.

E_m and E_c represent fitness landscapes with different degree of difficulty for the GAs. E_m's landscape is smoother than E_c's and the GAs are expected to converge faster in E_m's direction. This is corroborated by the results obtained by the GAs. In **Table 2** we can see that for ω^k with $k \geq 6$, E_m's directions, the algorithms need less time to converge. It should be specially noticed that moGA-SRM for those directions finds high quality images in less than $0.2T$. This behavior and the results by the last experiment mentioned above suggest that it may be worth trying dynamic configurations so that more resources could be assigned to those directions that require more time to converge accelerating the overall time needed to generate images simultaneously.

Table 3. Actual percentage of evaluations expended in each search direction

Algorithm	$W = \{\omega^1, \omega^2, \cdots, \omega^{11}\}$										
	ω^1	ω^2	ω^3	ω^4	ω^5	ω^6	ω^7	ω^8	ω^9	ω^{10}	ω^{11}
$cGA(200)$	100.0	100.0	100.0	100.0	100.0	100.0	100.0	100.0	100.0	100.0	100.0
$moGA(18, 198)$	13.0	22.1	15.0	11.5	11.0	9.1	7.8	6.9	6.4	5.9	6.5
$moGA(4, 44)$	10.2	20.9	13.1	12.4	10.9	9.3	7.7	7.2	6.6	6.0	7.2
$GA\text{-}SRM(2, 4)$	40.0	23.0	15.0	13.0	12.0	11.0	10.0	9.0	9.0	8.0	8.0
$moGA - SRM(9, 198)$	10.2	9.7	5.3	4.0	2.7	2.5	2.2	2.1	2.0	1.9	1.9
$moGA - SRM(2, 44)$	14.2	9.4	4.5	2.7	1.8	1.5	1.4	1.2	1.1	1.1	1.1
$moGA - SRM^*(2, 44)$	8.7	8.4	3.6	2.8	2.0	1.5	1.4	1.3	1.2	1.2	1.2

In **Table 2** moGA's and moGA-SRM's rows show the evaluations expended by the algorithm in all search directions. The actual percentage of the evaluations expended in each search direction is shown in **Table 3**. From this table it can be seen that with the multiobjective algorithms there is a substantial reduction of the actual number evaluations for each search direction. These reductions are explained by the information sharing induced by the selection process. As mentioned in **4.2** and indicated by **Eq. (19)**, the individuals with higher fitness in a specific direction are selected as parents. Thus, the individuals chosen to be parents for the k-th search direction at generation t may have been created for neighboring directions at generation t-1. To verify this point we also observe the composition of the parent population for each search direction. **Fig. 2** shows the average distribution for some of the ω^k directions after $0.1T$ and T evaluations, respectively. For example, in **Fig. 2(a)**, the parent population of ω^4 is in average composed by 18% of individuals coming from ω^3, 30% from ω^4 itself, and 13% from ω^5. From these figures we can see that each search direction benefits from individuals that initially were meant for other neighboring directions. This information sharing pushes forward the search reducing convergence times. Looking at **Fig. 2(a)** and **Fig. 2(b)** we can see that the information sharing is higher during the initial stages of the search.

Fig. 3 illustrates typical transitions of the non-normalized combined error $e(x)$ over the number of evaluations for some of the search directions by the GAs. The plots are cut after T evaluations. From these figures it can be visually appreciated the higher convergence velocity and higher convergence reliability

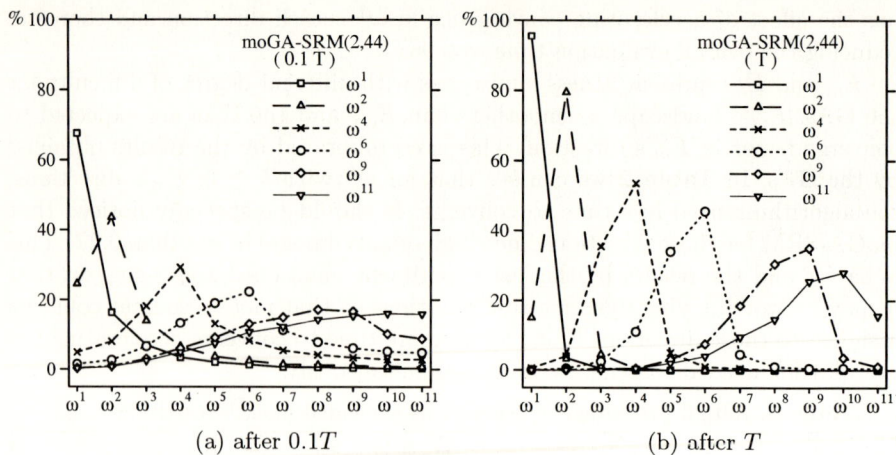

Fig. 2. moGA-SRM's average parent population distribution

(lower errors) by the algorithms that include SRM, GA-SRM and moGA-SRM. In general, moGA is faster than the cGA, but their final image quality tends to be the same. Also, it should be noticed that results by moGA and moGA-SRM are achieved simultaneously in one run (thus, T for these algorithms indicates the evaluations expended in all search directions).

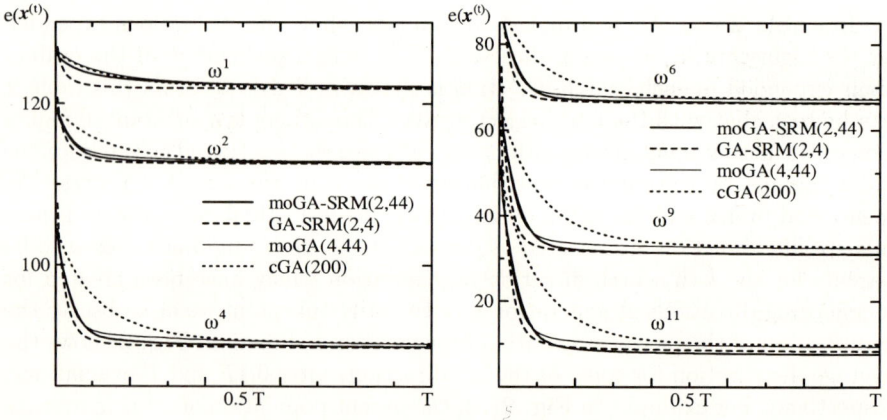

Fig. 3. Error transition for various ω^k

Fig. 4 show the original image "Lenna" and the images generated by two conventional halftoning techniques: ordered dithering (screen) and error diffusion[8]. **Fig. 5** show some of the simultaneously generated images by moGA-SRM. From these figures we can see that moGA-SRM generates more pleasant images to the human observer than traditional techniques. Another point to be remarked is that traditional halftoning techniques can generate only one image. On the other hand, among the images generated by moGA-SRM there is a gradual difference according to spatial and gray level resolution, which makes the GA based

(a) original image (b) ordered dithering (screen) (c) error diffusion

Fig. 4. Lenna's original and generated images by two conventional techniques

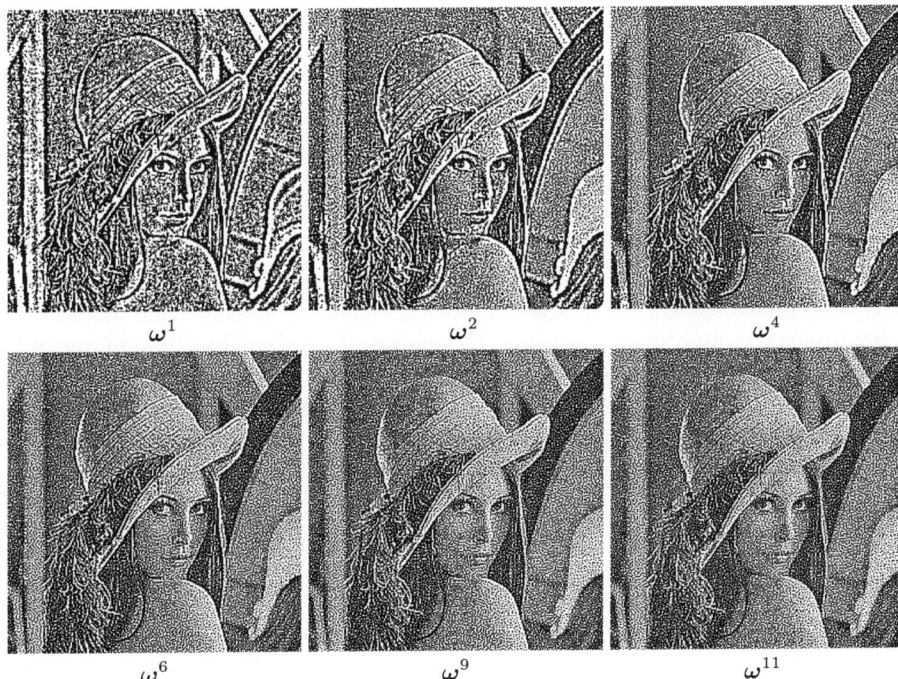

ω^1 ω^2 ω^4

ω^6 ω^9 ω^{11}

Fig. 5. Lenna's simultaneously generated images by moGA-SRM*(2,44) after $0.96T$

halftoning technique more flexible to users' requirements as well as more robust to constraints imposed by displaying and printing devices.

With regards to processing time, running software implementations of the algorithms in a Pentium III processor (600 MHz), to generate one image conventional techniques need only few seconds while GA-SRM (also implemented in software) needs about 8 minutes. Note that GA based techniques in this study process one block at a time always starting with random initial populations. Due to the high correlation among neighbor blocks of an image, reductions on processing time are expected by using previously generated image blocks in the initial populations of the subsequent blocks. However it is clear that, from a processing time standpoint, in order to apply GA based halftoning techniques on-line they must be improved further to reduce as much as possible the number of evaluations needed to generate higher quality images. Also, the GA's final implementation for industrial application must be in hardware.

Finally, we should also say that similar results were obtained for other SIBDA's benchmark images.

6 Conclusions

In this work we have extended an improved GA (GA-SRM) to a multiobjective optimization GA (moGA-SRM) for the image halftoning problem aiming to simultaneously generate halftone images with various combinations of gray level and spatial resolution.

GA-SRM is based on an empirical model of GA that puts parallel genetic operators in a cooperative-competitive stand with each other. To extend GA-SRM we follow a cooperative population search with aggregation selection preserving the fundamental features of the cooperative-competitive model. We compare the performance of four genetic algorithms generating halftone images: (i) a single objective simple GA (cGA), (ii) a single objective GA-SRM, (iii) a multiobjective simple GA (moGA), (iv) the proposed multiobjective GA-SRM (moGA-SRM).

From our experimental results we observe that multiobjective techniques benefit from information sharing and can greatly reduce processing time to generate simultaneously high quality images. To generate 11 images moGA requires only about 21% of the evaluations used by cGA. The cooperative-competitive model for parallel operators helps to increase the performance of single and multi objective GAs in this problem reducing even further processing time. GA-SRM requires about 15% and moGA-SRM about 9% of the evaluations used by cGA.

As future works, important issues to be explored related to the halftoning problem are (i) the effect of the definition of the weights set on the algorithm's stability and convergence, (ii) dynamic and parallel hierarchical configurations for moGA-SRM in order to accelerate the overall time needed to generate images simultaneously. Also, we are planning to continue studying moGA-SRM's behavior in a wider range of problems that include more than two objectives[18] and use it in other real world applications.

References

1. C. Fonseca and P. Fleming, "An overview of Evolutionary Algorithms in Multiobjective Optimization", *Evolutionary Computation*, 3(1):1-16, 1995.
2. J. Horn, "Multicriterion Decision Making", *Handbook of Evolutionary Computation*, Oxford University Press, Volume 1, pp.F1.9:1-F1.9:15, 1997.
3. C. Coello, "A Comprehensive Survey of Evolutionary-Based Multiobjective Optimization Techniques", *Knowledge and Information Systems*, 1(3):269-308, 1999.
4. D. Van Veldhuizen and G. Lamont, "Multiobjective Evolutionary Algorithms: Analyzing the State-of-the-Art", *Evolutionary Computation*, 8(2):125-147, 2000.

5. K. S. Tang, K. F. Man, S. Kwong and Q. He, "Genetic Algorithms and Their Applications", *IEEE Signal Processing Magazine*, vol.13, no. 6, pp.22-37, Jun. 1996.
6. N. Kobayashi and H. Saito, "Halftoning Technique Using Genetic Algorithm", *Proc. IEEE ICASSP'94*, vol.5, pp.105-108, Apr. 1994.
7. N. Kobayashi and H. Saito, "Halftone Algorithm Using Genetic Algorithm", *IEICE Trans.*, vol.J78-D-II, no. 10, pp.1450-1459, Oct. 1995 (in Japanese).
8. R. Ulichney, *Digital Halftoning*, MIT Press, Cambridge, 1987.
9. H. Aguirre, K. Tanaka and T. Sugimura, "Accelerated Image Halftoning Technique Using Improved Genetic Algorithm with Tiny Populations", *Proc. IEEE SMC'99*, vol.IV, pp.905-910, Oct. 1999.
10. H. Aguirre, K. Tanaka and T. Sugimura, "Accelerated Halftoning Technique Using Improved Genetic Algorithm", *IEICE Trans.*, vol.E83-A, no. 8, pp.1566-1574, Aug. 2000.
11. H. Aguirre, K. Tanaka and T. Sugimura, "Cooperative Crossover and Mutation Operators in Genetic Algorithms ", *Proc. GECCO'99*, p.772, Jul. 1999.
12. H. Aguirre, K. Tanaka and T. Sugimura, "Cooperative Model for Genetic Operators to Improve GAs", *Proc. IEEE ICIIS'99*, pp.98-106, Nov. 1999.
13. H. Aguirre, K. Tanaka, T. Sugimura and S. Oshita, "Cooperative-Competitive Model for Genetic Operators: Contributions of Extinctive Selection and Parallel Genetic Operators", *Proc. Late Braking Papers GECCO 2000*, pp.6-14, Jul. 2000.
14. H. Aguirre, K. Tanaka and T. Sugimura, "Empirical Model with Cooperative-Competitive Genetic Operators to Improve GAs: Performance Investigation with 0/1 Multiple Knapsack Problems", IPSJ Journal, vol.41, no.10, pp.2837-2851, Oct. 2000.
15. J. H. Holland, *Adaptation in Natural and Artificial Systems*, Univ. of Michigan Press, 1975.
16. D. E. Goldberg, *Genetic Algorithms in Search, Optimization and Machine Learning*, Addison-Wesley, Reading, 1989.
17. T. Bäck, *Evolutionary Algorithms in Theory and Practice*, Oxford Univ. Press, 1996.
18. H. Ishibashi, H. Aguirre, K. Tanaka and T. Sugimura, "Multi-objective Optimization with Improved Genetic Algorithm", *Proc. IEEE SMC2000*, pp.3852-3857, Oct. 2000.
19. J. Schaffer and J. Grefenstette, "Multiobjective Optimization with Vector Evaluated Genetic Algorithms", *Proc. 1st Int. Conf. on Genetic Algorithms*, pp.93-100, Jul. 1985.
20. M. Fourman, "Compaction of Symbolic Layout Using Genetic Algorithms", *Proc. 1st Int. Conf. on Genetic Algorithms*, pp.141-153, Jul. 1985.
21. F. Kurwase, "A Variant of Evolution Strategies for Vector Optimization", *Lecture Notes in Computer Science*, vol.496, pp.193-197, Springer-Verlag, 1991.
22. T. Murata and H. Ishibuchi, "MOGA:multi-objective genetic algorithms", *Proc. 2nd IEEE Int. Conf. on Evolutionary Computation*, pp.289-294, Nov. 1995.
23. J. Horn, N. Nafpliotis and D. E. Goldberg, "A Niched Pareto Genetic Algorithm for Multiobjective Optimization", *Proc. 1st Int. IEEE Conf. on Evolutionary Computation*, IEEE Service Center, vol.1, pp.82-87, 1994.
24. N. Srinivas and K. Deb, "Multiobjective Optimization Using Nondominated Sorting in Genetic Algorithms", *Evolutionary Computation*, 2(3):221-248, 1994.
25. R. C. Gonzalez and R. E. Wood, *Digital Image Processing*, Addison-Wesley, 1992.

then increase the step size in order to accelerate convergence. This adaptation is done every $N \cdot L_R$ generations where N is the number of parameters (i.e. dimension of search space) and L_R is a constant, usually equal to one. Selection is done out of the set union of parent and offspring, i.e. the better one of the two is chosen to become the parent of the next generation.

2.2 The (μ, λ)-ES

A slightly more advanced method is to take one or more parents and even more offspring, i.e. $\mu \geq 1$ and $\lambda > \mu$. Mutation is accomplished in a sihyperellipsoidmilar way as with the (1+1)-ES. Besides the 1/5 rule, another method for step size adaptation becomes available which is called *self-adaptive mutation* ([Bäck (1997a)]). In this method, the mutation steps are adapted every generation. They are either increased, decreased or kept the same, each with a probability of 1/3. On the average, 1/3 of the offspring will now be closer to their parents than before, 1/3 keeps progressing at the same speed and 1/3 explores further areas. Depending on how far away from the optimum we currently are, one of these three groups will do better than the others and therefore, more individuals out of it will be selected to the next generation, where their step sizes are inherited. The algorithm adapts the step size by itself, i.e. by means of mutation and selection.

2.3 The $(\mu/\mu_I, \lambda)$-CMA-ES

The Covariance Matrix Adaptation is a sophisticated method for online adaptation of step sizes in (μ, λ)-ES with intermediate recombination (i.e. averaging of parents). It was first described by [Hansen & Ostermeier (1996)] and further improved and evaluated by [Hansen & Ostermeier (1997)]. For a complete description of the algorithm, the reader is referred to the latter publication. The basic idea is to adapt step sizes and covariances in such a way, that the longest axis of the of mutation distribution always aligns in the direction of greatest estimated progress. This is done by accumulating information about former mutation steps and their success (*evolution path*) and searching it for correlations. Besides this very sophisticated method for step size adaptation, a CMA-ES also includes mutation (with Σ now being a full matrix) and selection.

3 Multiobjective Evolutionary Algorithms

As soon as there are many (possibly conflicting) objectives to be optimized simultaneously, there is no longer a single optimal solution but rather a whole set of possible solutions of equivalent quality. Consider for example the design of an automobile. Possible objectives could be: minimize cost, maximize speed, minimize fuel consumption and maximize luxury. These goals are clearly conflicting and therefore there is no single optimum to be found. Multiobjective EAs can yield a whole set of potential solutions - which are all optimal in some sense - and give the engineers the option to assess the trade-offs between different designs. One then could for example choose to create three different cars according to different marketing needs: a slow low-cost model which consumes least fuel, an intermediate solution and a luxury sports car where speed is clearly the primer objective. Evolutionary Algorithms are well suited to multiobjective optimization problems as they are fundamentally based on biological processes, which are inherently multiobjective.

After the first pioneering work on multiobjective evolutionary optimization in the eighties ([Schaffer (1984)], [Schaffer (1985)]), several different algorithms have been proposed and successfully applied to various problems. For comprehensive overviews and discussions, the reader is referred to [Fonseca & Fleming (1995)], [Horn (1997)], [Van Veldhuizen & Lamont (1998)] and [Coello (1999)].

3.1 Dominance and Pareto-Optimality

In contrast to fully ordered scalar search spaces, multidimensional search spaces are only partially ordered, i.e. two different solutions are related to each other in two possible ways: either one dominates the other or none of them is dominated.

DEFINITION 1: *Consider without loss of generality the following multiobjective optimization problem with m decision variables x (parameters) and n objectives y:*

$$\text{Maximize } \mathbf{y} = \mathbf{f}(\mathbf{x}) = (f_1(x_1,\ldots,x_m),\ldots,f_n(x_1,\ldots,x_m))$$
$$\text{where} \quad \mathbf{x} = (x_1,\ldots,x_m) \in X \quad (2)$$
$$\mathbf{y} = (y_1,\ldots,y_n) \in Y$$

and where **x** *is called* decision (parameter) vector, X parameter space, **y** objective vector *and* Y objective space. *A decision vector* $\mathbf{a} \in X$ *is said to* dominate *a decision vector* $\mathbf{b} \in X$ *(also written as* $\mathbf{a} \succ \mathbf{b}$*) if and only if:*

$$\forall i \in \{1,\ldots,n\} : f_i(\mathbf{a}) \geq f_i(\mathbf{b}) \\ \wedge \; \exists j \in \{1,\ldots,n\} : f_j(\mathbf{a}) > f_j(\mathbf{b}) \quad (3)$$

Additionally, we say **a** covers **b** *(*$\mathbf{a} \succeq \mathbf{b}$*) if and only if* $\mathbf{a} \succ \mathbf{b}$ *or* $\mathbf{f}(\mathbf{a}) = \mathbf{f}(\mathbf{b})$*.*

Based on this convention, we can define nondominated, *Pareto-optimal* solutions as follows:

DEFINITION 2: *Let* $\mathbf{a} \in X$ *be an arbitrary decision (parameter) vector.*

1. *The decision vector* **a** *is said to be* nondominated *regarding a set* $X' \subseteq X$ *if and only if there is no vector in* X' *which dominates* **a**; *formally:*

$$\nexists \mathbf{a}' \in X' : \mathbf{a}' \succ \mathbf{a} \quad (4)$$

2. The decision (parameter) vector **a** is called Pareto-optimal *if and only if* **a** *is nondominated regarding the whole parameter space X.*

If the set X' is not explicitly specified, the whole parameter space X is implied.

Pareto-optimal parameter vectors cannot be improved in any objective without causing a degradation in at least one of the other objectives. They represent in that sense *globally optimal* solutions. Note that a Pareto-optimal set does not necessarily contain all Pareto-optimal solutions in X. The set of objective vectors $\mathbf{f}(\mathbf{a}'), \mathbf{a}' \in X'$, corresponding to a set of Pareto-optimal parameter vectors $\mathbf{a}' \in X'$ is called *"Pareto-optimal front"* or *"Pareto-front"*.

3.2 Difficulties in Multiobjectve Optimization

In extending the ideas of single objective EAs to multiobjective cases, two major problems must be addressed:

1. How to accomplish fitness assignment and selection in order to guide the search towards the Pareto-optimal set.
2. How to maintain a diverse population in order to prevent premature convergence and achieve a well distributed, wide spread trade-off front.

Note that the objective function itself no longer qualifies as fitness function since it is vector valued and fitness has to be a scalar value. Different approaches to relate the fitness function to the objective function can be classified with regard to the first issue. For further information, the reader is referred to [Horn (1997)]. The second problem is usually solved by introducing elitism and intermediate recombination. *Elitism* is a way to ensure that good individuals do not get lost (by mutation or set reduction), simply by storing them away in a external set, which only participates in selection. *Intermediate recombination* on the other hand averages the parameter vectors of two parents in order to generate one offspring according to:

$$\begin{aligned} \mathbf{x}'_j &= \alpha \mathbf{x}^g_{j1} + (1-\alpha)\mathbf{x}^g_{j2}, j, j_1, j_2 \in \{1, \ldots, \mu\} \\ \mathbf{x}^{g+1}_i &= \mathbf{x}'_j + \mathcal{N}(0, \Sigma) \quad , i = 1, \ldots, \lambda \quad\quad , j \in \{1, \ldots, \mu\} \end{aligned} \quad (5)$$

Arithmetic recombination is a special case of intermediate recombination where $\alpha = 0.5$.

4 The Strength Pareto Approach

For this work, the *Strength Pareto Approach* for multiobjective optimization has been used. Comparative studies have shown for a large number of test cases that, among all major multiobjective EAs, the *Strength Pareto*

Evolutionary Algorithm (SPEA) is clearly superior ([Zitzler & Thiele (1999)] [Zitzler, Thiele & Deb (2000)]). It is based on the above mentioned principles of Pareto-optimality and dominance. The algorithm as proposed by [Zitzler & Thiele (1999)] was implemented in a restartable, fully parallel code as follows:

Step 1: Generate random initial population P and create the empty external set of nondominated individuals P'.
Step 2: Evaluate objective function for each individual in P in parallel.
Step 3: Copy nondominated members of P to P'.
Step 4: Remove solutions within P' which are covered by any other member of P'.
Step 5: If the number of externally stored nondominated solutions exceeds a given maximum N', prune P' by means of clustering.
Step 6: Calculate the fitness of each individual in P as well as in P'.
Step 7: Select individuals from $P + P'$ (multiset union), until the mating pool is filled.
Step 8: Adapt step sizes of the members of the mating pool.
Step 9: Apply recombination and mutation to members of the mating pool in order to create a new population P
Step 10: If maximum number of generations is reached, then stop, else go to Step 2.

4.1 Fitness Assignment

In Step 6, all individuals in P and P' are assigned a scalar fitness value. This is accomplished in the following two-stage process. First, all members of the nondominated set P' are ranked. Afterwards, the individuals in the population P are assigned their fitness value.

Step 1: Each solution $i \in P'$ is assigned a real value $s_i \in [0, 1)$, called *strength*. s_i is proportional to the number of population members $j \in P$ for which $i \succeq j$. Let n denote the number of individuals in P that are covered by i and assume N to be the size of P. Then s_i is defined as: $s_i = \frac{n}{N+1}$. The fitness f_i of i is equal to its strength: $f_i = s_i \in [0, 1)$.
Step 2: The fitness of an individual $j \in P$ is calculated by summing the strengths of all external nondominated solutions $i \in P'$ that cover j. Add one to this sum to guarantee that members of P' always have better fitness than members of P (note that the fitness is to be minimized):

$$f_i = 1 + \sum_{i, i \succeq j} s_i \, , f_i \in [1, N) \tag{6}$$

4.2 Selection and Step Size Adaptation

Step 7 requires an algorithm for the selection of individuals into the mating pool and Step 8 includes some method for dynamical adaptation of step sizes (i.e. mutation variances). For this paper, selection was done using the following binary tournament procedure:

Step 1: Randomly (uniformly distributed random numbers) select two individuals out of the population P.
Step 2: Copy the one with the better (i.e. lower for SPEA) fitness value to the mating pool.
Step 3: If the mating pool is full, then stop, else go to Step 1

Adaptation of the step sizes was done using the self-adaptive mutation method (c.f. section 2.3). Each element of P and P' is assigned an individual step size for every parameter, i.e. $\Sigma = diag(\sigma_i^2)$ is a diagonal matrix for each individual. The step sizes of all members of the mating pool are then either increased by 50%, cut to half or kept the same, each at a probability of 1/3.

4.3 Reduction by Clustering

In Step 5, the number of externally stored nondominated solutions is limited to some number N'. This is necessary because otherwise, P' would grow to infinity since there always is an infinite number of points along the Pareto-front. Moreover, one wants to be able to control the number of proposed possible solutions, because from a decision maker's point of view, a few points along the front are often enough. A third reason for introducing clustering is the distribution of solutions along the Pareto-front. In order to explore as much of the front as possible, the nondominated members of P' should be equally distributed along the Pareto-front. Without clustering, the fitness assignment method would probably be biased towards a certain region of the search space, leading to an unbalanced distribution of the solutions. For this work, the *average linkage method*, a clustering algorithm which has proven to perform well on Pareto optimization, has been chosen. The reader is referred to [Morse (1980)] or [Zitzler & Thiele (1999)] for details.

5 Strength Pareto Approach with Targeting

Compared to other methods like for example the Energy Minimization Evolutionary Algorithm (EMEA) (c.f. [Jonathan, Zebulum, Pacheco & Vellasco (2000)]), the SPEA has two major advantages: it finds the whole Pareto-front and not just a single point on it and it converges faster. The latter is a universal advantage, whereas the former is not. There are applications where a target value can be specified. One then wants to find the point on the Pareto-front which is closest to the user-specified target (in objective space). This eliminates the need to analyze all the points found by SPEA in order to make a decision. EMEA offers such a possibility, but it converges slower than SPEA and it is unable to find more than one point per run. Hence we wish to extend SPEA with some targeting facility that can be switched on and off depending on whether one is looking for a single solution or the whole front, respectively. We added this capability to SPEA by the following changes to the algorithm:

1. Between Step 6 and Step 7 the fitness values of all individuals in P and P' are scaled by the distance D of the individual from the target (in objective space) to some power q:

$$f_i = f_i \cdot D_i^q$$

This ensures that enough nondominated members close to the target will be found so that the one with minimal distance will appear at higher probability. The parameter q determines the sharpness of the concentration around the target.

2. Another external storage P_{best} is added, which always contains the individual out of P' which is closest to the target. Therefore, between steps 4 and 5, the algorithm calculates the distances of all members of P' to the target and picks the one with minimal distance into P_{best}. At all times, P_{best} only contains one solution.

3. At the end of the algorithm, not only the Pareto-front is output, but also the solution stored in P_{best}. Note that due to clustering and removal in P', the solution in P_{best} is not necessarily contained in P'. It is therefore an optimal solution which otherwise would not have appeared in the output.

The algorithm has been implemented and tested for convex and nonconvex test functions. Figures 1 to 4 show some results for the nonconvex test function T_2 as proposed in [Zitzler, Thiele & Deb (2000)]:

$$\text{Minimize } T_2(\mathbf{x}) = (f_1(x_1), f_2(\mathbf{x}))$$
$$\text{subject to } f_2(\mathbf{x}) = g(x_2, \ldots, x_m) h(f_1(x_1), g(x_2, \ldots, x_m))$$

where
$$\mathbf{x} = (x_1, \ldots, x_m)$$
$$f_1(x_1) = x_1$$
$$g(x_2, \ldots, x_m) = 1 + 9 \cdot \sum_{i=2}^{m} x_i/(m-1)$$
$$h(f_1, g) = 1 - (f_1/g)^2$$
(7)

where m is the dimension of the parameter space and $x_i \in [0, 1]$. The exact Pareto-optimal front is given by $g(\mathbf{x}) = 1$. The parameters of the algorithm were set as summarized in table 1.

The chosen target value is slightly off-front. Therefore, the targeting error will never be zero. Figure 1 shows the final population after 250 generations without targeting. The diamonds indicate members of the external nondominated set (Pareto-optimal front), whereas members of the regular population are denoted by crosses. In figure 2 the same run has been repeated with targeting. Figure 3

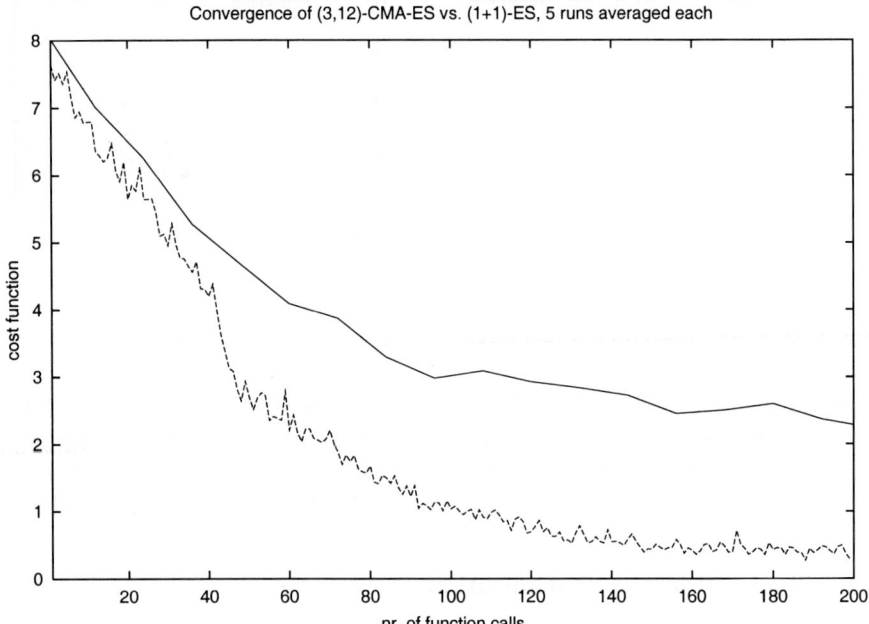

Fig. 5. Convergence of (3,12)-CMA-ES [solid line] and (1+1)-ES [dashed line] vs. number of evaluations of the objective function

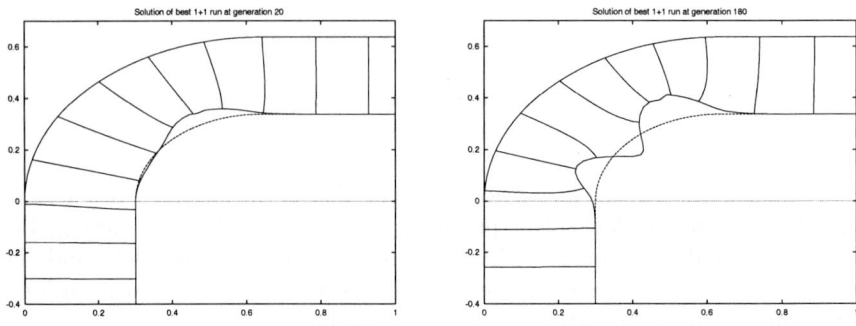

Fig. 6. Solution at generation 20

Fig. 7. Solution at generation 180

6.2 Multiobjective Optimization

We then introduced the total deformation of the channel contour as a second objective to be minimized simultaneously in order to minimize manufacturing costs. The second objective thus reads:

$$K = \sum_{i=1}^{11} p_i^2 \qquad (9)$$

where p_i are the shape parameters of the channel as introduced by [Mohammadi et al. (2000)]. The first objective remained unchanged. The algorithm used for this optimization was a SPEA with a population size of 20, a maximum size of the external nondominated set of 30 and a mating pool of size 10.

Figure 8 shows the Pareto-optimal trade-off front after 80 generations of the algorithm and figures 9 and 10 show the corresponding solutions, i.e. optimized shapes of the channel. One is now free to choose whether to go for minimal skewness at the expense of a higher deformation (c.f. figure 9), choose some intermediate result or minimize deformation in order to minimize manufacturing costs and still get the lowest skewness possible with the given amount of deformation (c.f. figure 10).

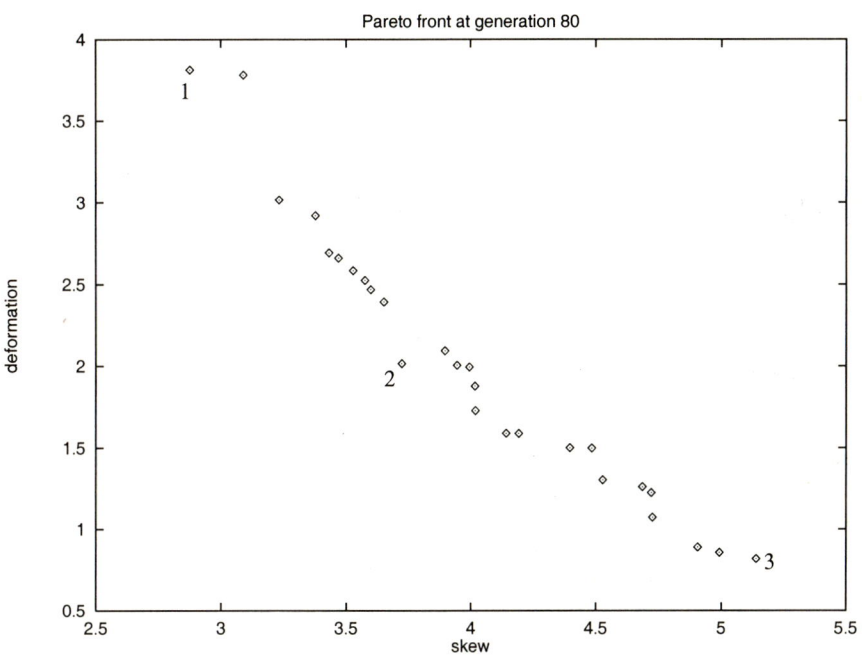

Fig. 8. Pareto-front of nondominated solutions after 80 generations

6.3 Comparison with Gradient Based Methods

Figures 11 and 12 show two classes of optimized shapes obtained by [Mohammadi et al. (2000)]. using gradient methods. It is interesting that the gradient technique offers the same two designs, namely the single-dented (fig. 11) and the double-dented (fig. 12) shapes, which we found with the evolution strategy after 40 or 180 generations, respectively. Therefore, we obtain qualitatively similar

[Schaffer (1985)] SCHAFFER, J. D. 1985 Multiple objective optimization with vector evaluated genetic Algorithms, *Proceedings of an International Conference on Genetic Algorithms and Their Applications*, sponsored by Texas Instruments and the U.S. Navy Center for Applied Research in Artificial Intelligence (NCARAI), 93-100.

[Van Veldhuizen & Lamont (1998)] VAN VELDHUIZEN, D. A. & LAMONT, G. B. 1998 Multiobjective evolutionary algorithm research: A history and analysis, Technical Report TR-98-03, Department of Electrical and Computer Engineering, Graduate School of Engineering, Air Force Institute of Technology, Wright-Patterson AFB, Ohio.

[Zitzler & Thiele (1999)] ZITZLER, E. & THIELE, L. Nov. 1999 Multiobjective Evolutionary Algorithms: A Comparative Case Study and the Strength Pareto Approach *IEEE Transactions on Evolutionary Computation*, **3(4)**, 257-271.

[Zitzler, Thiele & Deb (2000)] ZITZLER, E., DEB. K. & THIELE L. 2000 Comparison of Multiobjective Evolutionary Algorithms: Empirical Results *Evolutionary Computation*, **8(2)**, 173-195.

Multi-objective Optimisation of Cancer Chemotherapy Using Evolutionary Algorithms

Andrei Petrovski and John McCall

School of Computer and Mathematical Sciences, The Robert Gordon University,
St. Andrew Street, ABERDEEN, AB25 1HG, United Kingdom
{ap,jm}@scms.rgu.ac.uk

Abstract. The main objectives of cancer treatment in general, and of cancer chemotherapy in particular, are to eradicate the tumour and to prolong the patient survival time. Traditionally, treatments are optimised with only one objective in mind. As a result of this, a particular patient may be treated in the wrong way if the decision about the most appropriate treatment objective was inadequate. To partially alleviate this problem, we show in this paper how the multi-objective approach to chemotherapy optimisation can be used. This approach provides the oncologist with versatile treatment strategies that can be applied in ambiguous cases. However, the conflicting nature of treatment objectives and the non-linearity of some of the constraints imposed on treatment schedules make it difficult to utilise traditional methods of multi-objective optimisation. Evolutionary Algorithms (EA), on the other hand, are often seen as the most suitable method for tackling the problems exhibiting such characteristics. Our present study proves this to be true and shows that EA are capable of finding solutions undetectable by other optimisation techniques.

1 Introduction

Cancer chemotherapy is a highly complex process which controls tumour development by the administration of a cocktail of chemicals in a series of doses over a course of treatment. There is a wide variety of anti-cancer drugs available to oncologists. Due to their high toxicity, these drugs give rise to a variety of side-effects, ranging from cosmetically undesirable through debilitating through to the effects that are themselves life threatening. The oncologist therefore is faced with a complex task of designing a therapy which achieves certain treatment goals whilst limiting the toxic side-effects of the treatment to an acceptable level.

In the treatment of most common cancers multi-drug combinations are usually used. Traditionally, combination treatments are developed through empirical trials of different combinations, dosing, schedules and sequencing. However, since around 35 drugs are in common clinical use nowadays [17], it is evident that an almost infinite number of treatment schedules are conceivable and that the need for the optimisation of chemotherapeutic treatment is indisputable. The number of combinatorial possibilities for multi-drug schedules, coupled to the conflicting nature [13] and non-linearity of the constraints imposed on cancer treatments, make it difficult to solve the problem of cancer chemotherapy optimisation by means of empirical clinical experimentation or by means of traditional optimisation methods [6]. An alternative approach is to use evolutionary methods of computational optimisation to search for multi-drug treatment schedules that achieve certain treatment objectives and satisfy a number of simultaneous constraints.

A body of work has been established by the authors [7], [8], [10] and [11], where they have applied Genetic Algorithms to find the best (or at least suitable) treatment strategies given a single optimisation objective. In this paper, however, we endeavour to develop this approach further and to address the problem of finding treatment strategies that show a good performance with respect to more than one treatment objective. Thus, the evaluation of different treatment strategies will involve multiple measures (objectives) of performance, which should be optimised simultaneously, even though they may be conflicting in nature. The presence of conflicting objectives gives rise to a set of optimal solutions, known as the Pareto-optimal set. If all objectives are equally important, the conflict between them requires a compromise to be reached. A good solution to such problems involving conflicting objectives and therefore multiple evaluation criteria, should offer suitable, though possibly sub-optimal in the single-objective sense, performance in all objective dimensions [14]. Generally, there exists a multitude of such solutions; hence, the algorithm used to solve a multi-objective optimisation problem should find a wide variety of them, instead of just one.

Evolutionary Algorithms (EA) are a promising choice for solving the multi-objective optimisation problem of cancer chemotherapy for a number of reasons. Firstly, a set of Pareto-optimal solutions can, in principle, be captured in an EA population, thereby approximating the Pareto-optimal set in a single simulation run [2]. Secondly, in general Evolutionary Algorithms are less susceptible to the shape or continuity of the Pareto front than other techniques of multi-objective optimisation [16]. Thirdly, it has been shown by the authors (see [10] and [11]) that the problem of optimising cancer chemotherapy treatment belongs to the class of complex optimisation problems involving such features as discontinuity, multi-modality, non-connected, non-convex feasible regions, and inaccuracy in establishing model parameters. This is precisely the problem area where the methods of evolutionary computation really distinguish themselves from their competitors, thereby reinforcing the potential effectiveness of Evolutionary Algorithms in multi-objective optimisation of chemotherapeutic treatment.

The remaining sections are organised as follows. In section 2 we provide the background information on optimisation of chemotherapeutic treatment, which includes medical aspects of chemotherapy, the formulation of treatment design as a constrained multi-objective optimisation problem, and a description of salient features of Evolutionary Algorithms used in multi-objective optimisation. Section 3 explains implementation details of the evolutionary search for Pareto-optimal treatment schedules. The results of chemotherapy optimisation and their analysis are given in Section 4. Finally, Section 5 summarises the contribution of the present study to cancer chemotherapy and outlines possible directions for its further development.

2 Optimisation of Chemotherapeutic Treatment

Amongst the modalities of cancer treatment, chemotherapy is often considered as inherently the most complex [17]. As a consequence of this, it is extremely difficult to find effective chemotherapy treatments without a systematic approach. In order to realise such an approach, we need to take into account the medical aspects of cancer treatment.

2.1 Medical Aspects of Chemotherapy

Drugs used in cancer chemotherapy all have narrow therapeutic indices. This means that the dose levels at which these drugs significantly affect a tumor are close to those levels at which unacceptable toxic side-effects occur. Therefore, more effective treatments result from balancing the beneficial and adverse effects of a combination of different drugs, administered at various dosages over a treatment period.

The beneficial effects of cancer chemotherapy correspond to treatment objectives which oncologists want to achieve by means of administering anti-cancer drugs. A cancer chemotherapy treatment may be either curative or palliative. Curative treatments attempt to eradicate the tumour. It is believed that chemotherapy alone cannot eradicate cancer, but if the overall tumour burden is held below a certain level, other mechanisms (e.g. immune system or programmed cell death) will remove remaining tumour cells. Palliative treatments, on the other hand, are applied only when a tumour is deemed to be incurable. Here the objective is to maintain a reasonable quality of life for as long as possible.

The adverse effects of cancer chemotherapy stem from the systemic nature of this treatment: drugs are delivered via the bloodstream and therefore affect all body tissues. Since most anti-cancer drugs are highly toxic, they inevitably cause damage to sensitive tissues elsewhere in the body. In order to limit this damage, toxicity constraints need to be placed on the amount of drug applied at any time interval, on the cumulative drug dosage over the treatment period, and on the damage caused to various sensitive tissues [17]. In addition to toxicity constraints, the tumour size (i.e. the number of cancerous cells) must be maintained below a lethal level during the whole treatment period for obvious reasons.

The goal of cancer chemotherapy therefore is to achieve the beneficial effects of treatment objectives without violating any of the abovementioned constraints. This problem would not be much different from that of a general class of constrained optimisation problems, was it not for the conflict between treatment objectives. The objectives of curative and palliative treatments conflict with each other in the sense that drug schedules which tend to minimise tumour size are highly toxic and therefore have a negative effect on the quality of patient's life. Moreover, it has been shown that a severe treatment schedule that fails to cure can result in a shorter patient survival time (PST) than a milder palliative treatment [6].

Previously, the conflict between objectives was resolved by addressing each of them separately, that is, treatment strategies were sought which optimised only one of the objectives without considering the other [7]. The choice of the best strategy was left to the decision maker, i.e. the practicing oncologist who treats the patient; the role of the optimiser was to provide the alternatives to choose from. Although this cancer, as well as the values of maximum instantaneous and cumulative doses, can be found in [4], [8] or [11].

Regarding the objectives of cancer chemotherapy, we focus our study on the following two. The primary objective is to eradicate the tumour (curative treatment). We define eradication to mean a reduction of the tumour from the initial size to a size below 10^3 cells. Clinical experience shows that other mechanisms (e.g. programmed cell death, a.k.a. apoptosis) are capable of removing remaining tumour cells at this point.

In order to simulate the response of a tumour to chemotherapy, a number of mathematical models can be used [10]. The most popular is the Gompertz growth model with a linear cell-loss effect [17], which has been extensively validated in clinical trials:

$$\frac{dN}{dt} = N(t) \cdot \left[\lambda \ln\left(\frac{\Theta}{N(t)}\right) - \sum_{j=1}^{d} \kappa_j \sum_{i=1}^{n} C_{ij} \{H(t-t_i) - H(t-t_{i+1})\} \right] \quad (5)$$

where $N(t)$ represents the number of tumour cells at time t; λ, Θ are the parameters of tumour growth, $H(t)$ is the Heaviside step function; κ_j are the quantities representing the efficacy of anti-cancer drugs, and C_{ij} denote the concentration levels of these drugs. One advantage of the Gompertz model from the computational optimisation point of view is that the equation (5) yields an analytical solution after the substitution $u(t) = \ln(\Theta/N(t))$ [5]. Since $u(t)$ increases when $N(t)$ decreases, the primary optimisation objective of tumour eradication can be formulated as follows [9]:

$$\underset{\mathbf{c}}{\text{maximise}} \quad f_1(\mathbf{c}) = \int_{t_1}^{t_n} \ln\left(\frac{\Theta}{N(\tau)}\right) d\tau \quad (6)$$

subject to the state equation (5) and the constraints (1)-(4).

The second objective of cancer chemotherapy is to prolong the patient survival time (PST) maintaining a reasonable quality of life during the palliation period. If we denote the PST as T, then the second objective becomes:

$$\underset{\mathbf{c}}{\text{maximise}} \quad f_2(\mathbf{c}) = \int_{t_1}^{T} d\tau = T \quad (7)$$

again subject to (1)-(5).

Therefore, the evaluation function of the multi-objective optimisation problem of cancer chemotherapy takes the form of a two-dimensional vector function $F(\mathbf{c}) = [f_1(\mathbf{c}), f_2(\mathbf{c})]^T$, which maps the decision vectors $\mathbf{c} \in \Omega$ to the objective function space $\Lambda \subset \Re^2$ using the objectives (6) and (7).

As we mentioned in the previous section, these objectives are conflicting in nature. The conflict between objectives manifests itself in the fact that small tumours are more likely to be successfully eliminated, whereas it is much easier to palliate a large tumour [6]. Thus, in order to pursue the first treatment objective the maximum tolerable amount of drugs has to be administered at the start of treatment. The best palliative strategy, on the other hand, is to allow the tumour to grow up to the maximum size and then to maintain it at that level using only a necessary amount of drugs.

Taking this into account and considering the number of constraints imposed on chemotherapeutic treatment, it is not difficult to see that the traditional approaches to multi-objective optimisation of cancer chemotherapy (such, for example, as the weighting or constraint methods) are likely to fail. Our previous experiments with traditional optimisation methods (the complex and Hooke & Jeeves techniques) showed a lack of robustness in finding feasible solutions even in the case of single-objective optimisation [11]. Moreover, all traditional methods require several optimisation runs to obtain an approximation of the Pareto-optimal set. As the runs are performed independently from each other, synergies between them cannot be easily exploited, which may lead to substantial computational overhead [16].

Therefore, the necessity of a specialised optimisation technique to deal with the cancer chemotherapy MOP is evident. Recently, Evolutionary Algorithms (EA) have become established as an alternative to traditional methods. The major advantages of EA are: 1) the ability to effectively search through large solution spaces; 2) the ability to overcome the difficulties faced by the traditional methods mentioned above; and 3) the ability to approximate the Pareto-optimal set in a single run. In the following section we briefly discuss the salient features of Evolutionary Algorithms.

2.3 Evolutionary Multi-objective Optimisation

Evolutionary Algorithms entail a class of stochastic optimisation methods that simulate the process of natural selection. Although the underlying principles are quite simple, these algorithms have proven to be in general robust and powerful [1]. A large number of applications of EA to hard, real-world MOPs, the survey of which is given in [2], suggest that multi-objective optimisation of cancer chemotherapy is the problem set where Evolutionary Algorithms might excel.

As with any MOP, the problem of cancer chemotherapy optimisation involves two independent processes. The first process is the search through the solution space for the Pareto-optimal set. The search space of cancer chemotherapy MOP is very large [7], which makes the multi-directional and synergetic features of EA extremely helpful. The second process is decision-making, i.e. the selection of a suitable compromise solution from the Pareto-optimal set.

Depending on the order of performing these processes, the preferences of the decision maker (the oncologists in our case) can be made known either before, during or after the search process [14]. In the case of a priori preference articulation, the objectives of the given MOP are aggregated into a single objective that implicitly includes preference information (in the form of objective weights for example). This approach requires profound domain knowledge, which is not available for the optimisation problem of cancer chemotherapy [3].

If the search process is performed without any preference information given by the oncologist, then we are applying a posteriori preference articulation. Here, the search results in a set of candidate treatment schedules (ideally the Pareto-optimal set of treatments), from which the final choice is made by the oncologist. The main drawback of the latter approach is that it entirely excludes the domain knowledge, which in some cases might substantially reduce the size of the search space or/and its complexity. However, in a general case of cancer chemotherapy optimisation such a reduction is not advisable [8], which supports the suitability of the a posteriori approach.

Also, the process of decision-making may overlap with that of search. This means that after each optimisation step, a number of alternative treatment schedules (temporary Pareto-optimal set) are presented, on the basis of which the oncologist specifies further preference information, thereby guiding the search process. Such an approach is known as progressive preference articulation [14] and is a promising way to combine the advantages of the previous two. One example of how it can be used in the context of cancer chemotherapy is to optimise the modification of an existing treatment schedule rather than a schedule itself [11]. However, in this paper we concentrate our efforts on the optimisation of treatment schedules themselves as this is a more general problem. In solving this problem we do not wish to restrict the search process in any way, since a priori information on whereabouts of the Pareto-optimal set in the search space is unavailable. Therefore, hereafter we need to resort to a posteriori preference articulation approach to multi-objective optimisation.

Having established the strategic aspects of the method that is to be utilised for solving the cancer chemotherapy MOP, we now need to specify the implementation details. A general Evolutionary Algorithm can be presented as follows.

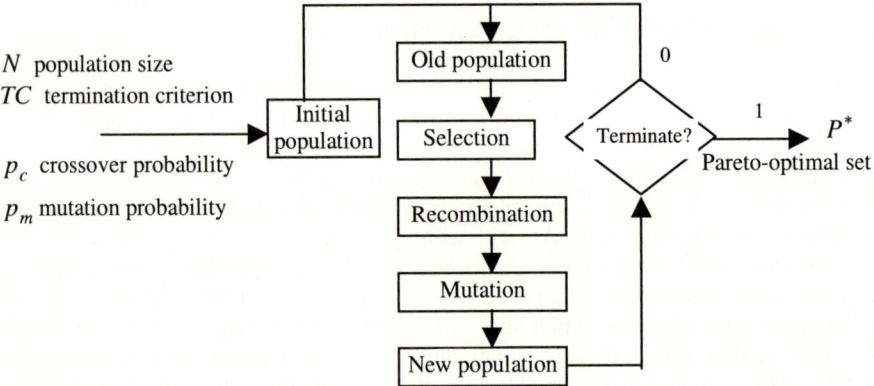

Fig. 1. Input, Output, and Internal Structure of a Generic Evolutionary Algorithm

This general structure holds for most EA implementations. The distinctive feature of Evolutionary Algorithms applied to multi-objective optimisation, however, is that they require addressing the following specific issues [15]. The first issue is how to accomplish fitness assignment, and consequently selection, given a vector-valued evaluation function $F : \Omega \to \Lambda$. In contrast to single-objective optimisation, where the fitness function takes into account only one optimisation objective, the fitness function of a multi-objective EA needs to map a k-dimensional objective function space to scalar numbers in such a way as to guide the search process to the Pareto-optimal set. Secondly, the diversity of an EA population has to be maintained more than ever in order to achieve a well distributed and well spread set of non-dominated solutions, in addition to preventing premature convergence.

A body of work has been established setting up various fitness assignment methods, selection techniques, and population diversifying schemes [2], [5], [14], [16]. As a consequence of this, many implementations of multi-objective EA are now available. In spite of this variety, however, there is no clear guideline on which EA implementation is suited to which sort of problem in the sense of ensuring that the derived solutions are the best available [15]. Thus, the choice is subjective and is often based on the developer's attempt to integrate the domains of the optimisation problem and that of the implementation algorithm [14].

Among the different implementation algorithms that have been proposed in the literature and have been used by EA practitioners, we have chosen and will base our study on the Strength Pareto Evolutionary Algorithm (SPEA) thoroughly described in [16]. This algorithm combines promising aspects of various multi-objective EA and has shown a superior performance on a number of test problems [15]. In the next section we describe how it can be applied to the multi-objective optimisation problem of cancer chemotherapy.

3   Evolutionary Search for Optimal Treatment Schedules

The search process aiming at finding non-dominated (with respect to the treatment objectives specified in Section 2.1) chemotherapy schedules is the main part of computational optimisation of chemotherapeutic treatment. The decision-making process is, of course, based on the results of this search, but it is left to oncologists and therefore lies outside the scope of the present paper.

The search for non-dominated treatment schedules is accomplished using the SPEA approach. Multi-drug chemotherapy schedules, represented by decision vectors $\mathbf{c} = (C_{ij}), i \in \overline{1,n}, j \in \overline{1,d}$, are encoded as binary strings. Using the EA terminology, the individual space $\mathbf{I}$ (a discretized version of Ω) can then be expressed as a Cartesian product

$$\mathbf{I} = A_1^1 \times A_1^2 \times \ldots \times A_1^d \times A_2^1 \times A_2^2 \times \ldots \times A_2^d \times \ldots \times A_n^1 \times A_n^2 \times \ldots \times A_n^d \qquad (8)$$

of allele sets A_i^j. Each allele set uses a 4-bit representation scheme

$$A_i^j = \{a_1 a_2 a_3 a_4 : a_k \in \{0,1\} \forall k \in \overline{1,4}\} \qquad (9)$$

so that each concentration level C_{ij} takes an integer value in the range of 0 to 15 concentration units. In general, with n treatment intervals and up to 2^p concentration levels for d drugs, there are up to 2^{npd} individual elements. Henceforth we assume that $n = 10$ and that the number of available drugs in restricted to three, one of which is strong but highly toxic, another is medium, and the last one

Application of Multi Objective Evolutionary Algorithms to Analogue Filter Tuning

Dr. Mark Thompson
Electronics Research Group,
Sheffield Hallam University
Sheffield,
United Kingdom
S1 1WB
M.Thompson@shu.ac.uk

Abstract. This paper discusses and compares the methods of Multi Objective Genetic Algorithm and Multi Objective Simulated Annealing applied to LC filter tuning. Specifically, the paper is concerned with the application and implementation of these methods to the design of an antenna tuning unit, providing the facility to adapt to changes in load impedance, temperature or environmental effects, ensuring maximum power transfer and harmonic rejection. A number of simulations were carried out to evaluate the relative performance of these algorithms.

1 Introduction

For radio antenna transmission systems it is of great importance to transmit maximum power to the antenna to achieve maximum transmission efficiency. In many cases, for example mobile and fixed tactical applications, these systems must deal with changing load and environmental aspects. The goal of obtaining fast antenna tuning systems that are capable of offering impedance matching whilst maintaining good harmonic rejection properties has become increasingly significant. One of the most popular impedance matching configurations used is the Pi-Network, which is simple in structure, can accommodate a wide range of load impedances and offers high harmonic rejection capabilities. Although the Pi-Network may be tuned manually, an automatic tuning system offers the advantages of improved tuning times and unsupervised tuning of a transmission system. In addition, if an automatic tuner can be developed which is capable of re-tuning quickly enough, then the transmission system may become dynamic, capable of adjusting to changes in antenna impedance, operating frequency or environment in order to maintain optimum performance. The Pi-Network's versatility allows additional criteria to be considered for optimisation i.e. harmonic rejection, parasitic effects, component costs etc. However, the contribution each component makes to the Pi-Network impedance characteristic is complex. Current tuning algorithms in commercially available equipment use a step by step approach, where the tuning network is adjusted iteratively, until impedance matching is achieved [1-5]. This method is slow and focuses only on impedance matching, rather than optimising the full capabilities of the Pi-Network. There have been reports in the literature of global optimisation techniques that have been applied to a range of applications, which consist of multiple objectives and where time constraints are also of importance. Previously, the Genetic Algorithm and Simulated Annealing [6] optimisation algorithms have been applied to this problem, however an aggregate fitness function is used, leading to the full capabilities of the system not being realised. The following paper attempts to use a pareto based fitness function to investigate the characteristics of the system, whilst also investigating MOEA ability to provide adequate network solutions.

2 The Pi-Network

Fig. 1 shows the arrangement of the pi-network between load and source impedances, R_s represents the transmitter source impedance, (typically 50 ohms resistive) Z_L represents the complex load impedance, while Z_1, Z_2 and Z_3 represent the impedance of the network at each stage from source to load respectively. In order to achieve conjugate impedance matching and hence maximum real power transfer from the transmitter to the load, Z_3 must equal the complex conjugate of the load impedance, Z_L.

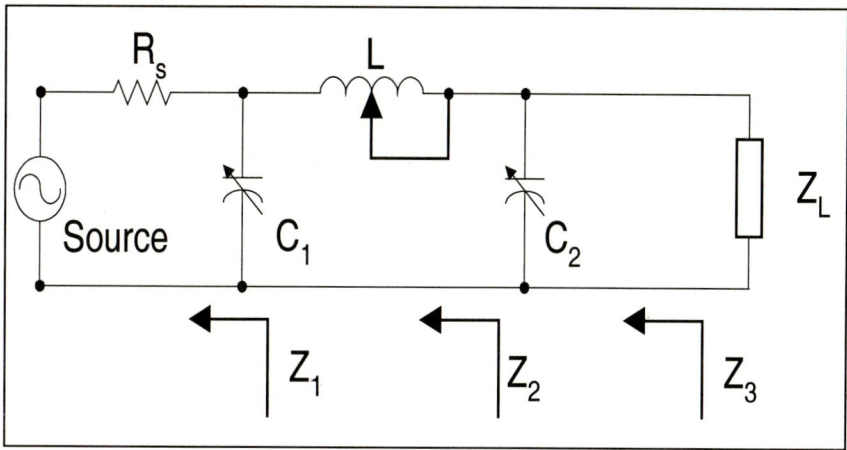

Fig. 1. Pi-Network used for Impedance Matching between Source and Load Impedances

Simple circuit analysis may be used to formulate the real and imaginary components of the Pi-Network output impedance, $Z_3 = R + jX$.

$$R = \frac{R_s}{\left(1-\omega^2 LC_2\right)^2 + \left(\omega C_1 R_s + \omega C_2 R_s - \omega^3 LC_1 C_2 R_s\right)^2} \quad (1)$$

$$X = \frac{\omega^3 LC_1^2 R_s^2 + 2\omega^3 LC_1 C_2 R_s^2 + \omega L - \omega C_1 R_s^2 - \omega C_2 R_s^2 - \omega^3 L^2 C_2 - \omega^5 L^2 C_1^2 C_2 R_s^2}{\left(1-\omega^2 LC_2\right)^2 + \left(\omega C_1 R_s + \omega C_2 R_s - \omega^3 LC_1 C_2 R_s\right)^2} \quad (2)$$

From these equations it is clear that each component influences the Pi-Network output impedance in a non-linear manner.

Initially the components are to be considered to be continuous in the range {0,1}; this will provide insight into the theoretical ideal characteristics of the network. Further work will then explore the issues of discrete components

3 Fitness Criteria

Impedance Matching Criteria: The purpose of the Pi-Network is to create a conjugate match between the transmitter and antenna. When an impedance mismatch between the source and load impedances exists, a proportion of the electromagnetic wave energy transmitted by the source is reflected back from the load impedance. The ratio of the reflected wave energy to forward wave energy is termed reflection coefficient, ρ. This relationship between the impedance mismatch and reflection coefficient allows the reflection coefficient to be used as a measure of impedance matching and may be expressed by the following equation:

$$|\rho| = \sqrt{\frac{(R_L - R_s)^2 + X_L^2}{(R_L + R_s)^2 + X_L^2}}$$

Where R_L represents the real component of the load impedance and X_L represents the imaginary component of the load impedance. The source impedance is assumed to be resistive of value R_s.

By measuring the reflection coefficient between the source and the input of the Pi-Network, the effectiveness of the Pi-Network to produce a conjugate match may be assessed.

As the reflection coefficient is always positive (0 when all of the power is absorbed by the load, 1 when the load reflects all of the power), the error function for maximum power absorption can be expressed simply as;

$$\text{Matching Fitness} = (1 - |\rho|)$$

By defining the error function as simply proportional to 1-ρ, small changes in ρ will in turn give rise to small changes in the value of that solution's fitness. Experimentation revealed that by increasing the rate of change in fitness as ρ tended to zero, an improvement in the speed of convergence may be achieved, this may be obtained through the fitness function shown;

$$\text{Matching Fitness} = (1-|\rho|)^2$$

Harmonic Rejection: Impedance matching is not the only parameter of importance, another extremely important consideration is that of harmonic rejection of the system. If harmonic rejection is not considered, then this can permit harmonics of the transmission frequency to also be transmitted, the result being the generation of interference at these frequencies for other radio users.

By modelling the transmitter to which the Pi-Matching Network is connected as an equivalent Thevinin voltage source and impedance (Fig. 2), the harmonic rejection characteristics of the network may be obtained.

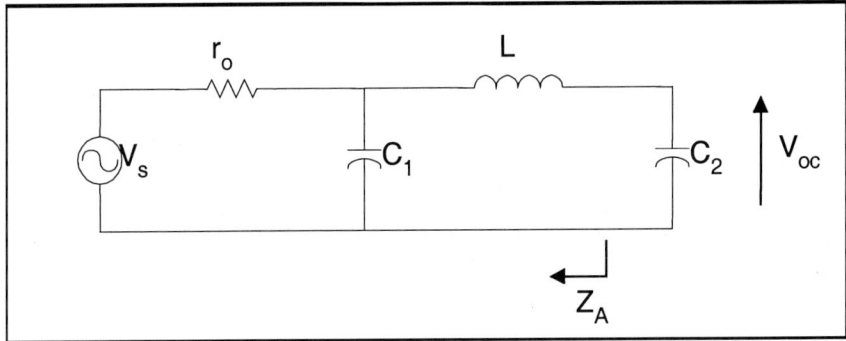

Fig. 2. The vinin equivalent circuit

If V_s is set to 1 volt then V_{oc} may be expressed as follows;

$$V_{oc} = \frac{1/j\omega_H C_2}{Z_A + 1/j\omega_H C_2} = \frac{r_o + j\omega_H C_1 r_o}{(1 + j\omega_H C_1 r_o) + j\omega C_2 (r_o - \omega_H^2 LC_1 r_o + j\omega L)}$$

Where

$$Z_A = \frac{r_o}{1 + j\omega_H C_1 r_o} + j\omega_H L = \frac{r_o - \omega_H^2 LC_1 r_o + j\omega_H L}{1 + j\omega_H C_1 r_o}$$

Denoting the real and imaginary components of V_{oc} as V_{Re} and V_{Im} respectively, these may be obtained from the above expression as;

$$V_{Re} = \frac{1-\omega_H^4 LC_1^2 C_2 r_o^2 + \omega_H^2 C_1 C_2 r_o^2 + \omega_H^2 C_1^2 r_o^2 - \omega_H^2 LC_2}{\left(1-\omega_H^2 LC_2\right)^2 + \left(\omega_H C_1 r_o + \omega_H C_2 r_o - \omega_H^3 LC_1 C_2 r_o\right)^2}$$

$$V_{Im} = \frac{-j\omega_H C_2 r_o}{\left(1-\omega_H^2 LC_2\right)^2 + \left(\omega_H C_1 r_o + \omega_H C_2 r_o - \omega_H^3 LC_1 C_2 r_o\right)^2}$$

The expressions for the real and imaginary components of the Thevinin equivalent impedance of the Pi-Network and source impedance may be expressed as follows:

$$Z_{eq}\, \text{Re} = \frac{r_o}{\left(1-\omega_H^2 LC_2\right)^2 + \left(\omega_H C_1 r_o + \omega_H C_1 r_o - \omega^3 LC_1 C_2 r_o\right)^2}$$

$$Z_{eq}\, \text{Im} = \frac{\omega_H^3 LC_1^2 r_o^2 + 2\omega_H^3 LC_1 C_2 r_o^2 + \omega_H L - \omega_H C_1 r_o^2 - \omega_H C_2 r_o^2 - \omega_H^3 L^2 C_2 - \omega_H^5 L^2 C_1^2 C_2 r_o^2}{\left(1-\omega_H^2 LC_2\right)^2 + \left(\omega_H C_1 r_o + \omega_H C_2 r_o - \omega_H^3 LC_1 C_2 r_o\right)^2}$$

For harmonic rejection, the need is to minimise the power delivered to the load at the harmonic frequency. Therefore the worst possible case would occur should the load impedance be the complex conjugate of the Thevinin equivalent impedance at the harmonic frequency. Minimising the power delivered to the load under these circumstances ensures minimum power transfer to the load at the harmonic frequency. During impedance matching, the complex impedances resonate causing the imaginary impedances of the circuit to cancel. Therefore the total impedance of the circuit connected to the Thevinin equivalent source is $2*Z_{eq}\text{Re}$ and consequently the modulus of the current flowing through the circuit may be expressed as follows:

$$|I| = \frac{|V_{oc}|}{2*Z_{eq}\,\text{Re}} \Rightarrow |I|^2 = \frac{|V_{oc}|^2}{4*Z_{eq}\,\text{Re}^2}$$

And the power dissipated into the load is then

$$P = |I|^2 Z_{eq}\,\text{Re} = \frac{|V_{oc}|^2}{4*Z_{eq}\,\text{Re}}$$

Therefore by minimising this function ensures minimisation of the harmonic distortion to the desired signal. From a 'fitness' criteria for the algorithms this may be described as follows;

$$\text{Rejection Fitness} = \frac{1}{P^2}$$

Therefore the fitness of the circuit improves as the power dissipated into the load at the harmonic frequency reduces. In the ideal case the power dissipated into the load at the harmonic frequency should tend toward zero, consequently the fitness function should reflect this.

It was discussed earlier how the Pi-Network's versatility creates problems when attempting to control it for optimal performance. When multiple objectives are considered, conflicting requirements may occur i.e. for a particular situation impedance matching may be achieved through low inductor values, while maximum harmonic rejection is achieved through maximum inductance. These conflicts between objectives compound the tuning problem by creating more complex error functions, containing greater numbers of local minima in which a search algorithm may become entrapped. By considering the objectives separately, using pareto optimality this problem may be eliminated and allows the interaction between objectives to be examined.

4 The Multi Objective Genetic Algorithm

Multi objective genetic algorithms (MOGA) allow the solution of problems consisting of more than one objective to be realised, typically using pareto optimality. The result of this approach is a set of possible solutions, covering a range of emphasis on the objectives, effectively searching the objective space for every objective weighting. This offers significant benefits over conventional genetic algorithms where traditionally multiple objectives are realised through the simple aggregation of the various objectives. By optimising the system in the pareto sense, a true appreciation of the interaction between objectives may be realised.

Each individual is evaluated in terms of pareto optimality [7], where each solution is evaluated for each objective separately and each solution is then compared to find the non dominated solutions.

For two points in n-dimensional objective space, where, $X=(x_1,x_2,x_3...x_n)$ and $Y=(y_1,y_2,y_3,..y_n)$, X is said to dominate Y when the following conditions are met;

$$(\forall i), X_i \leq Y_i \wedge (\exists i) X_i < Y_i$$

Consider a problem where it is required to minimise two cost functions, Fig. 3, shows a sample set of solutions. As the requirement is to minimise both costs, each point evaluates each objective separately, points A, B and C are dominant solutions as there are no solutions which provide lower cost values for both objectives. Similarly, point D is dominated by all of the other solutions as every other solution provides lower costs for both objectives. For the GA fitness evaluation, an approach proposed by (Fonseca and Fleming [8]) is used where each solution is ranked in proportion to the number of individuals that dominate that point. Therefore for a point, x, dominated by p solutions the rank of the solution would be calculated as:

$$rank(x) = 1 + p$$

front of the system. More rigorous quantitative evaluation methods adapted from Zitzler et al [16-17] were then used to inspect the algorithms' repeatability as well as provide tools for direct comparison of the algorithms performance. The original method proposed by Zitzler et al has been modified such that the proportion of the obtained pareto front solutions that are found to be dominant, non-dominated or dominated by a competitor's pareto front may be obtained. Further investigation into the assessment of the degree of dominance of the obtained pareto solutions may be pursued [18-19]; this will provide further tools for algorithm comparison and assessment. In addition, further work may explore the ability of these algorithms to provide solutions for non-ideal circuits considering such characteristics as internal resistance or other parasitic effects. These characteristics could then be incorporated into the circuit model. Furthermore, additional characteristics / constraints may be placed upon the system again to meet various issues such as: broadband impedance matching, minimal component values or restricted "off the shelf" component values.

References

[1] SUN, Y., FIDLER, J.K., "High Speed Automatic Antenna Tuning Units", IEE 9th Int. Conf. on Antennas and Propagation, 1995.
[2] GALKIN, V. A., "Automatic Antenna Matching Devices for Portable Radio Systems", Radiotekhnika, No. 11, pp71-73, 1991.
[3] SHAW, A.K., "Optimal Estimation of the Parameters of All-Pole Transfer Functions", IEEE Trans. on Circuits and Systems, Vol. 41, No. 2, February 1994.
[4] PETOVIC, P, J. MILEUSIC, TODOROVIC, J., " Fast Antenna Tuners For High Power HF Radio Systems", IEE Conf. Publication No. 308, European Conf. Circuit Theory and Design, Brighton, UK, 1989.
[5] THOMPSON, M., FIDLER, J.K. "Tuning The Pi-Network Using the Genetic Algorithm and Simulated Annealing", Proceedings of the 1997 European Conference on Circuit Theory and Design, pp 949-954, 1997.
[6] THOMPSON, M., FIDLER, J.K.," A novel approach for fast antenna tuning using transputer based simulated annealing", Electronic Letters, Vol. 36 No. 7, 2000.
[7] GOLDBERG, D.E., "Genetic Algorithms in Search, Optimization and Machine Learning", Addison-Wiley, 1989.
[8] FONSECA, C.M., FLEMING, P.J., "Genetic Algorithms for Multiobjective Optimization: Formulation, Discussion and Generalization, Proc. Fifth International Conference on Genetic Algorithms, ed. S. Forrest, Morgan Kaufman, 1993.
[9] METROPOLIS, N., ROSENBLUTH, A. W., ROSENBLUTH, M.N. AND TELLER, A.H.: "Equation of State Calculations by Fast Computing Machines", Journal of Chemical Physics 21 p1087, 1953.
[10] BOHACHEVSKY, I.O., JOHNSON, M.E., STEIN, M.L., "Generalized Simulated Annealing for Function Optimization", Technometrics, Vol. 28, No. 3, 1986.
[11] KIRKPATRICK S., GELATT C.D., VECCHI M.P., "Optimization by Simulated Annealing", Science Vol. 220, p671, 1983.
[12] LAARDHOVEN P.J.M. VON AND AARTS, E.H.L., "Simulated Annealing Theory and Applications", D. Reidel Publishing Company, pp40-138, 1988.

[13] SZU, H., HARTLEY, R., "Fast Simulated Annealing", Physics Letters A, Vol. 122, Number 3,4, pp157-162, 1987.
[14] WHIDBORNE, J.F., GU, D.W., POSTLETHWAITE, I., "Simulated Annealing for Multi-Objective Control System Design" UKACC International Conference on CONTROL '96, pp.376-381, 1996.
[15] ECCLESTONE, J., WHITAKER, D., "On the Design of optimal change-over experiments through multiobjective simulated annealing", Statistics and Computing Vol. 9 pp.37-42, 1999.
[16] ZITZLER, E., DEB, K., THIELE, L., "Comparison of Multiobjective Evolutionary Algorithms: Empirical Results", Evolutionary Computation Vol. 8 No. 2 pp.173-195, 2000.
[17] ZITZLER, E., THIELE, L., "Multiobjective Evolutionary Algorithms: A Comparative Case Study and the Strength Pareto Approach." , IEEE Transactions on Evolutionary Computation, Vol. 3 No. 4, pp 257-271, 1999.
[18] VAN VELDHUIZEN, D.A., LAMONT, G.B., "Multiobjective Evolutionary Algorithms: Analyzing the State of the Art", IEEE Transactions on Evolutionary Computation, Vol. 8 No. 2 pp125-147, 2000
[19] SHAW, K.J., FONSECA, C.M., FLEMING, P.J., "A Simple Demonstration of a Quantitative Technique for Comparing Multiobjective Genetic Algorithm Performance", Proceedings of the Genetic and evolutionary Computation Conference pp.119-120, 1999.
[20] DAVIS, L., "Handbook Of Genetic Algorithms", Van Nostrand Reinhold, pp1-53, 1991.
[21] DAVIS, L., "Genetic Algorithms and Simulated Annealing", Morgan Kaufmann, 1987.
[22] BUCKLES, B.P., PETRY, F.E., "Genetic Algorithms", IEEE Computer Society Press, pp5-19, pp30-47, 1994.
[23] GREFENSTETTE, J.J., "Optimization of Control Parameters for Genetic Algorithms", IEEE Transactions on Systems, Man, and Cybernetics, pp 122-128, Jan /Feb. 1986.
[24] GREFENSTETTE, J.J., BAKER, J.E., " How Genetic Algorithms Work: A Critical Look at Implicit Parallelism", Proceedings of the Third International Conference on Genetic Algorithms, Morgan Kaufmann, pp70-79, 1989.
[25] DE JONG, K., " Learning with Genetic Algorithms: An Overview", Machine Learning 3, Kluwer Academic, pp121-138, 1988.
[26] GOLDBERG, D.E., " Zen and the Art of Genetic Algorithms", Proceedings of the Third International Conference on Genetic Algorithms, pp80-85, 1989.
[27] SCHAFFER J. DAVID (ED) "Proceedings of the Third International Conference on Genetic Algorithms", Morgan Kaufman, 1989.
[28] MITCHELL, M., " An introduction to Genetic Algorithms", MIT Press, 1996.
[29] MICHALEWICZ, Z., "Genetic Algorithms+Data Structures=Evolution Programs", Springer, 1992.
[30] WHITLEY D. L., "Foundations Of Genetic Algorithms 2", Morgan Kaufmann, pp22-239, 1993.

Multiobjective Design Optimization of Real-Life Devices in Electrical Engineering: A Cost-Effective Evolutionary Approach

P. Di Barba, M. Farina, and A. Savini

Department of Electrical Engineering, University of Pavia, I-27100 Pavia, Italy
di_barba@etabeta.unipv.it

Abstract. When tackling the multicriteria optimization of a device in electrical engineering, the exhaustive sampling of Pareto optimal front implies the use of complex and time-consuming algorithms that are unpractical from the industrial viewpoint. In several cases, however, the accurate identification of a few non-dominated solutions is often sufficient for the design purposes. An evolutionary methodology of lowest order, dealing with a small number of individuals, is proposed to obtain a cost-effective approximation of non-dominated solutions. In particular, the algorithm assigning the fitness enables the designer to pursue either shape or performance diversity of the device. The optimal shape design of a shielded reactor, based on the optimization of both cost and performance of the device, is presented as a real-life case study.

1 Introduction

Optimal design in electromagnetism has a long history, from Maxwell (1869) on. In other fields of engineering like structural mechanics the history of optimal design is even longer, dating back to Lagrange (1770). In the latter area the modern development has taken place over the past three decades, anticipating the analogue development in electromagnetism and, to some extent, fostering it. In more recent years it has been possible to integrate the analysis of electromagnetic field with optimization techniques, so moving from computer-aided design (CAD) to automated optimal design (AOD).

The essential goal of AOD in electromagnetics is that of identifying, in a completely automatic way, the system or the device that is able to provide some prescribed performance, *e.g.* to minimize weight and materials cost or to maximize some output, taking into account physical constraints and geometrical bounds. This is actually an inverse problem and implies the simultaneous minimization of conflictual objectives.

In real-life engineering the presence of a single criterion or objective is somewhat an exception or a simplification. Therefore, the future of computational electromagnetics seems to be oriented towards, and conditioned by, the development of efficient methodologies and robust algorithms for solving multicriteria design problems.

From a formal viewpoint, a multicriteria problem is cast as follows:

$$\min_{x} F(x) \tag{1}$$

subject to $\quad g(x)<0$
$\quad\quad\quad\quad\quad\quad h(x)=0$

where $F(x)=(f_1(x),...,f_m(x))$ is a vector of m criteria or objectives, $x=(x_1,...,x_n)$ is the vector of n design variables defining the device or the system, $g(x)$ and $h(x)$ are inequality and equality constraints, respectively. In general, the utopia solution x^*, *i.e.* that minimizing all F_i simultaneously, does not exist and the so-called Pareto solutions are accepted, *i.e.* those for which no decrease in any of the criteria is obtained without a simultaneous increase in at least one of the other criteria.

Traditionally, multicriteria problems are reduced to singlecriterion problems, for instance by means of one of the following procedures:
  i) the use of a penalty function composed of the various criteria;
  ii) the separate solutions of singlecriterion problems and their trade-off;
  iii) the solution of a singlecriterion problem, taking the other criteria as constraints.
This approach leads to classical methods of multiobjective optimization and gives a solution which is supposed to be the optimum.

Often in the design of electromagnetic devices a satisfactory way to tackle the problem of multicriteria optimisation consists of applying the Pareto optima theory in connection with a suitable minimization algorithm. The result is a set of non-dominated solutions: in principle, all of them are optimal; in practice, each of them corresponds to a different degree of minimization of the single objectives.

Moreover, though looking attractive, the non-dominated approach often results to be unaffordable from the computational viewpoint; in fact, the evaluation of each objective may imply heavy non-linear field analyses in three-dimensional geometries. Consequently, the aim of a reliable method of multicriteria optimization should be to approximate the Pareto optimal front by fulfilling three requirements:
  - convergence to the front independent on the number, even very low, of non-dominated solutions;
  - remarkable diversity among non-dominated solutions;
  - moderate computational cost.
An attempt towards this goal is here presented.

2 EMO Strategy: Methodological Aspects

The aim of a stochastic multiobjective optimiser based on non-dominated sorting is to obtain as many solutions as possible lying on the Pareto optimal set while preserving diversity among them.

GA-based strategies [1],[2],[6] typically require some hundreds individuals for ensuring convergence. Moreover, when dealing with real-life optimization problems in electrical engineering, the evaluation of each objective often requires a FEM solution lasting several minutes [7],[10]. This difficulty often makes the use of GA-based strategies computationally unaffordable or highly unpractical from an industrial point of view.

Therefore we have decided to adopt a (1+1) ES algorithm as the optimization engine of the multiobjective strategy shown in Fig. 1 and Fig. 2 because, in our experience, it is robust and gives good convergence even when few individuals are considered. It should be noted that generation, mutation and annealing steps are implemented in parallel; this is possible because in our implementation individuals do not interact each other during the whole process, apart from the steps of Pareto ranking and fitness evaluation. In practice, the general structure of the algorithm is the same as NSGA whereas genetic operators have been replaced with the evolution strategy ones.

Two criteria must be pursued when assigning the fitness value to each individual:
1) forcing global convergence to the Pareto optimal set;
2) forcing diversity among solutions belonging to the same set.

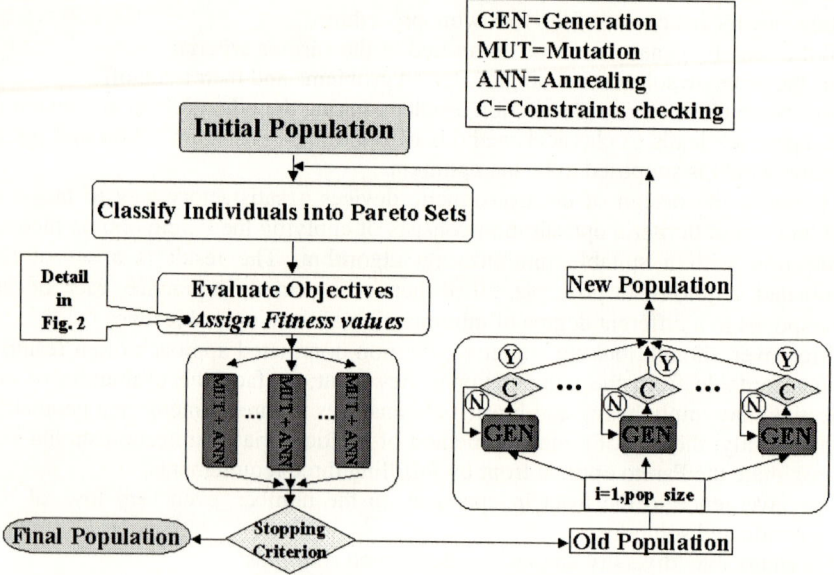

Fig. 1. Non-dominated Sorting Evolution Strategy Algorithm (NSESA): flowchart

To the first purpose, the fitness of each individual is evaluated according to the Pareto set, which it belongs to. To the second purpose, a sharing procedure is implemented within the current set in order to favour isolated solutions and prevent clustering. This step is particularly delicate when using a small number of individuals (say 5 to 10) and some changes with respect to classical sharing procedures [5],[8] are here proposed.

In general, when implementing a fitness sharing procedure, diversity of individuals in either the design space or the objective space can be considered. Moreover solutions with strong diversity in shape can be characterised by weak diversity in objective value (the opposite as well). Both procedures can lead to results useful for the device designer, who is interested in both shape and performance diversity of optimal solutions. This is why a sharing procedure in only one of the two spaces cannot guarantee a satisfactory approximation of the Pareto optimal front in the other space.

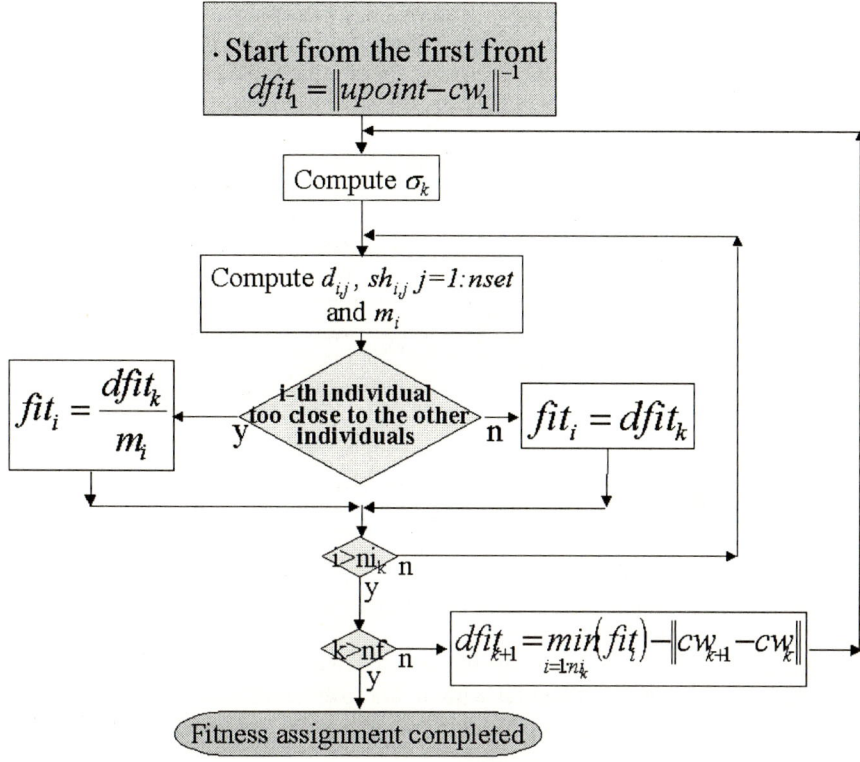

Fig. 2. Fitness assignment algorithm: flowchart

More into details, given a population sorted into Pareto sets, we at first consider the first set and assign a dummy fitness *dfit* to each individual as shown in Fig. 2, where cw_1 is the center of weight of the first front and *upoint* the utopia-point.

In order to set up the sharing procedure we then evaluate the normalized average distances $d_{i,j}$ among elements, in both design and objectives domain. Afterwards we implement the standard sharing formulas [3],[4] for the calculation of the sharing parameter $sh_{i,j}$ and the penalty coefficient m_i; we evaluate the niche radius σ in the following way:

$$\sigma_x = \frac{1}{(nset-1)}\sqrt{\sum_{p=1}^{ndof}(xmax_p - xmin_p)^2} \qquad (2)$$

$$\sigma_f = \frac{1}{(nset-1)}\sqrt{\sum_{p=1}^{obif}(fmax_q - fmin_q)^2}$$

Finally, the fitness value of the i-th individual is evaluated and assigned. Before moving to the $k+1$-th front a new dummy fitness $dfit_{k+1}$ has to be evaluated, as shown in Fig. 2. In order to increase the convergence rate the new value of dummy fitness depends on the center of weights cw_k, cw_{k+1} of current and next set, respectively. The procedure is repeated for all successive sets.

We point out that convergence towards the optimal front is always controlled in the objective space, while sharing procedures can be performed in either design space or objective space.

The following convergence indexes have been defined:

$$C_x(iter) = \frac{1}{npop}\sum_{i=1}^{npop}\sqrt{\sum_{j=1}^{ndof}\left(x_{ij}^{iter} - x_{ij}^{iter-1}\right)^2} \qquad (3)$$

$$C_f(iter) = \frac{1}{npop}\sum_{i=1}^{npop}\sqrt{\sum_{k=1}^{nobjf}\left(f_{ik}^{iter} - f_{ik}^{iter-1}\right)^2}$$

Finally, three stopping criteria have been implemented:
a) maximum number of iterations;
b) minimum value of convergence index in the objective space;
c) maximum number of iterations with no improvement found.

Results of previous investigations [9], [11] on simplified test problems have validated the effectiveness of the strategy proposed.

3 EMO Strategy: Numerical Aspects

Several test cases on real-valued analytical functions have been carried out for validating the code implemented. Here we show results for one of them, namely the Deb's t_3 problem. It is characterized by two variables and two objectives, giving rise to a non-connected Pareto front; the problem can be defined as follows:

$$\min_{(x_1,x_2)} (f_1, f_2) \qquad (4)$$

$$\begin{cases} f_1(x_1) = x_1 \\ f_2(x_1, x_2) = 1 + 9x_2 - \sqrt{x_1(1+9x_2)} - x_1 \sin(10\pi x_1) \end{cases}$$

where $(x_1, x_2) \in (0,1) \times (0,1)$

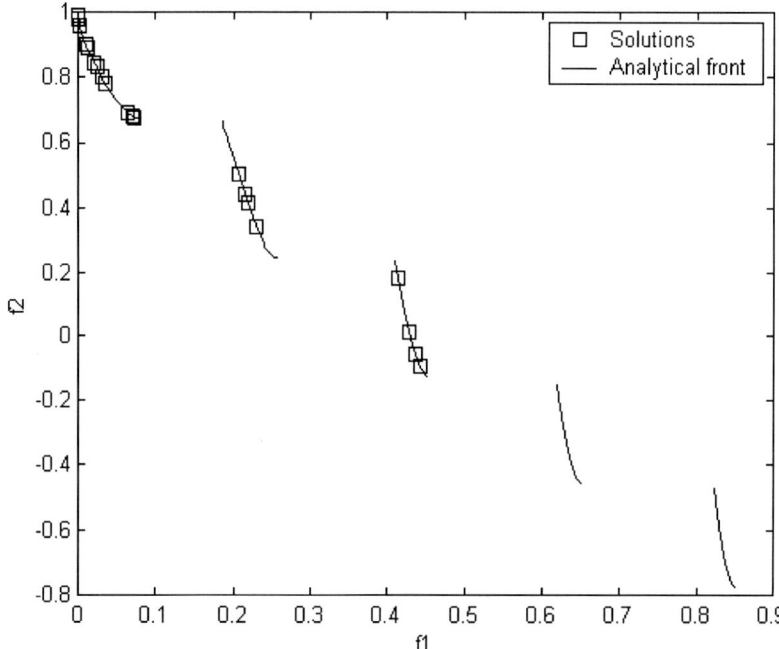

Fig. 3. NSESA: 20 individuals solution for validation test

As can be seen from Fig. 3, a solution composed of twenty individuals has been found; individuals are distributed along three of the five branches the POF is composed of. The starting population was chosen in a random way in the design space. Given the ik-th individual at niter-th iteration, the following two expressions have been used in design space and in objective space, respectively, in order to quantify the POF approximation error all along the evolution:

$$errorx(ik, iter) = x_2(ik, iter) \tag{5}$$

$$errorf(ik, iter) = f_2(ik, iter) - 1 + \sqrt{f_1(ik, iter)} + \\ + f_1(ik, iter)\sin(10\pi f_1(ik, iter))$$

The log value of both errors is plotted in Fig. 4 with reference to a single individual.

4 An Industrial Case Study

4.1 The Device

The shape optimization of a single-phase series reactor for power applications is considered [12]; the reactor is employed to reduce the peak value of short-circuit current and so to mitigate its electrodynamical effects.

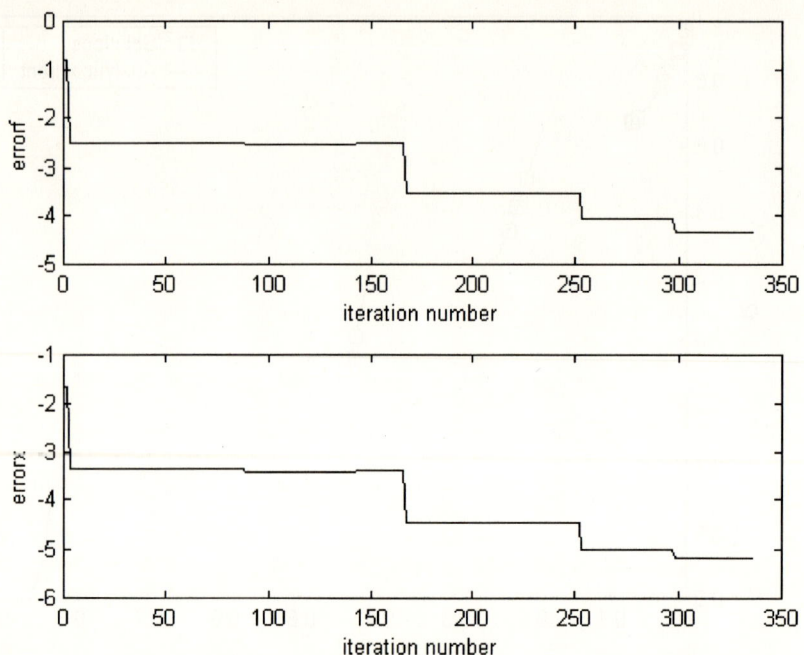

Fig. 4. History of approximation errors for the solution shown in Fig. 3

The log value of both convergence indexes is plotted in Fig. 5.

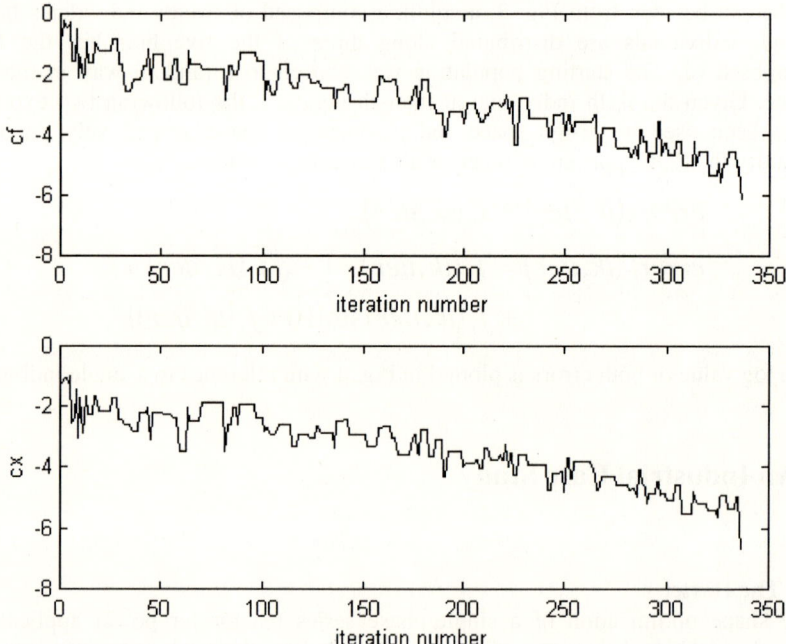

Fig. 5. History of convergence indexes for the solution shown in Fig. 3

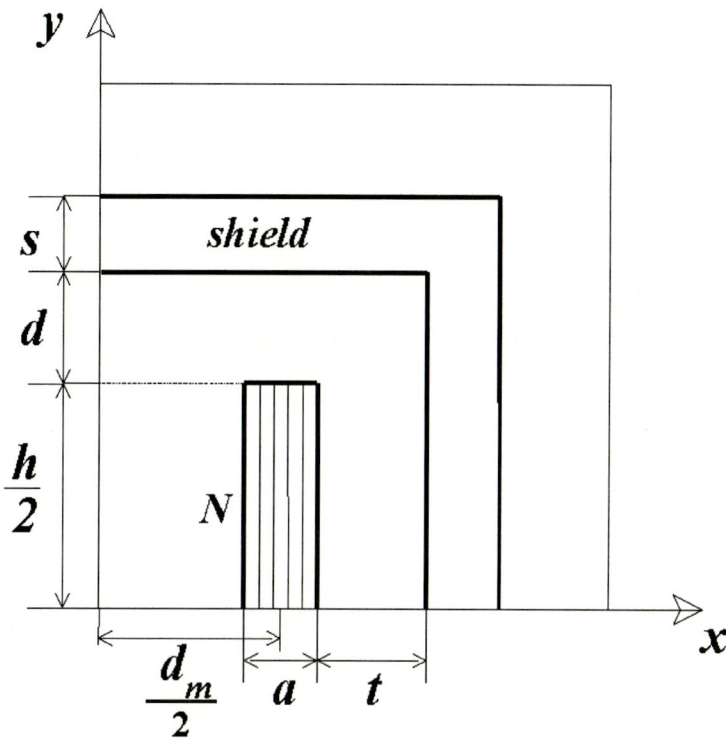

Fig. 6. Cross-section of the reactor (one quarter) and design variables

The reactor, the cross-section of which is shown in Fig. 6, is characterised by a coreless winding with cylindrical shape (foil winding); it is boxed in a laminated magnetic shield with rectangular shape in order to protect the surrounding environment from the strong stray field. The latter, in turn, gives rise to power losses in the winding that limit the operation of the device. The higher the winding, the lesser the stray field; on the other hand, the realization of a higher winding and shield, though reducing the effect of leakage, causes an increase of volume and cost of the reactor so that a conflict of design criteria is originated. For a prototype reactor rating 5.9 MVA at a nominal current of 893 A the following values hold: h= 500 mm, dm=590 mm, a=210 mm, d=80 mm, t=40 mm, N=212, filling factor of the winding k_s= 0.504.

4.2 Analysis

The distribution of magnetostatic field in the reactor, for which the rectangular symmetry is assumed, is governed by the Poisson's equation in terms of vector potential $\mathbf{A}=(0,0,A)$

$$-div\left(\frac{1}{\mu}gradA\right) = J \qquad (6)$$

subject to boundary conditions A=0 along x=0 and elsewhere; J=3.57 Amm^{-2} is the current density in the winding while μ_r=1 and μ_r=10^4 are the values assumed for relative permeability of non-magnetic materials and iron, respectively. To solve (6) numerically, the two-dimensional field region shown in Fig.6 has been discretized by means of a regular grid of finite elements, namely triangles with quadratic variation of potentials; the total number of elements is ne=950 approximately. The evolutionary optimizer calls the MagNet code [13] for performing the field analysis and then updates the finite element grid at each iteration.

4.3 Design

In general, up to seven design variables defining the shape of the device can be considered: geometric height h, mean diameter d_m, radial thickness of the winding a, number of turns N, axial distance d between winding and magnetic shield, thickness s of the shield, radial distance t between winding and shield.

Two conflictual criteria can be defined:
- the material cost f_1 of the reactor, namely the weighted sum of copper and iron weights, to be minimized:

$$f_1 = 4k_i w_i \left[s\left(\frac{d_m + a}{2} + t\right)l + s\left(\frac{h}{2} + d + s\right)l \right] + k_c w_c k_s lah \qquad (7)$$

with k_i=1, k_c=3 while w_i=7860 kgm^{-3} and w_c=8930 kgm^{-3} are mass densities of iron and copper, respectively;
- the fringing field f_2 inside the winding, i.e. the mean radial component of magnetic induction in the cross-section of the winding, to be minimized as well:

$$f_2 = \frac{1}{NW}\sum_{i=1}^{NW}|B_x(i)| \qquad (8)$$

where NW=64 is the number of points of a grid sampling the radial induction in the winding.

The minimisation of the fringing field has two important benefits: from a global point of view it leads to a strong reduction of additional losses in the winding and thus increases the efficiency of the reactor; on the other hand, the probability of local overheating inside the coil and its consequent failure is reduced.

The following constraints have been prescribed:
- the rated value of inductance L=23.57 mH;
- the induction in the core, not exceeding 0.8 T, when the current is equal to $\sqrt{2}I_{nom}$ with I_{nom}=893 A;

- the current density in the winding;
- the insulation gaps d and t between winding and core.

Consequently, three independent design variables have been selected, *i.e.* height h, mean diameter d_m and number of turns N of the winding, respectively. Finally, a set of bounds preserves the geometrical congruency of the model, namely:

$$0.5 \leq h \leq 1.5 \quad m \quad 0.1 + 2a \leq d_m \leq 1.8 \quad m \quad 162 \leq N \leq 262 \quad (9)$$

The sensitivity surfaces of both f_1 and f_2 against (h, d_m) for given number of turns $N=200$ are reported in Fig. 7 and Fig. 8, respectively.

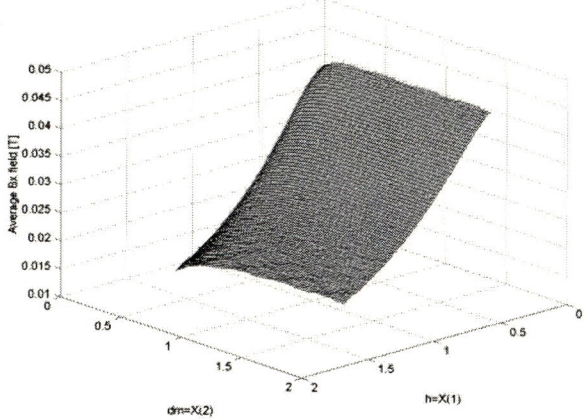

Fig. 7. Average B_x field in the winding as a function of mean diameter d_m and height h of the winding itself

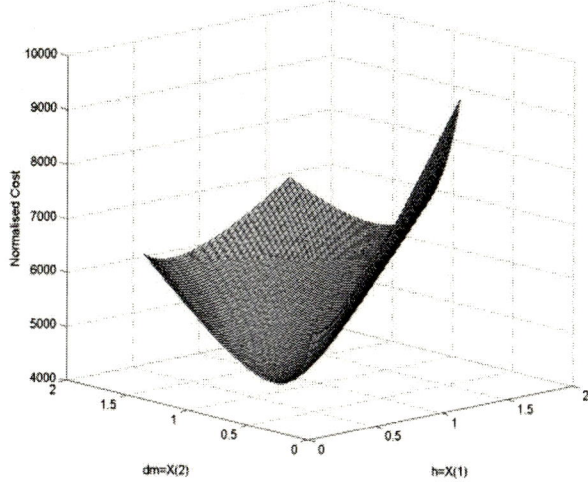

Fig. 8. Normalized cost of the reactor as a function of mean diameter d_m and height h of the winding

The conflict between the two objectives is evident from the comparison of both surfaces.

In order to estimate the maximum cost *totcost* of the EMO strategy implemented, the following formula holds

$$tot\cos t = niter \times npop \times nobj \times femtime \qquad (10)$$

where: number of objective functions *nobj*=2 to 3, maximum number of iterations *niter*=300. In our experience of real-life problems, typical number of individuals is *npop*=5 to 20, while the cost of a single FEM analysis is *femtime*=1 to 5 min.
As for the case study developed, due to the linear magnetostatic analysis and the inexpensive evaluation of f_2, we had *femtime*×*nobj*=0.8 min thus requiring some 48 hours for the stopping criterion to be satisfied.

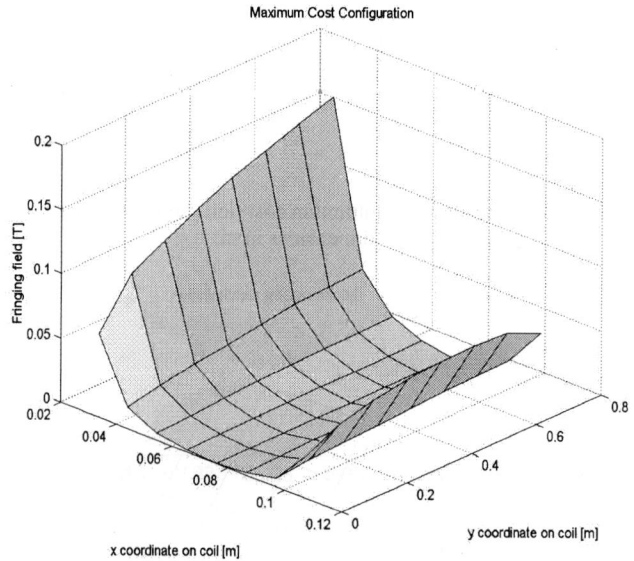

Fig. 12. Fringing field in the winding for the maximum cost configuration

6 Conclusion

In real-life engineering, when adopting an algorithm of multiobjective optimisation based on Pareto optimality, it is of primary importance to reduce the number of calls to the objective function, often requiring a FEM analysis. In the paper, a cost-effective EMO strategy has been developed and applied to the shape design of a realistic electromagnetic device.

From the methodological viewpoint, the results show that a lowest-order evolution strategy algorithm with a small number of individuals (5 to 10) can be conveniently used as the engine of the multiobjective optimization. Nevertheless, the procedure of fitness assignment should be modified with respect to classical formulas. In fact, the latter refer to large number of individuals (50 to 100) and depend on some tuning parameters, usually defined by means of empirical formulas. In the paper a fitness assignment procedure is proposed, making the formulas forcing diversity of individuals univocal and easy-to-implement.

Turning to the case study, the cost and performance optimisation of a shielded reactor has been achieved. A wide number of configurations belonging to the Pareto optimal front have been identified, so offering the designer an effective choice among devices that rank from the best performing one to the less expensive one. From an industrial point of view, having a set of Pareto-optimal solutions makes it easy to fulfil *a posteriori* technology-related constraints that are typical of real-life engineering, whereas in scalar optimization they have to be carefully prescribed *a priori* in order the only solution be feasible.

Finally, the proposed strategy allows the designer to pursue either shape or performance diversity of Pareto-optimal devices.

References

[1] E. Zitzler, E., Thiele, L.: Multiobjective Evolutionary Algorithms: A Comparative Case Study and the Strength Pareto Approach. IEEE Trans. Evol. Comp., Vol. 3 no. 4, pp. 257-271, 1999

[2] Srinivans, N., Deb, K.: Multiobjective Optimization using Non-dominated Sorting in Genetic Algorithms. Evol. Comput., Vol. 2 no. 3, pp. 221-248, 1994

[3] Deb, K.: Multi-Objective Algorithm: Problem Difficulties and Construction of Test Problems. Technical Report no. CI-60/98, Department of Computer Science, XI University of Dortmund GE, 1998

[4] Deb, K.: Non-linear Goal Programming Using Multi-Objective Genetic Algorithm. Technical Report no. CI-60/98, Department of Computer Science, XI University of Dortmund GE, 1998

[5] Rahmat-Samii, Y., Michielseen, E. (ed.s): Electromagnetic Optimization by Genetic Algorithms. John Wiley & Sons, USA 1999

[6] Goldberg, D. E.: Genetic Algorithms in Search, Optimization and Machine Learning. Addison Wesley, USA 1989

[7] Kim, M. K., Lee, C., Jung, H.: Multiobjective Optimal Design of Three-phase Induction Motor using Improved Evolution Strategy. IEEE Trans. Mag., vol. 34, no. 5, pp. 2980-2983, 1998

[8] Sereni, B., Krahenbuhl, L., Nicolas, A.: Niching Genetic Algorithm for Optimization in Electromagnetics (parts I and II). IEEE Trans. Mag., vol. 34, no. 5, pp. 2984-2987, 1998

[9] Di Barba, P., Farina, M., Savini, A.: Vector Shape Optimization of an Electrostatic Micromotor using a Genetic Algorithm. COMPEL, vol.19, no.12, pp. 576-581, 2000

[10] Battistetti, M., Di Barba, P., Dughiero, F., Farina, M., Lupi, S., Savini, A.: Multiobjective Design Optimisation of an Inductor for Surface Heating: an Innovative Approach. Presented at CEFC2K, June 4-7, 2000, Milwaukee (USA). Submitted to IEEE Trans. on Magnetics

[11] Di Barba, P., Farina, M., Savini, A.: An Improved Technique for Enhancing Diversity in Pareto Evolutionary Optimization of Electromagnetic Devices. Presented at 9th Intl IGTE Symposium, September 11-13, 2000, Graz (Austria). Submitted to COMPEL

[12] Di Barba, P.: A Fast Evolutionary Method for Identifying Non-inferior Solutions in Multicriteria Shape Optimization of a Shielded Reactor. Presented at 6th Intl OIPE Workshop, September 25-27, 2000, Torino (Italy). Submitted to COMPEL

[13] MagNet Version 6, Getting Started Guide, Infolytica Corporation, 1999. http://www.infolytica.com/

[14] Miettinen, K. M. Nonlinear Multiobjective Optimization Kluwer Academic Publishers

Application of Multiobjective Evolutionary Algorithms for Dose Optimization Problems in Brachytherapy

Michael Lahanas[1], Natasa Milickovic[1], Dimos Baltas[1,2], and Nikolaos Zamboglou[1,2]

[1] Department of Medical Physics and Engineering, Strahlenklinik, Klinikum Offenbach, 63069 Offenbach, Germany.
[2] Institute of Communication and Computer Systems, National Technical University of Athens, 15773 Zografou, Athens, Greece.

Abstract. In High Dose Rate (HDR) brachytherapy the conventional dose optimization algorithms consider the multiple objectives in form of an aggregate function which combines individual objectives into a single utility value. As a result, the optimization problem becomes single objective, prior to optimization. Up to 300 parameters must be optimized satisfying objectives which are often competing. We use multiobjective dose optimization methods where the objectives are expressed in terms of quantities derived from dose-volume histograms or in terms of statistical parameters of dose distributions from a small number of sampling points. For the last approach we compare the optimization results of evolutionary multiobjective algorithms with deterministic optimization methods. The deterministic algorithms are very efficient and produce the best results. The performance of the multiobjective evolutionary algorithms is improved if a small part of the population is initialized by deterministic algorithms.

1 Introduction

High dose rate brachytherapy is a treatment method for cancer where empty catheters are inserted within the tumor volume. Once the correct position of these catheters is verified, a single ^{192}Ir source is moved inside the catheters at discrete positions (dwell positions) using a computer controlled machine. The problem that we consider is the determination of the n dwell times (which sometimes are called as well dwell position weights or simply weights) for which the source is at rest and delivers radiation at each of the n dwell positions, resulting in a three-dimensional dose distribution which fulfills the defined quality criteria. In modern brachytherapy, the dose distribution has to be evaluated with respect to the irradiated normal tissues and the Planning Target Volume (PTV) which includes besides the Gross Tumor Volume (GTV) an additional margin accounting for position inaccuracies, patient movements, etc. Additionally, for all critical structures, either located within the PTV or in its immediate vicinity or otherwise within the body contour, the dose should be smaller than a critical dose D_{crit}. In practice it is difficult, if not impossible to meet all these objectives. Usually, the above mentioned objectives are mathematically quantified separately, using different objective functions and then added together in various proportions to define the overall treatment objective function [1],[2].

The number of source positions varies from 20 to 300. It is therefore a high dimensional problem with competing objectives. The use of a single weighted sum leads to information loss and is not generally to be recommended, especially for non convex problems and for those cases where objectives have not the same dimensions and in addition maybe competing. An understanding of which objectives are competing or non-competing is valuable information. We therefore use multiobjective evolutionary algorithms in HDR brachytherapy. One algorithm is based on the optimization of dose-volume histograms (DVH), which describes the distribution of the dose within an object, or from these derived distributions. These distributions are evaluated for the PTV, the surrounding tissue and organs at risk from a set of up to 100000 sampling points [3]. The calculation of the DVH requires a considerable amount of time and for implants with 300 sources the optimization requires a few hours. Another limitation of this method is that a comparison with deterministic algorithms is not possible. We have therefore considered the optimization of the dose distribution using as objectives the variance of the dose distribution on the PTV surface and within the PTV obtained from a set of 1500–4000 sampling points. These functions are convex and a unique global minimum exists.

In the past comparisons of the effectiveness of evolutionary algorithms have been made with either other evolutionary algorithms [4] or with manually optimized plans [1], [2]. We have compared the Pareto fronts obtained by multiobjective evolutionary algorithms with the Pareto fronts obtained by a weighted sum approach using deterministic optimization methods such as quasi-Newton algorithms and Powells modified conjugate gradient algorithm which does not requires derivatives of the objective function [5].

2 Methods

2.1 Calculation of the Dose Rate

The dose rate around each of the small cylindrical shaped sources is dominated by the $1/r^2$ term with modifications due to absorption and scattering in the surrounding material. The dose value $d(r)$ at $r = (x, y, z)$ is:

$$d(x) = \sum_{i=1}^{N_S} w_i K(r - r_i) \qquad (1)$$

In (1) r_i is the position of the ith source and N_s the total number of sources. $K(r-r_i)$ is the dosimetric kernel describing the dose rate per unit source strength at r from a source positioned at r_i. The dwell position weight $w_i = S_k t_i$ is proportional to the strength S_k of the of the single stepping source, where t_i is the dwell time of the ith source dwell position [6]. Because of the high dose gradients a dose specification at a single point inside the PTV is not possible in interstitial brachytherapy. For this reason we use as a reference dose D_{ref} the average dose value at the PTV surface.

2.2 Dose-Volume Histogram Based Optimization Using the Conformal Index

In the paper of Baltas et al. [7] a conformal Index (COIN) was proposed as a measure of implant quality and dose specification in brachytherapy. This index takes into account patient anatomy, both of the tumor and normal tissues and organs, see Fig.1.

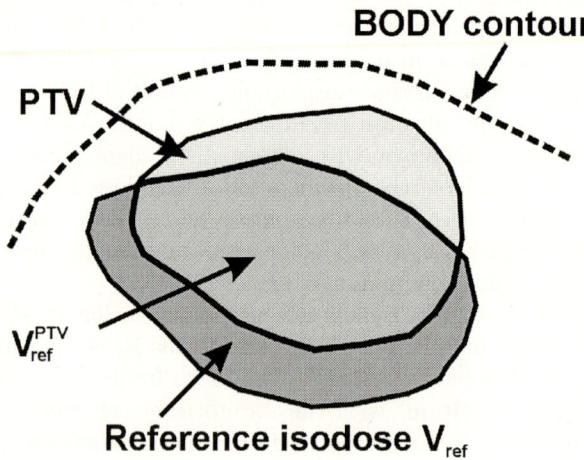

Fig. 1. Two-dimensional schematic diagram of the COIN = $c_1 c_2$ based optimization. The coefficients c_1 and c_2 consider the coverage of the PTV by the isosurface with the prescription dose D_{ref} and parts of the tissue surrounding the PTV.

COIN is defined as:

$$\text{COIN} = c_1 \cdot c_2 \qquad (2)$$

where $c_1 = V_{ref}^{PTV}/V_{PTV}$ and $c_2 = V_{ref}^{PTV}/V_{ref}$. The coefficient c_1 is the fraction of the PTV, V_{ref}^{PTV}, that is enclosed by D_{ref} and is a measure of how accurately the PTV is covered by D_{ref}. The coefficient c_2 is the fraction of the volume of the reference dose, V_{ref}, that is covered by PTV. It is also a measure of how much normal tissue outside the PTV is covered by D_{ref}. COIN can be calculated from the cumulative DVHs of the PTV and the body at the reference dose D_{ref} i.e. $\text{DVH}_{PTV}(D_{ref})$ and $\text{DVH}_{body}(D_{ref})$ respectively:

$$\text{COIN} = V_{PTV} \cdot \text{DVH}_{PTV}(D_{ref})^2/100 \cdot V_{body} \cdot \text{DVH}_{body}(D_{ref}) \qquad (3)$$

V_{body}, V_{PTV} are the volumes of the body and the PTV, respectively. We describe the dependence of the conformal index COIN on the choice of the reference dose value as the COIN distribution, see Fig. 2(b). Usually the dose values are normalized to D_{ref} and are given either as fractions or percentages of D_{ref}.

The "ideal" dose distribution is characterized by the following:

- $c_1 = c_2 = 1$ i.e. COIN=1 at $D = D_{ref}$, which means that the reference dose value isodose 3D envelope is identical with the PTV.
- For $D < D_{ref}$, an extremely rapid fall-off of the COIN value which corresponds to a rapid fall-off of the dose outside the PTV (normal tissues).
- COIN ≈ 0 for $D > D_{ref}$, that means that there are negligible volumes with dose values higher than $D < D_{ref}$.

The cumulative dose volume histograms of the PTV and the body for a rib implant is shown in Fig. 2(a). Due the rapid decrease of the DVH of the body a large number of sampling points is necessary in order to calculate with a high accuracy the DVH, the COIN distribution and the COIN integral at dose values close to the reference dose value and above. The COIN distribution from the DVHs of Fig. 2(a) and the COIN integral is shown in Fig. 2(b).

2.3 Dose Statistics Based Optimization

The DVH based optimization method requires a large number of sampling points for the computation of the histograms and the COIN distribution and therefore is computational expensive. We have developed a stratified sampling approach where the sampling points are non uniform distributed and which reduces the number of required sampling points by a factor of 5–10. Even then for implants with 200–300 sources the optimization time can reach 1–2 hours. A comparison of the performance with deterministic and gradient based algorithms is not practical or not even possible. Therefore we consider another set of two objectives: For the conformity objective we use the variance f_S of the dose distribution of sampling points uniformly distributed on the PTV. In order to avoid excessive high dose values inside the PTV we require a small as possible dose distribution variance f_V inside the PTV. Due to the source characteristics these two objectives are competing. We use normalized variances for the two objectives:

$$f = \frac{1}{m^2 N} \sum_{i=1}^{N} (d_i - m)^2 \qquad (4)$$

Where m is the average dose value and N the corresponding number of sampling points.

2.4 Multiobjective Optimization with Deterministic Algorithms

These objectives allow us to use deterministic gradient based algorithms. We use a weighted sum approach for the multiobjective optimization, where for a

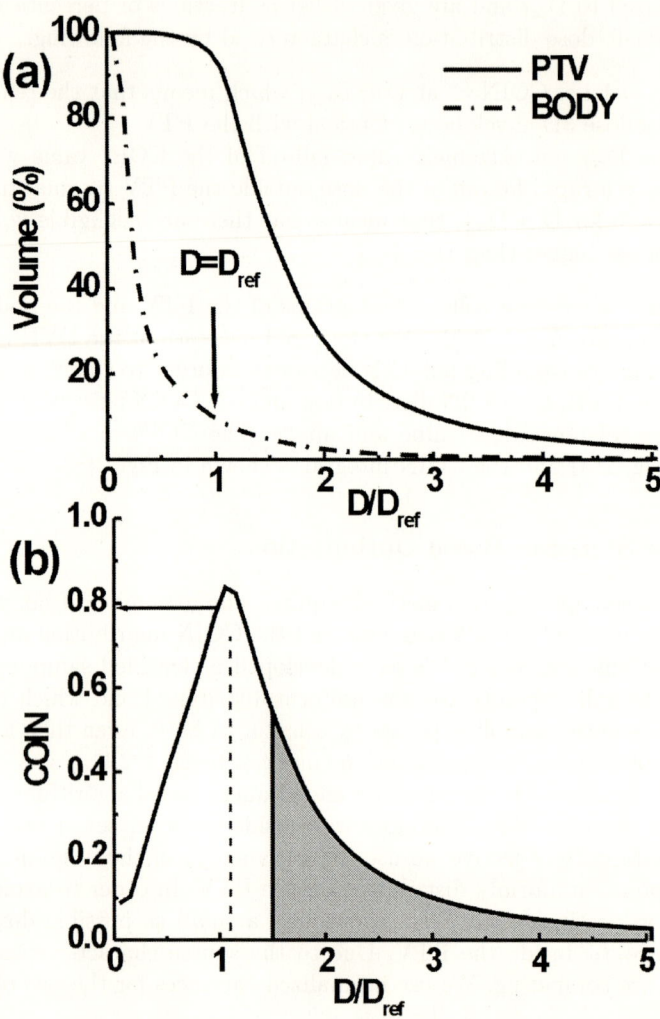

Fig. 2. (a)Dose-volume histograms of the PTV and the body as a function of dose. (b)The corresponding COIN distribution. The shaded area to the right of $D/D_{ref} = 1.5$ is the COIN integral. The objectives are maximum COIN value at $D = D_{ref}$ and minimum COIN integral for the avoidance of high dose values in the PTV and the surrounding tissue.

set of weights for the volume and surface variance we perform a single objective optimization of f_w:

$$f_w = w_S f_S + w_V f_V \tag{5}$$

where $w_S, w_V \geq 0$ are the surface and volume importance factors, respectively and $w_S + w_V = 1$. We used 21 optimization runs where w_S varied from 0 to 1 in steps of 0.05 to determine the shape of the trade-off curve. A problem in using deterministic optimization methods is that the solution contains a large number of dwell weights with negative values. This is a non physical solution. In the past either constrained optimization methods were used or a correction was applied by setting to 0 all negative weights in each optimization step. A constrained optimization method increases the number of parameters by a factor of two. The correction method for the negative weights reduces the quality of the optimization results. We use a simple technique by replacing the decision variables, the weights w_k, with the parameters $w'_k = w_k^{1/2}$. Using this mapping technique we avoid non feasible solutions. For this unconstrained optimization we use the Polak-Ribiere variant of Fletcher-Reeves algorithm or the Broyden-Fletcher-Goldfarb-Shanno quasi-Newton based algorithm [5]. These require the first derivative of the objective function with respect to the decision variables to be calculated. The derivative of the normalized variance f used by the gradient based optimization methods is:

$$\frac{\partial f}{\partial w'_k} = \frac{2}{Nm^3} \sum_{i=1}^{N} (d_i - m) \left[m \frac{\partial d_i}{\partial w'_k} - d_i \frac{\partial m}{\partial w'_k} \right] \tag{6}$$

As a gradient free method we used the modified Powell method of Numerical Recipes [5].

2.5 Multiobjective Optimization with Evolutionary Algorithms

The population of our multiobjective evolutionary algorithm consists of strings storing a set of weights for each source dwell position. The weights are initially produced randomly distributed in the interval [0, 1]. A part of the population can be initialized, if this is possible, by solutions of deterministic algorithms.

Three selection mechanism can be used. The niched Pareto algorithm (NPGA) proposed by Horn and Nafpliotis [8], the strength evolutionary approach algorithm (SPEA) by Zitzler and Thiele [9] and the non dominated ranking algorithm (NRGA) by Fonseca and Fleming [10],[11].

After a new population is formed, the strings of randomly selected pairs undergo a crossover operation with a probability P_c and mutation with a probability P_m. We have found that P_c must be larger than 0.7 and P_m should be smaller than 0.1. The size of the population should be larger than 50. Various crossover types can be selected such as single point, two point, and arithmetic crossover. For the mutation operation also we have used various forms: uniform or non-uniform mutation. We use a real representation for the gene values. A detailed description of the genetic operators is given in reference [12].

For the NPGA algorithm we use a tournament selection, the tournament population size is a free parameter and can be used for the modification of the selective pressure. Tests have shown that it should be normally 10% of the population size. For much smaller values the genetic algorithm is sensitive to fluctuations, while much larger values can lead to a premature convergence. We applied special genetic operators for decision variables as described by Michalewicz [13]. Some of them offer the possibility for a better performance of the genetic algorithms in the late stage of the optimization process. For NPGA we use a sharing mechanism described by S. Deb [14]. The sharing parameter σ_{share} is given by:

$$\sigma_{share} \approx \frac{0.5}{\sqrt[P]{q}} \qquad (7)$$

where q is the desired number of distinct Pareto-optimal solutions and P is the number of variables in the problem.

Selecting the Solution from the Pareto Set. After the last generation is processed by the SPEA, NRGA or NPGA algorithm, members of the population are expected to be close to the Pareto frontier. A member of the non dominated set is selected which has a minimum Euclidean distance to the ideal optimum. The ideal point is defined by the minimum values (f_1^{min}, f_2^{min}) of each objective function. The distance is calculated by normalizing each objective to a maximum value of 1 using the corresponding largest objective value found in the population. This member is presented as the solution of the optimization process. Additionally members are selected each with the best result in each objective. A list is produced with the objective values for all the members of the Pareto set. Additionally the user can examine the dose distributions and the dose-volume histogram and isodose contours of every member of the population. Based on this information of the trade-off surface of the various objectives a decision maker can select the best result. In our current implementation each objective has equal priority.

3 Results

The dose variances are calculated from 1000–4000 quasi-randomly distributed sampling points. For the COIN based optimization ≈ 100000 points are generated. The distances of these points to each source dwell position r, more precisely the inverse square distances $1/r^2$, are stored for speed maximization in look-up tables. We assume a invariant kernel $K(r) = 1/r^2$ and ignore any spatial anisotropy, namely attenuation and scattering effect. This dosimetric simplification has no measurable influence on the results of the optimization.

All calculations presented in our study have been made by using for the mutation probability P_m a value of 0.0065 and for the crossover probability P_c

a value of 0.85. Furthermore a uniform mutation option has been selected and a two point crossover has been used. The selection of a two point crossover means that the string representation of a member is cut at two random positions and the two end parts are interchanged. This increases the efficiency of the exploitation [15].

The optimization time depends mainly on the number of dwell positions and the population size. For 200 dwell positions and up to 200 generations it can take 1 hour with an Intel Pentium III 700 MHz processor with 512 MB RAM.

The flowchart for the COIN based optimization algorithm is shown in Fig. 3. For each member of the population for a given generation a renormalization is carried out according to the resulting COIN distribution, so that the maximum COIN value is observed at $D = D_{ref}$ [7]. The dose prescription is realized at the D_{ref}, the isodose value resulting in the maximal conformity. This results generally in mean normalized dose values at the surface of PTV different from 1.0.

The multiobjective genetic algorithm, which uses dose-volume based constraints, produces equivalent or even better results than algorithms which were based on phenomenological methods and used in the majority of treatment planning systems [16], [17], [18].

As an example in Fig. 4 the multiobjective genetic algorithm provides a solution with a more homogeneous dose distribution inside the PTV than by conventional optimization algorithms of a treatment planning system. Due to the large computational time for the COIN based optimization we used only the NPGA algorithm.

For the variance based objectives we used 22 different implant cases from various anatomic regions. For these implants different number of catheters were used and their topology differed from case to case. The study aimed to assess the dose homogeneity and conformity and to determine if a common set of importance factors exists, allowing a single objective function to be used with these weights.

An example of the geometry of a PTV is shown in Fig.5(a) including the catheters, the source dwell positions and the sampling points on the PTV surface which define the surface variance. In Fig.5(b) the isosurface for the prescription dose is shown, which should have the same shape as the PTV.

The deterministic gradient based algorithms are very effective in generating the Pareto front using a summed weights approach. Powells algorithm which does not require derivatives is efficient only for implants with a small number of sources. For implants with 250-300 sources the optimization time can reach a few hours for a single objective run, whereas the gradient based algorithms require only 1-2 minutes. Gradient based algorithms are limited by the fact that they can be trapped in local minima, or that non convex regions are not accessible using the weighted sum method [19].

From the evolutionary algorithms SPEA has been found to produce the best results, since it applies an elitism and sharing mechanism. Therefore the Pareto fronts are more uniformly distributed as compared with NPGA. For implants fronts which in most cases are much better than the Pareto front obtained by evolutionary multiobjective algorithms.

If the number of objectives increases then the number of combinations using a weighted sum approach with deterministic algorithms increases. Deterministic methods are not efficient for non analytic complex objectives such as used by the COIN based method. When more objectives are included then a non convex feasible space could be the result [20]. A combination of deterministic and evolutionary multiobjective algorithms seems to be the best choice for a robust and efficient multiobjective dose optimization in HDR brachytherapy. The targets of the dose optimization cannot be expressed uniquely by a single set of objective functions. This is because conformity and homogeneity can be expressed with various functional forms and for the complex geometry of the PTV and the variety of topological configurations it is not known which set is the best. Is the COIN based dose optimization approach better than the dose-statistics approach using variances and if yes how much better?

We are currently studying for various sets of objectives the Pareto fronts using multiobjective evolutionary algorithms and if possible in combination with deterministic algorithms. We expect to understand their limitations and their robustness and performance for the complex problem of the dose-optimization in brachytherapy.

Acknowledgments. We would like to thank Dr. E. Zitzler and P. E. Sevinç for the FEMO library. This investigation was supported by a European Commission Grant (IST-1999-10618, Project: MITTUG).

References

1. Yu, Y., Schell, M. C.: A genetic algorithm for the optimization of prostate implants. Med. Phys. **23** (1996) 2085–2091
2. Yang, G., Reinstein, L. E., Pai, S., Xu, Z.: A new genetic algorithm technique in optimization of permanent ^{125}I prostate implants. Med. Phys. **25** (1998) 2308–2315
3. Lahanas, M., Baltas, D., Giannouli, S., Milickovic, N., Zamboglou, N.: Generation of uniformly distributed dose points for anatomy-based three-dimensional dose optimization methods in brachytherapy. Med. Phys. **27** (2000) 1034–1046
4. Zitzler, E., Deb, K., Thiele, L.: Comparison of Multiobjective Evolutionary Algorithms: Empirical Results. Evolutionary Computation. **8** (2000) 173–195
5. Press,W. H., Teukolsky,S. A., Vetterling, W.T., Flannery,B. P.: Numerical Recipes in C. 2nd ed. Cambridge University Press, Cambridge, England. 1992
6. Nath, R., Anderson, L. L., Luxton, G., Weaver, K. A., Williamson, J. F., Meigooni, A. S.: Dosimetry of interstitial brachytherapy sources: Recommendations of the AAPM Radiation Therapy Committee Task Group No. 43. Med. Phys. **22** (1995) 209–234
7. Baltas D., Kolotas, C., Geramani, K., Mould, R. F., Ioannidis, G., Kekchidi, M., Zamboglou, N.: A Conformal Index (COIN) to evaluate implant quality and dose specifications in brachytherapy. Int. J. Radiat. Oncol. Biol. Phys., **40** (1998) 512–524

8. Horn, J., Nafpliotis, N.: Multiobjective optimization using the niched Pareto genetic Algorithm. IlliGAL Report No.93005. Illinois Genetic Algorithms Laboratory. University of Illinois at Urbana-Champaign, 1993
9. Zitzler, E., Thiele, L.: Multiobjective Evolutionary Algorithms: A Comparative Case Study and the Strength Pareto Approach. IEEE Transactions on Evolutionary Computation. **37** (1999) 257–271
10. Fonseca, M., Fleming, P. J.: Multiobjective optimization and multiple constraint handling with evolutionary algorithms I: A unified formulation. Research report 564, Dept. Automatic Control and Systems Eng. University of Sheffield, Sheffield, U.K., Jan. 1995
11. Fonseca, M., Fleming, P. J.: An overview of evolutionary algorithms in multiobjective optimization. Evolutionary Computation **3** (1995) 1–16
12. Milickovic, N., Lahanas, M., Baltas, D., Zamboglou, N.: Comparison of evolutionary and deterministic multiobjective algorithms for dose optimization in brachytherapy. These proceedings
13. Michalewicz, Z.: Genetic Algorithms + Data Structures = Evolution Programs. Springer Verlag. 1996
14. Deb, K.: Non-linear goal programming using Multi-objective Genetic Algorithms. Technical Report CI-60/98, Department of Computer Science /LS11. University of Dortmund, Germany. (1999)
15. Goldberg, D. E., Richardson, J.: Genetic Algorithms with Sharing for Multimodal Function Optimization. J.J. Grefenstette (Editor), Genetic Algorithms and Their Applications: Proceedings of the Second International Conference on Genetic Algorithms. Lawrence Erlbaum Associated. (1987) 41–49
16. Lahanas, M., Baltas, D., Zamboglou, N.: Anatomy-based three-dimensional dose optimization in brachytherapy using multiobjective genetic algorithms. Med. Phys. **26** (1999) 1904–1918
17. Edmundson, K.: Geometry based optimization for stepping source implants, in: Brachytherapy HDR and LDR, A. A. Martinez, C. G. Orton and R. F. Mould eds., Nucleotron: Columbia. (1990) 184–192
18. Van der Laarse, T. P. E. Prins.: Introduction to HDR brachytherapy optimisation, In: R. F. Mould, J. J. Battermann, A. A. Martinez and B. L . Speiser eds. Brachytherapy from Radium to Optimization. Veenendaal, The Netherlands: Nucleotron International. (1994) 331–351
19. Das, I. Dennis, J.: A Closer Look at Drawbacks of Minimizing Weighted Sums of Objectives for Pareto Set Generation in Multicriteria Optimization Problems. Structural Optimization **14** (1997) 63–69
20. Deasy, J. O.: Multiple local minima in radiotherapy optimization problems with dose-volume constraints. Med. Phys. **24** (1997) 1157–1161

2 Formulation of Linguistic Rule Extraction

In this section, we formulate linguistic rule extraction from numerical data as a three-objective combinatorial optimization problem for pattern classification. Our basic idea is to simultaneously maximize classification ability and interpretability of extracted rule sets.

2.1 Assumptions

We assume that m training patterns (i.e., labeled patterns) are given as numerical data for an n-dimensional c-class pattern classification problem. We denote those training patterns as $\mathbf{x} = (x_{p1}, ..., x_{pn})$, $p = 1, 2, ..., m$. For simplicity of explanation, each attribute value x_{pi} is assumed to be a real number in the unit interval [0, 1], i.e., $x_{pi} \in [0, 1]$. This means that the pattern space of our pattern classification problem is the n-dimensional unit hypercube $[0, 1]^n$. In computer simulations of this paper, all attribute values are normalized into real numbers in the unit interval [0, 1].

We also assume that a set of linguistic values is given for describing each attribute. That is, we assume that a fuzzy partition of the pattern space $[0, 1]^n$ is given. For simplicity of explanation, we use five linguistic values in Fig. 1 for all the n attributes. Of course, our approaches described in this paper are applicable to more general cases where a different set of linguistic values is given to each attribute. In such a general case, membership functions are not necessary to be triangular. They are specified according to domain knowledge and intuition of human experts.

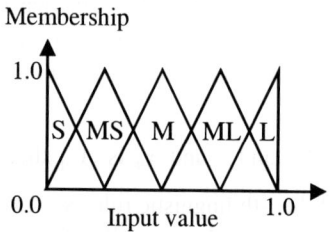

Fig. 1. Membership functions of five linguistic values (S: *small*, MS: *medium small*, M: *medium*, ML: *medium large*, and L: *large*).

As Jin [17] pointed out, it is not easy to understand fuzzy rules with many antecedent conditions. Thus we use "DC: *don't care*" as an additional linguistic value in the antecedent part of our linguistic rule in (1). Linguistic rules with many "*don't care*" conditions can be concisely written even for high-dimensional pattern classification problems. The following are some examples of such linguistic rules for a 13-dimensional wine classification problem used in Section 5.

R_1: If x_7 is *medium* and x_{11} is *medium* then Class 1 with 0.56. (2)

R_2: If x_{10} is *small* then Class 2 with 0.94. (3)

R_3: If x_7 is *small* then Class 3 with 0.85. (4)

These linguistic rules are very simple and easily understood. In the above linguistic rules, "*don't care*" conditions are omitted.

When we use the five linguistic values in Fig. 1 and "DC: *don't care*" as antecedent linguistic values in our linguistic rule in (1), we have $(5+1)^n$ combinations of antecedent linguistic values as follows:

$$\text{Rule } R_j : \text{If } x_1 \text{ is } \begin{Bmatrix} S \\ MS \\ M \\ ML \\ L \\ DC \end{Bmatrix} \text{ and ... and } x_n \text{ is } \begin{Bmatrix} S \\ MS \\ M \\ ML \\ L \\ DC \end{Bmatrix} \text{ then Class } C_j \text{ with } CF_j. \quad (5)$$

As shown in Appendix, the consequent class C_j and the certainty grade CF_j of each linguistic rule can be easily specified from training patterns when its antecedent conditions are given. Thus our linguistic rule extraction problem can be viewed as finding a small number of combinations of antecedent linguistic values among the above $(5+1)^n$ combinations. The total number of possible rule sets is 2^N where $N = (5+1)^n$. It is not easy to examine all the possible rule sets even for a three-dimensional pattern classification problem. In this case, the total number of possible rule sets is $2^{6\times6\times6} \cong 1.05\times10^{65}$. In the case of high-dimensional problems, we can examine only a tiny portion of possible rule sets because the search space is huge.

2.2 Three-Objective Combinatorial Optimization Problem

Let S_{ALL} be the set of $(5+1)^n$ linguistic rules for our n-dimensional pattern classification problem. They correspond to the $(5+1)^n$ combinations of the five linguistic values in Fig. 1 and "*don't care*". Our linguistic rule extraction problem is to find a small number of simple linguistic rules with high classification ability from the rule set S_{ALL}, i.e., to find a compact and high-performance rule set. Because there is a tradeoff between compactness and performance, we try to find non-dominated rule sets with respect to conflicting criteria.

We measure the classification performance of a rule set S ($S \subset S_{ALL}$) by the number of correctly classified training patterns. Compactness of S is measured by two criteria: the number of linguistic rules in S and the total number of antecedent conditions in S. Of course, "*don't care*" conditions are not counted among antecedent conditions. The number of antecedent conditions in a linguistic rule is referred to as its length. Thus the total number of antecedent conditions is the same as the total length of linguistic rules. Based on these discussions, our linguistic rule extraction problem is formulated as follows:

$$\text{Maximize } f_1(S), \text{ minimize } f_2(S), \text{ and minimize } f_3(S), \qquad (6)$$

where

$f_1(S)$: The number of correctly classified training patterns by S,

$f_2(S)$: The number of linguistic rules in S,

$f_3(S)$: The total length of linguistic rules in S.

For example, when a rule set S consists of the three linguistic rules R_1, R_2 and R_3 in (2)-(4), $f_2(S)$ and $f_3(S)$ are calculated as $f_2(S) = 3$ and $f_3(S) = 4$, respectively. The first objective $f_1(S)$ is calculated by classifying all the m training patterns by the rule set S. In Appendix, we show how each pattern is classified by linguistic rules (see [2] for various fuzzy reasoning methods for pattern classification).

The third objective $f_3(S)$ is not the average length of extracted rules but the total length. Let $f_{3*}(S)$ be the average length. For example, $f_{3*}(S)$ is calculated as $f_{3*}(S) = 1.33$ for the rule set S with R_1, R_2 and R_3. Let us construct another rule set S^+ by adding a linguistic rule R_4 of the length one to S. For the new rule set S^+ with $R_1 \sim R_4$, $f_{3*}(S^+)$ is calculated as $f_{3*}(S^+) = 1.25$. That is, the average length is improved by adding R_4 to the rule set S while the complexity of the rule set is increased. Even if the added linguistic rule R_4 does not improve the classification performance of the rule set (i.e., $f_1(S) = f_1(S^+)$), the new rule set S^+ is not dominated by the rule set S when we use the average length as the third objective. This simple example shows that the average length is not an appropriate objective for measuring the simplicity of extracted linguistic rules in the framework of multi-objective optimization. Thus we use the total length as the third objective $f_3(S)$. In [7], the average length was used for rule selection. Since three objectives were combined into a scalar fitness function in [7] for obtaining a single optimal rule set, the above-mentioned difficulty of the average length can be ignored. The difficulty of the average length is crucial only when this objective is used in the framework of multi-objective optimization for obtaining non-dominated rule sets.

2.3 Simple Numerical Example

Let us consider a simple numerical example in Fig. 2 (a) where 121 training patterns are given in the unit square $[0, 1] \times [0, 1]$. If we use a standard grid-type fuzzy partition in Fig. 2 (b), we can generate 5×5 linguistic rules with no "*don't care*" conditions. All the given training patterns are correctly classified by those 25 linguistic rules. On the other hand, a much simpler rule set can be extracted if we consider linguistic rules with "*don't care*" conditions.

As shown in Subsection 2.1, the total number of possible combinations of antecedent linguistic values is 36 for the two-dimensional pattern classification problem when we use "*don't care*" as an additional linguistic value. By examining subsets of those 36 linguistic rules, we can find that the following three linguistic rules correctly classify all the given training patterns:

R_A : If x_1 is *medium* then Class 2 with 0.75. (7)

R_B : If x_2 is *large* then Class 2 with 0.84 . (8)

R_C : (x_1, x_2) is Class 1 with 0.19. (9)

The last linguistic rule R_C has no linguistic condition (i.e., it has two "*don't care*" conditions). Since R_C has a small certainty grade (i.e., 0.19), this rule is used for pattern classification only when the other two linguistic rules R_A and R_B are not applicable. In this manner, inconsistency is resolved through certainty grades.

All the non-dominated rule sets of our linguistic rule extraction problem for this numerical example are shown in Table 1. Since the numerical example in Fig. 2 is very simple, we can find all non-dominated solutions by examining all rule sets with one, two or three linguistic rules. Usually we can not find all non-dominated solutions by such an enumeration method especially for high-dimensional pattern classification problems. In the rest of this paper, we explain how genetic algorithms can be used for finding non-dominated solutions of our three-objective linguistic rule extraction problem.

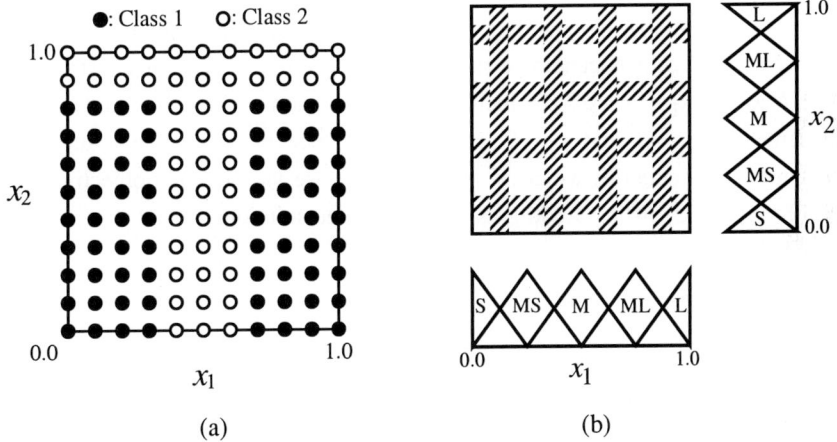

Fig. 2. A numerical example and a grid-type fuzzy partition.

Table 1. All the non-dominated solutions for the numerical example in Fig. 1.

Rule set: S	# of patterns: $f_1(S)$	# of rules: $f_2(S)$	Total length: $f_3(S)$
{ }	0 (0%)	0	0
{ R_C }	72 (60%)	1	0
{ R_A, R_C }	105 (87%)	2	1
{ R_A, R_B, R_C }	121 (100%)	3	2

3 Rule Selection

3.1 Basic Idea of Rule Selection

We have already proposed a GA-based rule selection method [7, 15] where multiple objectives were combined into a scalar fitness function for applying standard single-objective genetic algorithms. We have also proposed two-objective genetic algorithms for finding non-dominated rule sets with respect to the classification performance and the number of linguistic rules [8]. In this paper, the total rule length is added to the two-objective rule selection method in [8].

It is not difficult to extend our former two-objective rule selection method to the case of three objectives. Let N be the number of linguistic rules that can be generated from given training patterns. Those N linguistic rules are used as candidate rules in rule selection. In low-dimensional pattern classification problems, all the $(5+1)^n$ linguistic rules in (5) are considered as candidate rules. All linguistic rules, however, are not always generated (e.g., when training patterns are not evenly distributed in the entire pattern space). In high-dimensional pattern classification problems, the number of candidate rules should be much smaller than $(5+1)^n$ because the string length is the same as the number of candidate rules in our rule selection method.

Any subset S of N candidate rules can be represented by a binary string of the length N as $S = s_1 s_2 \cdots s_N$. The inclusion and exclusion of the j-th candidate rule are represented by $s_j = 1$ and $s_j = 0$, respectively. Since every feasible solution of our problem is represented by a binary string, we can use various multiobjective genetic algorithms [28, 33, 34] for finding its non-dominated solutions.

3.2 Candidate Rule Prescreening

In high-dimensional pattern classification problems, we can not handle all the $(5+1)^n$ linguistic rules in (5) as candidate rules. Thus a prescreening procedure of candidate rules is necessary for applying our rule selection method to high-dimensional problems. A simple trick is to examine only short linguistic rules with a few antecedent linguistic conditions [7]. This trick, which has also a good effect on the third objective of our rule extraction problem, can significantly decrease the number of candidate rules. Even when the total number of linguistic rules is huge, the number of short rules is not so large. For example, the number of linguistic rules of the length one with a single antecedent condition is calculated as $_{13}C_1 \times 5 = 65$ for a 13-dimensional problem such as wine data. In our computer simulation on the wine data, we examine one rule of the length zero, 65 rules of the length one, and $_{13}C_2 \times 5 \times 5 = 1950$ rules of the length two for generating candidate rules in our rule selection method. In [7], the certainty grade of each linguistic rule was also used in addition to the length for prescreening candidate rules.

3.3 Domain-Specific Heuristic Procedures

The performance of genetic algorithms for rule selection can be improved by incorporating domain-specific heuristic procedures. One procedure is for eliminating unnecessary rules. As shown in Appendix, we use a fuzzy reasoning method based on a single winner rule in the classification phase. This means that each pattern is classified by a single winner rule in a rule set. If a linguistic rule is not used as a winner rule for any training pattern, that rule can be removed from the rule set with no deterioration of its classification performance. This elimination improves the second and third objectives of our linguistic rule extraction problem. Our rule elimination procedure removes such a linguistic rule before each rule set is evaluated in three-objective genetic algorithms.

Another trick is to bias the mutation probability. Even when we use an appropriate prescreening procedure, usually the number of selected rules is much smaller than that of candidate rules. That is, binary strings should consist of a small number of 1's and a large number of 0's. The standard mutation tends to increase the number of 1's when binary strings have much more 0's than 1's. For efficiently searching for binary strings with a small number of 1's, we use biased mutation probabilities where the mutation probability from 1 to 0 is much larger than that from 0 to 1.

4 Genetics-Based Machine Learning

4.1 Basic Idea of Genetics-Based Machine Learning

The quality of non-dominated rule sets obtained by rule selection strongly depends on the choice of a prescreening procedure. While some studies [4, 12] showed high performance of short rules with only a few antecedent conditions, this is not always the case. Some pattern classification problems may need long rules as well as short rules. In this case, the search among $(5+1)^n$ linguistic rules is necessary for finding good rule sets. Genetics-based machine learning (GBML) algorithms are promising tools for finding non-dominated rule sets in the huge search space.

We have already proposed Michigan-style GBML algorithms for generating linguistic rules for high-dimensional pattern classification problems [9, 14]. In our Michigan-style algorithm, a single linguistic rule was coded as a string. A population with a fixed number of linguistic rules was evolved by genetic operations for finding good linguistic rules. Since our objectives in this paper are not only classification performance but also the number of linguistic rules and the total rule length, the number of linguistic rules should not be fixed. It is, however, difficult to directly optimize the number of linguistic rules in the framework of Michigan-style algorithms because a fitness value is assigned to each linguistic rule. Thus we use a Pittsburgh-style algorithm with variable sting length for finding non-dominated rule sets.

4.2 Genetic Operations

As in our Michigan-style algorithms in [9, 14], each linguistic rule is coded by its antecedent linguistic values as "$A_{j1}A_{j2} \cdots A_{jn}$". Since the consequent class and the certainty grade are easily specified by training patterns, they are not coded as a part of the string. A rule set is denoted by a concatenated string. Each substring of the concatenated string corresponds to a single linguistic rule.

A new rule set (i.e., a new string) is generated by crossover and mutation. In our computer simulation, we use a kind of one-point crossover shown in Fig. 3, which changes the number of linguistic rules in each rule set. This crossover operation randomly selects a different cutoff point for each parent to form an offspring. A mutation operation randomly replaces each element (i.e., each antecedent linguistic value) of the string with another linguistic value. Elimination of existing rules and addition of new rules can be also used as mutation operations. Such mutation operations change the number of linguistic rules in each string.

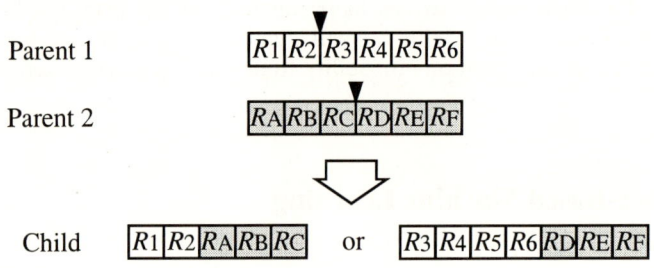

Fig. 3 Crossover operation.

4.3 Hybridization with Michigan-Style Algorithm

The search ability of Pittsburgh-style algorithms to find good linguistic rules is somewhat inferior to that of Michigan-style algorithms when they are applied to high-dimensional pattern classification problems [11, 13]. Michigan-style algorithms, however, can not directly optimize rule sets because a fitness value is assigned to each linguistic rule (not to each rule set). On the other hand, Pittsburgh approach can directly optimize rule sets. A natural idea for utilizing advantages of these two kinds of GBML approaches is to combine them into a single hybrid algorithm [10, 11]. A Michigan-style algorithm, which is used as a mutation operation in our Pittsburgh-style algorithm, partially modifies each rule set by generating new rules. By this hybridization, search ability of our Pittsburgh-style algorithm to efficiently find good linguistic rules is significantly improved. In our computer simulation, a single iteration of a Michigan-style algorithm was applied with a prespecified probability to every rule set generated by the genetic operations of our Pittsburgh-style algorithm.

5 Computer Simulations

5.1 Test Problem and Simulation Conditions

We applied the rule selection method and the hybrid GBML algorithm to wine data (available from UC Irvine Database: http://kdd.ics.uci.edu/) for finding non-dominated rule sets of our linguistic rule extraction problem. These two algorithms were implemented in the framework of three-objective genetic algorithms. The three objectives were combined into the following fitness function, which was used in a roulette wheel selection with a linear scaling.

$$fitness(S) = w_1 \cdot f_1(S) - w_2 \cdot f_2(S) - w_3 \cdot f_3(S). \qquad (10)$$

For finding various non-dominated solutions, weights w_1, w_2 and w_3 were not fixed but randomly updated whenever a pair of parent strings were selected as in our two-objective genetic algorithm in [8]. Non-dominated solutions were separately stored from the current population. Some of the stored non-dominated solutions were added to the current population for maintaining its diversity and quality.

Both the rule selection method and the hybrid GBML algorithm were executed under the following parameter specifications:

> Population size: 50 rule sets.
> Stopping conditions: 1000 generations.

That is, a population of 50 rule sets was evolved until the 1000th generation in both algorithms. We applied each algorithm to wine data 20 times. In the following subsections, we report non-dominated solutions obtained from those 20 trials.

5.2 Simulation Results by Rule Selection

Since wine data involve 13 attributes, the number of possible combinations of antecedent linguistic values is $(5+1)^{13} \cong 1.3 \times 10^{10}$ (i.e., about 13 billion). It is impossible to apply our rule selection method to candidate rules generated from all those combinations. We examined only short linguistic rules of the length two or less. By this prescreening procedure, 1834 linguistic rules were generated as candidate rules. Thus each rule set was represented by a binary string of the length 1834. The task of our rule selection method is to find non-dominated rule sets from those candidate rules. From 20 trials, we obtained 17 non-dominated rule sets. Due to the space limitation, we show 12 rule sets with high classification rates. Too small rule sets are not shown in this table (e.g., those with only two rules). From Table 2, we can see that our rule selection method found various non-dominated rule sets. Some are very compacts, and others have high classification rates. We can observe a tradeoff between the classification performance and the compactness of rule sets.

candidate rules is necessary when it is applied to high-dimensional pattern classification problems. Then we explained how non-dominated rule sets can be found by a genetics-based machine learning algorithm in a huge search space with all possible linguistic rules. These two schemes were applied to wine data with 13 attributes. Simulation results showed that compact rule sets with high classification performance were found. Finally we modified our linguistic rule extraction problem for applying it to function approximation.

Acknowledgement. This study was partially supported by Saneyoshi Scholarship Foundation.

Appendix: Rule Generation and Pattern Classification

The consequent class C_j and the certainty grade CF_j of our linguistic rule in (1) can be determined by the following heuristic procedure (for example, see [15]):

Step 1) Calculate the compatibility grade $\mu_{R_j}(\mathbf{x}_p)$ of each training pattern $\mathbf{x}_p = (x_{p1}, ..., x_{pn})$ with the linguistic rule R_j by the product operation as

$$\mu_{R_j}(\mathbf{x}_p) = \mu_{j1}(x_{p1}) \times \cdots \times \mu_{jn}(x_{pn}), \quad p = 1, 2, ..., m, \tag{A1}$$

where $\mu_{ji}(\cdot)$ is the membership function of the antecedent linguistic value A_{ji}.

Step 2) For each class, calculate the total compatibility grade of the training patterns with the linguistic rule R_j:

$$\beta_{\text{Class } h}(R_j) = \sum_{\mathbf{x}_p \in \text{Class } h} \mu_{R_j}(\mathbf{x}_p), \quad h = 1, 2, ..., c. \tag{A2}$$

Step 3) Find the consequent class C_j that has the maximum value of $\beta_{\text{Class } h}(R_j)$:

$$\beta_{\text{Class } C_j}(R_j) = \text{Max}\{\beta_{\text{Class } 1}(R_j), ..., \beta_{\text{Class } c}(R_j)\}. \tag{A3}$$

If the consequent class can not be uniquely determined, we do not extract the linguistic rule R_j. For example, if $\beta_{\text{Class } h}(R_j) = 0$ for all classes, we do not generate R_j.

Step 4) Specify the certainty grade CF_j as follows:

$$CF_j = \{\beta_{\text{Class } C_j}(R_j) - \overline{\beta}\} / \sum_{h=1}^{c} \beta_{\text{Class } h}(R_j), \tag{A4}$$

where

$$\overline{\beta} = \sum_{\substack{h=1 \\ h \neq C_j}}^{c} \beta_{\text{Class } h}(R_j)/(c-1). \tag{A5}$$

Let S be a set of generated linguistic rules. A new pattern $\mathbf{x}_p = (x_{p1},...,x_{pn})$ is classified by S using a fuzzy reasoning method based on a single winner rule [15]. The winner rule R_{j*} for $\mathbf{x}_p = (x_{p1}, ..., x_{pn})$ is determined as follows:

$$\mu_{R_{j*}}(\mathbf{x}_p) \cdot CF_{j*} = \max\{\mu_{R_j}(\mathbf{x}_p) \cdot CF_j \mid R_j \in S\}. \tag{A6}$$

The new pattern $\mathbf{x}_p$ is classified by the winner rule R_{j*}. That is, $\mathbf{x}_p$ is assigned to the consequent class of R_{j*}. If multiple linguistic rules with different consequent classes have the same maximum value in (A6), the classification of $\mathbf{x}_p$ is rejected.

References

1. Abe, S., Lan, M.-S., Thawonmas, R.: Tuning of a Fuzzy Classifier Derived from Data, *Int. J. Approximate Reasoning* 14 (1996) 1-24.
2. Cordon, O., del Jesus, M. J., Herrera, F.: A Proposal on Reasoning Methods in Fuzzy Rule-Based Classification Systems, *Int. J. Approximate Reasoning* 20 (1999) 21-45.
3. Herrera, F., Lozano, M., Verdegay, J. L.: Tuning Fuzzy Logic Controllers by Genetic Algorithms, *Int. J. Approximate Reasoning* 12 (1995) 299-315, 1995.
4. Holte, R. C.: Very Simple Classification Rules Perform Well on Most Commonly Used Dataset, *Machine Learning* 11 (1993) 63-91.
5. Horikawa, S., Furuhashi, T., Uchikawa, Y.: On Fuzzy Modeling Using Fuzzy Neural Networks with the Back-Propagation Algorithm, *IEEE Trans. on Neural Networks* 3 (1992) 801-806.
6. Ishibuchi, H.: Fuzzy Reasoning Method in Fuzzy Rule-based Systems with General and Specific Rules for Function Approximation, *Proc. of 8th IEEE Int. Conference on Fuzzy Systems* (1999) 198-203.
7. Ishibuchi, H., Murata, T., Nakashima, T.: Linguistic Rule Extraction from Numerical Data for High-Dimensional Classification Problems, *Int. J. Advanced Computational Intelligence* 3 (1999) 386-393.
8. Ishibuchi, H., Murata, T., Turksen, I. B.: Single-Objective and Two-Objective Genetic Algorithms for Selecting Linguistic Rules for Pattern Classification Problems, *Fuzzy Sets and Systems* 89 (1997) 135-149.
9. Ishibuchi, H., Nakashima, T.: Improving the Performance of Fuzzy Classifier Systems for Pattern Classification Problems with Continuous Attributes, *IEEE Trans. on Industrial Electronics* 46 (1999) 1057-1068.
10. Ishibuchi, H., Nakashima, T.: Linguistic Rule Extraction by Genetics-Based Machine Learning, *Proc. of Genetic and Evolutionary Computation Conference* (2000) 195-202.
11. Ishibuchi, H., Nakashima, T., Kuroda, T.: A Hybrid Fuzzy GBML Algorithm for Designing Compact Fuzzy Rule-Based Classification Systems, *Proc. of 9th IEEE International Conference on Fuzzy Systems* (2000) 706-711.
12. Ishibuchi, H., Nakashima, T., Morisawa, T.: Simple Fuzzy Rule-Based Classification Systems Perform well on Commonly Used Real-World Data Sets, *Proc. of 16th Annual Meeting of the North American Fuzzy Information Processing Society* (1997) 251-256.
13. Ishibuchi, H., Nakashima, T., Murata, T.: Genetic-Algorithm-Based Approaches to the Design of Fuzzy Systems for Multi-Dimensional Pattern Classification Problems, *Proc. of 3rd IEEE Int. Conference on Evolutionary Computation* (1996) 229-234.
14. Ishibuchi, H., Nakashima, T., Murata, T.: Performance Evaluation of Fuzzy Classifier Systems for Multi-Dimensional Pattern Classification Problems, *IEEE Trans. on Systems, Man, and Cybernetics - Part B: Cybernetics* 29 (1999) 601-618.

15. Ishibuchi, H., Nozaki, K., Yamamoto, N., Tanaka, H.: Selecting Fuzzy If-Then Rules for Classification Problems using Genetic Algorithms, *IEEE Trans. on Fuzzy Systems* 3 (1995) 260-270.
16. Jang, J. -S. R.: ANFIS: Adaptive-Network-Based Fuzzy Inference System, *IEEE Trans. on Systems, Man, and Cybernetics* 23 (1993) 665-685.
17. Jin, Y.: Fuzzy Modeling of High-Dimensional Systems: Complexity Reduction and Interpretability Improvement, *IEEE Transactions on Fuzzy Systems* 8 (2000) 212-221.
18. Jin, Y., von Seelen, W., Sendhoff, B.: On Generating FC^3 Fuzzy Rule Systems from Data using Evolution Strategies, *IEEE Transactions on Systems, Man and Cybernetics - Part B: Cybernetics* 29 (1999) 829-845.
19. Karr, C. L., Gentry, E. J.: Fuzzy Control of pH using Genetic Algorithms, *IEEE Trans. on Fuzzy Systems* 1 (1993) 46-53.
20. Kosko, B.: Fuzzy Systems as Universal Approximators, *Proc. of 1st IEEE Int. Conference on Fuzzy Systems* (1992) 1153-1162.
21. Leondes, C. T. (ed.): *Fuzzy Theory Systems: Techniques and Applications*, Academic Press, San Diego (1999).
22. de Oliveira, V.: Semantic Constraints for Membership Function Optimization, *IEEE Trans. on Systems, Man, and Cybernetics - Part A: Systems and Humans* 29 (1999) 128-138.
23. Pedrycz, W., de Oliveira, V.: Optimization of Fuzzy Models, *IEEE Trans. on Systems, Man, and Cybernetics - Part B: Cybernetics* 26 (1996) 627-637.
24. Setnes, M., Babuska, R., Kaymak, U., van Nauta Lemke, H. R.: Similarity Measures in Fuzzy Rule Base Simplification, *IEEE Trans. on Systems, Man, and Cybernetics - Part B: Cybernetics* 28 (1998) 376-366.
25. Setnes, M., Babuska, R., Verbruggen, B.: Rule-Based Modeling: Precision and Transparency, *IEEE Trans. on Systems, Man, and Cybernetics - Part C: Applications and Reviews* 28 (1998) 165-169.
26. Setnes, M., Roubos, H.: GA-Fuzzy Modeling and Classification: Complexity and Performance, *IEEE Trans. on Fuzzy Systems* 8 (2000) 509-522.
27. Takagi, T., Sugeno, M.: Fuzzy Identification of Systems and Its Applications to Modeling and Control, *IEEE Trans. on Systems, Man, and Cybernetics* 15 (1985) 116-132.
28. van Veldhuizen, D. A., Lamont, G. B.: Multiobjective Evolutionary Algorithms: Analyzing the State-of-the-Art, *Evolutionary Computation* 8 (2000) 125-147.
29. Wang, L. -X.: Fuzzy Systems are Universal Approximators, *Proc. of 1st IEEE Int. Conference on Fuzzy Systems* (1992) 1163-1170.
30. Wang, L. X., Mendel, J. M.: Generating Fuzzy Rules by Learning from Examples, *IEEE Trans. on Systems, Man, and Cybernetics* 22 (1992) 1414-1427.
31. Yen, J., Wang, L.: Simplifying Fuzzy Rule-Based Models using Orthogonal Transformation Methods, *IEEE Trans. on Systems, Man, and Cybernetics - Part B: Cybernetics* 29 (1999) 13-24.
32. Yen, J., Wang, L., Gillespie, W.: Improving the Interpretability of TSK Fuzzy Models by Combining Global Learning and Local Learning, *IEEE Trans. on Fuzzy Systems* 6 (1998) 530-537.
33. Zitzler, E., Deb, K., Thiele, L.: Comparison of Multiobjective Evolutionary Algorithms: Empirical Results, *Evolutionary Computation* 8 (2000) 173-195.
34. Zitzler, E., Thiele, L.: Multiobjective Evolutionary Algorithms: A Comparative Case Study and the Strength Pareto Approach, *IEEE Transactions on Evolutionary Computation* 3 (1999) 257-271.

Determining the Color-Efficiency Pareto Optimal Surface for Filtered Light Sources

Neil H. Eklund[1] and Mark J. Embrechts[2]

[1] Oak Grove Scientific, 11A Manchester Drive, Clifton Park NY 12065
eklund@acm.org
[2] Department of Decision Sciences and Engineering Systems, Rensselaer Polytechnic Institute, Troy NY 12180
embrem@rpi.edu

Abstract. While there are many factors that determine the commercial potential of an electric light source, color and efficiency are arguably the most important. Tradeoffs between color and efficiency are frequently made in lighting applications, typically by moving between different light source technologies. However, the potential exists to change position in color-efficiency space by filtering a light source. Because color is specified in two dimensions, and efficiency in one, the Pareto-optimal color and efficiency front defines a surface. This paper presents a method for determining color-efficiency Pareto optimal surface for a filtered light source.

1 Introduction

There are many factors that determine the commercial potential of an electric light source, including color, efficiency, color rendering index, color temperature, and the reliability and lifetime of the source. While the success or failure of novel light source technologies in the marketplace is dependent on all of these factors, color and efficiency are arguably the most important factors.

Tradeoffs between color and efficiency are made in different applications. For example, low pressure sodium lamps, which have "bad" color but high efficiency are used in many outdoor applications; meanwhile, incandescent lamps, which have "good" color but low efficiency are used in many indoor applications. These examples switch between light source technology to move around the color-efficiency Pareto optimal surface; however, by filtering a light source, it is also possible to move about on the Pareto optimal surface. In cases where the efficiency of a light source is high, but the color is undesirable, it might be possible to filter the light such that the efficiency is reduced by a small amount, but the color becomes acceptable. The goal of this research is to develop a technique for determining the color-efficiency Pareto optimal surface for filtered light sources.

2 Characterization of Light Sources

Various radiometric, photometric, and colorimetric properties of a light source can be determined from the source's spectral power distribution (SPD) [1], the radiant power per unit wavelength as a function of wavelength. The SPDs of the three light sources used in this paper are plotted in Figure 1. Note that SPDs can be smooth and continuous (e.g., incandescent lamps, sulfur lamps), or more spiky, with power either spread throughout the visible spectrum (e.g., metal halide lamps, fluorescent lamps), or concentrated principally in one portion of the visible spectrum (e.g., high pressure sodium lamps, low pressure sodium lamps).

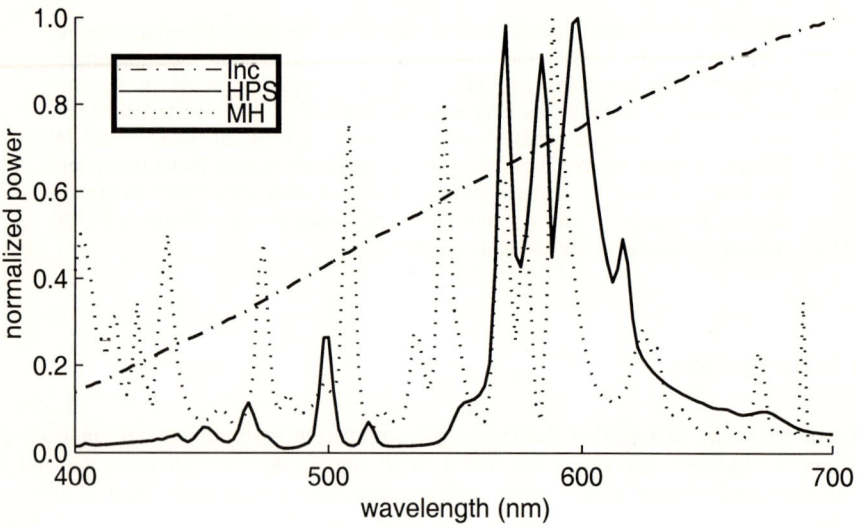

Fig. 1. SPD for metal halide (MH), high pressure sodium (HPS) and incandescent (Inc) lamps.

2.1 The CIE System

The International Commission on Illumination (CIE[1]) is the ISO recognized body for all matters regarding the science and art of lighting. The CIE has developed [2] a method to represent the color and brightness of a light source from its SPD. The SPD is weighted by the color matching functions and integrated over the visible spectrum to provide the tristimulus values, X, Y and Z:

---

[1] After the French, 'Commission Internationale de l'Eclairage'.

$$X = k \int_\lambda P_\lambda \bar{x}(\lambda) d\lambda \qquad (1)$$

$$Y = k \int_\lambda P_\lambda \bar{y}(\lambda) d\lambda \qquad (2)$$

$$Z = k \int_\lambda P_\lambda \bar{z}(\lambda) d\lambda \qquad (3)$$

where $\bar{x}(\lambda)$, $\bar{y}(\lambda)$, and $\bar{z}(\lambda)$ are color matching functions (weighting factors, Fig. 2), $P(\lambda)$ is the power at wavelength λ, and k is an application specific constant (set to 1 for the work presented here). The Y tristimulus value is proportionate to the brightness of the light source. Efficiency is defined as the ratio of the Y tristimulus value of a filtered light source to the Y tristimulus value of the same unfiltered light source (i.e., "how bright is the filtered light source compared to the original light source"). Note that because this is a measure of relative efficiency (rather than absolute efficiency), efficiency as defined here should not be directly compared between lamps.

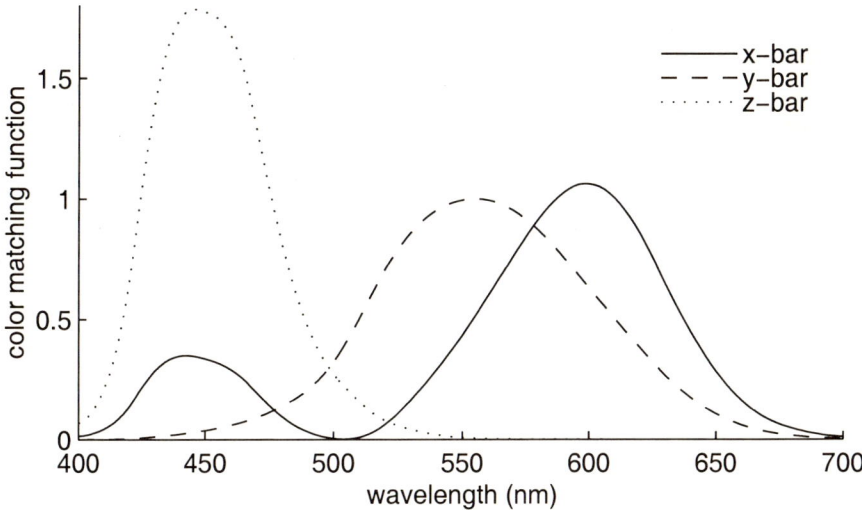

Fig. 2. CIE color matching functions.

4 Results

Figures 3-5 are contour plots of the color-efficiency Pareto-optimal surface for the three different sources. The heavy, sail shaped line is the spectrum locus - the set of chromaticity coordinates for monochromatic lights (and the straight line at the bottom of the "sail", connecting two end points of the spectrum locus, known as the purple line [1]). The spectrum locus defines color space - no light source can ever plot outside the spectrum locus. The heavy dashed line is the blackbody locus - the set of chromaticity coordinates for a blackbody radiator at various temperatures. Light sources near the blackbody locus in color space are frequently considered more desirable in the lighting industry, as they tend to appear more "natural". The heavy dot denotes the chromaticity coordinates of the unfiltered light source (and the only point where efficiency == 1.0).

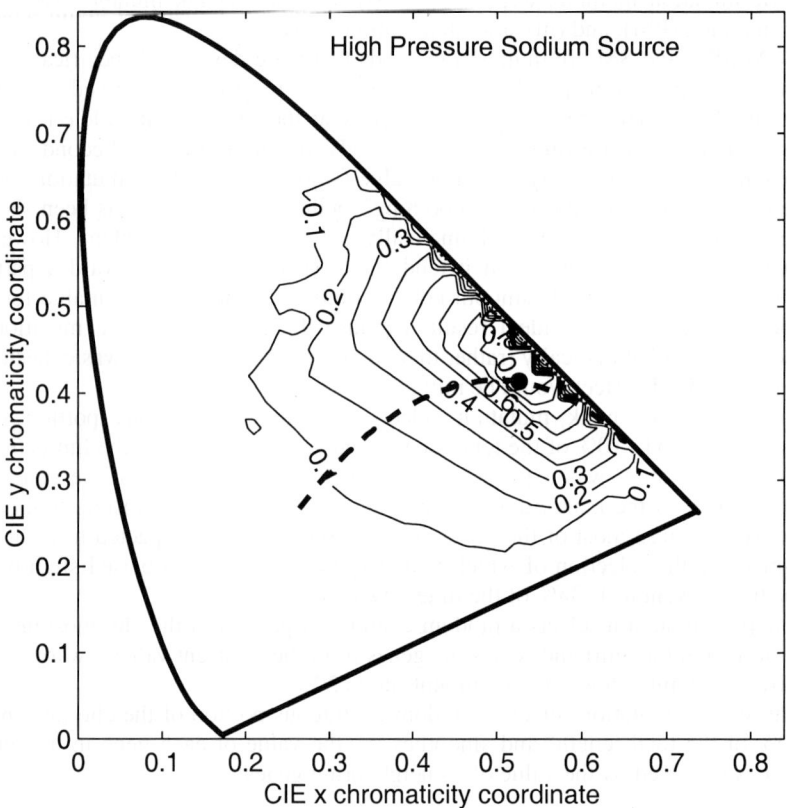

Fig. 3. Efficiency contours for the high pressure sodium source.

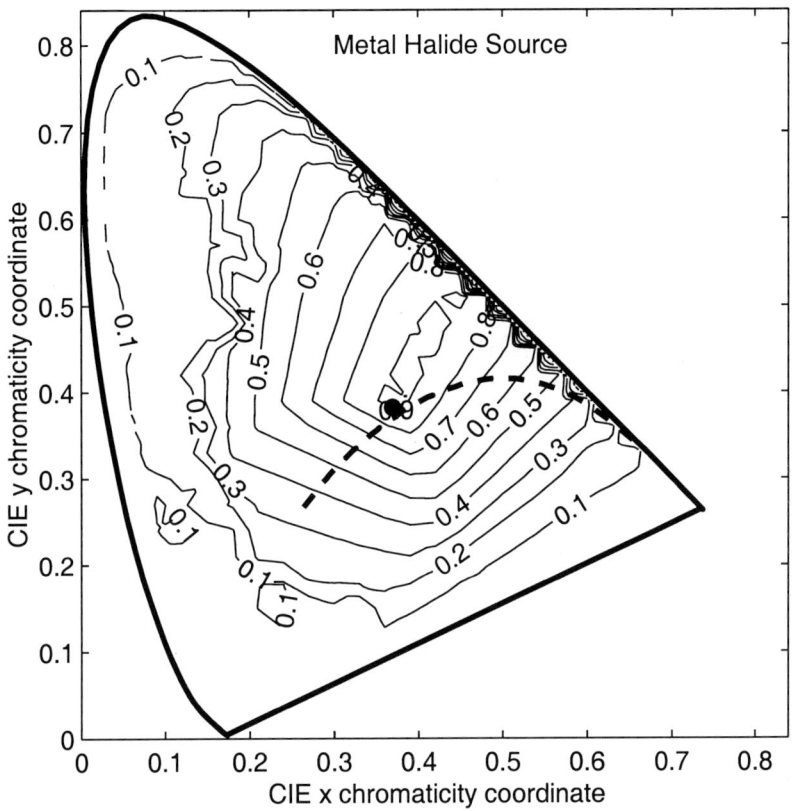

Fig. 4. Efficiency contours for the metal halide source.

The shape of the color-efficiency surface is consistent with what one might expect from the SPDs of the three sources. The amount of color space achievable at any level of efficiency by the HPS source is much smaller than the area achievable by the other two sources, due to the HPS lamp having most of its power in a relatively small region of the visible spectrum. Although both the HPS and incandescent sources have energy spread throughout the spectrum, the HPS source has a relatively narrow (compared to the incandescent) band of high efficiency, related to the relative spiky spectrum of the HPS lamp.

5 Discussion

This paper presents a method for determining the color-efficiency Pareto optimal surface for filtered light sources, which has two major applications. First, as novel light sources are developed which are extremely efficient, but have a color that is unacceptable for some applications, filtering may adjust the color to an acceptable region while maintaining a relatively high level of efficiency, permitting a wider range of application for the lamp technology. For example the sulfur lamp [5] has a very high luminous efficacy, but its chromaticity coordinates (x=0.33, y=0.41) are relatively distant from the blackbody locus, which makes it unsuitable in appearance for many indoor applications. This lamp could be filtered to bring it much closer to the blackbody locus while still maintaining high luminous efficacy, which would make it much more desirable for indoor applications (and, consequently, more marketable).

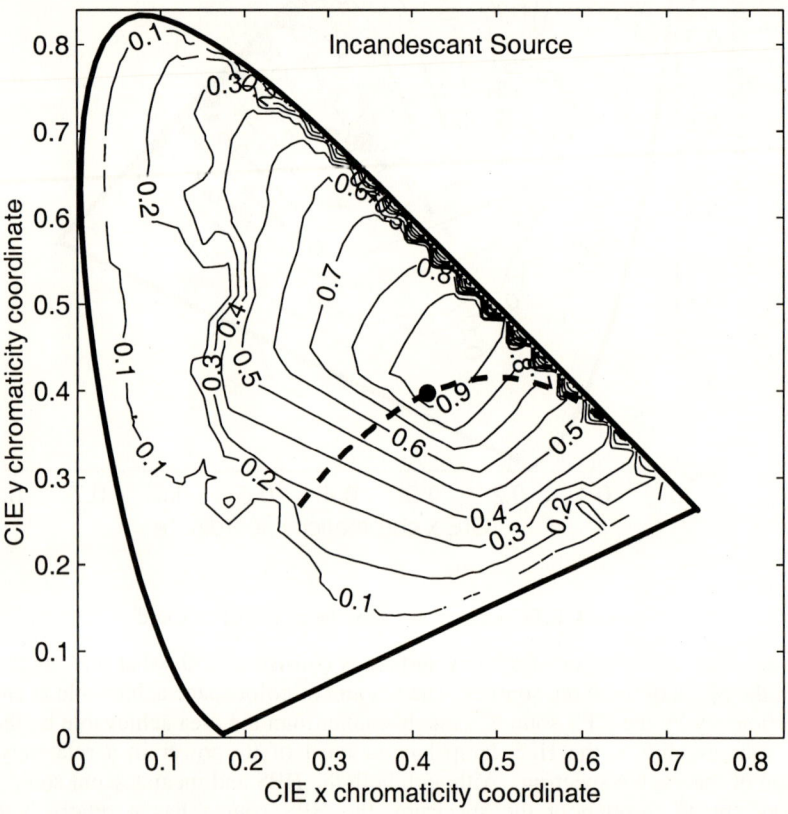

Fig. 5. Efficiency contours for the incandescent source.

A second major application is filtering existing light source technologies for use in new applications. For example, metal halide lamps are used extensively in theatre lighting where colored lighting is desired. Currently, the shift in color is achieved by modifying the composition of metal salts in the lamp, which tends to reduce the lamp life. A filtered standard (i.e., longer life) metal halide lamp might be an attractive alternative. The technique described here gives manufactures the ability to determine where in color-efficiency space filtered lamps might be competitive.

References

1. Wyszecki, G. Stiles, W.: Color Science: Concepts and Methods, Quantitative Data and Formulae. 2nd edn. John Wiley & Sons, New York (1982)
2. Colorimetry (Official Recommendations of the International Commission on Illumination). CIE Publication No. 15 (E-1.3.1), Bureau Central de la CIE, Paris
3. Houck, C., Joines, J., Kay, M.: A Genetic Algorithm for Function Optimization: A MATLAB Implementation. North Carolina State University - Industrial Engineering Technical Report 95-09 (1995)
4. Eklund, N., Embrechts, M.: Multi-Objective Optimization of Spectra Using Genetic Algorithms. J. Illuminating Eng. Soc. Am. (in press) (2001)
5. Siminovitch, M., Gould, C., Page,E.: A High-Efficiency Indirect Lighting System Utilizing the Solar 1000 Sulfur Lamp. Proc. of the Right Light 4 Conf., Nov. 19-21, 1997, Copenhagen, Denmark.

Multi-objective Design Space Exploration of Road Trains with Evolutionary Algorithms

Nando Laumanns[1], Marco Laumanns[2], and Dirk Neunzig[1]

[1] RWTH Aachen, Institut für Kraftfahrwesen (ika)
D-52074 Aachen, Germany
{laumanns,neunzig}@ika.rwth-aachen.de
[2] ETH Zürich, Institut für Technische Informatik und Kommunikationsnetze (TIK)
CH-8092 Zürich, Switzerland
laumanns@tik.ee.ethz.ch

Abstract. This paper examines the road train concept as a new alternative in long-distance freight traffic. The design of such a system is a difficult task since many different and conflicting criteria arise depending on the application spectrum, the legal conditions and the preferences of the carrier. Furthermore the evaluation of each decision alternative relies on a time consuming and sophisticated simulation. Evolutionary algorithms (EAs) have shown to be a useful tool for multi-objective optimization in engineering design. Based on a unified model, we develop a problem-specific evolutionary algorithm which features strong elitism, an unlimited archive of non-dominated solutions and density dependent selection. This EA is able to create alternatives which dominate previous manually engineered solutions as well as those derived from exhaustive search.

1 Introduction

In freight traffic the importance of trucks grows constantly. Based on a forecast by the Prognos-Institute experts expect an increase of transportation performance on German roads by 55 % until the year 2015 [7]. This would cause a proportional rise of the mileage with a higher traffic load unless transportation regulations are changed. One possibility to avoid this effect is to extend the vehicle load in freight transportation.

In Germany the vehicle load is limited to a maximum weight of 40 t with a maximum vehicle length of 18.75 m. This makes it possible to carry for example two C 782 containers with a load capacity of max. 25 t. In order to avoid road damages, the maximum load per axle of today has to be kept constant.

Based on concepts currently used in other countries, the ika (Institut für Kraftfahrwesen, RWTH Aachen) developed a road train concept, consisting of a semi-trailer truck and two semi-trailers, connected by a dolly (see Figure 1).

Fig. 1. Example of a road train with two semi-trailers.

First simulations of the longitudinal dynamics of those road trains showed a considerable decrease of fuel consumption in connection with an increase of traffic flow on German highways, although their power trains were not adapted to the new demands. Furthermore a rise in maximum speed by 25% from 80 km/h to 100 km/h resulted in a significantly lower increase of the fuel consumption of 5%. Therefore two versions of the road train are to be developed [6]. Maneauverability and transversal dynamical behaviour of road trains were investigated in [9].

This paper now focuses on the design space exploration of road trains concerning many different optimization criteria. The next section discusses the issues raised in road train design, states the resulting optimization problem in terms of decision variables and objectives and explains how each decision alternative is evaluated by extensive numerical simulation of different driving conditions. In section 3 we motivate the use of evolutionary algorithms for this design space exploration task and describe which algorithmic modifications were necessary to deal with the large number of objectives. The results are presented in section 4 and compared to previously known solutions. In section 5 we conclude with the implications of our results from the engineering as well as from the algorithmic point of view.

2 Optimization of a Road Train

The optimization of a whole new vehicle concept with respect to fuel consumption and driving dynamics is a very complex subject, because the lack of existing data and knowledge leaves a wide open space for experiments concerning the power train and the overall weight of the road train.

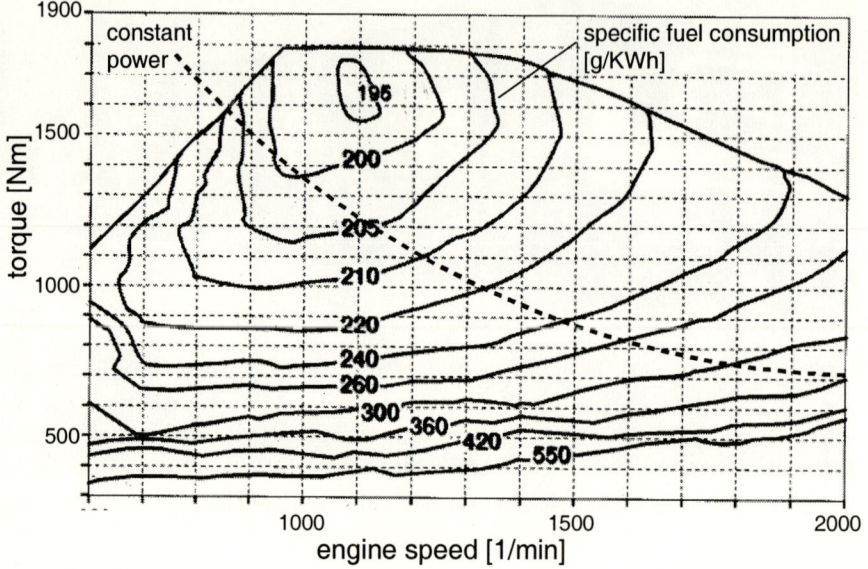

Fig. 2. Engine characteristic graph.

An increase of weight leads to an increase of road and climbing resistance. That changes the engine operating point and therefore its efficiency and fuel consumption.

Every single driving condition defines a point in the engine characteristic graph. The number of revolutions is determined by the velocity of the vehicle and the total gear ratio, consisting of rear-axle ratio and transmission ratio. The necessary torque is a result of necessary power output (influenced by velocity, efficiency of the gear box, acceleration, and road gradient) and revolutions. A lower total gear ratio reduces the engine speed. Under the presumption of constant running resistance due to constant velocity and road gradient the required power remains unchanged. The line of constant power indicates this relationship in Figure 2.

Long-distance transport vehicles usually drive rather statically, operating at maximum authorized speed. This leads to the assumption that the gear ratio should be low enough to cause an engine operating point in the area of lowest specific fuel consumption. However, this area is close to the line of maximum torque. Small increases of the running resistance, resulting from headwind or road gradient, cannot be compensated by requesting more torque from the engine, but force the driver to shift gears or to go at a lower speed.

Since the drivability of the vehicle requires a big distance between the most frequent engine operating point and the line of maximum torque, resulting in powerful engines and high gear ratios, it opposes the attempt to reduce the fuel consumption.

The goal of the optimization is to find a combination of overall weight, gear box, engine and driving strategy minimizing fuel consumption, optimizing the driving performance and increasing driving convenience.

Another difficulty in the design process of a long-distance freight vehicle is the large application spectrum. Some carriers operate only in a rather even area, like the Netherlands for example. It is obvious that they would prefer a road train version different from one a carrier would choose whose standard route crosses the Alps. The latter puts much more emphasis on the climbing capacity than the other.

2.1 Overall Weight

Based on current semi-trailer trucks the overall weight of an average prime mover with a nominal power of more than 310 kW can be set to about 8 tons. An ordinary semi-trailer weighs about 7 tons, leaving a load capacity of 25 tons. Therefore an upper limit for the overall weight of the road train can be set at 72 tons, representing a prime mover with two full-size semi-trailers.

Assuming a constant ratio of load capacity and overall trailer weight of 0.78125, the load capacity can be varied from 25 tons to 50 tons.

2.2 Power Train

The simulated road train has a power train consisting mainly of a combustion engine (diesel), a clutch, a manually shifted transmission and a rear-axle differential. The engine used as basis for the simulation is the Mercedes-Benz OM 442 LA. It was slightly modified to represent an average modern truck engine. It has a nominal power of 314 kW at 2000 rpm and an optimal specific fuel consumption of 193 g/kWh.

Two standard gear-boxes were chosen for the simulation. Both of them provide 16 gears, with the direct gear being 15th and 16th respectively. The efficiency of the gear box in the direct gear is assumed to be 2% higher than in the other gears, because the flow of power avoids the toothed wheels, which cause the loss of efficiency.

2.3 Resulting Design Variables

One part of the optimization process is the choice of a suitable engine. The engine characteristic graph of the modified OM 442 LA mentioned above is the basis for the engines used in the simulations. The creation of more powerful engines is achieved by multiplying the engine torque with $1+x_0$, $x_0 \in [0,1]$, in every point of the engine characteristic graph. The relating efficiencies remain unchanged.

The second design variable x_1, $x_1 \in [0,1]$, defines the overall weight and therefore the load capacity as well.

The ratios of the manual gear-box remain unchanged, the total gear ratio is varied through x_2, $x_2 \in [0,1]$, responsible for the rear-axle ratio.

individuals randomly). In the selection part, the better individuals (in respect to the objective function) are kept for the production of offspring, while the worse alternatives are deleted.

In multi-dimensional objective spaces the partial order makes the selection decision more complicated. Modern multi-objective evolutionary algorithms use selection rules which avoid to aggregate the multiple objectives into a surrogate scalar function and thus retain the true multi-objective nature of the problem [2].

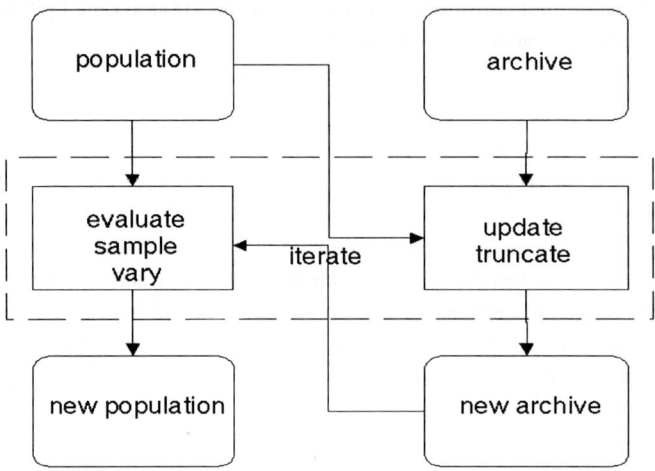

Fig. 4. Flow chart of a universal multi-objective evolutionary algorithm (UMMEA).

3.2 Choice of the Algorithm

Though a great variety of (multi-objective) evolutionary algorithms has emerged so far, many engineering design problems still require to define or to customize an application-specific implementation. Here, the Unified Model for Multi-objective Evolutionary Algorithms (UMMEA [4]) is used. This model allows to systematically combine the different operators that have been proposed and discussed in the literature and to include own problem-specific instances, a schematic view is given in Figure 4.

Like in many other algorithms, an archive is used besides the normal population to store all non-dominated solutions offline during the run. In our case the archive must be very large because the possible size of the trade-off surface increases with the objective space dimension. Practically we even do not have to bound its size at all because the long duration of the simulation (about 30 seconds) already limits the total number of alternatives that can be generated in a reasonable amount of computing time.

The parents for the next generation are selected exclusively from the archive. In order to avoid genetic drift and an oversampling of easily accessible objective space regions, it is necessary to employ density dependent selection: In each iteration the density is estimated for every point represented by the individuals, and the individuals are selected with a probability reciprocal to this density. This leads to a more uniform distribution of alternatives in the approximated trade-off surface.

Each individual represents a vector of design variables. The variation operator for this study only applies mutation, which is carried out by adding a normal distributed random variable

$$x_i^{(t+1)} = x_i^{(t)} + x_i', \quad x_i' \sim N(0, \sigma)$$

to each decision variable x_i, while a constant σ=0.02 was used. Recombination turned out not to be of use here since the interdependence of the design variables in every part of the objective space seems to be very high.

4 Results

In order to analyze the quality of the vehicles developed by the evolutionary algorithm a road train version is designed in a traditional way, based on simple rules for optimizing a power train of a truck [8,11].

In addition, two grid searches over the whole area of possible combinations are performed, each with a total number of 2160 elements. One of them was restricted to a maximum authorized speed of 80 km/h, the other to 100 km/h.

For the design space exploration with the evolutionary algorithm a hierarchical approach was used. The first run of the evolutionary algorithm is performed to narrow the design variable intervals. An analysis of the trade-offs between the different objectives leads to the conclusion that a focus on reducing the fuel consumption would not necessarily worsen the other objective values to an unacceptable amount. Furthermore, this goal is the main factor for the profitability of a vehicle concept and deserves special attention. Therefore we chose the average fuel consumption on the highway (y_7) as the objective value that defines a ranking of the solutions; y_3 and y_4 can be used to represent the second main part of the driving performance, the required climbing ability. In this case the reduction of the maximum velocity must not exceed 5 km/h. The solutions that did not meet this criterion were removed from the ranking. The remaining individuals were ranked according to the fuel consumption on highways. The top solution was considered as the best version.

According to Figures 5 and 6 the modified design variable intervals are defined as follows: $x_0 \in [0.4, 0.6]$; $x_1=1$; $x_2 \in [0.3, 0.4]$; $x_3 \in [0, 0.5]$; $x_4=0$ for the 100 km/h road train and $x_0 \in [0.0, 0.4]$; $x_1=1$; $x_2 \in [0.55, 0.85]$; $x_3 \in [0, 0.5]$; $x_4=0$ for the 80 km/h road train.

Limited to those intervals a second run of the same evolutionary algorithm then fulfilled a more exact approximation of the Pareto set in the region of interest. Of course, there are other ways to cope with the large number of incomparable alternatives in the presence of many objectives. These typically rely on preference information, for instance aggregating (or dropping) objectives, lexicographic ordering or the transformation of objectives into constraints. In many cases, however, it is very diffi- cult to derive an exact numerical representation of the preferences, even if the designer certainly has some fuzzy preferences in mind. Moreover, since we had different decision makers with different preferences in mind, the aim is first to explore the Pareto set as broad as possible with a minimum number of simulations before exploiting interesting regions through restriction of the decision variable space as described above. Finally it should be mentioned that even dropping highly correlated objectives does not help since these correlations are usually not known in advance, can differ much in different regions of the search space, and they do not contribute to the dimensionality of the Pareto set.

Final results show a huge advantage of road trains with respect to fuel consumption. A decrease of 23% (80 km/h-version) respectively 26% (100 km/h-version) is achieved on highways in spite of the rather tough gradients. In steady-state operation fuel consumption advantages of up to 35% are accomplished. With acceleration being at a sensible level the road trains have no disadvantages in climbing ability and required gear shifts.

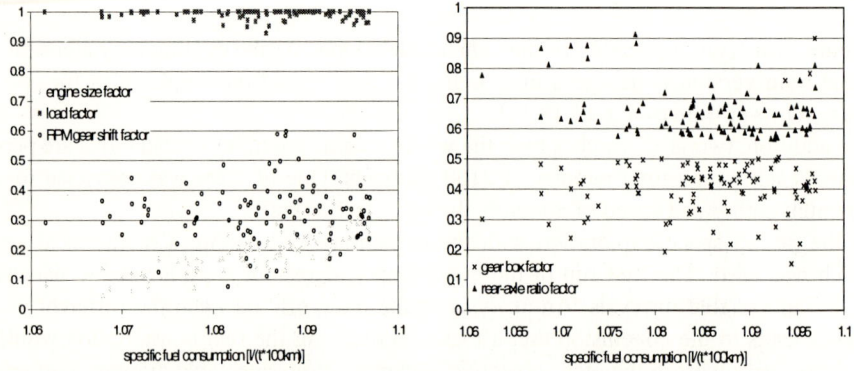

Fig. 5. Design variables and specific fuel consumption for the 100 km/h road train.

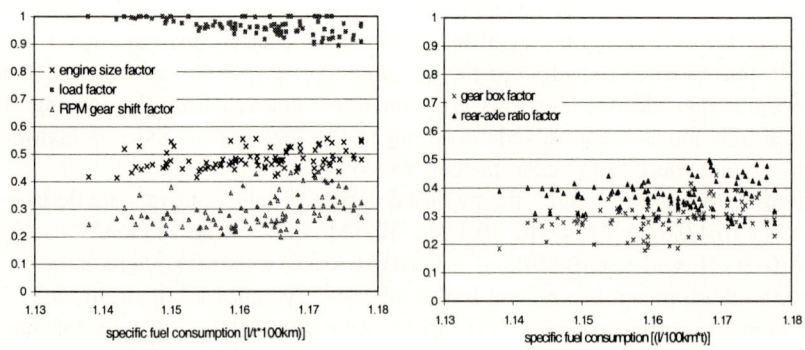

Fig. 6. Design variables and specific fuel consumption for the 80 km/h road train.

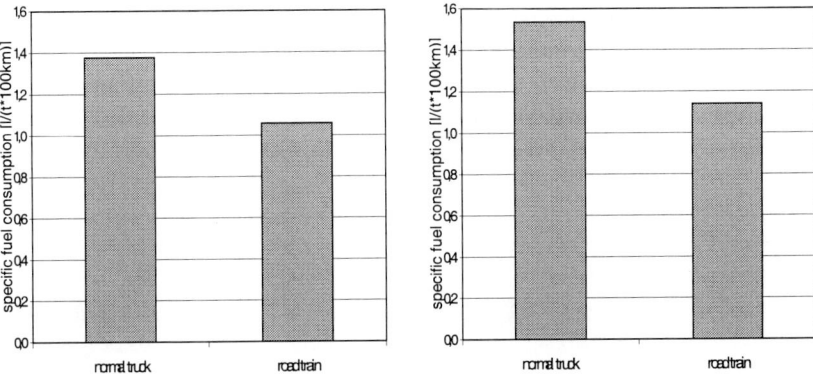

Fig. 7. Comparison between a normal truck and a road train concerning fuel consumption on a highway, maximum authorized speed of 100 km/h (left) and 80 km/h (right).

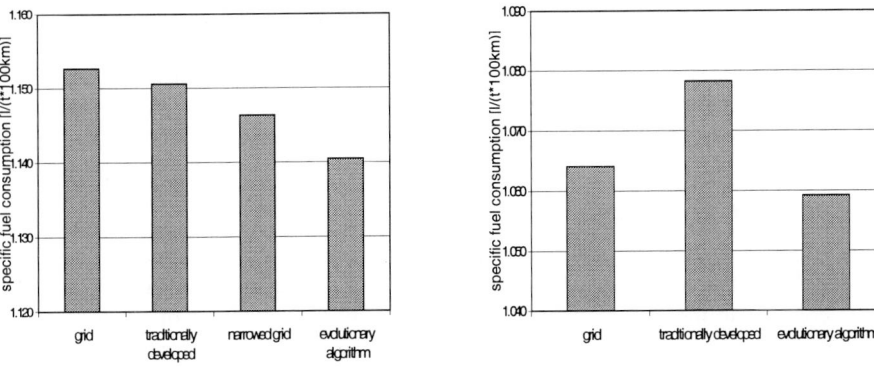

Fig. 8. Specific fuel consumption on a highway for the 100 km/h road train (left) and the 80 km/h road train (right).

The comparison of the different road train versions indicates that the evolutionary algorithm is able to generate solutions which dominate all other results. Showing the same climbing ability and acceleration as the traditionally developed versions and the ones gained by a grid-scan over the whole parameter area, the EA-solution needs about 1% less fuel on the highway. The 100 km/h version is even better than the best version found out by a grid-scan of 1000 elements distributed over the narrowed intervals.

Fig. 9 shows the relation between the objective function y_3 (maximal velocity in 14th gear with 1.5% road gradient) and y_8 (specific fuel consumption on highway). This relation provides information about the trade-off between drivability and fuel economy.

The creation of 1300 individuals already produces a rather large number of solutions, which have to be considered better than any solution found out without the evolutionary algorithm. This advantage in efficiency will become even more important when more sophisticated driving scenarios – and thus more time consuming simulations – will be used, which is subject to further research.

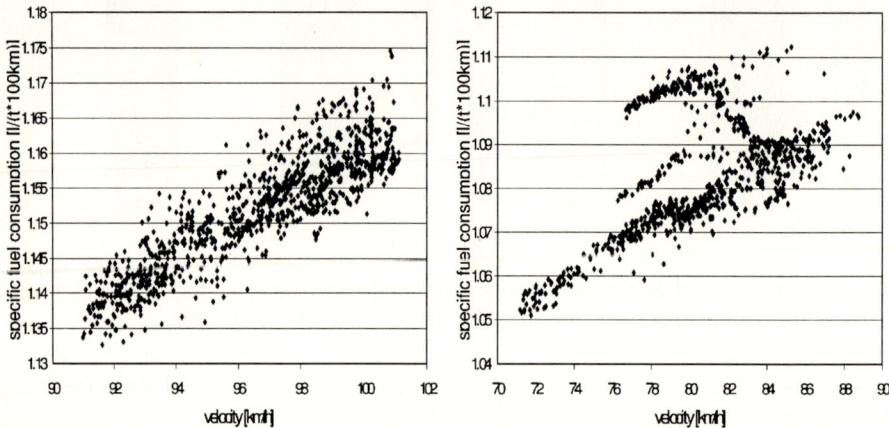

Fig. 9. velocity (1,5% road gradient) and specific fuel consumption on a highway for the 100 km/h road train (left) and the 80 km/h road train (right).

5 Conclusion and Outlook

The results of the design space exploration showed a number of interesting new aspects concerning the optimization of the road train concept. First of all road trains have huge advantages compared to standard trucks concerning fuel consumption per load. This efficiency is certainly the most dominating aspect when developing vehicles for long-distance transport. Furthermore the use of rather powerful engines considerably increases the climbing ability of the road trains without worsening the fuel consumption massively.

The approximation of the Pareto set enables engineers and carriers to choose the right configuration for a special application. A hand-made customization for every application spectrum would require a huge amount of time and work. Thus the design space exploration performed in this paper is a powerful tool in vehicle development. Other possible future fields of operation are the analysis and optimization of vehicles equipped with driver assistance systems or collision avoidance, which need extensive human-machine-interaction.

From the algorithmic point of view the optimization problem turned out to be a challenging task for the evolutionary algorithm because of its high-dimensional objective space. Since on average more than 30 % of the generated solutions were non-dominated, a huge archive was needed to reflect the whole range of possible efficient solutions. In this situation it is difficult to maintain an appropriate selection pressure towards the real trade-off surface. Therefore we chose a strong elitist approach, i.e. all parents were drawn from the archive. This was not harmful in respect to premature convergence as the archive always exhibits enough diversity. On the contrary, the solutions have even shown to be too diverse to make recombination of distant individuals viable. Finally, a density based selection was necessary to reach a good distribution of alternatives to all regions of the potential trade-off surface, where the parents were sampled with a probability reciprocal to their estimated density instead of just applying a rank-based selection scheme based on the densities. Future enhancements of the algorithm could include a scheme to combine both density and preference information in the selection process.

References

1. Bäck, T., Hammel, U., and Schwefel, H.-P.: Evolutionary computation: Comments on the history and current state. IEEE Transactions on Evolutionary Computation 1(1), pp. 3-17, 1997.
2. Fonseca, C. M. and Fleming, P. J.: An overview of evolutionary algorithms in multiobjective optimization, Evolutionary Computation 3(1),1-16.
3. Horn, J.: Multicriteria decision making. In T. Bäck, D. B. Fogel, and Z. Michalewicz (eds.), Handbook of Evolutionary Computation. Institute of Physics Publishing, Bistol, 1997.
4. Laumanns, M., Zitzler, E., and Thiele, L.: A Unified Model for Multi-objective Evolutionary Algorithms with Elitism. In A. Zalzala et al., eds., Congress on Evolutionary Computation (CEC 2000), volume 1, pages 46-53. IEEE Press, Piscataway NJ, 2000.
5. Ludmann, J.: Beeinflussung des Verkehrsablaufs auf Straßen: Analyse mit dem fahrzeugorientierten Verkehrssimulationsprogramm PELOPS, Schriftenreihe Automobiltechnik, Forschungsgesellschaft Kraftfahrwesen mbH Aachen, 1998
6. Ludmann, J., Neunzig, D., Weilkes, M., Wallentowitz, H.: The effectivity of new traffic-technologies and transportation-systems in suburban areas and on motorways. International Transactions in Operational Research, Res. 6, Pergamon, 1999
7. Prognos AG: Prognos European Transport Report 2000 - 22 Western and Eastern European Countries 1998-2000-2010. Basel, August 2000.
8. Rubi, V.: Industrielle Nutzfahrzeugentwicklung, Schriftenreihe Automobiltechnik, Forschungsgesellschaft Kraftfahrwesen mbH Aachen, 1993
9. Sandkühler, D.: Simulationstechnische Untersuchung des Fahrverhaltens von mehrgliedrigen Road-Train Sattelzügen, Studienarbeit am Institut für Kraftfahrwesen, Aachen, 1998
10. Sen, P. and Yang, J.-B.: Multiple criteria decision support in engineering design. London, 1998.
11. Wallentowitz, H.: Longitudinal Dynamics of Motor Vehicles, Forschungsgesellschaft Kraftfahrwesen mbH Aachen, 2000

Multiobjective Optimization of Mixed Variable Design Problems

Johan Andersson and Petter Krus

Department of Mechanical Engineering
Linköping University, SE-581 83 Linköping, Sweden
{johan, petkr}@ikp.liu.se

Abstract. In this paper, a new multiobjective genetic algorithm is employed to support the design of a hydraulic actuation system. First, the proposed method is tested using benchmarks problems gathered from the literature. The method performs well and it is capable of identifying multiple Pareto frontiers in multi-modal function spaces. Secondly, the method is applied to a mixed variable design problem where a hydraulic actuation system is analyzed using simulation models. The design problem constitutes of a mixture of determining continuous variables as well as selecting components from catalogs. The multi-objective optimization results in a discrete Pareto front, which illustrate the trade-off between system cost and system performance.

1 Introduction

Most engineering design problems consist of several often conflicting objectives. In many cases, the multiple objectives are aggregated into one single overall objective function. Optimization is then conducted with one optimal design as the result. The result is then strongly dependent on how the objectives are aggregated. To avoid this difficulty and in order to explore a broader set of optimal solutions the concept of Pareto optimality is employed. Valuable insight about the trade-off between the objectives could be gained by investigating the set Pareto optimal solutions. Vilfredo Pareto defined Pareto optimality as the set where every element is a problem solution for which no other solutions can be better in all design attributes. A solution in a Pareto optimal set cannot be deemed superior to the others in the set without including preference information to rank competing attributes.

This paper develops a Pareto optimization method for use in multiobjective, multi-modal design spaces. For a general design problem, the design space consists of continuous variables as well as selection of individual components from catalogs or databases. Furthermore, numerical simulations and other CAE tools are often employed to evaluate design solutions; i.e. simulation is employed to transform solutions from the design space to the attribute space. As the attributes or objectives are calculated using numerical simulations, there is no simple way of obtaining derivatives of the objective functions. Therefore genetic algorithms are well suited for such applications—they do not need derivatives of the objective functions and they have been shown to be effective in optimizing mixed variable problems [11] in multi-modal search spaces [7].

The paper first defines a general multiobjective optimization problem and reviews related work on multiobjective genetic algorithms. Then, a new method is proposed and validated using a problem gathered from the literature. Later the method is applied to a real design problem containing a mixture of continuous design variables and discrete selections of components from catalogs. The problem is solved by connecting the optimization strategy to a simulation program.

1.1 The Multiobjective Design Problem

A general multiobjective design problem could be expressed by equations (1) and (2).

$$\min \mathbf{F}(\mathbf{x}) = (f_1(\mathbf{x}), f_2(\mathbf{x}), ..., f_k(\mathbf{x}))^T \quad (1)$$

$$s.t. \ \mathbf{x} \in S$$

$$\mathbf{x} = (x_1, x_2, ..., x_n)^T \quad (2)$$

where $f_1(x), f_2(x), ..., f_k(x)$ are the k objectives functions, $(x_1, x_2, ..., x_n)$ are the n optimization parameters, and $S \in R^n$ is the solution or parameter space. Obtainable objective vectors, $\{\mathbf{F}(\mathbf{x}) | x \in S\}$ are denoted by Y. $Y \in R^k$ is usually referred to as the attribute space.

The Pareto set consists of solutions that are not dominated by any other solutions. Considering a minimization problem and two solution vectors $\mathbf{x}, \mathbf{y} \in S$. $\mathbf{x}$ is said to dominate $\mathbf{y}$, denoted $\mathbf{x} \succ \mathbf{y}$, if:

$$\forall i \in \{1,2,...,k\}: f_i(\mathbf{x}) \le f_i(\mathbf{y}) \ \text{and} \ \exists j \in \{1,2,...,k\}: f_j(\mathbf{x}) < f_j(\mathbf{y}) \quad (3)$$

The space in R^k formed by the objective vectors of Pareto optimal solutions is known as the Pareto optimal front.

1.2 Multiobjective Genetic Algorithms

Genetic algorithms are modeled after mechanisms of natural selection. Each optimization parameter (x_n) is encoded by a gene using an appropriate representation, such as a real number or a string of bits. The corresponding genes for all parameters $x_1...x_n$ form a chromosome capable of describing an individual design solution. A set of chromosomes representing several individual design solutions comprise a population where the most fit are selected to reproduce. Mating is performed using crossover to combine genes from different parents to produce children. The children are inserted into the population and the procedure starts over again, thus creating an artificial Darwinian environment. For a general introduction to genetic algorithms, see [6].

Additionally, there are many different types of multiobjective genetic algorithms. Literature surveys and comparative studies on multiobjective genetic algorithms could be found in for example [3, 10 and 12].

Most multiobjective genetic algorithms use either the selection mechanism or some sort of Pareto based ranking to produce non-dominated solutions. In the proposed method, the ranking scheme presented by Fonseca and Fleming [5] is employed.

In the multiobjective GA (MOGA) [5] each individual is ranked according to their degree of dominance. The more population members that dominate an individual, the higher ranking the individual is given. Here an individual's ranking equals the number of individuals that it is dominated by plus one, see Figure 1. Individuals on the current Pareto front will have a rank of 1, as they are non-dominated. The rankings are then scaled to score individuals in the population. In MOGA both sharing and mating restrictions are employed in order to maintain population diversity. Fonseca and Fleming also introduce preference information and goal levels to reduce the Pareto set to those that simultaneously meet certain attribute values.

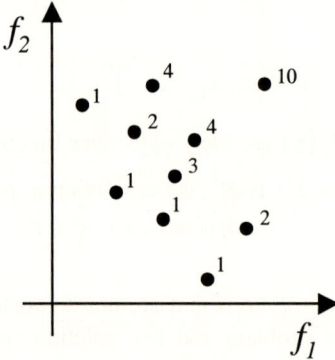

Fig. 1. Population ranking according to Fonseca and Fleming.

Although there is a substantial body of research on multiobjective genetic algorithms, there are still important issues that current methods address with only partial success. The methods typically require extensive genetic algorithm parameter tuning on a problem-by-problem basis in order for the algorithm to perform well. However, in a real-world problem there is little knowledge about the shape of the attribute space, which makes it difficult to assess problem specific parameters. Additionally, existing methods do not handle consistently the location of multiple Pareto frontiers in multi-modal problem spaces. The method presented in this paper is capable of identifying multiple frontiers without any problem specific parameter tuning.

2 The Proposed Method

The multiobjective struggle genetic algorithm (MOSGA) [1] combines the struggle crowding genetic algorithm [7] with Pareto-based ranking as devised in [5].

In the struggle algorithm, a variation of restricted tournament selection [8], two parents are chosen randomly from the population, and crossover/mutation are performed to create a child. The child then has to compete with the most similar individual in the entire population, and replaces it if the child has a better fitness. This replacement strategy counteracts genetic drift that can spoil population diversity. The struggle genetic algorithm has been demonstrated to perform well in multi-modal function landscapes where it successfully identifies and maintains multiple peaks.

There is no single objective function to determine the fitness of the different individuals in a Pareto optimization. Therefore, the ranking scheme presented by Fonseca and Fleming is employed, and the "degree of dominance" is used to rank the population. Each individual is given a rank based on the number of individuals in the population that are preferred to it, i.e. for each individual the algorithm loops through the population counting the number of preferred individuals. "Preferred to" could be implemented in a strict Pareto optimal sense or extended to include goal levels on the objectives in order to limit the frontier.

The principle of the MOSGA algorithm is outlined below.

Step 1: Initialize the population.

Step 2: Select parents randomly from the population.

Step 3: Perform crossover and mutation to create a child.

Step 4: Calculate the rank of the child, and a new ranking of the population that considers the presence of the child.

Step 5: Find the most similar individual, and replace it with the new child if the child's ranking is better.

Step 6: Update the ranking of the population if the child has been inserted.

Step 7: Perform steps 2-6 until the mating pool is filled.

Step 8: If the stop criterion is not met go to step 2 and start a new generation.

The similarity between of two individuals is measured using a distance function. The method has been tested with distance functions based upon the Euclidean distance in both the attribute as well as the parameter space. A mixed distance function combining both the attribute and the parameter distance has been evaluated as well.

2.1 Genome Representation

The genome encodes design variables in a form suitable for the GA to operate upon. Design variables may be values of parameters (real or integer) or represent individual components selected from catalogs or databases. Thus, the genome is a hybrid list of real numbers (for continuous parameters), integers and references to catalog selections, see Figure 2.

A catalog could be either a straight list of elements, or the elements could be arranged in a hierarchy. Each element of a catalog represents an individual component. The characteristics of catalogs would be discussed further on and exemplified by the design example.

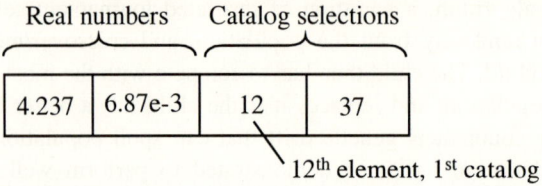

Fig. 2. Example of the genome encoding. The first two elements represent real variables and the last two elements catalog selections.

2.2 Similarity Measures

Speciating GAs require a measure of likeness between individuals, a so called similarity measure. Here the similarity measure is based on a distance function calculating the distance between two genomes. The similarity could be based on the distance in either the attribute space (between the objectives), the phenotype space (between the design parameters) or the genotype space (in the genome encoding). As direct encoding is used (not a conversion to a string of bits), a phenotype and a genotype distance function would yield the same result. It is shown that the choice between an attribute based and a parameter based distance function might have a great influence on the outcome of the optimization.

Attribute Based Distance Function
One way of comparing two individual designs is to calculate their distance in attribute space. As we want the population to spread evenly on the Pareto front (in attribute space) it seems to be a good idea to use an attribute based distance measure. The distance between two solutions (genomes) in attribute space is calculated using the normalized Euclidean distance (4).

$$\text{Distance}(a,b) = \sqrt{\sum_{i=1}^{k} \left(\frac{f_{ia} - f_{ib}}{f_{i\max} - f_{i\min}} \right)^2 \frac{1}{k}} \quad (4)$$

Where f_{ia} and f_{ib} are the objective values for the i:th objective for a and b respectively. $f_{i\max}$ and $f_{i\min}$ is the maximum and the minimum of the i:th objective in the current population, and k is the number of objectives. Thus, the distance function will vary between 0, indicating that the individuals are identical, and 1 for the very extremes.

Phenotype Based Distance Function
Another way of calculating the distance between solutions is to use the distance in parameter (phenotype) space. As the genome is a hybrid mixture of real numbers and catalog selections, we have to define different distance functions to work on different type of elements. The methods described here build on the framework presented by

Senin et al. [11]. In order to obtain the similarity between two individuals the distance between each search variable is calculated. The overall similarity is then obtained by summing up the distances for each search variable.

Real Number Distance

A natural distance measure between two real numbers is the normalized Euclidean distance, see equation (5).

$$\text{Distance}(a,b) = \sqrt{\left(\frac{a-b}{\max \text{ distance}}\right)^2} \quad (5)$$

Where a and b are the values for the two real numbers and max distance is the maximum possible distance between the two values (i.e. the search boundaries).

Catalog Selection Distance

Distance between two catalog selections could be measured through relative position in a catalog or a catalog hierarchy. The relative position is only meaningful if the catalog is order, see Figure 3.

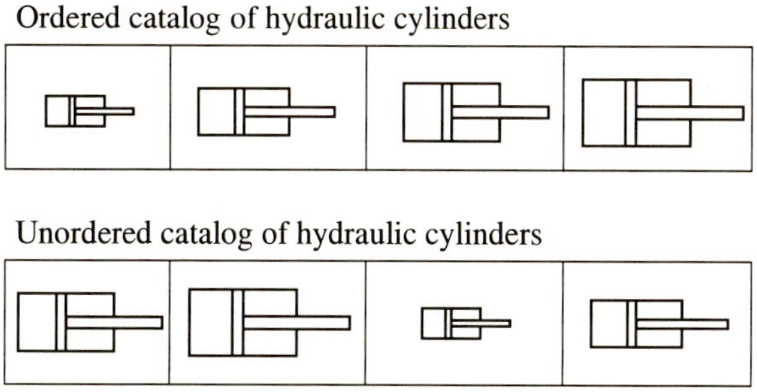

Ordered catalog of hydraulic cylinders

Unordered catalog of hydraulic cylinders

Fig. 3. Examples of ordered and unordered catalogs.

The dimensionless distance between two elements within the same catalog is expressed by equation (6) and exemplified in Figure 4.

$$\text{Distance}(a,b) = \frac{pos(a) - pos(b)}{\max \text{ distance}} \quad (6)$$

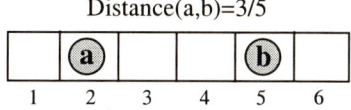

Fig. 4. Distance evaluation for two elements of an ordered catalog.

For catalog hierarchies equation (6) has to be generalized. For elements belonging to the same sub-catalog, the distance is evaluated using the relative position within that sub-catalog. Otherwise, the maximum length of the path connecting the different sub-catalog is used. This implies that for two given sub-catalogs an element in one catalog

If the function g is multi-modal, the corresponding multiobjective problem will have global and local Pareto-optimal frontiers. A multi-modal g function is defined in equation (11).

$$g(x_2) = 2 - \exp\left\{-\left(\frac{x_2 - 0.2}{0.004}\right)^2\right\} - 0.8\exp\left\{-\left(\frac{x_2 - 0.6}{0.4}\right)^2\right\} \qquad (11)$$

Figure 8(a) shows the g function for $0 \leq x_2 \leq 1$ with the global optima located at $x_2=0.2$ and a local optima at $x_2=0.6$. Figure 8(b) shows a plot of f_1 and f_2 in the attribute space with the global and local Pareto optimal solutions. 10,000 randomly chosen solutions are generated and plotted in Figure 8(b) to illustrate that the problem is biased—the solution density is higher towards the local Pareto-optimal front.

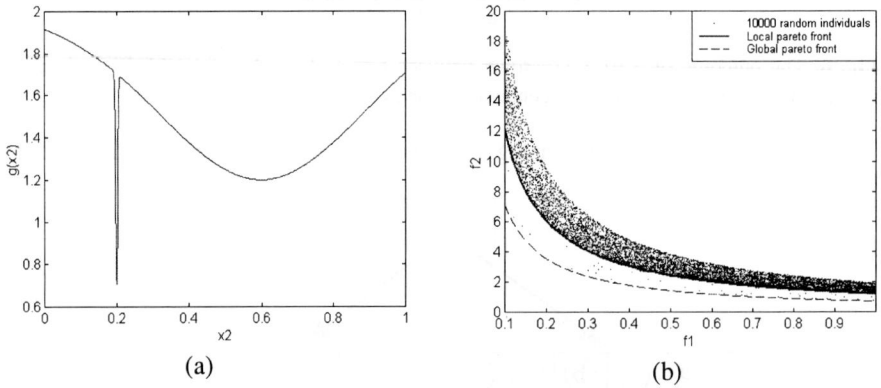

Fig. 8. Figure (a) shows the multi-modal function $g(x_2)$, where the global optima is situated at $x_2=0.2$ and the local optima at $x_2=0.6$. For the multiobjective problem, a f_1-f_2 plot for 10000 random solutions is shown in (b). Notice the low solution density at the global Pareto optimal front.

The optimization was conducted with a population size of 60 individuals and ran for 200 generations. The variables are real encoded, and BLX crossover is employed to produce offspring. Deb reported that the NSGA was trapped in the local Pareto front in 59 out of 100 runs.

The original MOSGA algorithm used an attribute based distance function resulting in the algorithm converging to the local Pareto frontier in only 7% of 100 optimizations. The algorithm found the preferred global Pareto optimal front in 86% of the optimizations, as shown in figure 9 (a) and (b). In 7% of the optimizations, it converged to both frontiers. Thus, the MOSGA seems more robust in locating the global Pareto optimal frontier.

However, one whishes that the algorithm should be capable of identifying both frontiers in every optimization run. By changing to a parameter based distance function this can be achieved. However, the parameter based distance function was slower and less exact in its convergence to the frontier.

In the MOSGA, the new child has to compete with the individual most similar to itself. When the comparison is done in parameter space, a portion of the population will find and maintain local optima, where solutions close in the parameter space are all dominated. When using an attribute based distance function, solutions at local optima might have to compete with solutions at the global optima, as they might be close in attribute space. Therefore, local optima would not be maintained.

By combining equally weighted attribute-based and a parameter-based distance functions to form a mixed distance measure, the advantages of fast convergence and the ability of finding multiple frontiers were realized. Figure 9 shows how the algorithm spreads the population evenly on both frontiers when using the mixed distance function. To summarize, the attribute distance function performs well on problems with one Pareto frontier. For problems with multiple frontiers, a mixed distance function is preferred. A more detailed discussion about the properties of the algorithm is given in [1 and 2].

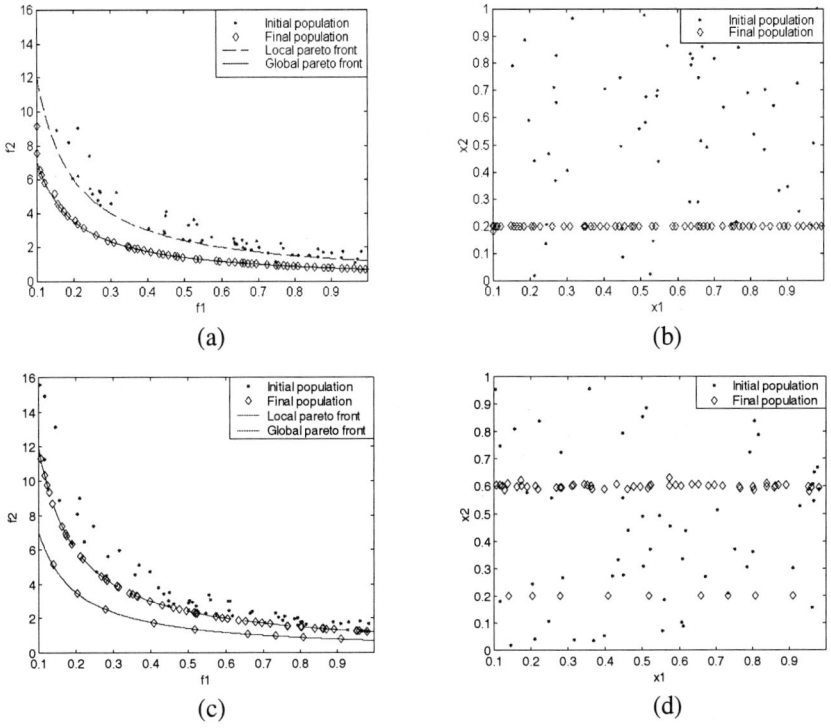

Fig. 9. Optimization results using different distance functions. In (a) and (b) an attribute based distance function is used and the population has converged to the global Pareto front. In (c) and (d) the mixed distance function is used and the population converges to both the global and the local frontier. (a) and (c) show the result in attribute space, whereas (b) and (d) show the result in parameter space.

Thus, the method is capable of reliably identifying multiple Pareto frontiers in a single optimization run, outperforming other techniques. Another advantage is that the method does not require problem specific parameter settings. The only GA parameters that have to be determined are population size, number of generations and the distance function. The method has been successfully tested on several benchmark problems proposed by Deb, see [1].

3 Design Example

The object of study for the design example is a hydraulic actuation system. The system consists of a hydraulic cylinder that is connected to a mass. The motion of the mass is controlled by a directional valve, which in turn is controlled by a proportional controller. The system is powered from a constant pressure hydraulic supply system.

In order to investigate the properties of different designs, the system has been modeled in the simulation package Hopsan [9]. For every new genome, the optimization strategy calls the simulation program to evaluate that particular design. Each component in the simulation model consists of a set of algebraic and differential equations taking aspects such as friction, leakage and non-linearities into account. A graphical representation of the system model is depicted in figure 10.

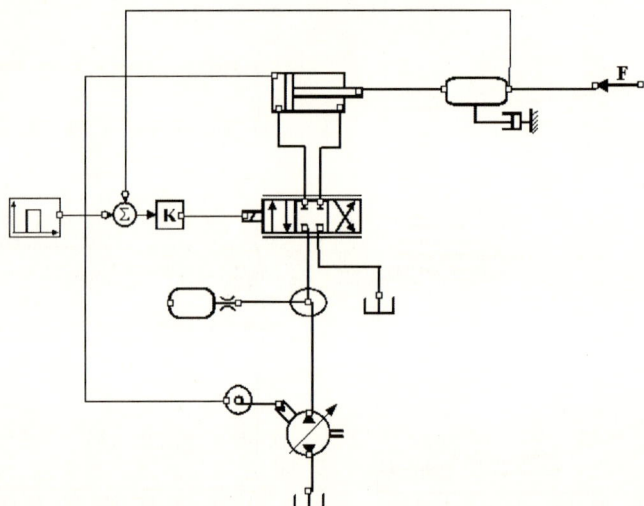

Fig. 10. The simulation model of the hydraulic actuation system. The main components are; (from the upper left) cylinder, mass, pulse generator, p-controller, directional valve, accumulator and constant pressure pump.

The objective of the study is to design a system that has good controllability to a low cost. Naturally, these two objectives are in conflict with each other. To achieve good controllability we can choose a fast servo valve, which is more expensive, then a slower proportional valve. Therefor, there is a trade-off between cost and controllability. The cost for a particular design is composed of the cost for the individual components as well as the cost induced by the energy consumption.

The system has been studied for a pulse in the position command. The control error and the energy consumption are calculated based on the simulation result.

When designing the system cylinders and valves are selected from a catalog of existing components. Other parameters such as the control parameter, a leakage coefficient and the maximal flow of the supply system have to be determined as well. Thus the problem is multiobjective with two objectives and five optimization variables of which two are discrete catalog selections ant three are continuous variables.

3.1 Component Catalogs

For the catalog selections, catalogs of valves and cylinders have been created. For the directional valve, the choice is between a slow but cheap proportional valve or an expensive and fast servo valve. Valves from different suppliers have been arranged in two ordered sub-catalogs as depicted in figure 11. The same structure applies to the cylinders as they are divided into sub-catalogs based on their maximal pressure level. The pressure in the system has to be controlled so that the maximum pressure for the cylinder is not exceeded. A low-pressure system is cheaper but has inferior performance compared to a high-pressure system.

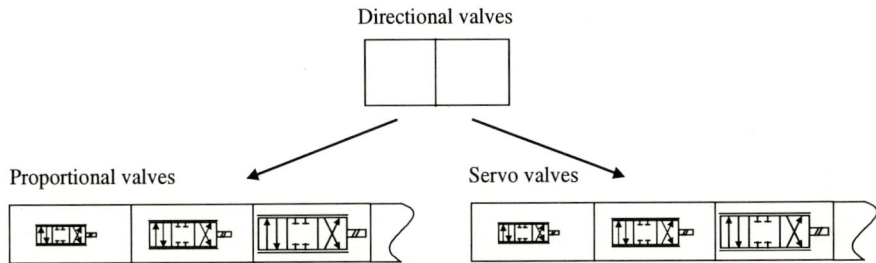

Fig. 11. The catalog of directional valves is divided into proportional valves and servo valves. Each sub-catalog is ordered based on the valve size. For each component, a set of parameters describing the component is store together with information on price and weight.

Naturally, the component catalog is connected to the simulation program. The optimization strategy however needs information about the topology of the catalog in order for the genetic operators to work.

3.2 Optimization Results

The system has been optimized using a population of 40 individuals and 400 generations. In order to limit the Pareto frontier a goal level on the control error was introduced. The goal level corresponds to the highest acceptable control error. Without such a goal level, the result would include very cheap designs that do not follow the position command at all. The introduction of goal levels therefore focuses the population on the most interesting parts of the Pareto frontier.

The result could be divided into four distinct regions depending on valve type and pressure level, see figure 12.

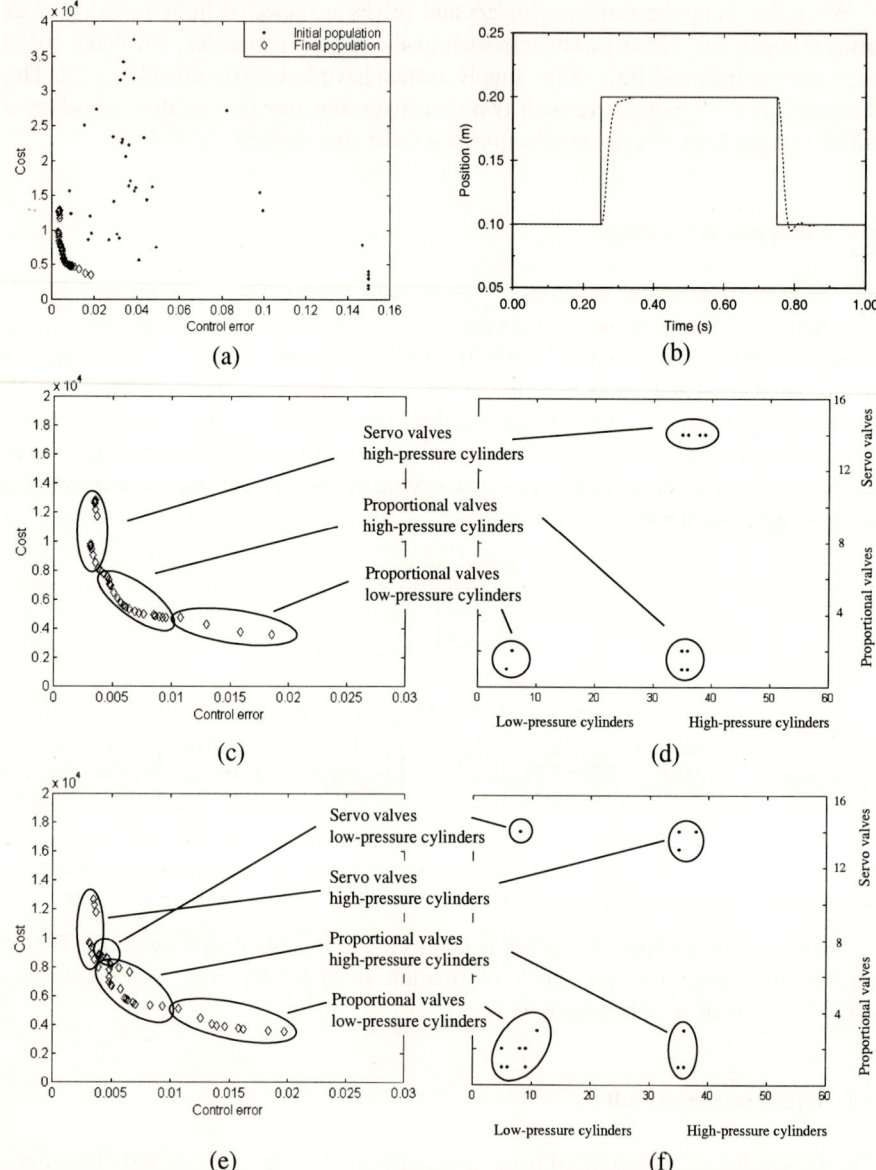

Fig. 12. Optimization results. In (a) the initial and final population of the optimization is shown. In (b) the simulated pulse response for a reasonably fast solution is depicted. Figure (c) shows an enlargement of the Pareto front where different regions have been identified based on valve and cylinder selections, as shown in (d). The graphs (c) and (d) are obtained using an attribute based distance function, whereas (e) and (f) are the corresponding graphs obtained using the mixed distance function.

As can be seen from figure 12 there is a trade-off between system performance (control error) and system cost. By accepting a higher cost, better performance could be achieved. The cheapest designs consist of small proportional valves and low-pressure cylinders. By choosing larger proportional valves and high-pressure cylinders, the performance could be increased on the expense of higher cost. If a still better performance is desired, a servo valve has to be chosen, which is more expensive but has better dynamics.

The continuous parameters, such as the control parameter, tend to smoothen out the Pareto front. For a given valve and cylinder, different settings on the continuous parameters affect the pulse response. A faster response results in a lower control but also a higher energy consumption and thereby higher cost. Therefore, there is a local trade-off between cost and performance for each catalog selection.

4 Discussion

In the proposed method, new solutions have to compete with the most similar individual before they are inserted into the population. Therefore, the similarity measure has a great influence on the optimization result. When using the attribute based distance function as a similarity measure, the true Pareto optimal front is identified, as shown in figure 12 (c) and (d). When using the mixed distance function some dominated solutions survive, for example servo valves with low-pressure cylinders, see 12 (e) and (f). These solutions represent local optima, as they dominate the solutions that are close in parameter space.

The obtained results are in accordance with the results from the mathematical test functions. An attribute based distance function gives fast convergence to the Pareto optimal front, whereas a mixed distance function is a little slower in convergence but is capable of finding and maintaining multiple Pareto frontiers, see figure 9.

For an engineering problem, the optimization formulation is often a simplification of the real world problem, which in part requires human or inquantifiable judgment. When deciding upon the final design there are usually more criteria to consider then just the optimization objectives. Therefore, knowledge of the existents of local Pareto optimal solutions is very valuable. For example, aspects such as robustness, product portfolio, maintenance and quality might be important but hard to include in the optimization. A local Pareto optimal solution might therefore be preferred to a solution at the global Pareto optimal front. Hence, a method that identifies and maintains local Pareto optimal solutions is valuable from an engineering perspective.

5 Conclusions

In this paper, a new multiobjective genetic algorithm has been presented and applied to solve a mathematical test problem as well as a mixed variable design problem. The method is capable of finding and maintaining multiple Pareto optimal fronts with a minimum of problem specific parameter settings. For the design problem, a hydraulic actuation system has been studied with the help of a simulation program. The optimi-

Adaptive Range Multiobjective Algorithms (ARMOGAs) developed from ARGAs for multiobjective optimization are applied to the aerodynamic multiobjective design optimization.

Aerodynamic design for supersonic transport (SST) is considered in this study. A next-generation SST is required to improve the supersonic cruising performance and to prevent the sonic boom. However, there is a severe tradeoff between lowering the drag and boom. Therefore, the next-generation SST may cruise at a supersonic speed only over the sea. This means that it is important to improve not only supersonic performance but also transonic performance, and thus the multipoint aerodynamic optimization is needed. In addition to the reduction of both aerodynamic drags, structural constraints should be considered to keep the wings from having impractically large aspect ratios.

Three-objective optimization for supersonic wings, which minimized transonic and supersonic drag coefficients and the bending moment at the wing root, were reported in [6-7]. In order to consider the viscous effect, a Navier-Stokes solver was used to evaluate the aerodynamic performances at both transonic and supersonic conditions [7]. Successful results were obtained by the multiobjective optimization. There were Pareto solutions that outperformed the NAL's second design in all three objectives, and those wings were similar to the "arrow wing" planform. Although the arrow wing is known to be good for supersonic aerodynamics, it is known to have aeroelastic and control problems due to a large sweep angle. The primary concern is a pitching (twist) moment of the wing. The design results also showed that the second derivative of the wing thickness distribution was discontinuous. This lead to another concern of the designer for the possible boundary layer separation at the maximum thickness location. Therefore the minimization of the pitching moment is added as the present fourth objective with an improved wing thickness parameterization.

National Aerospace Laboratory (NAL) in Japan is working on the scaled experimental supersonic airplane project (NEXST-I) [8]. A scaled experimental supersonic airplane without a propulsion system will be launched with a rocket in 2002. The airplane will be separated from the rocket after launch and will glide down to sample the flight data in the supersonic region. The flight data will be compared with the CFD results to validate the reliability and accuracy of CFD predictions. NAL designed several configurations for the experimental aircraft. The present Pareto solutions obtained are compared with the NAL's design. In order to verify the present optimization method, the present Pareto solutions are also compared with the Pareto solutions obtained before.

2 Adaptive Range Multiobjective Genetic Algorithms

To reduce the large computational burden, the reduction of the total number of evaluations is needed. On the other hand, a large string length is necessary for real parameter problems. ARGAs, which originally proposed by Arakawa and Hagiwara, are a quite unique approach to solve such problems efficiently [9-10]. Oyama developed real-coded ARGAs and applied them to the transonic wing optimization [4].

2.1 Real-Coded Adaptive Range Genetic Algorithms

The main difference between ARGAs and conventional GAs is the introduction of the range adaptation. The flowchart of ARGAs is shown in Fig. 1. Population is reinitialized every M generations for the range adaptation so that the population advances toward promising regions. Another difference is the elimination of the range limits because design variables are encoded into the normal distribution.

In the real-coded ARGAs, the real value of i-th design variable P_i is encoded to a real number r_i defined in (0,1) so that r_i is equal to the integrations of the normal distribution form $-\infty$ to Pn_i.

$$r_i = \int_{-\infty}^{pn_i} N(0,1)(z)dz \tag{1}$$

$$P_i = \sigma_i \cdot pn_i + \mu_i \tag{2}$$

where the average μ_i and the standard deviation σ_i of i-th design variable are calculated by sampling the top half of the previous population to promote the population toward search regions of high fitness. A schematic view of this coding is illustrated in Fig. 2.

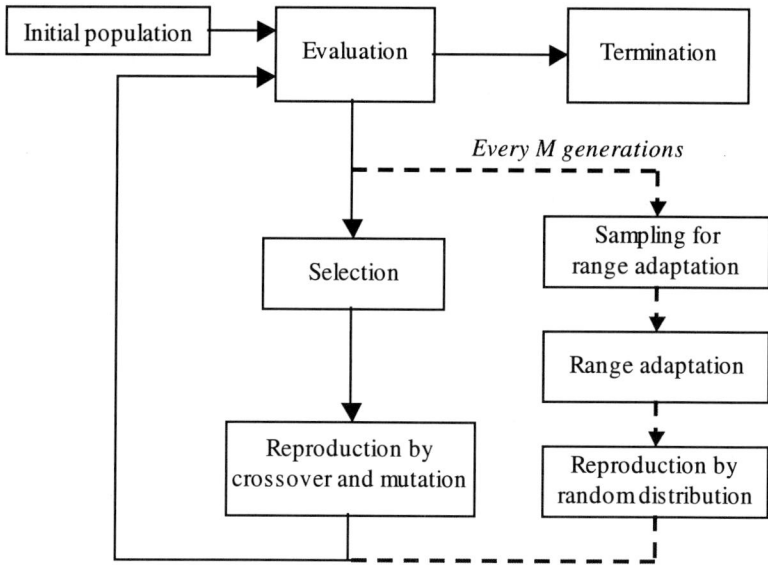

Fig. 1. Flowchart of ARGAs

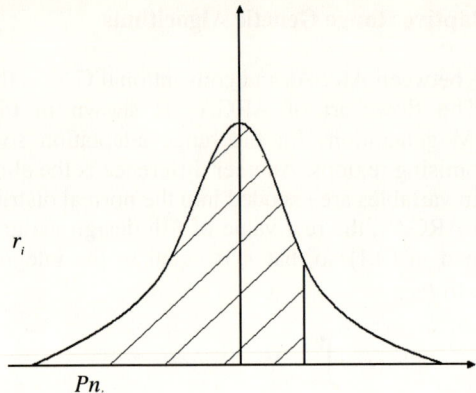

Fig. 2. Normal distribution is used for encoding in ARGAs

2.2 Extension of ARGAs to the Multiobjective Problem

In this study, ARGAs have to deal with multiple Pareto solutions for the multiobjective optimization. The basis of ARMOGAs is the same as ARGAs, but a straightforward extension may cause a problem in the diversity of the population. To better preserve the diversity of solution candidates, the normal distribution for encoding is changed as shown in Fig. 3. The searching region is partitioned into three parts (i, ii, iii). The region i and iii make use of the same encoding method as ARGAs. In contrast, the region ii adopts the conventional real-number encoding method.

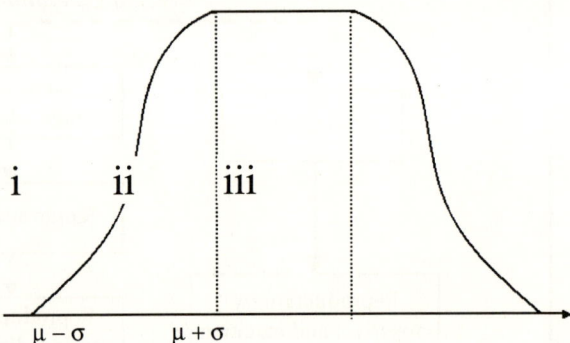

Fig. 3. Normal distribution used in ARGAs' encoding is extended to maintain the diversity of the population for ARMOGAs

3 Problem Definitions

3.1 Objective Functions

The objective functions used in this study can be stated as follows:

1. Drag coefficient at transonic cruise, $C_{D,t}$
2. Drag coefficient at supersonic cruise, $C_{D,s}$
3. Bending moment at the wing root at supersonic cruise condition, M_B
4. Twisting moment at supersonic cruise condition, M_T

In the present optimization, all four objective functions are to be minimized. Both the transonic and supersonic drag coefficients are evaluated by a Navier-Stokes solver. Both the bending and twisting moments are calculated by directly integrating the computed pressure load at the supersonic condition. The bending moment represents the lateral moment that acts at the wing root. The twisting moment is the pitching moment measured at the leading edge of the root along the line normal to the root. The present optimization is performed at two design points for the transonic and supersonic cruises. Each flow conditions and the target lift coefficients are described below.

1. Transonic cruising Mach number, $M_{\infty,t}=0.9$
2. Supersonic cruising Mach number, $M_{\infty,s}=2.0$
3. Target lift coefficient at transonic cruising condition, $C_{L,t}=0.15$
4. Target lift coefficient at supersonic cruising condition, $C_{L,s}=0.10$
5. Reynolds number based on the root chord length at both conditions, $Re=1.0 \times 10^7$

To maintain constant lift constraints, the angle of attack is predicted by using C_{L_α} obtained from the finite difference. Thus, three Navier-Stokes computations per evaluation are required.

3.2 Design Parameters

Design variables are categorized to planform, warp shape and the thickness distribution. The definitions of design parameters are same as the previous optimization except for the thickness distribution. As mentioned earlier, the previous thickness definition has the lack of smoothness at the maximum thickness as shown in Fig. 4 [11]. To improve it, two more control points, which are symmetrical with respect to maximum thickness location, are added as shown in Fig. 5. Therefore, the present definition makes the second derivative continuous at the maximum thickness. As a result, 11 control points are used to represent the thickness distribution by a Bezier curve at three spanwise sections (root, kink and tip). Linear interpolation is used to interpolate the thickness distribution in spanwise direction. Table 1 describes the constraints for the thickness definition.

The wing planform is determined by six design variables as shown in Fig. 6. Since the wing area is fixed, the chord length at the wing tip is determined automatically. Constraints and the range of design variables are described in Tab. 2. The warp shape is composed of camber and twist. The camber surface is defined from the airfoil camber lines at the inboard and outboard of the wing separately. Each surface is represented by the Bezier surface which has four polygons in the chordwise direction and three in the spanwise direction. In case of the wing twist, a B-spline curve with six polygons is used. The total number of design parameters becomes 72.

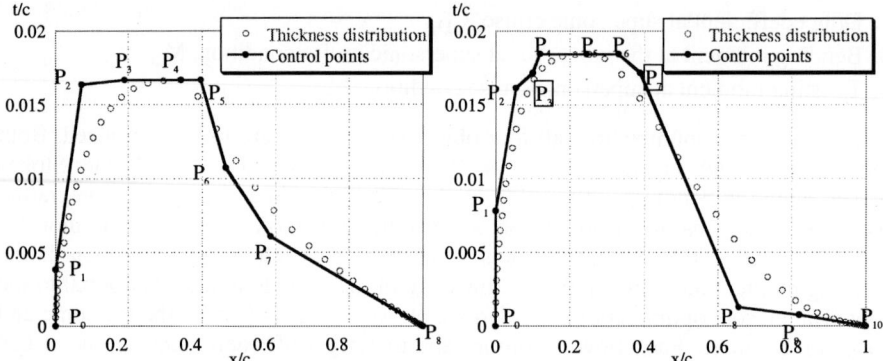

Fig. 4. Previous thickness definition   **Fig. 5.** Present thickness distribution

Table 1. Constraints for thickness distribution

Maximum thickness	$3 < Z_{P_5} < 4$
Maximum thickness location	$15 < X_{P_5} < 70$
First derivative constant at P_5	$Z_{P_4} = Z_{P_5} = Z_{P_6}$
Second derivative constant at P_5	$X_{P_5} - X_{P_3} = X_{P_7} - X_{P_5}$, $Z_{P_3} = Z_{P_7}$
First derivative constant at leading edge	$X_{P_0} = X_{P_1}$

Table 2. Constraints for planform shape

Chord length at root	$10 < C_{root} < 20$
Chord length at kink	$3 < C_{kink} < 15$
Inboard span length	$2 < b_1 < 7$
Outboard span length	$2 < b_2 < 7$
Inboard sweep angle (deg)	$35 < \alpha_1 < 70$
Outboard sweep angle (deg)	$35 < \alpha_2 < 70$
Wing area	$S = 60$

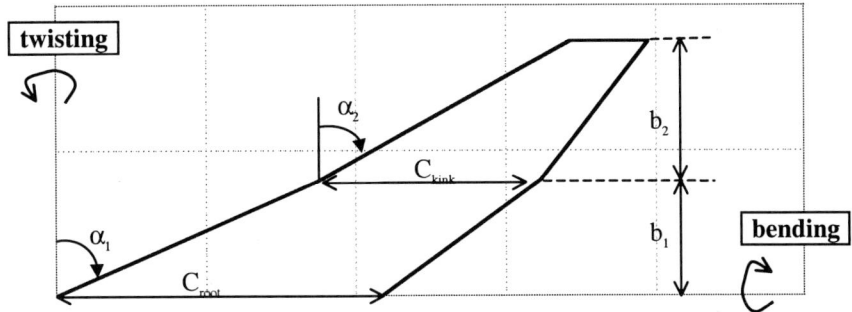

Fig. 6. Wing planform is defined by six design variables. Schematic view of bending and twisting moments are also shown

3.3 Evaluation by CFD

Previous results showed the importance of the viscous effect for wing designs. Thus, the three-dimensional, compressible, thin layer Navier-Stokes code is again used to evaluate aerodynamic performance of a three-dimensional wing at both transonic and supersonic conditions. This Navier-Stokes code employs total-variation-diminishing type upwind differencing and the lower-upper factored symmetric Gauss-Seidel scheme [12]. An algebraic mixing length turbulence model by Baldwin and Lomax is adopted [13]. To accelerate the convergence, the multigrid method is also used [14].

Taking advantage of the characteristic of GAs, the present optimization is parallelized on SGI ORIGIN 2000 at the Institute of Fluid Science, Tohoku University. The system has 640 PE's with peak performance of 384 GFLOPS and 640 GB of memory. The master PE manages the optimization process, while the slave PE's compute the Navier-Stokes code. The population size used in this study was set to 64 so that the process was parallelized with 32-128 PE's depending on the availability, because the transonic and supersonic computations can be processed separately. It should be noted that the parallelization was almost 100% because almost all the CPU time was dominated by Navier-Stokes computations. The present optimization requires about six hours per each generation parallelized on 128 PEs.

3.4 Details of the Present ARMOGA

In this study, the design variables are encoded in the real numbers. Blended crossover (BLX-α) is adopted as a crossover operator. This crossover method produces children on a segment defined by two parents and user specified parameter α. Parameter α is set to 0.5 except for the planform definition design variables. In the case of the six planform design variables, α is set to 0.0 to prevent the computational divergence of new candidates. After the crossover, mutation takes place at a probability of 20% based on a uniform random number selected over 10% of the range.

Selection is based on the Pareto ranking method and fitness sharing. Each individual is assigned to its rank according to the number of individuals that dominate it. A standard fitness sharing function is used to maintain the diversity of the population. The so-called best-N selection is also employed.

A population is set to 64, and the range adaptation is performed every 10 generations starting from the 15th generation.

4 Optimization Results

4.1 Overview of Pareto Solutions

The evolution was computed for 75 generations. After the computation, all the solutions evolved were sorted again to find the final Pareto solutions. The Pareto solutions were obtained in four-dimensional objective function space. To understand the distribution of Pareto solutions, all Pareto solutions are projected into two-dimensional objective function space between transonic and supersonic drag coefficients as shown in Fig. 7. In this figure, the Pareto solutions obtained from the previous optimization with three design objectives are also plotted. The present Pareto front is larger than before, in particular, better tradeoff solutions appear in the tradeoff surface I. The planform shapes of the extreme Pareto solutions, which minimize respective objective functions, appear physically reasonable as shown in Fig. 8. A wing, which minimizes the transonic cruising drag, has a less leading-edge sweep and a large aspect ratio. On the contrary, a wing with the lowest supersonic drag coefficient has a large leading-edge sweep to remain inside the Mach cone.

4.2 Influences of the Bending and Twisting Moments to Drag Coefficients

To examine influences of the bending and twisting moments, all the present Pareto solutions in Fig. 7 are labeled by the bending and twisting moments, respectively, as shown in Fig. 9. The wings, which locate near the tradeoff surface between transonic and supersonic drag coefficients (tradeoff surface I, Fig. 7), have impractically large bending moments as shown in Fig.9 (a). The bending moment is closely related to both transonic and supersonic drag coefficients. On the other hand, the twisting moment has an influence only on supersonic drag coefficient. As a consequence, the region II in Fig. 7 was primarily corresponding the minimization of the bending moment, not to the new objective function of the twisting moment minimization. The planform shapes, which have the lowest bending moment obtain/ed by the present and previous optimization respectively, are plotted in Fig. 10. Since these planform shapes are supposed to be indifferent, the present minimum wing and the wings belonged to the region II are found thanks to ARMOGA. Similarly, the improvement of the present tradeoff surface I from the previous result (Fig. 7) is due to ARMOGA.

Pareto solutions are also projected into the two-dimensional plane with the supersonic drag coefficient and the twisting moment in Fig. 11. A clear tradeoff is found. Figure 11 is also labeled by aspect ratios but there is no trend in performance based on the aspect ratios.

4.3 Comparison with NAL's Second Design and the Previous Design

To examine the quality of the present Pareto solutions, two wings are compared with NAL's second design wing as well as the previous wing obtained by three-objective optimization. NAL SST Design Team already finished the fourth aerodynamic design for the scaled experimental supersonic airplane to be launched in 2002 (NEXST-I). To summarize their concepts briefly, the first design determined the planform shapes among 99 candidates, then the second design was performed by the warp optimization based on the linearized theory. The third design aimed a natural-laminar-flow (NLF) wing by an inverse method using a Navier-Stokes code. Finally, the fourth design was performed for a wing-fuselage configuration. Because a fully developed turbulence is assumed in the present Navier-Stokes computations, it is improper to compare the present Pareto solutions to NAL's NLF wing design. Therefore, the NAL second design is chosen for a comparison.

Table 3 summarizes the aerodynamic performances of four wings compared: two present Pareto solutions (A, B), a previous Pareto solution (OBJ-3) and NAL's second design. The aerodynamic calculation of NAL's and the previous design is performed by using the same Navier-Stokes solver. All three Pareto solutions are superior to NAL's second design in all four objectives. The wing planform shapes are compared as shown in Fig. 12. The present and the previous planform shapes are similar to the "arrow wing" type. On the other hand, NAL's planform is similar to the conventional "delta wing" planform. These results indicate that the present arrow wing doesn't necessarily have a large pitching moment because NAL's design has a higher pitching moment.

The thickness distributions at the wing root of three Pareto solutions (A, B, OBJ-3) are presented in Fig. 13. In this figure, Pareto solutions A and B have much smoother thickness distributions than a previous Pareto solution of OBJ-3. The present wings do not have a kink in the thickness distribution thanks to the improved parameterization, and less likely to cause boundary layer separation.

Table 3. Aerodynamic performances of selected four wings

	$C_{D,t}$	$C_{D,s}$	M_B	M_T
Pareto (A)	0.00998863	0.01085439	18.15	62.35
Pareto (B)	0.01007195	0.01093646	17.39	60.60
OBJ-3	0.01004036	0.01093742	18.21	61.00
NAL2nd	0.01010175	0.01097646	18.23	63.31

5 Conclusion

Four-objective aerodynamic optimization of the wings for SST was performed by ARMOGA. In addition to the previous objective functions, which are to minimize the transonic and supersonic drag coefficients and the bending moment at the wing root, the minimization of the twisting moment is added. The number of design variables is increased from 66 to 72 to improve the thickness distribution. A Navier-Stokes solver is used to evaluate the aerodynamic performances.

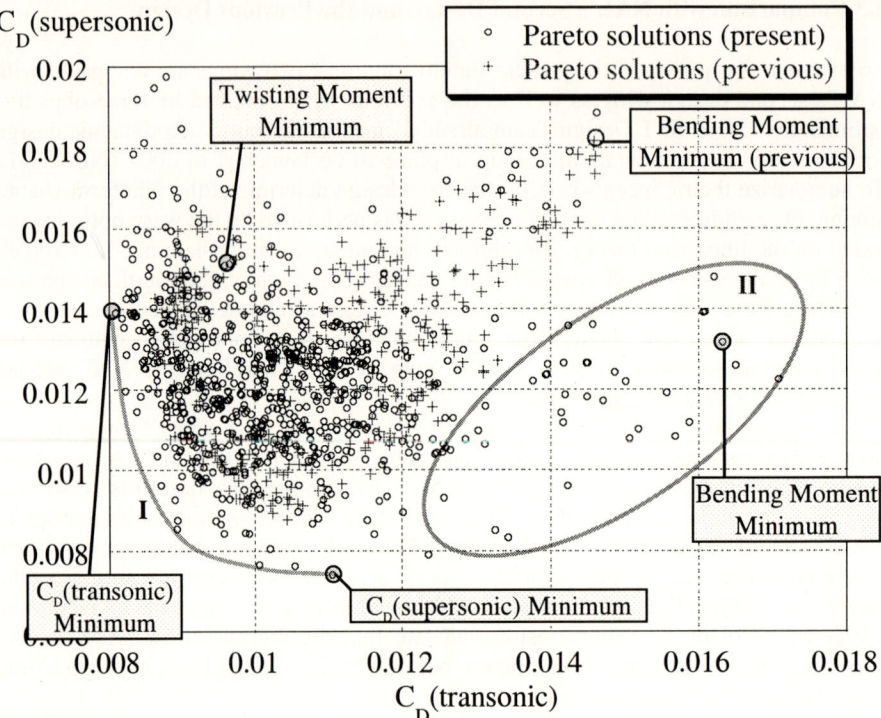

Fig. 7. Projection of present 4-objective Pareto front to transonic and supersonic drag tradeoffs. Pareto solutions obtained by previous 3-objective optimization are also plotted here. Extreme Pareto solutions are indicated. A previous Pareto solutions with the minimum bending moment is also indicated

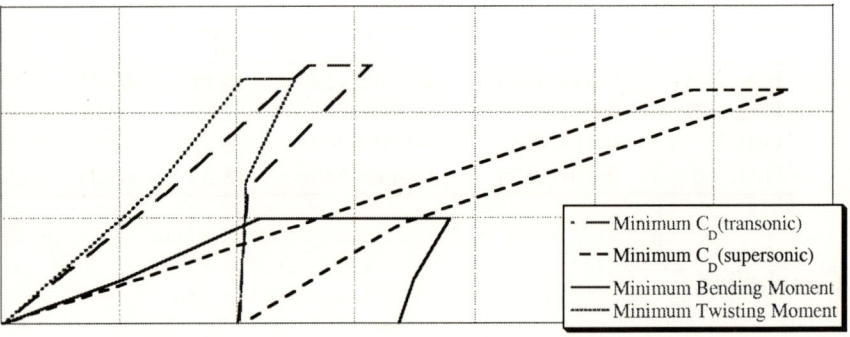

Fig. 8. Typical planform shapes of the extreme Pareto solutions

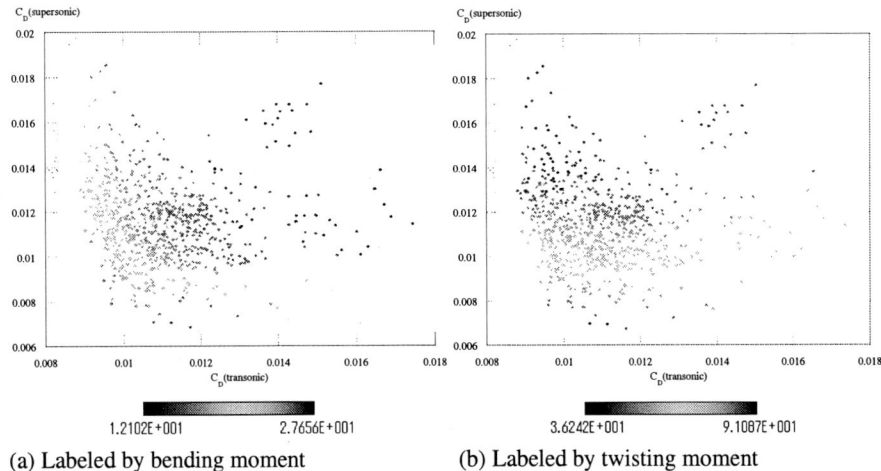

(a) Labeled by bending moment   (b) Labeled by twisting moment

Fig. 9. Projection of Pareto front to transonic and supersonic drag tradeoffs labeled by bending and twisting moments

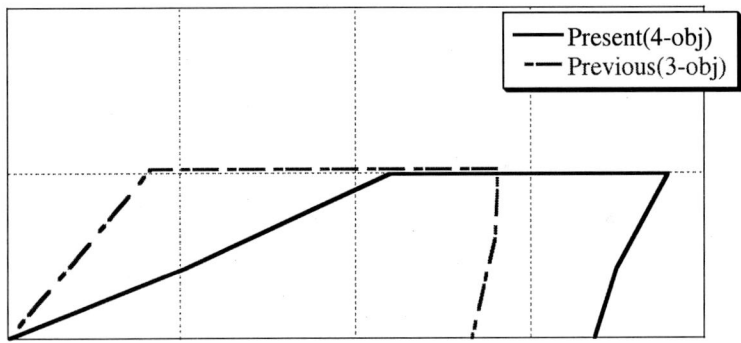

Fig. 10. Comparison of planform shapes having lowest bending moment obtained by the present and previous optimizations

As a result of the optimization, reasonable Pareto solutions were successfully obtained. The planform configurations of the extreme Pareto solutions are found physically reasonable. The resulting Pareto front appeared better than the previous case thanks to the range adaptation. ARMOGA is confirmed to work well in a large search space. By improving the definition of the thickness distributions, more realistic thickness distributions are obtained.

The present Pareto solutions, which are superior to NAL's second design in all four objective functions, are compared with NAL's wing and an optimal wing obtained before. As for the planform, optimal wings are similar to the "arrow wing" type. On the other hand, the NAL's design is similar to the conventional "delta wing" type. It also shows that even the arrow wing can reduce the pitching moment below that of the NAL second design. The present arrow wing is a good design candidate for the next-generation SST.

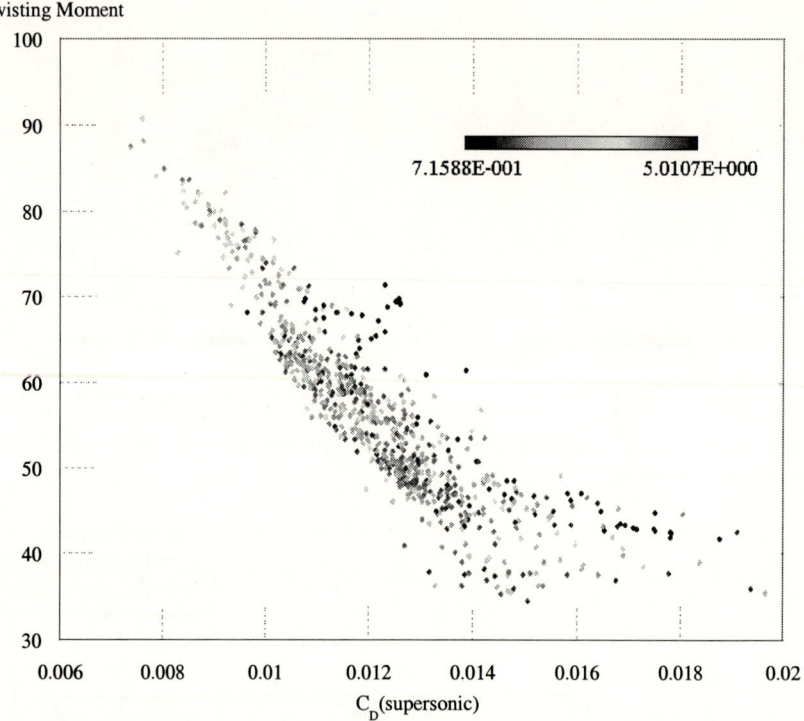

Fig. 11. Projection of Pareto front to supersonic drag and twisting moment tradeoffs labeled by aspect ratios

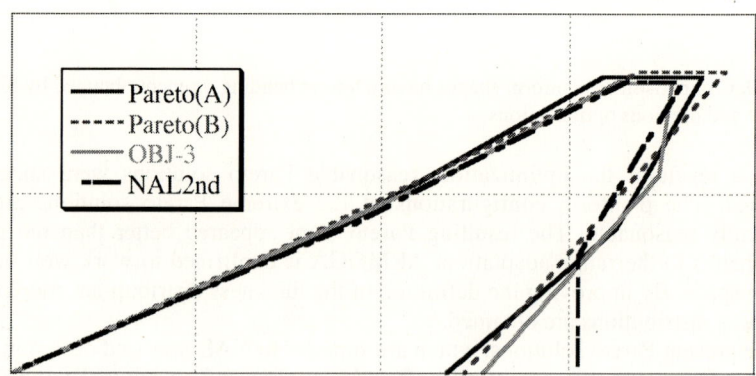

Fig. 12. Comparison of planform shapes among selected Pareto solutions and NAL's design. Planform shapes of the present (A, B) and previous results (OBJ-3) are similar to the "arrow wing"

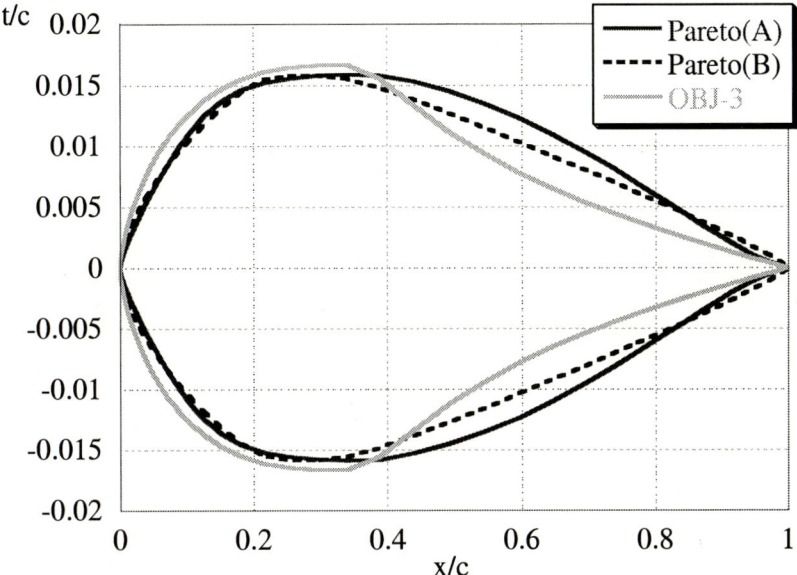

Fig. 13. Comparison of thickness distributions at the wing root among selected Pareto solutions. Thickness distributions of Pareto solutions (A, B) are much smoother at the maximum thickness location than that of the previous result (OBJ-3) is

Acknowledgements. To perform the wing design for SST, the computation was carried out in parallel using ORIGIN2000 in the Institute of Fluid Science, Tohoku University. This research was partly funded by Japanese Government's Grants-in-Aid for Scientific Research, No. 10305071. The third author's research has been partly supported by Bombardier Aerospace, Toronto, Canada. The authors would like to thank National Aerospace Laboratory's SST Design Team for providing many useful data.

References

1. Yamamoto, K., Inoue, O.: Applications of Genetic Algorithm to Aerodynamic Shape Optimization. AIAA paper 95-1650 (1995)
2. Quagliarella, D., Periaux, J., Poloni, C., Winter, G. (Eds.): Genetic Algorithms and Evolution Strategies in Engineering and Computer Science. John Wiley and Sons, Chichester (1998)
3. Doorly, D.: Parallel Genetic Algorithms for Optimisation in CFD. In: Winter, G., et al. (Eds.): Genetic Algorithms in Engineering and Computer Science. John Wiley and Sons, Chichester (1995) 251-270
4. Oyama, A., Obayashi, S., Nakamura, S.: Real-Coded Adaptive Range Genetic Algorithm Applied to Transonic Wing Optimization. Lecture notes in Computer Science, Vol. 1917. Springer-Verlag, Berlin Heidelberg New York (2000) 712–721

5. Fonseca, C. M., Fleming, P. J.: Genetic algorithms for multiobjective optimization: formulation, discussion and generalization. Proc. of the 5th Int. Conference on Genetic Algorithms, Morgan Kaufmann Publishers (1993) 416-423
6. Obayashi, S., Sasaki, D., Takeguchi, Y., Hirose, N.: Multiobjective Evolutionary Computation for Supersonic Wing-Shape Optimization. IEEE Transactions on Evolutionary Computation, Vol. 4, No. 2 (2000) 182-187
7. Sasaki, D., Obayashi, S., Sawada, K., Himeno, R.: Multiobjective Aerodynamic Optimization of Supersonic Wings Using Navier-Stokes Equations. Proc. of ECCOMAS 2000 [CD-ROM] (2000)
8. Iwamiya, T.: NAL SST Project and Aerodynamic Design of Experimental Aircraft. Proc. of 4th ECCOMAS Computing Fluid Dynamics Conference, Vol. 2 (1998) 580-585
9. Arakawa, M., Hagiwara, I.: Development of Adaptive Real Range (ARRange) Genetic Algorithms, JSME Int. J., Series C, Vol. 41, No. 4 (1998) 969-977
10. Arakawa, M., Hagiwara, I.: Nonlinear Integer, Discrete and Continuous Optimization Using Adaptive Range Genetic Algorithms, Proc. of 1997 ASME Design Engineering Technical Conferences (1997)
11. Grenon, R.: Numerical Optimization in Aerodynamic Design with Application to a Supersonic Transport Aircraft. Proc. of Int. CFD Workshop for Super-Sonic Transport Design (1998) 83-104
12. Obayashi, S., Grurswamy, G. P.: Convergence Acceleration of a Navier-Stokes Solver for Efficient Static Aeroelastic Computations. AIAA Journal, Vol. 33 (1995) 1134-1141
13. Baldwin, B. S., Lomax, H.: Thin Layer Approximation and Algebraic Model for Separated Turbulent Flows. AIAA paper 78-257 (1978)
14. Jameson, A., Caughey, D. A.: Effect of Artificial Diffusion Scheme on Multigrid Convergence. AIAA Paper 77-635 (1977)

Accurate, Transparent, and Compact Fuzzy Models for Function Approximation and Dynamic Modeling through Multi-objective Evolutionary Optimization

Fernando Jiménez[1], Antonio F. Gómez-Skarmeta[1],
Hans Roubos[2], and Robert Babuška[2]

[1] Dept. Informática, Inteligencia Artificial y Electrónica
University of Murcia, Spain
{fernan,skarmeta}@dif.um.es
[2] Control Engineering Laboratory
Delft University of Technology, the Netherlands
{J.A.Roubos,R.Babuska}@ITS.TUDelft.NL

Abstract. Evolutionary algorithms to design fuzzy rules from data for systems modeling have received much attention in recent literature. Many approaches are able to find highly accurate fuzzy models. However, these models often contain many rules and are not transparent. Therefore, we propose several objectives dealing with transparency and compactness besides the standard accuracy objective. These objectives are used to find multiple Pareto-optimal solutions with a multi-objective evolutionary algorithm in a single run. Attractive models with respect to compactness, transparency and accuracy are the result.

Keywords: Takagi-Sugeno fuzzy model, Pareto optimality, multi-objective evolutionary algorithm.

1 Introduction

This paper deals with fuzzy model parameter estimation and structure selection. In fuzzy model identification, we can, in general, take into account three criteria to be optimized: compactness, transparency and accuracy. Different measures for these criteria are proposed here. Compactness is related to the size of the model, i.e. the number of rules, the number of fuzzy sets and the number of inputs for each rule. Transparency is related to linguistic interpretability [1,2] and locality of the rules. Often one is interested in the local behavior of the global nonlinear model. Such information can be obtained by constraining the model-structure during identification. Transparency and model interpretability for data-based fuzzy models received a lot of interest in recent literature [3,4,5,6].

Evolutionary Algorithms (EA) [7,8] have been recognized as appropriate techniques for multi-objective optimization because they perform a search for multiple solutions in parallel [9,10,11]. EAs have been applied to learn both the this way, *preference articulation* implicitly defines a *utility function* which discriminates between candidate solutions. Approaches based on weights, goals and priorities have been used more often. Moreover, preference articulation can be achieved in different ways depending on how the computation and the decision processes are combined in the search for compromise solutions. Three broad classes can be identified, *a priori*, *a posteriori*, and *progressive* articulation of preferences.

2.3 Rule Set Simplification Techniques

Automated approached to fuzzy modeling often introduce redundancy in terms of several similar fuzzy sets that describe almost the same region in the domain of some variable. According to some similarity measure, two or more similar fuzzy sets can be merged to create a new fuzzy set representative for the merged sets [22]. This new fuzzy set substitutes the ones merged in the rule base. The merging process is repeated until fuzzy sets for each model variable cannot be merged, i.e., they are not similar. This simplification may results in several identical rules, which are removed from the rule set.

We consider the following similarity measure between two fuzzy sets A and B:

$$S(A, B) = \frac{|A \cap B|}{|A \cup B|} \tag{5}$$

If $S(A, B) > \theta_S$ (we use $\theta_S = 0.6$) then fuzzy sets A and B are merged in a new fuzzy set C as follows:

$$\begin{aligned} a_C &= min\{a_A, a_B\} \\ b_C &= \alpha b_A + (1 - \alpha) b_B \\ c_C &= \alpha c_A + (1 - \alpha) c_B \\ d_C &= max\{d_A, d_B\} \end{aligned} \tag{6}$$

where $\alpha \in [0, 1]$ determine the influence of A and B on the new fuzzy set C.

3 Criteria for Fuzzy Modeling

We consider three main criteria to search for an acceptable fuzzy model: (i) accuracy, (ii) transparency, and (iii) compactness. It is necessary to define quantitative measures for these criteria by means of appropriate objective functions which define the complete fuzzy model identification.

The accuracy of a model can be measured with the *mean squared error*:

$$MSE = \frac{1}{K} \sum_{k=1}^{K} (y_k - \hat{y}_k)^2 \tag{7}$$

where y_k is the true output and $\hat{y}_k$ is the model output for the kth input vector, respectively, and K is the number of data samples.

Many measures are possible for the second criterion, transparency. Nevertheless, in this paper we only consider one of most significant, *similarity*, as a first starting point. The similarity S among distinct fuzzy sets in each variable of the fuzzy model can be expressed as follows:

$$S = \max_{\substack{i,j,k \\ A_{ij} \neq B_{ik}}} S(A_{ij}, B_{ik}), \ i = 1, \ldots, n, \ j = 1, \ldots, M, \ k = 1, \ldots, M \qquad (8)$$

This is an aggregated similarity measure for the fuzzy rule-based model with the objective to minimize the maximum similarity between the fuzzy sets in each input domain.

Finally, measures for the third criterion, the compactness, are the number of rules M and the number of different fuzzy sets L of the fuzzy model. We assume that models with a small number of rules and fuzzy sets are compact.

In summary, we have considered three criteria for fuzzy modeling, and we have defined the following measures for these criteria:

Criteria	Measures
Accuracy	MSE
Transparency	S
Compactness	M, L

4 Multi-objective Evolutionary Algorithm

The main characteristics of the Multi-Objective Evolutionary Algorithm are the following:

1. The proposed algorithm is a Pareto-based multi-objective EA for fuzzy modeling, i.e., it has been designed to find, in a single run, multiple non-dominated solutions according to the Pareto decision strategy. There is no dependence between the objective functions and the design of the EA, thus, any objective function can easily be incorporated. Without loss of generality, the EA minimizes all objective functions.
2. Constraints with respect to the fuzzy model structure are satisfied by incorporating specific knowledge about the problem. The initialization procedure and variation operators always generate individuals that satisfy these constraints.
3. The EA has a variable-length, real-coded representation. Each individual of a population contains a variable number of rules between 1 and max, where max is defined by a decision maker. Fuzzy numbers in the antecedents and the parameters in the consequent are coded by floating-point numbers.
4. The initial population is generated randomly with a uniform distribution within the boundaries of the search space, defined by the learning data and model constraints.

Rule Set Level Variation Operators

- **Crossover 1**: Given two parents $I_1 = (R_1^1 \ldots R_{M_1}^1)$ and $I_2 = (R_1^2 \ldots R_{M_2}^2)$, this operator exchanges information about the number of rules of the parents and information about the rules of the parents, but no rule is internally crossed. Two children are produced: $I_3 = (R_1^1 \ldots R_a^1 R_1^2 \ldots R_b^2)$ and $I_4 = (R_{a+1}^1 \ldots R_{M_1}^1 R_{b+1}^2 \ldots R_{M_2}^2)$, where $a = round(\alpha \cdot M_1 + (1 - \alpha) \cdot M_2)$ and $b = round((1 - \alpha) \cdot M_1 + \alpha \cdot M_2)$. The number of rules of the children is between M_1 and M_2.
- **Crossover 2**: This operator increases the number of rules of the two children as follows: the first child contains all M_1 rules of the first parent and $\min\{\max - M_1, M_2\}$ rules of the second parent; the second child contains all M_2 rules of the second parent and $\min\{\max - M_2, M_1\}$ rules of the first parent.
- **Mutation 1**: This operator deletes or adds, both with equal probability, one rule in the rule set. For deletion, one rule is randomly deleted from the rule set. For rule-addition, one rule is randomly generated, according to the initialization procedure described, and added to the rule set.

Rule Level Variation Operators

- **Crossover 3**: Given two parents $I_1 = (R_1^1 \ldots R_i^1 \ldots R_{M_1}^1)$ and $I_2 = (R_1^2 \ldots R_j^2 \ldots R_{M_2}^2)$, this operator produces two children $I_3 = (R_1^1 \ldots R_i^3 \ldots R_{M_1}^1)$ and $I_4 = (R_1^2 \ldots R_j^4 \ldots R_{M_2}^2)$, with $R_i^3 = \alpha R_i^1 + (1 - \alpha) R_j^2$ and $R_j^4 = \alpha R_j^2 + (1-\alpha) R_i^1$, where i, j are random indexes from $[1, M_1]$ and $[1, M_2]$ respectively.
- **Crossover 4**: Given two parents $I_1 = (R_1^1 \ldots R_i^1 \ldots R_{M_1}^1)$ and $I_2 = (R_1^2 \ldots R_j^2 \ldots R_{M_2}^2)$, this operator produce two children $I_3 = (R_1^1 \ldots R_i^3 \ldots R_{M_1}^1)$ and $I_4 = (R_1^2 \ldots R_j^4 \ldots R_{M_2}^2)$, where R_i^3 and R_j^4 are obtained with the *uniform* crossover.
- **Mutation 2**: This operator removes a randomly chosen rule and inserts a new one which is randomly generated by the rule-initialization procedure.

Parameter Level Variation Operators

- **Crossover 5**: Given two parents, and one rule of each parent randomly chosen, this operator crosses the fuzzy numbers corresponding to a random input variable or the consequent parameters. The crossover is arithmetic.
- **Mutation 3**: This operator mutates a random fuzzy number or the consequent of a random rule. The new fuzzy number or consequent is generated at random.
- **Mutation 4**: This operator changes the value of one of the antecedent fuzzy sets a, b, c or d of a random fuzzy number, or a parameter of the consequent ζ, of a randomly chosen rule. The new value of the parameter is generated at random within the constraints by a non-uniform mutation.

5 Optimization Models and Decision Making

After preliminary experiments in which we have checked different optimization models, the following remarks can be maded:

1. The minimization of the number of rules M of the individuals has negative influence on the evolution of the algorithm. The reason is than this parameter is not an independent variable to optimize, as the amount of information in the population decreases when the average number of rules is low, which is not good for exploration. Then, we do not minimize the number of rules during the optimization, but we will take it into account at the end of the run, in a posteriori articulation of preferences applied to the last population.
2. It is very important to note that a very transparent model will be not accepted by a decision maker if the model is not accurate. In most fuzzy modeling problems, excessively low values for similarity hamper accuracy, for which these models are normally rejected. Alternative decision strategies, as *goal programming*, enable us to reduce the domain of the objective functions according to the preferences of a decision maker. Then, we can impose a goal g_S for similarity, which stop minimization of the similarity in solutions for which goal g_S has been reached.
3. The measure L (number of different fuzzy sets) is considerably reduced by the rule set simplification technique. So, we do not define an explicit objective function to minimize L.

According to the previous remarks, we finally consider the two following optimization models:

Optimization Model 1:

$$\begin{aligned} Minimize\ f_1 &= MSE \\ Minimize\ f_2 &= S \end{aligned} \quad (9)$$

Optimization Model 2:

$$\begin{aligned} Minimize\ f_1 &= MSE \\ Minimize\ f_2 &= max\{g_S, S\} \end{aligned} \quad (10)$$

At the end of the run, we consider the following a posteriori articulation of preferences applied to the last population to obtain the final compromise solution:

1. Identify the set $X^* = \{x_1^*, \ldots, x_p^*\}$ of non-dominated solutions according to:

$$\begin{aligned} Minimize\ f_1 &= MSE \\ Minimize\ f_2 &= S \\ Minimize\ f_3 &= M \end{aligned} \quad (11)$$

2. Choose from X^* the most accurate solution x_i^*; remove x_i^* from X^*;
3. If solution x_i^* is not accurate enough or there is no solution in the set X^* then STOP (no solution satisfies);

4. If solution x_i^* is not transparent or compact enough then go to step 2;
5. Show the solution x_i^* as output.

Computer aided inspection shown in Figure 3 can help in decisions for steps 2 and 3.

6 Experiments and Results

Consider the 2nd order nonlinear plant studied by Wang and Yen in [24,25]:

$$y(k) = g(y(k-1), y(k-2)) + u(k) \qquad (12)$$

with

$$g(y(k-1), y(k-2)) = \frac{y(k-1)y(k-2)(y(k-1) - 0.5)}{1 + y^2(k-1)y^2(k-2)} \qquad (13)$$

The goal is to approximate the nonlinear component $g(y(k-1), y(k-2))$ of the plant with a fuzzy model. As in [24], 400 simulated data points were generated from the plant model (12). Starting from the equilibrium state $(0,0)$, 200 samples of identification data were obtained with a random input signal $u(k)$ uniformly distributed in $[-1.5, 1.5]$, followed by 200 samples of evaluation data obtained using a sinusoidal input signal $u(k) = \sin(2\pi k/25)$. The resulting signals and the real surface are shown in Figure 1.

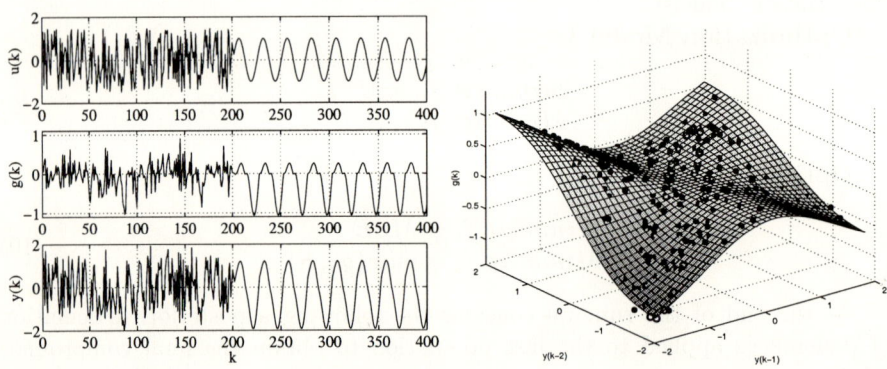

Fig. 1. *Left*: Input $u(k)$, unforced system $g(k)$, and output $y(k)$ of the plant in (12). *Right*: Real surface.

The following values for the parameters of the EA were used in the simulations: population size 100, crossover probability 0.8, mutation probability 0.4, number of children for the preselection scheme 10, minimum number of individuals for each number of rules 5, and maximum number of individuals for each number of rules 20. All crossover and mutation operators are applied with the same probability. The EA stops when the solutions satisfy the decisor maker.

We show results obtained with the EA by using the optimization models (9) ($max = 5$) and (10) ($max = 5$, $g_S = 0.25$). Figure 2 shows the non-dominated solutions in the last population according to (11) for both optimization models (9) and (10). One can appreciate the effectiveness of the preselection technique and the added explicit niche formation technique to maintain diversity in the populations. The main differences between the optimization models (9) and (10) are that with (9) the EA obtains more diversity but the fuzzy models are less accurate. Goal-based model have the disadvantage that it is necessary to choose, a priori, a good goal for the problem, although this value is representative of the maximum degree of overlapping of the fuzzy sets allowed by a decisor.

According to the described decision process, we finally choose a compromise solution showed in Figure 3 by means of different graphics for the obtained model. Figure 3(a) shows the local model. The surface generated by the model is shown in Figure 3(b), fuzzy sets for each variable are showed in Figure 3(c), and finally, the identification and validation results as well as the prediction error are shown in Figure 3(d).

We compared our results, with those obtained by the four different approaches proposed in [25] and [26]. The best results obtained for in each case are summarized in Table 1, with an indication of the number of rules, number of different fuzzy sets, consequent type, and obtained MSE for training and evaluation data. In [25], the low MSE on the training data is in contrast with the MSE for the evaluation data which indicates overtraining. The solution in [26] is similar to the solutions in this paper with respect to the accuracy, transparency and compactness, but hybrid techniques (initial fuzzy clustering and a sequence of specific genetic algorithms) were required in [26]. Solutions in this paper are obtained with a single EA and they have been chosen among different alternatives, which is an advantage for an appropriate decision process.

Table 1. Fuzzy models for the dynamic plant. All models are of the Takagi-Sugeno type.

Ref.	No. of rules	No. of sets	Consequent	MSE train	MSE eval
[25]	36 rules (initial)	12 (B-splines)	Linear	$1.9 \cdot 10^{-6}$	$2.9 \cdot 10^{-3}$
	24 rules (optimized)	-	Linear	$2.0 \cdot 10^{-6}$	$6.4 \cdot 10^{-4}$
[26]	7 rules (initial)	14 (triangular)	Linear	$1.8 \cdot 10^{-3}$	$1.0 \cdot 10^{-3}$
	5 rules (optimized)	5 (triangular)	Linear	$5.0 \cdot 10^{-4}$	$4.2 \cdot 10^{-4}$
This paper[1]	5 rules	5 (trapezoidal)	Linear	$2.0 \cdot 10^{-3}$	$1.3 \cdot 10^{-3}$
This paper[2]	5 rules	6 (trapezoidal)	Linear	$5.9 \cdot 10^{-4}$	$8.8 \cdot 10^{-4}$

[1] Solution corresponds to the solution marked with * in Figure 2(left), and Figure 3(left)

[2] Solution corresponds to the solution marked with * in Figure 2(right), and Figure 3(right)

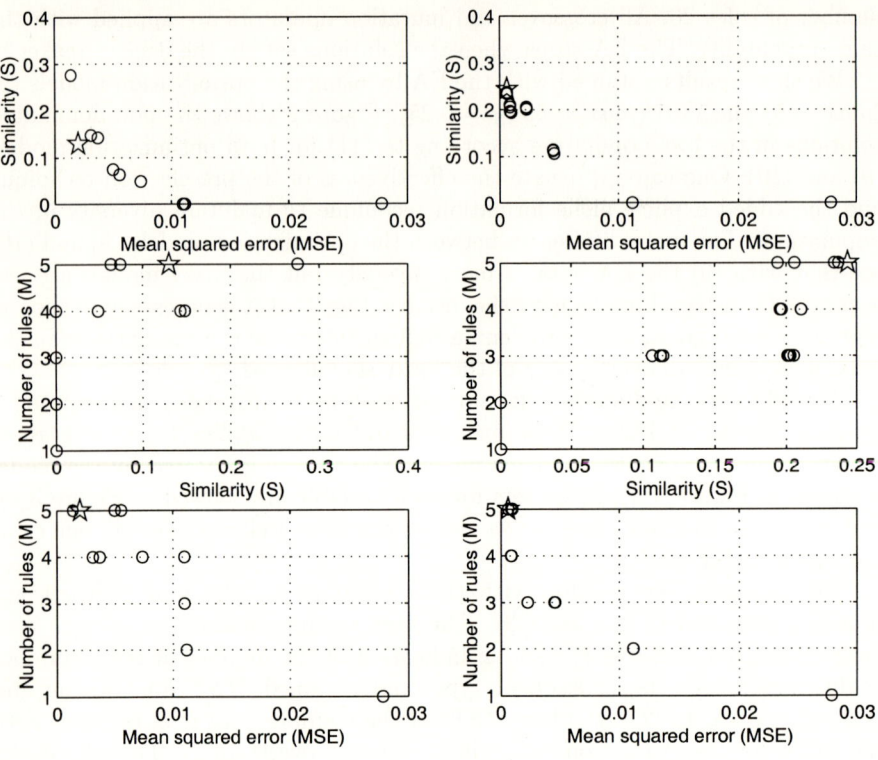

Fig. 2. *Left:* Non-dominated solutions according to (11) obtained with the Pareto-based multi-objective EA by using the optimization model (9). *Right:* Non-dominated solutions according to (11) obtained with the Pareto-based multi-objective EA by using the optimization model (10). Solution marked with * is the final compromise solution.

7 Conclusions and Future Research

This paper remarks some initial results in the combination of Pareto-based multi-objective evolutionary algorithms and fuzzy modeling. Criteria such as accuracy, transparency and compactness have been taken into account in the optimization process. Some of these criteria have been partially incorporated into the EA by means of ad hoc techniques, such as rule set simplification techniques. An implicit niche formation technique (preselection) in combination with other explicit techniques with low computational costs have been used to maintain diversity. These niche formation techniques are appropriate in fuzzy modeling if excessive amount of data are required. Excessive computational times would result if sharing function were used. Elitism is also implemented by means of the preselection technique. A goal based approach has been proposed to help to obtain more accurate fuzzy models. Results obtained are good in comparison with other more complex techniques reported in literature, with the advantage that the proposed

2 The Niched Pareto Genetic Algorithm

Since 1993, a number of researchers have implemented Goldberg's 1989 [6] general suggestion of combining a Pareto-based selection with some form of fitness sharing to achieve a selective pressure toward the non-dominated front while maintaining diversity along that front. [7] developed the NPGA which implements the basic suggestion of Goldberg [6] by using *Pareto domination tournaments* for selection, and fitness sharing in *objective space* to introduce speciation in a partially ordered search space. [8]'s Multiobjective GA (MOGA) used a similar strategy, with Pareto rank-based selection and objective space fitness sharing, while the non-dominated sorting GA (NSGA) [9] used Goldberg's [6] suggested ranking method and then used fitness sharing in the genotypic and phenotypic spaces (rather than in the objective space). Since then, other researchers have implemented their own variations on these similar strategies, all with relative success. Below we describe the NPGA2 and compare and contrast it with the original 1994 version, as well as with MOGA and with NSGA.

2.1 The Original NPGA

The extension of the traditional GA to the 1994 NPGA involved the addition of two specialized genetic operators: (1) Pareto domination tournaments and (2) continuously updated fitness sharing. These operators alter the traditional mechanism of selection by partially ordering the population and by maintaining diversity in the population through successive generations, respectively.

In the 1994 NPGA, tournament selection was used because of its simplicity, well-known effectiveness, and the ability to adjust selection pressure by changing the tournament size. Tournaments were modified to handle the Pareto criterion in the following manner. k competitors were chosen at random from the current generation's population. A set of t_{dom} individuals was also chosen at random as the Pareto sample set P. Each of the k competitors was compared to each member of P. If a competitor was dominated by a member of P it lost the competition. If all of the k competitors lost, then fitness sharing was used to break the "tie". Each competitor's *niche count* (see below) was calculated in the objective space (i.e., using its evaluated objective values, such as *cost* and *reliability*, as coordinates). The competitor with lowest niche count (i.e., the "least crowded") won the tournament. As with normal tournament selection, enough tournaments were held to fill the next generation's population with "winners". Crossover and mutation could then be applied to the new population.

2.2 The NPGA2

Although the original NPGA was successfully applied to several test problems, including a groundwater detection problem, we felt that the Pareto domination sampling used in the tournaments could be made less noisy. Indeed, the sampling can be seen as an approximation of Fonseca and Fleming's [8] Pareto ranking scheme, in which the rank of an individual is determined by the number of current population members that dominate the individual. In the NPGA, the more members that dominate a solution, the more likely it is that a random sample of the population will include a dominating member. To make such Pareto ranking deterministic in the new NPGA2, we use the *degree of domination* of an individual as its rank, as [8] and as used by others since. But unlike [8] we maintain the use of tournament selection, simply using Pareto rank to determine tournament winners. As in the original NPGA, tournament ties are resolved by fitness sharing, using niche counts as calculated in the objective space. A more detailed discussion of the new NPGA2 selection and niching mechanisms follows.

To initiate selection each individual in the population of designs is assigned a rank equal to the degree of Pareto domination experienced by that design. The degree of domination, or rank, of an individual design is the total number of designs in the population that dominate that design. A design is said to (Pareto) *dominate* another individual in the population if it is at least equal in all objectives to that individual and better in at least one. *Non-dominated* designs, or those that are not dominated by any individuals in the population, are assigned a rank of zero. In Fig. 1, where the objectives are to minimize both cost and mass remaining, an example of a Pareto domination ranked population of ten designs is shown. Designs of equal rank are designated by the same symbol.

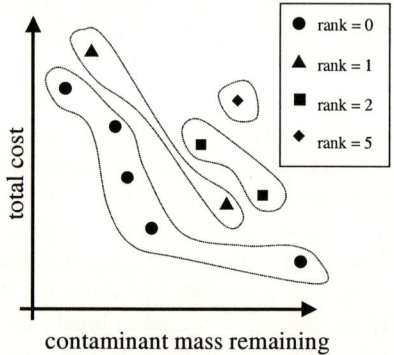

Fig. 1. Ranking by degree of Pareto domination

Once the entire population has been ranked according to the Pareto domination rank, candidate designs are chosen for reproduction via deterministic tournament selection [6]. The controlling variable in tournament selection competitions is the tournament size k. A group of k competitors is randomly selected from the population of ranked designs. If there is a single candidate with the lowest rank (i.e. less dominated), this candidate is the "clear winner" of the tournament and is selected for reproduction. If there is no clear winner, none of the candidates are preferred and the tournament selection ends in a tie.

In the case of a tie, the population density around each competitor is calculated within a specified Cartesian distance (in objective function-space), called the niche radius (see Figure 5). A niche count is calculated by summing the number of designs within the niche radius of each competitor, weighted by the radial distance between the competitor and the other designs. The niche count for competitor i is given by:

$$m_i = \begin{cases} \sum_{j \in pop} \left(1 - \dfrac{d_{ij}}{\sigma_{share}}\right) & \text{if } d_{ij} < \sigma_{share} \\ 0 & \text{if } d_{ij} \geq \sigma_{share} \end{cases} \quad (1)$$

where d_{ij} is the distance between competitor i and other population member j and σ_{share} is the niche radius. The winner of the tied tournament is the competitor with the lowest niche count.

As in the original NPGA, niche counts are calculated using individuals in the partially filled next generation population, rather than from the current generation population. The use of this *continuously updated fitness sharing* was suggested by [10]. They showed that the naïve combination of tournament selection and fitness sharing would lead to chaotic perturbations of the population composition. Note that the values of the objective functions should be scaled to equal ranges in order to determine the niche count, as in:

$$O'_i = \dfrac{O_i - O_{i,\min}}{O_{i,\max} - O_{i,\min}} \quad (2)$$

where O'_i, $O'_{i,\min}$, and $O'_{i,\max}$ are the scaled, minimum, and maximum values of objective O_i, respectively.

2.3 Comparison to Non-dominated Sorting

NPGA2 uses tournament selection based on the ranking given by [8]'s degree-of-domination measure, which has been implemented in a number of other EMO algorithms. Goldberg, in [6], suggested a different ranking scheme. Srinivas and Deb [9] have named this *non-dominated sorting*, and it has also been implemented in a number of other EMO algorithms. In a non-dominated sort of the population, the current non-dominated subset of the population is assigned rank 0 and is then temporarily removed from consideration. The remaining population is then evaluated to determine another non-dominated subset, which is then given rank 1 and removed from consideration. This process continues until the entire population has been ranked [6]. As with the degree-of-domination ranking, the non-dominated ranks can then be used for selection in a number of ways, such as tournaments. In the case of the NSGA, Srinivas and Deb use proportionate selection with fitness sharing.

How does the ranking generated by a non-domination sort compare to the ranking generated by the degree-of-domination? First, it is clear that both rankings treat the *on-line Pareto optimal set* (i.e., the non-dominated set from the current generation population) the same: every member of the Pareto set gets a rank of zero. As for the rest of the population, it can be shown that both types of ranking respect the Pareto ordering. That is, if an individual **A** Pareto dominates an individual **B**, then both types of ranking will rank **A** better (lower) than **B**. In the case of non-dominated sorting, this is quite clear, as **A** must be removed from consideration before **B** can be considered non-dominated, and therefore **A** must be ranked before **B**. In the case of degree-of-domination ranking, **B** must have a greater degree of domination than **A** because every individual that dominates **A** must also dominate **B** (by the transitive property of the Pareto domination relation), and by assumption **A** dominates **B** as well. Thus **B**'s degree of domination must be greater than that of **A**.

Non-dominated ranking can differ from degree-of-domination ranking however, when comparing individuals that are **not** ordered by the Pareto domination relation. Thus an individual **A** might be ranked better (lower) than an individual **B** by non-dominated ranking, because it is "closer" to the Pareto front, while being ranked worse (higher) by degree-of-domination ranking, because **A** is more "crowded" by solutions that dominate it[1].

3 The Groundwater Remediation Problem

We apply the NPGA2 to a new problem domain: cleanup of a hypothetical contaminated groundwater site. The site is given an initial contaminant plume, after which active remediation, in the form of pump-and-treat (PAT) technology, is applied. A fixed number of wells in fixed locations are used to pump out contaminated water to be treated on the surface. The goal of the NPGA2 is to find the optimal pumping rates for the wells such that total cost (e.g., power to the pumps, volume of water treated, etc.) is minimized while also minimizing the amount of contaminant remaining after the treatment period (ten years). For each set of decoded pumping rates generated by the NPGA2, we run a computer simulation of the resulting groundwater and contaminant flow over the ten year remediation horizon, to determine a total cost of treatment and total contaminant mass remaining.

Although our specific test cases are artificial and somewhat idealized, the overall approach is realistic. In real-world situations, contamination is detected, an initial contaminant plume is estimated (based on known mass of missing contaminant and/or sampling of concentrations from test wells), the hydrology flow field is modeled (again based on known and sampled geological conditions), and various pumping strategies (e.g., locations, depths, and pumping rates) are simulated. Remediation strategies exhibiting the best tradeoffs between total cost and contaminant mass remaining, under simulation, are chosen for implementation. The rest of this section provides details of the simulation model.

---

[1] But this can only be true if both individuals are not Pareto optimal, i.e., of rank 0.

3.1 Description of the Contaminated Site

The site is modeled as a confined homogeneous aquifer 1000 by 1010 meters in area and 30 meters in thickness (see Fig. 2). The model is discretized in two dimensions using 10-m square grid blocks. Constant-head boundaries are imposed on the east and west sides of the domain while no-flow boundaries are imposed on the north and south. (Thus the "background flow" of groundwater is a constant and uniform east-west flow.) The aquifer is modeled as having a homogeneous, isotropic hydraulic conductivity (i.e., uniform porosity, etc.). Extraction wells are modeled as being open over the entire thickness of the confined aquifer. Removal of contaminant by the treatment system is simulated as equilibrium, nonlinear adsorption. The adsorbent concentration, q, is given by the Freundlich isotherm equation: $q = K_{AB} C_{i,j}^{1/n}$.

For a hypothetical contaminant, we chose Trichloroethlyene (TCE), a commonly observed and studied groundwater contaminant. TCE is treated as a conservative, dissolved species. To generate the initial plume of contaminant shown in Fig. 2, a constant source of approximately 750 ppm is applied until a plume of approximately 1,000 kg of TCE is released into the confined aquifer. At that point, the source is removed and active remediation begins, continuing for a ten year period. The *ex-situ* treatment technology is granular activated carbon (GAC).

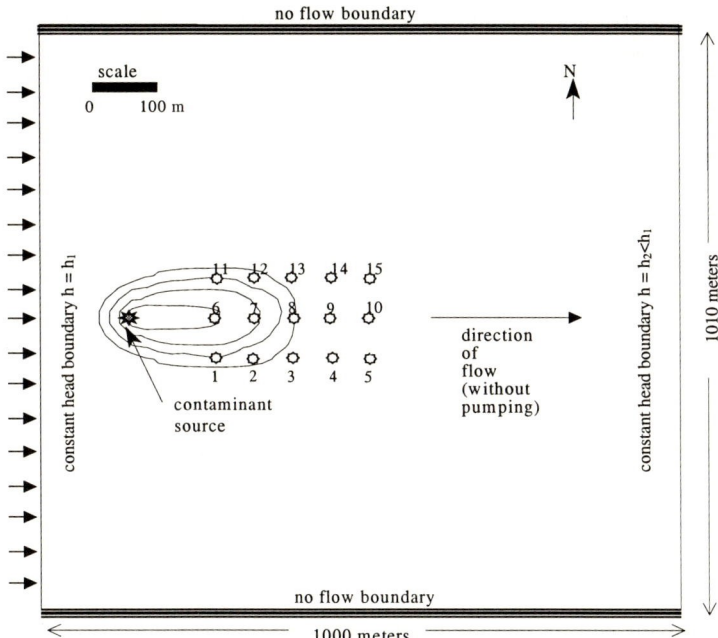

Fig. 2. The initial contaminant plume to be remediated

span of the tradeoff curve covered by these solutions. We determine the percentage of Pareto optimal solutions found by a method m by aggregating all of the Pareto optimal solutions found by the methods and calculating P_m / P_M, where P_M is the total number of Pareto optimal solutions found by all methods and P_m is the number of Pareto optimal solutions found by method m. The endpoints of the span are defined as the maximum MR' (~100%) and the minimum MR' achievable, given the maximum pumping rate constraint.

We implement archiving to eliminate redundant objective function evaluations. If a design is found to have been evaluated previously, the previous evaluation is used. Only new designs are evaluated, so that only new designs count against computational effort.

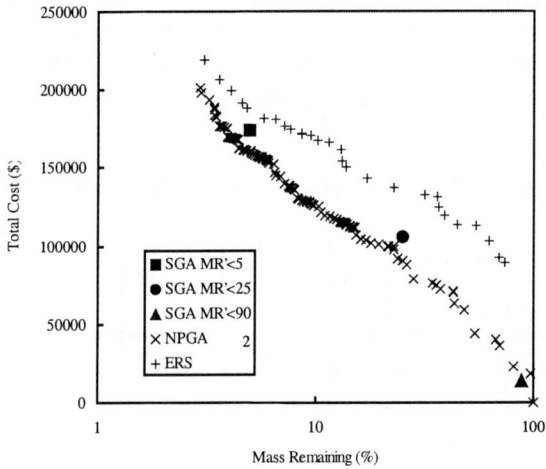

Fig. 3. Performances on the 15 well problem

5   Results and Discussion

Table 3 summarizes the results of all the experiments, in terms of percentage of Pareto optimal solutions found by each algorithm for each scenario. We aggregate all non-dominated solutions found by each algorithm, for a particular scenario, and take the Pareto optimal subset of the aggregation. We then credit each algorithm for its contribution to that set. Table 3 indicates that as the problem becomes harder (number of wells increases), simpler approaches like ERS are outperformed by the more sophisticated heuristics.

If we look at the 15 well case in more detail (Fig. 3), in which each algorithm's Pareto optimal set is shown plotted in objective space, we can see that the genetic algorithms "dominate" ERS. The SGA comes close to the aggregated optimal front with two of its solutions, while its third solution is actually on the front.

Finally, we take a look at a couple of individual solutions to the 15 well scenario, picked from two different parts of the aggregated Pareto optimal tradeoff curve. In Fig. 4, right, we graph the well pumping rates for a solution that costs less than $50,000 but leaves approximately 60% of the contaminant behind. Note how the NPGA2 found that only two wells need to be turned on, and these are directly downflow of the source. On the left of Fig. 4 we show contour plots illustrating the distribution of the remaining contaminant. In Fig. 5, we see another tradeoff solution, with MR' = 15%, but this one must apparently spend more power and therefore money, pumping out more of the contaminant, in order to achieve a 15 percentage mass remaining. Fig.s 4 and 5 together demonstrate that the NPGA2 is designing reasonable strategies: (a) as MR' increases, fewer wells are needed, (b) the pumping rates are more or less symmetric about the mean direction of groundwater flow, and (c) the remaining contaminant plume is mostly symmetric about the direction of flow.

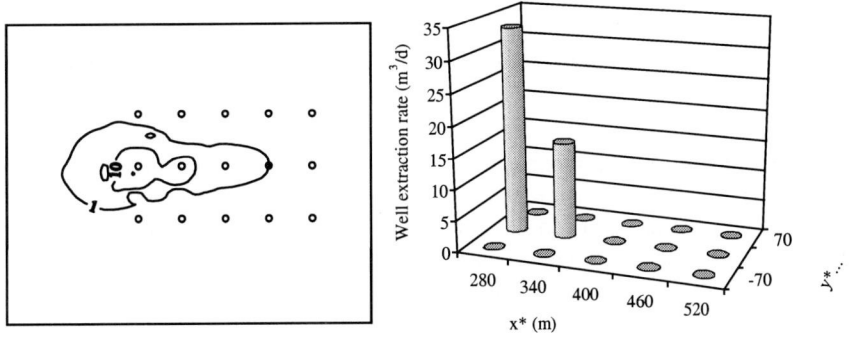

Fig. 4. NPGA2 solution for 60% MR' design is reasonable

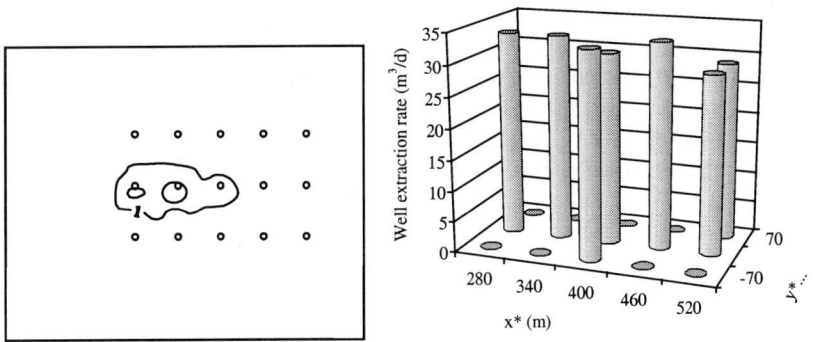

Fig. 5. NPGA2 solution for 15% MR' design

6 Conclusions

In summary, we have applied the new NPGA2 to a groundwater quality management problem consisting of active remediation by pump-and-treat. Through the addition of fitness sharing, we have overcome the problem described by [4] in finding Pareto optimal solutions over the entire tradeoff curve.

To gain some confidence in the ability of the algorithm to scale with problem size and difficulty, we ran a series of test problems where the NPGA2 was compared to two other methods, SGA and ERS. As the problems increased in complexity, by considering additional decision variables, the NPGA2 was more effective and efficient than either the SGA or the ERS in finding more Pareto optimal designs that span the entire tradeoff curve.

We conclude that EMO algorithms like the NPGA2 can be an effective method for producing tradeoff curves for subsurface remediation problems. Tradeoff curves such as those presented here may give decision makers the capability of making better informed decisions.

In future we think we can consider additional objective functions, such as maximizing reliability, minimizing drawdown, or minimizing remediation time. We also plan to try other decision variables, such as time-variable extraction rates. Further tests of the applicability of this approach should consider realistic, contaminated sites, especially those with a significant degree of heterogeneity. Future comparative studies should include more sophisticated algorithms than ERS or SGA, such as other successful EMO methods, as well as multi-objective simulated annealing, and other multiobjective and classic optimization approaches.

Acknowledgements. The majority of this work has been made possible by an EPA Office of Research and Development grant under the contract number CR826614-01-0. A portion of the computational work was conducted on MTU Computational Science & Engineering's High Performance Computing Platform, funded by NSF. Any views presented herein by the authors are not necessarily reflective of the views of the EPA or NSF. The authors wish to thank Reed M. Maxwell, from the Lawrence Livermore National Laboratory, for the gracious use of his numerical groundwater flow and transport models. Dr. Horn's Fall 1998 Genetic Algorithms class co-wrote the NPGA2 code used in the experiments.

References

1. Wagner, B. J., Recent advances in simulation-optimization groundwater management modeling, *Rev. Geophys., Supplement, U.S. Natl. Rep. Int. Union Geod. Geophys., 33*, 1021-1028, 1995.
2. Freeze, R. A., and S. M. Gorelick, Convergence of stochastic optimization and decision analysis in the engineering design of aquifer remediation, *Ground Water, 37*(6), 934-954, 1999.

evaluate each rule efficiency and a mechanism to allow the formation of hierarchies, although there is not an explicit pressure towards the formation of general rules. XCS [21][22] is a classifier system which has shown a strong tendency to achieve the stated objectives. One of its main strengths is the definition of fitness, based on the prediction accuracy of each rule rather than on the prediction itself. Although it does not use an explicit bias to favour the generalization of rules, this is achieved by the application of the GA to environmental niches [22].

MOLeCS is a recent CS [2][3] designed to solve classification tasks by supervised learning. Its main contribution is the definition of the *accuracy* and *generality* measures for each rule, and the use of multiobjective solution techniques for optimizing them simultaneously. In this sense, the learning task is represented explicitly as a multiobjective problem: the optimization of the accuracy and the generality of each rule.

This paper studies, under the MOLeCS architecture, the performance of different multiobjective evolutionary algorithms in achieving the learning goals, which can be summarized as: obtaining the *minimum* rule set that *covers accurately* all the examples. Different niching strategies are also considered in order to maintain a parallel set of rules and thus, achieve covering. First, we place the MOLeCS system into the CSs and MOEAs (multiobjective evolutionary algorithms) frameworks. In section 3 we describe MOLeCS and discuss different multiobjective approaches and niching methods. Next, we compare them using some benchmark problems often tested in the research community. In section 5, we show the application of MOLeCS to a real classification task based on the prediction of breast cancer. Finally, we give our conclusions and future work.

2 Background

In this section, a brief overview of classifier systems is made, remarking on the main differences between previous CSs and MOLeCS. Next, we review the similarities and differences between a typical multiobjective evolutionary algorithm and a classifier system as MOLeCS.

2.1 Classifier Systems

Typically, learning classifier systems codify each individual as one rule, while the solution that must be obtained is a complete set of rules (that is, all the population). Two major issues arise from this approach: (a) the evaluation of each rule (fitness) and (b) the maintenance of a group of rules.

The fitness evaluation method must provide a scalar measure that weighs the efficiency of each rule. In traditional classifier systems, this fitness measure was based on the payoff prediction; that is, the payoff that the classifier would receive from the environment if its action was selected (e.g. Holland's CS [11]). Recently, XCS [21][22] has migrated the fitness from the payoff prediction to the accuracy of the prediction, which results in better performance. Horn's study

[13] also addresses the classifier's accuracy, which is defined as the percentage of correctly classified examples over the covered training examples.

Different niching mechanisms are proposed in the research community to ensure covering by the co-evolution of different rules: sharing payoff between active classifiers (see [13]), performing restricted replacement [9], or translating the panmictic GA to the active classifiers (match set or action set) which can be classified as a kind of restricted mating [21],[22].

The classifier system's ability to evolve generalizations is a major issue that has recently received a growing interest. In XCS, the application of the GA to the action sets has resulted in a pressure towards generalizations. The generalization hypothesis [21] states that given two classifiers (rules) C1 and C2 (both equally accurate, where C2 is a generalization of C1) the more general rule (C2) tends to displace the more specific one (C1). This is due to the fact that the more general rule participates in more action sets, having more reproductive opportunities and achieving thus more copies. Frey and Slate [8] proposed a CS where fitness was based on accuracy. A pressure towards generalization was induced using an "utility" measure. This was computed as the number of correctly classified examples over the total number of examples seen by the system. Rules with a low "utility" measure were deleted in order to favour those rules more used.

From the **accuracy** perspective, our approach is more related to Horn's study, since fitness is based on accuracy, computed as the percentage of correctly classified examples. However, our system is taking account of the classifier **generality** too, which is a more complex task. Generality in MOLeCS is considered explicitly in the fitness evaluation stage, in contrast to XCS where generality is enforced in an implicit way. Frey and Slate also proposed the use of two different measures similar to ours, although they did not try to optimize them with multiobjective techniques. In this sense, the application of multiobjective techniques is new and offers promising perspectives for learning systems. **Covering** in MOLeCS is induced in the replacement stage and it is based on restricted replacement methods. Although Michigan style classifier systems are incremental, we have started with a non-incremental proposal. Our main reason is to understand the task of the multiobjective evaluation and niching, under a "classic GA" scheme. The migration to an incremental version will be performed in a near future, with the aim of improving the computational cost.

2.2 MOEAs and CSs

A multiobjective evolutionary algorithm (MOEA) and a learning classifier system (CS) have many related points:

- *Maintenance of a group of different solutions.* The solution of an MOEA is usually a set of many points, approximating the set of non-inferior solutions (Pareto optimal set). Similarly, the solution returned by a CS is a set of many rules, solving together a concept description. In order to find and maintain such multiple optima, some niching mechanism is required. MOEAs usually implement sharing, with the aim of obtaining a uniform distribution of the Pareto front or the Pareto optimal set.

- *Evaluation of an individual.* In MOEAs, each solution fitness depends on the other solutions, as it happens to each rule in the CSs.

Nevertheless, we can highlight some differences:

- *Generational and Non-Generational schemes.* One desirable feature in CSs is the on-line performance. This is promoted with a non-generational scheme, which introduces slight changes into the population by replacing only a small fraction of the population in each generation. This is not a very common scheme in MOEAs, although some proposals do exist [17]. Crowding is a natural way of performing niching in a non-generational scheme. For that reason, it has widely been used in classifier systems [11], [9].
- *Solution returned by the system.* Another important issue is the solution that MOEAs and CSs must obtain. MOEAs usually try to find a well-distributed population belonging to the Pareto front (or Pareto optimal set), from where the decision maker (DM) can perform a selection. If we translate this objective to MOLeCS, it should be expected to find the Pareto front corresponding to the *generality* and the *accuracy* goals. When we use the system in the exploitation mode, the DM has to select the best rules from all the available rules in the Pareto optimal set. Which rules should be used? If we want to perform accurate classifications, it seems obvious to choose always the most accurate rules. Therefore, we can state our learning problem as a particular multiobjective problem (MOP) where the DM's choices are known *a priori*. We can take profit of that by guiding the search towards the preferred areas of the DM.

3 Description of the System

Each individual in MOLeCS codifies a rule (classifier) of type: $rule : condition \rightarrow action$. The condition is the conjunction of tests over the problem attributes. It is represented by the ternary string: $\{0, 1, \#\}^k$, with length equal to the number of describing attributes. The symbol '#' (don't care) matches all values of an attribute, so it permits us to express generalizations. The action part of the rule is represented by the binary string: $\{0, 1\}^l$.

Each individual must have a fitness value in order to apply the appropriate selective pressure. This is not an easy task, since each classifier does not represent a complete solution to the overall problem. In fact, each rule r_i can match a different number of examples and can predict those examples in different degrees of accuracy. We compute these two features for each rule in the population:

$$generality(r_i) = \frac{\text{\# covered examples } (r_i)}{\text{\# examples in the training set}} \quad (1)$$

$$accuracy(r_i) = \frac{\text{\# correctly classified examples } (r_i)}{\text{\# covered examples } (r_i)} \quad (2)$$

If fitness is only based on *accuracy* the search will be biased towards accurate but too specific rules. This can result in an enhancement of the solution set, poor covering, etc. On the contrary, basing fitness on *generality* will result in low performance (in terms of classification accuracy). The solution is to balance these two characteristics (accuracy and generality) and optimize them simultaneously. Our hypothesis is that a multiobjective approach will lead the search towards general and accurate rules, resulting in a minimum, complete and accurate set of rules. We have tested and compared different multiobjective strategies, which are described in section 3.1.

Once the fitness assignment phase is performed, the GA proceeds to the selection and recombination stages. Selection is performed with stochastic universal sampling (SUS) [1].

As stated before, a concept description must be complete; that is, all the input examples must be covered. We use the term *covering* as the ratio of instances covered by the entire rule set RS to the size of the training set:

$$covering = \frac{\# \text{ examples covered by RS}}{\# \text{ examples in the training set}} \quad (3)$$

Promoting general classifiers is not sufficient to reach a 100% of covering. The genetic algorithm can tend, due to the genetic drift [10], to one general and accurate classifier and usually one classifier does not solve the overall problem. Therefore, we must enforce the *co-evolution* of a set of fit rules by niching mechanisms. Niching in MOLeCS is performed in the replacement stage (section 3.2).

Once the system has learned, it is used under an exploit or test phase. It works as follows. An example coming from the test set is presented. Then, the system finds the matching rules and applies the fittest rule to predict the associated action or class. As explained before, in case of equally fit rules, the most accurate rule is chosen.

3.1 Multiobjective Learning

In the previous section, we have defined our learning as a multiobjective problem (MOP). Now, we formalize the concepts mentioned before. Next, we consider different multiobjective algorithms to solve our MOP.

Definition 1 *The MOP evaluation function, $F : X \to Y$ maps decision variables $\mathbf{x} = (x_1, ..., x_n) \in X$ to vectors $\mathbf{y} = (y_1...y_k) \in Y$. In MOLeCS, the decision variables are the rules, while the objective vectors, of dimensionality k=2, are of type: $\mathbf{y} = F(\mathbf{x}) = (f_1(\mathbf{x}), f_2(\mathbf{x}))$, where $f_1(\mathbf{x}) = accuracy(\mathbf{x})$ computed from equation (2) and $f_2(\mathbf{x}) = generality(\mathbf{x})$, from equation (1).*

Definition 2 *Our multiobjective learning problem is defined as follows:*
  Maximize $\mathbf{y} = F(\mathbf{x}) = (accuracy(\mathbf{x}), generality(\mathbf{x}))$
  where $\mathbf{x}$ is the decision vector, and $accuracy(\mathbf{x})$ and $generality(\mathbf{x})$ are described by equations (2) and (1) respectively.

There are several MOEA techniques [7], [5], [18]. We have tested and compared four different algorithms, representative of three main algorithmic approaches [1]: a Pareto-approach, a non-Pareto approach and a plain aggregating approach.

Pareto-based Approach. We consider the Pareto approach proposed by Goldberg [9], which consists of ranking the population into non-dominated sets and then, assigns fitness according to this rank.

The Pareto approach gives the same selective pressure to non-dominated objective vectors. Suppose we have two objective vectors: $\mathbf{y}_1 = (1, 0.6)$ and $\mathbf{y}_2 = (0.5, 0.9)$, being non-dominated as depicted in figure 1(a). In this case, the Pareto approach assigns them the same rank. This means that the final solution set returned by the system can contain overgeneral rules (e.g., vector $\mathbf{y}_2$) as well as maximally general rules[2] (e.g., vector $\mathbf{y}_1$). Nevertheless, in the exploitation phase, these overgeneral rules will not be selected, because they can degrade the classification performance. The decision maker will always choose the best accurate rules (i.e. the maximally general rules).

Our hypothesis is that these overgeneral rules do not contribute significantly to our final solution. Indeed, they can degrade our search towards an accurate rule set, because they consume resources from the population and thus, they may prevent other accurate rules from being explored.

If we know the decision preferences in advance, we may guide the GA more efficiently towards these preferences. For that reason, we have designed a modification of the Pareto approach which gives a bias towards accurate rules. As shown in figure 1(b), inside each group of non-dominated classifiers a second level of ranking is performed, based on the accuracy of classifiers. In the following, we will refer to the Pareto original approach as PR (Pareto ranking) and to the modified algorithm with the accuracy bias as PAR (Pareto-accuracy ranking).

Population-based non-Pareto Approach. Using the idea of promoting the most accurate areas, we have designed a population-ranking method, based on the lexicographic ordering. The algorithm ranks the population according to the accuracy objective. When two or more individuals equally accurate are found, they are ordered by the generality objective. In this way, we state that the first goal to be achieved is accuracy (in order to obtain accurate classifiers) and second, these classifiers must be *as general as possible*. An example of this ranking, which we term *accuracy-generality ranking* (AGR), is depicted in figure 1(c).

---

[1] We consider here the MOEA classification scheme made by Fonseca and Fleming in [7].
[2] Kovacs [14] defines a maximally general rule as an accurate rule (accuracy=1.0) which cannot be more general without becoming inaccurate. A rule being inaccurate due to excessive generalizations, is called an overgeneral rule. A suboptimally general rule is an accurate rule that can be more general (have more '#') without losing its accuracy.

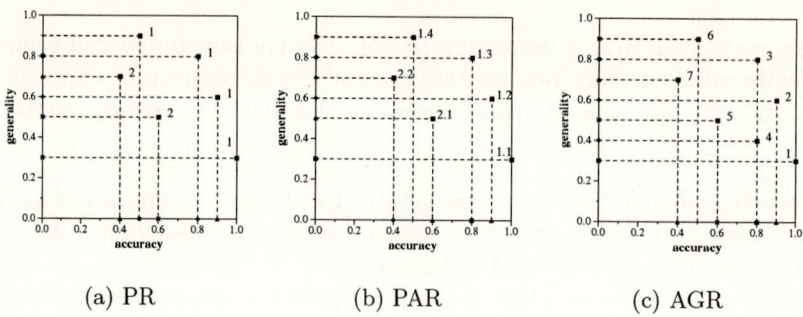

Fig. 1. Multiobjective evaluation methods: Pareto ranking (PR), Pareto-accuracy ranking (PAR) and accuracy-generality ranking (AGR).

Plain Aggregating Approach. The weighted sum (WS) approach weighs and sums up all the objectives obtaining directly a scalar fitness value. Our multiobjective problem is then solved as: $max \sum_{i=1}^{k} w_i y_i$ where **w** is the weight vector that must be set depending on the relative importance of the objectives. When we test MOLeCS with well-known problems (see section 4), we can tune these coefficients properly. In this case, we use the WS as a bound from where the other algorithms can be compared.

A Priori and A Posteriori Preference Articulation. From the DM's perspective, PAR, AGR and WS can be classified as *a priori preference articulation* methods [19], because they use the decision preferences in the fitness/selection stage. Our preferences are: "those accurate rules being as general as possible" or in other words, the *maximally general rules*.

On the contrary, the Pareto based approach (PR) does not use these preferences until the exploitation phase, when the learning process has finished. Therefore, it searches for a compromise between the accuracy and the generality goals. From the resulting set of non-dominated solutions, the DM has to select the maximally general rules *a posteriori*.

In sections 4 and 5, we will test if the methods based on *a priori preference articulation* can outperform the solution given by *a posteriori preference articulation* method as the Pareto approach.

3.2 Niching

Niching methods are the key point for classifier systems to evolve a population of diverse rules. We enforce niching in the replacement stage using crowding strategies. We have analysed crowding, two variants of crowding including selective pressure and deterministic crowding.

Crowding (or crowding factor model, CF) was introduced by De Jong [6]. The algorithm tries to preserve the diversity of population, by replacing each new individual (from the offspring population) by a similar one in the parents population. To be exact, the new individual is compared to a subpopulation of *cf* members and the member with the highest similarity is replaced.

In order to induce a convergence pressure in the CF model, two variants are tested: CIF and CC. Both try to replace a "low fitness and similar individual". The former differs from the CF model in the selection of the subpopulation of *cf* members. Instead of selecting them randomly, the selection is performed with a probability inversely proportional to fitness. The second variant, which we term CC, was used in the Simple Classifier System (SCS) [9]. The method consists of selecting each member going to the cf-subpopulation from a bucket of *csp* individuals. The worst individual from the csp-bucket is inserted into the cf-subpopulation. Then, the algorithm proceeds as the CF model.

Deterministic crowding (DC) makes a competition between each pair of {parent,offspring}, choosing the competing pairs with a minimum-distance criterion [15]. The child only replaces its parent when its fitness is greater.

4 Experimental Results

4.1 Design of Experiments

We first analyse the learning performance of the system with two well-known learning tasks, usually tested in the CSs' community [21], [16]: the multiplexer problem and the parity problem. This election is made for several reasons. First, for the simplicity to test our system, since the desired solution is known. Second, because they represent two different types of problems. The multiplexer allows generalizations in its rules; so the bias towards generality is desirable for the achievement of a minimal set. On the contrary, the parity problem does not need any generality pressure, since all the rules required to describe the problem must be specific. In this sense, we will test the ability of MOLeCS (and its multiobjective algorithms) to scale to the different levels of generalization.

The results are shown for the multiplexer with 11 inputs (11-mux) and the parity problem with 5 inputs (5-par). Each problem is tested with: four multiobjective strategies (PR, PAR, AGR and WS) and four different niching methods (CF, CIF, CC and DC). We also compare our results with a single-objective EA, optimizing only accuracy (i.e., using the WS approach with $\mathbf{w} = (1,0)$).

Each result is the average of five runs using different seed numbers. The parameters settings for the 11-mux problem are: population size = 800, Pgen=0.3 (probability of generalization in the initialization of population and in the mutation operator), G=0.2 (generation gap), $p_c = 0.9$ (probability of crossover), p_m=0.01 (probability of mutation per gene). If DC is used, $p_c = 1.0$ according to the algorithm definition. The crowding methods CF, CC and CIF require the subpopulation sizes. They are tuned previously, and here we only show the results obtained with the best parameters (whose values are reported in the corresponding figures). In case of the 5-par problem, the parameters settings are the same, except for the population size that is 250.

4.2 Metrics of Performance

In order to test the learning performance of the system, we will use the following metrics:

- *Covering.* This is the ratio of training instances covered by the population, as defined in equation (3). This measure is related to the ability of the niching methods to maintain multiple rules. Covering can also be improved if generalizations are found.
- *Accuracy.* Two metrics of accuracy can be performed on the overall rule set. The first one, termed as *crude accuracy* (CA) [12], is defined as the number of correctly classified examples over the covered examples. The second measure, or *corrected crude accuracy* (CCA), is the ratio between the correct classified examples and the total number of examples presented to the system.
- *Size of the solution set.* One desirable goal in learning is to minimize the size of the solution set. If the set is small, it is more explanatory and easier to understand by the human experts. We will consider the size of the solution set as the number of different rules.
- *Optimal Population.* Kovacs [14] defines an optimal population (denoted as [O]) as having three characteristics: it is complete, non-overlapping and minimal. In case of the 6-multiplexer problem, the optimal population in MOLeCS consists of 8 rules, of type: 01#0##:0. In the 11-multiplexer problem, [O] consists of 16 rules, while the 5-parity problem needs 32 rules. This measure is useful for testing if the developed rules have reached the optimal generality, in comparison to accuracy and covering that do not necessary give us this information. In the multiplexer problem, reaching a 100% of accuracy and covering does not imply directly that [O] has been reached.
- *Learning speed.* Although the speed measure is more important in the test epoch (exploit phase), the speed in the training epoch is also desirable, specially when the system has to learn from real-world applications.

4.3 Results

Figure 2 shows a summary of the results obtained in the 11-multiplexer problem. A graph represents a fixed niching method, and inside each one there is a curve for each multiobjective strategy. The curves plot the CCA (corrected crude accuracy), which is measured over all the training examples. Covering is not shown since all methods achieved the 100%. The WS approach was previously tuned to $\mathbf{w} = (0.75, 0.25)$.

The main differences arise between the crowding methods. CF is the method with the worst accuracy, which ranges from 0.70 to 0.80 (see figure 2(a)). Adding a selective pressure in the replacement stage improves the performance. This happens specially with CC -see figure 2(c)-, where accuracy is about 0.90. These results are obtained for subpopulation sizes of 30/30. Nevertheless, this method is very sensitive to the parameter settings. When the size of the csp-subpopulation raises up, increasing the selective pressure, the accuracy (not reported here)

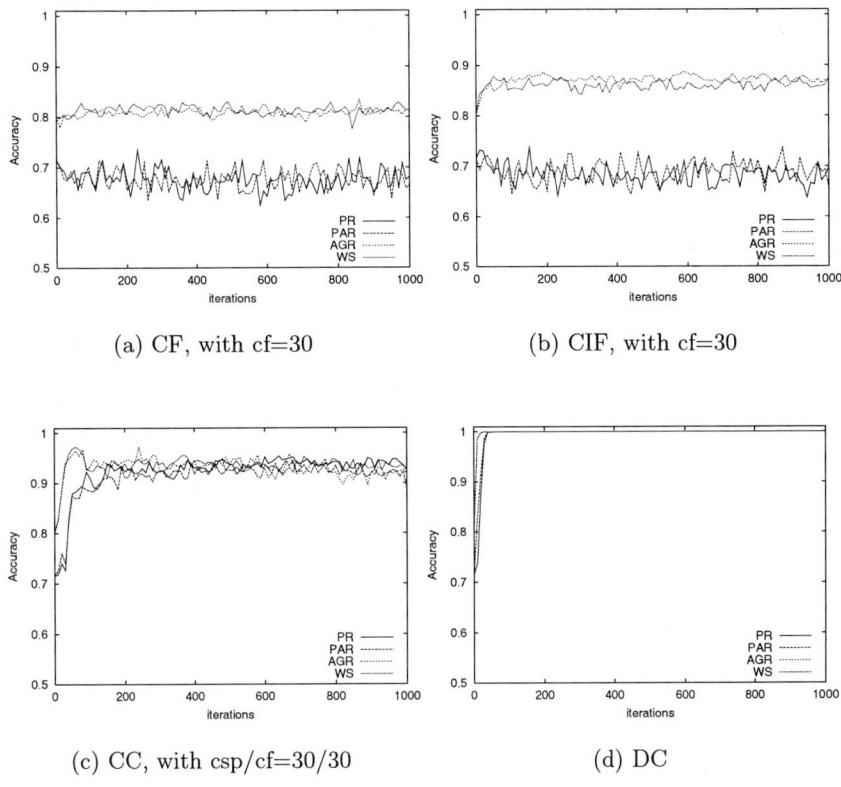

Fig. 2. Results in the 11-multiplexer problem. Comparison between four different crowding methods: CF, CIF, CC and DC. Each niching method is shown for each multiobjective evaluation method: PR, PAR, AGR and WS with $\mathbf{w} = (0.75, 0.25)$. Curves show the corrected accuracy average over five runs, traced along 1000 iterations.

decreases. The results achieved with deterministic crowding -figure 2(d)- outperform all previous results. Accuracy reaches 1.0 in the early generations. The method balances appropriately the selective pressure and the maintenance of niches, reaching the optimal performance. These results are consistent with other niching studies which demonstrate the superiority of DC on different test problems [16].

When the appropriate crowding method is used, there are no significant differences between the four multiobjective algorithms, in terms of accuracy and speed performance.

In the 5-par problem, the behaviour of the different niching methods is similar to the 11-mux problem (see figure 3). What is important to mention here is the difference in performance that arises between some multiobjective methods. Figure 3(d) shows that the Pareto approach has the poorest accuracy (with a value of 0.78). This is because PR does not establish any preference towards the rule accuracy, but towards a compromise between generality and accuracy. Thus, PR seeks for overgeneral rules as well as for maximally general rules.

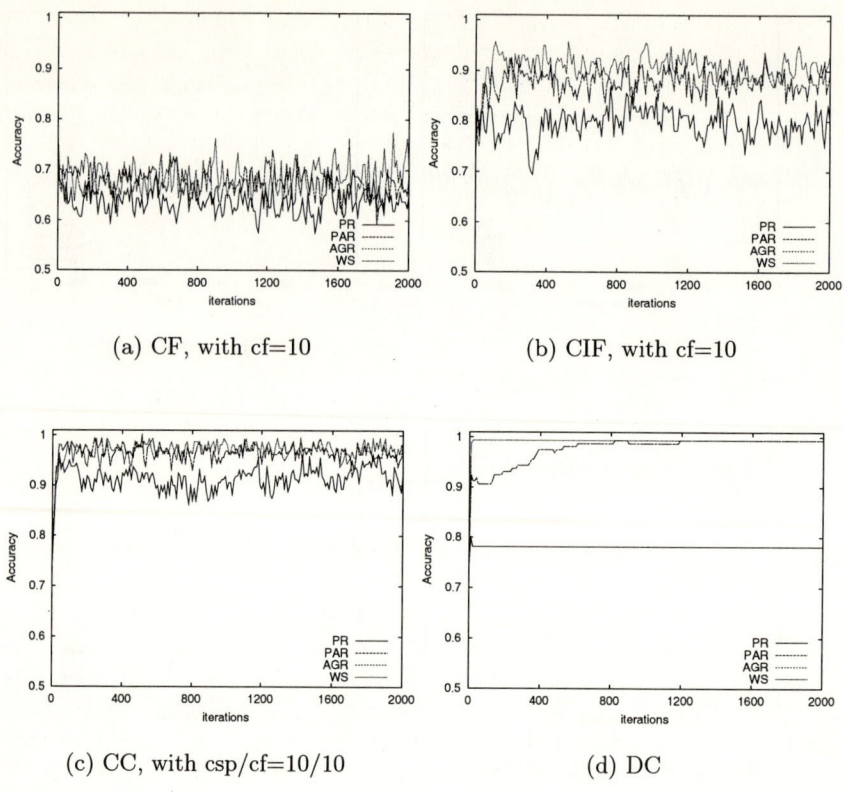

Fig. 3. Results in the 5-parity problem. Comparison between four different crowding methods: CF, CIF, CC and DC. Each niching method is shown for each multiobjective evaluation method: PR, PAR, AGR and WS with $\mathbf{w} = (1, 0)$. Curves show the corrected accuracy average over five runs, traced along 2000 iterations.

These results confirm our hypothesis about the presence of overgeneral rules in the population. They do not contribute to the expected solution, preventing other desirable accurate rules from being explored. Adding a preference towards accurate rules, as PAR and AGR do, improves the previous results. WS has the best speed, although its application depends on the appropriate knowledge about the weight vector settings. In these results, $\mathbf{w} = (1, 0)$, which is the same as a single objective optimization towards accuracy.

Table 1 reports the different performance measures obtained from the final rule set (after learning is performed). In the 11-mux problem, all the multiobjective approaches have achieved the same performance, except for PR which presents the highest rule set size (217). The last row in the table shows the results obtained by a single-objective learning (SO), optimizing accuracy. Covering achieves only 93% of the examples, while the final rule set is much more complex than the other approaches (with 791 different rules) and the optimal population is not reached at all.

Table 1. Results in the 11-mux and 5-par problems, summarized for the different multiobjective strategies and compared to a single optimization method (row labeled as SO). DC is used as the crowding method. In case of the 11-mux, $\mathbf{w} = (0.75, 0.25)$. In case of 5-par problem, $\mathbf{w} = (1, 0)$, which is the same as the single-objective algorithm. **Cov** is covering, **CCA** is the corrected accuracy, **Size** is the number of different rules and **%[O]** is the percentage of the optimal population reached by the final rule set. These measures are computed from the final rule set (obtained when the training epoch has finished).

	11-multiplexer				5-parity			
	Cov	CCA	Size	%[O]	Cov	CCA	Size	%[O]
PR	1	1	217	1	1	0.78	21	0.60
PAR	1	1	40	1	1	0.99	34	0.99
AGR	1	1	40	1	1	0.99	34	0.99
WS	1	1	40	1	1	0.99	32	0.98
SO	0.93	0.93	791	0				

In the 5-par problem, PR has converged with 0.78 of accuracy and only 60% of the expected optimal population. In fact, the population has collapsed to the non-inferior solutions, which represent only two points in the objective space. The first one, with objective vector $\mathbf{y}_1 = (0.03, 1)$, corresponds to rules of type: 00010 : 1. The second one, with vector $\mathbf{y}_2 = (0.5, 0.5)$, corresponds to the overgeneral rules: ##### : 1 and ##### : 0. Once the population has converged to these points, it is very difficult to increase the number of specific rules, because of the presence of too many #'s in the population schemata. This is also the reason for such a small set size.

5 Application to a Real-World Classification Problem

In this section, we apply MOLeCS to the Wisconsin breast cancer database, obtained from the UCI repository [4]. The database contains 699 instances, with 9 numerical attributes ranging from 1 to 10, and two classes (benign or malignant). There are 16 instances with missing attribute values. The class distributions are unbalanced, having 458 (65.5%) benign and 241 (34.5%) malignant instances.

As the describing attributes are numerical and not binary, we must consider again our rule representation. First, we can discretize each numerical feature into a string of bits. This allows us to maintain our binary representation in the rules. The second possibility is to represent a rule as a set of real-valued intervals, as proposed by Wilson in [23]. We have implemented and tested both representations, without significant differences for the Wisconsin database.

Each experiment is averaged for five different seed numbers. Accuracy is estimated using ten-fold cross-validation (for details see [20]). The results reported in table 2 show the covering, the crude accuracy and the corrected accuracy, measured on test sets. MOLeCS is run with DC and the four multiobjective strategies. We also ran the single-objective optimization algorithm, obtaining a corrected accuracy of 0.90. PR achieved a result of 0.65, while PAR and AGR

Table 2. Results using the Wisconsin database, obtained with a ten-fold cross-validation experiment. We compare the four multiobjective strategies and the single-objective algorithm, using DC as crowding method. The table reports: **Cov** (ratio of covered examples), **CA** (crude accuracy) and **CCA** (corrected crude accuracy).

	Cov	CA	CCA
PR	1	0.65	0.65
PAR	0.98	0.97	**0.95**
AGR	0.98	0.97	**0.95**
WS (.75,.25)	0.98	0.94	0.92
SO	0.94	0.95	0.90

reached the maximum value, with 0.95 of accuracy. This confirms the results obtained with the multiplexer and parity problems. First, it seems suitable to apply multiobjective methods in order to optimize each rule accuracy and generality. Enforcing only the accuracy leads the system to develop a high number of specific rules. This makes the learning more difficult, because more rules are needed to describe the problem. SO resulted in less covering and even less crude accuracy than PAR and AGR. If we give the same pressure (or preference) to accuracy than to generality (as PR does) we degrade the final rule set accuracy, as happened with 5-par problem. Therefore, the best learning performance is achieved when we optimize generality and accuracy, but with a preference towards accuracy as it is implemented by the methods PAR and AGR. The obtained accuracy is of 0.95 ± 0.016 with a 95% confidence interval, which is comparable to other learning classifier systems. XCS [24] also reached an accuracy of 0.95 in a similar experiment using the same database.

6 Conclusions and Future Work

This paper has studied the performance of MOLeCS using different MOEA techniques. The results are compared to a single-objective EA optimizing only the accuracy goal, in order to prove the suitability of the multiobjective approach. The experiments with single-objective optimization demonstrate that the system evolves too many specific rules. This produces an enhancement of the solution set, making the learning more difficult and achieving poor covering. If we optimize the accuracy and generality of each rule, we improve the learning performance. Nevertheless, giving the same importance to these attributes (as Pareto ranking does) makes the system evolve overgeneral rules in the search process, preventing other maximally general rules from being explored and maintained. The overall accuracy of the final rule set is thus degraded. In this sense, including the decision preferences in the search (e.g., with PAR or AGR) leads up to a better achievement of the learning goals. The results with the Wisconsin database have reached an accuracy of 0.95, performing as well as XCS.

The use of niching methods in MOLeCS is necessary to ensure the maintenance of a set of rules that covers the examples. The paper has studied different crowding algorithms under a non-generational scheme. In terms of covering and stability the best results are obtained with deterministic crowding.

As a future work, it is necessary to perform further investigation on the applicability of MOLeCS to real world databases. We can test the MOLeCS performance on more complex problems, having more describing attributes, performing multiple categorization rather than binary, etc.

When we deal with medical databases with two unbalanced classes, it is interesting to distinguish between the correct predictions made by the system when the true decision is "benign" and the correct predictions when the true decision is "malignant". In this case, the accuracy measure does not give enough information. Other measures as sensitivity, specificity and area under the ROC curve must be included in our further analysis.

Another important future research with MOLeCS is to study our approach with problems with highly unbalanced classes. Giving a pressure towards generalization might displace specific rules (that cover few examples from a certain class) by other general rules (covering examples from other more numerous classes). This can be prevented in the replacement stage or by measuring the generality of each rule relatively to its niche.

Acknowledgements. The results of this work were obtained with the equipment co-funded by *Direcció de Recerca de la Generalitat de Catalunya (D.O.G.C 30/12/1997)*. The authors acknowledge the support provided by Epson Iberica, under 1999 Rosina Ribalta Award, and the support of *Enginyeria i Arquitectura La Salle*.

References

1. J.E. Baker. Reducing bias and inefficieny in the selection algorithm. In J.J.Grefenstette, editor, *Genetic Algorithms and their Applications: Proceedings of the Second International Conference on Genetic Algorithms*, pages 14–21, 1987.
2. Ester Bernadó Mansilla and Josep Maria Garrell i Guiu. MOLeCS: A MultiObjective Classifier System. In *Proceedings of the Genetic and Evolutionary Computation Conference (GECCO)*, page 390, 2000.
3. Ester Bernadó Mansilla and Josep Maria Garrell i Guiu. MultiObjective Learning in a Genetic Classifier System (MOLeCS). In *Butlletí de l'ACIA, 22. 3r Congrés Català d'Intel.ligència Artificial*, 2000.
4. C.L. Blake and C.J. Merz. UCI Repository of machine learning databases, [http://www.ics.uci.edu/~mlearn/MLRepository.html]. University of California, Irvine, Dept. of Information and Computer Sciences, 1998.
5. Carlos A. Coello. A Comprehensive Survey of Evolutionary-Based Multiobjective Optimization Techniques. *Knowledge and Information Systems. An International Journal*, 1(3):269–308, August 1999.
6. K.A. De Jong. *An analysis of the behavior of a class of genetic adaptive systems. (Doctoral Dissertation)* . PhD thesis, University of Michigan, 1975.

7. Carlos M. Fonseca and Peter Fleming. An Overview of Evolutionary Algorithms in Multiobjective Optimization. *Evolutionary Computation*, 3(1):1–16, 1995.
8. P.W. Frey and D.J. Slate. Letter recognition using Holland-style adaptive classifiers. *Machine Learning*, 6:161–182, 1991.
9. David E. Goldberg. *Genetic Algorithms in Search, Optimization and Machine Learning*. Addison-Wesley Publishing Company, Inc., 1989.
10. D.E. Goldberg and P. Segrest. Finite Markov chain analysis of Genetic Algorithms. In J.J. Grefenstette, editor, *Proceedings of the Second International Conference on Genetic Algorithms*, pages 1–8. Lawrence Erlbaum, 1987.
11. John H. Holland. Escaping Brittleness: The Possibilities of General Purpose Learning Algorithms Applied to Parallel Rule-Based Systems. *Machine Learning: An Artificial Intelligence Approach, Vol. II*, pages 593–623, 1986.
12. John H. Holmes. Quantitative Methods for Evaluating Learning Classifier System Performance in Forced Two-Choice Decision Tasks. In *Second International Workshop on Learning Classifier Systems (IWLCS-99)*, 1999.
13. J. Horn, D.E. Goldberg, and K. Deb. Implicit Niching in a Learning Classifier System: Nature's Way. *Evolutionary Computation, 2(1)*, pages 37–66, 1994.
14. Tim Kovacs. XCS Classifier System Reliably Evolves Accurate, Complete and Minimal Representations for Boolean Functions. In Roy, Chawdhryand, and Pant, editors, *Soft Computing in Engineering Design and Manufacturing*, pages 59–68. Springer-Verlag, 1997.
15. Mahfoud, Samir W. Crowding and preselection revisited. In R.Maenner and B.Manderick, editors, *Parallel Problem Solving from Nature, 2*, pages 27–36. Elsevier:Amsterdam, 1992.
16. Mahfoud, Samir W. *Niching Methods for Genetic Algorithms*. PhD thesis, University of Illinois at Urbana-Champaign, 1995.
17. Manuel Valenzuela-Rendón and Eduardo Uresti-Charre. A Non-Generational Genetic Algorithm for Multiobjective Optimization. *Proceedings of the Seventh International Conference on Genetic Algorithms*, pages 658–665, 1997.
18. David A. Van Veldhuizen. *Multiobjective Evolutionary Algorithms: Classifications, Analyses, and New Innovations*. PhD thesis, Dept. of Electrical and Computer Engineering. Air Force Institute of Technology, Wright-Patterson AFB, Ohio, 1999.
19. David A. Van Veldhuizen and Gary B. Lamont. Multiobjective Evolutionary Algorithms: Analyzing the State-of-the-Art. *Evolutionary Computation*, 8(2):125–174, Summer 2000.
20. Sholom M. Weiss and Casimir A. Kulikowski. *Computer Systems That Learn. Classification and Prediction Methods from Statistics, Neural Nets, Machine Learning and Expert Systems*. Morgan Kaufmann, 1991.
21. Stewart W. Wilson. Classifier Fitness Based on Accuracy. *Evolutionary Computation*, 3(2):149–175, 1995.
22. Stewart W. Wilson. Generalization in the XCS Classifier System. In J.Koza et al., editor, *Genetic Programming: Proceedings of the Third Annual Conference*. San Francisco, CA: Morgan Kaufmann, 1998.
23. Stewart W. Wilson. Get Real! XCS with Continuous-Valued Inputs. In L. Booker, S. Forrest, Mitchell M., and Riolo R, editors, *Festschrift in Honor of John H. Holland*. Center for the Study of Complex Systems, University of Michigan, 1999.
24. Stewart W. Wilson. Mining Oblique Data with XCS. In *Third International Workshop on Learning Classifier Systems (IWLCS-2000)*, 2000.

Author Index

Obayashi, Shigeru 639
Okabe, Tatsuya 96
Oshita, Shinjiro 501
Osyczka, Andrzej 141

Parmee, I.C. 52
Petrovski, Andrei 531
Pratap, Amrit 284
Pulido, Gregorio T. 126

Rahoual, Malek 416
Ranjithan, S. Ranji 299
Roubos, Hans 653

Sait, Sadiq M. 400
Sannomiya, N. 443
Sasaki, Daisuke 639
Savini, A. 560
Sbalzarini, Ivo F. 516
Sendhoff, Bernhard 96
Steuer, Ralph E. 41

Sugimura, Tatsuo 501

Talbi, El-Ghazali 416
Tan, K.C. 111, 344
Tanaka, Kiyoshi 501
Teich, Jürgen 314
Thiele, Lothar 181
Thompson, Mark 546

Van Veldhuizen, David A. 226

Watson, Richard A. 269
White, Brian A. 668
Wright, Jonathan 256

Youssef, Habib 400

Zamboglou, Nikolaos 167, 574
Zhao, Y. 443
Zitzler, Eckart 181
Zydallis, Jesse B. 226

Author Index

Aguirre, Hernán E.   501
Andersson, Johan   624

Babuška, Robert   653
Bagchi, Tapan P.   458
Balicki, Jerzy   373
Baltas, Dimos   167, 574
Becker, Bernd   154
Bernadó i Mansilla, Ester   696
Blumel, Anna L.   668
Brizuela, C.   443

Carlyle, W. Matthew   472
Chetan, S. Kishan   299
Coello Coello, Carlos A.   21, 126
de Coligny, Marc   486
Corne, David W.   269

Dakshina, Harish K.   299
Deb, Kalyanmoy   67, 284, 385
Dhaenens, Clarisse   416
Di Barba, P.   560
Drechsler, Nicole   154
Drechsler, Rolf   154

Eklund, Neil H.   603
El Moudani, Walid   486
Embrechts, Mark J.   603
Erickson, Mark   681

Farina, M.   560
Fonseca, Carlos M.   213
Fowler, John W.   472

Gandibleux, Xavier   429
Garrell i Guiu, Josep M.   696
Gel, Esma S.   472
Gen, Mitsuo   82
Goel, Tushar   67, 385
Gómez-Skarmeta, Antonio F.   653
Grunert da Fonseca, Viviane   213

Hall, Andreia O.   213
Hanne, Thomas   197
Hapke, Maciej   241
Horn, Jeffrey   681

Hughes, Evan J.   329, 668

Iredi, Steffen   359
Ishibuchi, Hisao   82, 588

Jaszkiewicz, Andrzej   241
Jiménez, Fernando   653
Jin, Yaochu   96

Katoh, Naoki   429
Khan, Salman A.   400
Khor, E.F.   111, 344
Kim, Bosun   472
Kitowski, Zygmunt   373
Knowles, Joshua D.   269
Kominek, Paweł   241
Koumoutsakos, Petros   516
Krenich, Stanislaw   141
Krus, Petter   624

Lahanas, Michael   167, 574
Lamont, Gary B.   226
Laumanns, Marco   181, 612
Laumanns, Nando   612
Lee, T.H.   111, 344
Loosemore, Heather   256

Mabed, Mohamed Hakim   416
Mayer, Alex   681
McCall, John   531
Merkle, Daniel   359
Meyarivan, T.   284
Middendorf, Martin   359
Miettinen, Kaisa   1
Milickovic, Natasa   167, 574
Mora-Comino, Félix   486
Morikawa, Masashi   639
Morita, Hiroyuki   429
Müller, Sibylle   516
Murata, Tadahiko   82, 588

Nakahashi, Kazuhiro   639
Nakashima, Tomoharu   588
Neunzig, Dirk   612
Nunes Cosenza, Carlos A.   486

3. McKinney, D. C., and M. Lin, Genetic algorithm solutions of groundwater management models, *Water Resour. Res., 30*(6), 1897-1906, 1994.
4. Ritzel, B. J., J. W. Eheart, and S. Ranjithan, Using genetic algorithms to solve a multiple objective groundwater pollution containment problem, *Water Resour. Res., 30*(5), 1589-1603, 1994.
5. Cieniawski, S. E., J. W. Eheart, and S. Ranjithan, Using genetic algorithms to solve a multiobjective groundwater monitoring problem, *Water Resour. Res., 31*(2), 399-409, 1995.
6. Goldberg, D. E., *Genetic Algorithms in Search, Optimization, and Machine Learning*, Addison-Wesley, Reading, Mass., 1989.
7. Horn, J., N. Nafpliotis, and D. E. Goldberg, A niched Pareto genetic algorithm for multi-objective optimization, *The Proceedings of the First IEEE Conference on Evolutionary Computation (ICEC '94)*, Piscataway, NJ: IEEE Service Center, 82-87, 1994.
8. Fonseca, C. M., and Fleming, P. J., Genetic algorithms for multiobjective optimization: formulation, discussion, and generalization, *Proceedings of the Fifth International Conference on Genetic Algorithms (ICGA 5)*, San Mateo, CA: Morgan Kaufman, 416-423, 1993.
9. Srinivas, N., and Deb, K., Multiobjective optimization using nondominated sorting in genetic algorithms, *Intl. Journal of Evolutionary Computation*, 2, 221-248, 1994.
10. Oei, C., Goldberg, D. E., and Chang, Tournament Selection, Niching, and the Preservation of Diversity, *IlliGAL Technical Report 94002*, 1994.
11. Maxwell, R. M., Understanding the effects of uncertainty and variability on groundwater-driven health risk assessment, Ph.D. dissertation, Univ. of Calif., Berkeley, 1998.
12. LaBolle, E. M., G. E. Fogg, and A. F. B. Tompson, Random-walk simulation of transport in heterogeneous porous media: Local mass-conservation problem and implementation methods, *Water Resour. Res., 32*(3), 583-593, 1996.
13. Chan Hilton, A. B., and T. B. Culver, Constraint handling for genetic algorithms in optimal remediation design, *J. Water Resour. Plan. Manage., 126*(3), 128-137, 2000.

MOLeCS: Using Multiobjective Evolutionary Algorithms for Learning

Ester Bernadó i Mansilla and Josep M. Garrell i Guiu

Computer Science Department
Enginyeria i Arquitectura La Salle, Ramon Llull University
Passeig Bonanova, 8, 08022-Barcelona (Spain)
{esterb,josepmg}@salleURL.edu

Abstract. MOLeCS is a classifier system (CS) which addresses its learning as a multiobjective task. Its aim is to develop an optimal set of rules, optimizing the *accuracy* and the *generality* of each rule simultaneously. This is achieved by considering these two goals in the rule fitness. The paper studies four multiobjective strategies that establish a compromise between accuracy and generality in different ways. The results suggest that including the decision maker's preferences in the search process improves the overall performance of the obtained rule set. The paper also studies a third major objective: *covering* (the maintenance of a set of different rules solving together the learning problem), through different niching mechanisms. After a performance analysis using some benchmark problems, MOLeCS is applied to a real-world categorization task: the diagnosis of breast cancer.

1 Introduction

The learning task performed by a classifier system (CS) is itself multiobjective [13]: it has to find a concept description, usually represented by a set of rules, which should be: (a) complete, (b) accurate and (c) minimum. In terms of classification, a rule set is *complete* when it covers (satisfies) all the examples, whereas a rule set is *accurate* when it classifies the examples correctly (i.e., without misclassification errors). The third objective involves *minimizing* the number of rules, in order to obtain concise and comprehensible descriptions. Another related objective is the system's capability to express *generalizations*, that is, to generalize all the similar examples.

These multiple objectives are closely connected. Generalization of equivalent examples allows more concise representations and leads up to smaller rule sets, promoting covering. But these objectives are opposed to accuracy in some way. If the system performs an excessive generalization, the accuracy of classification will be degraded. Thus, optimizing these conflicting objectives simultaneously involves a weak equilibrium, which is difficult for a CS to reach and maintain.

The classifier systems' community has solved this multiobjective learning task in an implicit way. Holland's CS [11] uses a credit allocation algorithm to

Lecture Notes in Computer Science

For information about Vols. 1–1920
please contact your bookseller or Springer-Verlag

Vol. 1921: S.W. Liddle, H.C. Mayr, B. Thalheim (Eds.), Conceptual Modeling for E-Business and the Web. Proceedings, 2000. X, 179 pages. 2000.

Vol. 1922: J. Crowcroft, J. Roberts, M.I. Smirnov (Eds.), Quality of Future Internet Services. Proceedings, 2000. XI, 368 pages. 2000.

Vol. 1923: J. Borbinha, T. Baker (Eds.), Research and Advanced Technology for Digital Libraries. Proceedings, 2000. XVII, 513 pages. 2000.

Vol. 1924: W. Taha (Ed.), Semantics, Applications, and Implementation of Program Generation. Proceedings, 2000. VIII, 231 pages. 2000.

Vol. 1925: J. Cussens, S. Džeroski (Eds.), Learning Language in Logic. X, 301 pages 2000. (Subseries LNAI).

Vol. 1926: M. Joseph (Ed.), Formal Techniques in Real-Time and Fault-Tolerant Systems. Proceedings, 2000. X, 305 pages. 2000.

Vol. 1927: P. Thomas, H.W. Gellersen, (Eds.), Handheld and Ubiquitous Computing. Proceedings, 2000. X, 249 pages. 2000.

Vol. 1928: U. Brandes, D. Wagner (Eds.), Graph-Theoretic Concepts in Computer Science. Proceedings, 2000. X, 315 pages. 2000.

Vol. 1929: R. Laurini (Ed.), Advances in Visual Information Systems. Proceedings, 2000. XII, 542 pages. 2000.

Vol. 1931: E. Horlait (Ed.), Mobile Agents for Telecommunication Applications. Proceedings, 2000. IX, 271 pages. 2000.

Vol. 1658: J. Baumann, Mobile Agents: Control Algorithms. XIX, 161 pages. 2000.

Vol. 1756: G. Ruhe, F. Bomarius (Eds.), Learning Software Organization. Proceedings, 1999. VIII, 226 pages. 2000.

Vol. 1766: M. Jazayeri, R.G.K. Loos, D.R. Musser (Eds.), Generic Programming. Proceedings, 1998. X, 269 pages. 2000.

Vol. 1791: D. Fensel, Problem-Solving Methods. XII, 153 pages. 2000. (Subseries LNAI).

Vol. 1799: K. Czarnecki, U.W. Eisenecker, Generative and Component-Based Software Engineering. Proceedings, 1999. VIII, 225 pages. 2000.

Vol. 1812: J. Wyatt, J. Demiris (Eds.), Advances in Robot Learning. Proceedings, 1999. VII, 165 pages. 2000. (Subseries LNAI).

Vol. 1932: Z.W. Raś, S. Ohsuga (Eds.), Foundations of Intelligent Systems. Proceedings, 2000. XII, 646 pages. (Subseries LNAI).

Vol. 1933: R.W. Brause, E. Hanisch (Eds.), Medical Data Analysis. Proceedings, 2000. XI, 316 pages. 2000.

Vol. 1934: J.S. White (Ed.), Envisioning Machine Translation in the Information Future. Proceedings, 2000. XV, 254 pages. 2000. (Subseries LNAI).

Vol. 1935: S.L. Delp, A.M. DiGioia, B. Jaramaz (Eds.), Medical Image Computing and Computer-Assisted Intervention – MICCAI 2000. Proceedings, 2000. XXV, 1250 pages. 2000.

Vol. 1936: P. Robertson, H. Shrobe, R. Laddaga (Eds.), Self-Adaptive Software. Proceedings, 2000. VIII, 249 pages. 2001.

Vol. 1937: R. Dieng, O. Corby (Eds.), Knowledge Engineering and Knowledge Management. Proceedings, 2000. XIII, 457 pages. 2000. (Subseries LNAI).

Vol. 1938: S. Rao, K.I. Sletta (Eds.), Next Generation Networks. Proceedings, 2000. XI, 392 pages. 2000.

Vol. 1939: A. Evans, S. Kent, B. Selic (Eds.), «UML» – The Unified Modeling Language. Proceedings, 2000. XIV, 572 pages. 2000.

Vol. 1940: M. Valero, K. Joe, M. Kitsuregawa, H. Tanaka (Eds.), High Performance Computing. Proceedings, 2000. XV, 595 pages. 2000.

Vol. 1941: A.K. Chhabra, D. Dori (Eds.), Graphics Recognition. Proceedings, 1999. XI, 346 pages. 2000.

Vol. 1942: H. Yasuda (Ed.), Active Networks. Proceedings, 2000. XI, 424 pages. 2000.

Vol. 1943: F. Koornneef, M. van der Meulen (Eds.), Computer Safety, Reliability and Security. Proceedings, 2000. X, 432 pages. 2000.

Vol. 1944: K.R. Dittrich, G. Guerrini, I. Merlo, M. Oliva, M.E. Rodriguez (Eds.), Objects and Databases. Proceedings, 2000. X, 199 pages. 2001.

Vol. 1945: W. Grieskamp, T. Santen, B. Stoddart (Eds.), Integrated Formal Methods. Proceedings, 2000. X, 441 pages. 2000.

Vol. 1946: P. Palanque, F. Paternò (Eds.), Interactive Systems. Proceedings, 2000. X, 251 pages. 2001.

Vol. 1947: T. Sørevik, F. Manne, R. Moe, A.H. Gebremedhin (Eds.), Applied Parallel Computing. Proceedings, 2000. XII, 400 pages. 2001.

Vol. 1948: T. Tan, Y. Shi, W. Gao (Eds.), Advances in Multimodal Interfaces – ICMI 2000. Proceedings, 2000. XVI, 678 pages. 2000.

Vol. 1949: R. Connor, A. Mendelzon (Eds.), Research Issues in Structured and Semistructured Database Programming. Proceedings, 1999. XII, 325 pages. 2000.

Vol. 1950: D. van Melkebeek, Randomness and Completeness in Computational Complexity. XV, 196 pages. 2000.

Vol. 1951: F. van der Linden (Ed.), Software Architectures for Product Families. Proceedings, 2000. VIII, 255 pages. 2000.

Vol. 1952: M.C. Monard, J. Simão Sichman (Eds.), Advances in Artificial Intelligence. Proceedings, 2000. XV, 498 pages. 2000. (Subseries LNAI).

Vol. 1953: G. Borgefors, I. Nyström, G. Sanniti di Baja (Eds.), Discrete Geometry for Computer Imagery. Proceedings, 2000. XI, 544 pages. 2000.

Vol. 1954: W.A. Hunt, Jr., S.D. Johnson (Eds.), Formal Methods in Computer-Aided Design. Proceedings, 2000. XI, 539 pages. 2000.

Vol. 1955: M. Parigot, A. Voronkov (Eds.), Logic for Programming and Automated Reasoning. Proceedings, 2000. XIII, 487 pages. 2000. (Subseries LNAI).

Vol. 1956: T. Coquand, P. Dybjer, B. Nordström, J. Smith (Eds.), Types for Proofs and Programs. Proceedings, 1999. VII, 195 pages. 2000.

Vol. 1957: P. Ciancarini, M. Wooldridge (Eds.), Agent-Oriented Software Engineering. Proceedings, 2000. X, 323 pages. 2001.

Vol. 1960: A. Ambler, S.B. Calo, G. Kar (Eds.), Services Management in Intelligent Networks. Proceedings, 2000. X, 259 pages. 2000.

Vol. 1961: J. He, M. Sato (Eds.), Advances in Computing Science – ASIAN 2000. Proceedings, 2000. X, 299 pages. 2000.

Vol. 1963: V. Hlaváč, K.G. Jeffery, J. Wiedermann (Eds.), SOFSEM 2000: Theory and Practice of Informatics. Proceedings, 2000. XI, 460 pages. 2000.

Vol. 1964: J. Malenfant, S. Moisan, A. Moreira (Eds.), Object-Oriented Technology. Proceedings, 2000. XI, 309 pages. 2000.

Vol. 1965: Ç. K. Koç, C. Paar (Eds.), Cryptographic Hardware and Embedded Systems – CHES 2000. Proceedings, 2000. XI, 355 pages. 2000.

Vol. 1966: S. Bhalla (Ed.), Databases in Networked Information Systems. Proceedings, 2000. VIII, 247 pages. 2000.

Vol. 1967: S. Arikawa, S. Morishita (Eds.), Discovery Science. Proceedings, 2000. XII, 332 pages. 2000. (Subseries LNAI).

Vol. 1968: H. Arimura, S. Jain, A. Sharma (Eds.), Algorithmic Learning Theory. Proceedings, 2000. XI, 335 pages. 2000. (Subseries LNAI).

Vol. 1969: D.T. Lee, S.-H. Teng (Eds.), Algorithms and Computation. Proceedings, 2000. XIV, 578 pages. 2000.

Vol. 1970: M. Valero, V.K. Prasanna, S. Vajapeyam (Eds.), High Performance Computing – HiPC 2000. Proceedings, 2000. XVIII, 568 pages. 2000.

Vol. 1971: R. Buyya, M. Baker (Eds.), Grid Computing – GRID 2000. Proceedings, 2000. XIV, 229 pages. 2000.

Vol. 1972: A. Omicini, R. Tolksdorf, F. Zambonelli (Eds.), Engineering Societies in the Agents World. Proceedings, 2000. IX, 143 pages. 2000. (Subseries LNAI).

Vol. 1973: J. Van den Bussche, V. Vianu (Eds.), Database Theory – ICDT 2001. Proceedings, 2001. X, 451 pages. 2001.

Vol. 1974: S. Kapoor, S. Prasad (Eds.), FST TCS 2000: Foundations of Software Technology and Theoretical Computer Science. Proceedings, 2000. XIII, 532 pages. 2000.

Vol. 1975: J. Pieprzyk, E. Okamoto, J. Seberry (Eds.), Information Security. Proceedings, 2000. X, 323 pages. 2000.

Vol. 1976: T. Okamoto (Ed.), Advances in Cryptology – ASIACRYPT 2000. Proceedings, 2000. XII, 630 pages. 2000.

Vol. 1977: B. Roy, E. Okamoto (Eds.), Progress in Cryptology – INDOCRYPT 2000. Proceedings, 2000. X, 295 pages. 2000.

Vol. 1978: B. Schneier (Ed.), Fast Software Encryption. Proceedings, 2000. VIII, 315 pages. 2001.

Vol. 1979: S. Moss, P. Davidsson (Eds.), Multi-Agent-Based Simulation. Proceedings, 2000. VIII, 267 pages. 2001. (Subseries LNAI).

Vol. 1983: K.S. Leung, L.-W. Chan, H. Meng (Eds.), Intelligent Data Engineering and Automated Learning – IDEAL 2000. Proceedings, 2000. XVI, 573 pages. 2000.

Vol. 1984: J. Marks (Ed.), Graph Drawing. Proceedings, 2001. XII, 419 pages. 2001.

Vol. 1987: K.-L. Tan, M.J. Franklin, J. C.-S. Lui (Eds.), Mobile Data Management. Proceedings, 2001. XIII, 289 pages. 2001.

Vol. 1989: M. Ajmone Marsan, A. Bianco (Eds.), Quality of Service in Multiservice IP Networks. Proceedings, 2001. XII, 440 pages. 2001.

Vol. 1991: F. Dignum, C. Sierra (Eds.), Agent Mediated Electronic Commerce. VIII, 241 pages. 2001. (Subseries LNAI).

Vol. 1992: K. Kim (Ed.), Public Key Cryptography. Proceedings, 2001. XI, 423 pages. 2001.

Vol. 1993: E. Zitzler, K. Deb, L. Thiele, C.A.Coello Coello, D. Corne (Eds.), Evolutionary Multi-Criterion Optimization. Proceedings, 2001. XIII, 712 pages. 2001.

Vol. 1995: M. Sloman, J. Lobo, E.C. Lupu (Eds.), Policies for Distributed Systems and Networks. Proceedings, 2001. X, 263 pages. 2001.

Vol. 1998: R. Klette, S. Peleg, G. Sommer (Eds.), Robot Vision. Proceedings, 2001. IX, 285 pages. 2001.

Vol. 2000: R. Wilhelm (Ed.), Informatics: 10 Years Back, 10 Years Ahead. IX, 369 pages. 2001.

Vol. 2003: F. Dignum, U. Cortés (Eds.), Agent Mediated Electronic Commerce III. XII, 193 pages. 2001. (Subseries LNAI).

Vol. 2004: A. Gelbukh (Ed.), Computational Linguistics and Intelligent Text Processing. Proceedings, 2001. XII, 528 pages. 2001.

Vol. 2006: R. Dunke, A. Abran (Eds.), New Approaches in Software Measurement. Proceedings, 2000. VIII, 245 pages. 2001.

Vol. 2009: H. Federrath (Ed.), Designing Privacy Enhancing Technologies. Proceedings, 2000. X, 231 pages. 2001.

Vol. 2010: A. Ferreira, H. Reichel (Eds.), STACS 2001. Proceedings, 2001. XV, 576 pages. 2001.

Vol. 2024: H. Kuchen, K. Ueda (Eds.), Functional and Logic Programming. Proceedings, 2001. X, 391 pages. 2001.